1,000,000 Books

are available to read at

www.ForgottenBooks.com

Read online
Download PDF
Purchase in print

ISBN 978-0-259-97978-4
PIBN 10040324

This book is a reproduction of an important historical work. Forgotten Books uses state-of-the-art technology to digitally reconstruct the work, preserving the original format whilst repairing imperfections present in the aged copy. In rare cases, an imperfection in the original, such as a blemish or missing page, may be replicated in our edition. We do, however, repair the vast majority of imperfections successfully; any imperfections that remain are intentionally left to preserve the state of such historical works.

Forgotten Books is a registered trademark of FB &c Ltd.
Copyright © 2018 FB &c Ltd.
FB &c Ltd, Dalton House, 60 Windsor Avenue, London, SW19 2RR.
Company number 08720141. Registered in England and Wales.

For support please visit www.forgottenbooks.com

1 MONTH OF FREE READING

at
www.ForgottenBooks.com

By purchasing this book you are eligible for one month membership to ForgottenBooks.com, giving you unlimited access to our entire collection of over 1,000,000 titles via our web site and mobile apps.

To claim your free month visit: www.forgottenbooks.com/free40324

* Offer is valid for 45 days from date of purchase. Terms and conditions apply.

English
Français
Deutsche
Italiano
Español
Português

www.forgottenbooks.com

Mythology Photography **Fiction** Fishing Christianity **Art** Cooking Essays Buddhism Freemasonry Medicine **Biology** Music **Ancient Egypt** Evolution Carpentry Physics Dance Geology **Mathematics** Fitness Shakespeare **Folklore** Yoga Marketing **Confidence** Immortality Biographies Poetry **Psychology** Witchcraft Electronics Chemistry History **Law** Accounting **Philosophy** Anthropology Alchemy Drama Quantum Mechanics Atheism Sexual Health **Ancient History** **Entrepreneurship** Languages Sport Paleontology Needlework Islam **Metaphysics** Investment Archaeology Parenting Statistics Criminology **Motivational**

THE SOCIETY OF ENGINEERS
IN THE
UNIVERSITY OF MINNESOTA

OFFICERS, 1892-1893

J. FRANK CORBETT, '94, *President.*
GEORGE E. BRAY, '94, *Vice-President.*
CHARLES H. CHALMERS, '94, *Secretary.*
ARTHUR M. FRAZEE, '95, *Treasurer.*
NOAH JOHNSON, '94, *Business Manager.*

THE YEAR BOOK OF THE SOCIETY OF ENGINEERS
1892–1893

EDITORIAL COMMITTEE

HENRY B. AVERY, *Managing Editor.*
JOHN W. ERF, *Business Manager.*
J. FRANK CORBETT, *President of the Society.*

DEPARTMENTAL EDITORS

O. J. ANDERSON, '93	*Civil Engineering.*
H. B. AVERY, '93	*Mechanical Engineering.*
G. H. MORSE, '93	*Electrical Engineering.*
H. E. WHITE, '93	*Mining.*
L. L. LONG, '94	*Architecture.*

THE SOCIETY OF ENGINEERS

LIST OF MEMBERS

May, 1893

HONORARY MEMBERS

C. W. Hall, M. A.
Wm. R. Hoag, C. E.
J. E. Wadsworth, C. E.
Harry W. Jones, Arch.

Wm. A. Pike, B. S.
Geo. D. Shepardson, M. E.
Wm. R. Appleby, B. A.
H. E. Smith, M. E.

J. H. Gill, B. M. E.

ACTIVE MEMBERS

O. J. Anderson,
Henry B. Avery,
Wm. S. Abernethy,
Horace S. Andrews,
Norman B. Atty,
Geo. F. Adams,
Martin A. Anderson,
Chas. W. Arrick,
A. A. Adams,
W. F. Andrews,
Sidney K. Adams,
Chas. R. Aldrich,
Frank L. Batchelder,
Austin Burt,
Jno. Blackmer,
Jno. Bradford,
A. R. Bryan,
Frank E. Burch,
Jno. A. Bohland,
Wm. H. Beard,
Adam E. Bishman,
S. P. Burg,
Daniel Buck,
Carl Burghart,

Chas. M. Babcock,
Henry E. Byorum,
Adam C. Beyer,
Albert W. Burch,
G. E. Bray,
H. G. Cooley,
A. W. Chase,
A. O. Cunningham,
Roscoe Cramb,
Chas. H. Chalmers,
J. Frank Corbett,
G. A. Casseday,
L. H. Chapman,
C. H. Cross,
G. R. Caley,
R. E. Carswell,
G. L. Chesnut,
Lee Coleman,
Harry Cutler,
Geo. B. Couper,
G. W. Digen,
A. M. Dunton,
W. H. Dewey,
Hans Dahl,

Louis Dinsmore,
Fred G. Dustin,
J. W. Erf,
B. H. Esterly,
S. A. Ellis,
J. H. Evans,
L. J. Fuller,
C. R. Fallis,
A. M. Frazee,
R. E. Ford,
Benj. Grunberg,
James B. Gilman,
V. I. Gray,
A. E. Garland,
J. D. Guthrie,
H. P. Hoyt,
A. L. Hill,
P. E. Holt,
G. B. Huntington,
H. E. Hatch,
Jas. Hildreth,
H. E. Hjaidemal,
F. M. Hughes,
Victor Hugo,

W. W. Hyslop,	B. O'Brien,	W. W. Tanner,
Clive Hastings,	Clinton M. Perry,	G. W. Turner,
C. D. Hilferty,	Edward E. Pratt,	G. L. Thornton,
Lewis Iverson,	Levi B. Pease,	W. M. Tilderquist,
Noah Johnson,	C. L. Pillsbury,	B. E. Tunstad,
M. A. Joyslin,	C. S. Phelps,	H. L. Tanner,
C. P. Jones,	F. W. Reidhead,	R. W. Wentworth,
R. B. Kernohan,	T. A. Rockwell,	Wm. C. Weeks,
W. M. Knapp,	F. M. Rounds,	Otto Wolfrum,
A. G. Kinney,	H. W. Rutherford,	Geo. B. Woodford,
J. P. Kane,	B. W. Roberts,	R. M. Wheeler,
F. W. Long,	G. A. Rhame,	C. D. Wilkinson,
H. D. Lackor,	F. W. Springer,	F. B. Walker,
Perry Lewis,	Fred von Schlegell,	A. C. Weaver,
E. F. Lindmere,	E. E. Smith,	H. E. West,
Jas. H. Linton,	B. P. Shepherd,	E. C. Wells,
L. L. Long,	E. S. Savage,	C. P. Waller,
G. H. Morse,	C. B. Sprague,	W. C. Walthers,
A. D. McNair,	N. P. Stewart,	H. M. Wheeler,
Jas. N. Munro,	F. J. Savage,	H. N. Whittelsy,
C. E. Magnusson,	L. V. Smith,	Delos C. Washburn,
A. W. McCrea,	Jas. E. Spry,	H. E. White.
Robest Nesbitt,	Jno. L. Stackhouse,	M. A. York,
Robert Northway,	M. D. Staughton,	Washington Yale,
Victor A. Neil,	A. W. Strong,	Frank Zimmerman,
R. W. Nelson,	F. D. Sterling,	Clarence J. Zintheo.
John Norris,	Winfred O. Stout,	

CORRESPONDING MEMBERS

Anderson, Christian, B. C. E., '88 - - - -	Portland, Ore.
Andrews, George C., B. M. E., '87 - - - -	Minneapolis, Minn.
Aslakson, Baxter M., B. M. E., '91 - - -	Indianapolis, Ind.
Barr, John H., B. M. E., '83; M. S., '88 - - -	Ithaca, N. Y.
Burch, Edward P., B. E. E., '92 - - - -	Minneapolis, Minn.
Burt, John L., B. C. E., '90 - - - - -	Minneapolis, Minn.
Burtis, William H., B. E. E., '92 - - - -	Oshkosh, Wis.
Bushnell, Charles S., B. M. E., '78 - - -	Minneapolis, Minn.
Bushnell, Elbert E., B. M. E., '85 - - - -	New York, N. Y.
Chowen, Walter A., B. C. E., '91, - - -	Brown's Valley, Minn.
Coe, Clarence S., B. C. E., '89 - - - -	Wenatschee, Wash.
Crane, Fremont, B. S., '86; B. C. E., '87 - - -	Prescott, Ariz

Dann, Wilbur W., B. C. E., '90 - - - - Minneapolis, Minn.
Dawley, William S., B. C. E.,'79 - - - - - Danville, Ill.
Douglas, Fred L., B. C. E., '91 - - - - - Trenton, N. J.

Felton, Ralph P., B. M. E., '92 - - - - Lehigh, Wis.
Furber, Pierce P., B. C. E., '79 - - - - Saint Louis, Mo.

Gerry, Martin H., B. M. E., '90; B. E. E., '91 Minneapolis, Minn.
Gilman, Fred H., B. C. E., '90 - - - - Minneapolis, Minn.
Gill, James H., B. M. E., '92 - - - - Minneapolis, Minn.
Gillette, Lewis S., B. S., '76; B. C. E., '76 - Minneapolis, Minn.
Goodkind, Leo, B. Arch., '92 - - - - - Saint Paul, Minn.
Gray, William I., B. E. E., '92 - - - - Minneapolis, Minn.
Greenwood, Williston W., B. C. E., '90 - - - New York, N. Y.

Hankenson, John J., B. C. E., '92 - - - Minneapolis, Minn.
Hayden, John F., B. C. E., '90 - - - - - Fargo, N. D.
Hendrickson, Eugene A., B. S., '76; B. C. E., '76 Saint Paul, Minn.
Higgins, Elvin L., B. C. E.,'92 - - - - Hutchinson, Minn.
Higgins, John T., B. C. E., '90 - - - - Minneapolis, Minn.
Hoag, William R., B. C. E., '84; C. E., '88 - Minneapolis, Minn.
Howard, Monroe S., B. E. E., '92 - - - - Lake City, Minn.
Hoyt, William H., B. C. E., '90 - - - - - Duluth, Minn.
Huhn, George P., B. E. E., '91 - - - - Minneapolis, Minn.

Leonard, Henry C., B. C. E., '75; B. S., '78 - Minneapolis, Minn.
Loe, Erie H., B. M. E., '88 - - - - Minneapolis, Minn.
Loy, George J., B. C. E., '84 - - - - Spokane Falls, Wash.

Mann, Fred M., B. C. E., '92 - - - - - Boston, Mass.
Matthews, Irving W., B. C. E., '84 - - - Waterville, Wash.
Morris, John, B. M. E., '88 - - - - Minneapolis, Minn.

Nilson, Thorwald E., B. M. E., '90 - - - Minneapolis, Minn.

Pardee, Walter S., B. Arch., '77 - - - - Minneapolis, Minn.
Peters, William G., B. C. E., '83 - - - - - Tacoma, Wash.
Plowman, George T., B. Arch., '92 - - - Minneapolis, Minn.

Rank, Samuel A., B. S., '75; B. C. E., '75 - Central City, Col.
Reed, Albert I., B. C. E., '85 - - - - Hastings, Minn.

Smith, Louis O., B. C. E., '83 - - - - - LeSueur, Minn.
Smith, William C., B. C. E., '90 - - - Saint Cloud, Minn.
Stewart, Clark, B. S., '75; B. C. E. '75 - - Minneapolis, Minn.

Thayer, Charles E., B. C. E., '76 - - - Minneapolis, Minn.
Trask, Birney E., B. C. E., '90 - - - - Fort Plains, N. Y.

Woodmansee, Charles C., B. Arch., '86 - - Saint Paul, Minn.
Woodward, Herbert M., B. M. E., '90 - - - Milwaukee, Wis.

THE YEAR BOOK

OF THE

SOCIETY OF ENGINEERS

MAY, 1893.

THE COMMONWEALTH AND THE COLLEGE.

CHRISTOPHER W. HALL, M. A., *Dean of the College of Engineering, Metallurgy and the Mechanic Arts.*

Minnesota is supporting a School of Applied Science, called the College of Engineering, Metallurgy and the Mechanic Arts. Is that a profitable thing for the State to do?

Let this first be said: The College of Mechanic Arts, which was the Engineering College of the University during the twenty years from 1873 to 1892, instructed 1,123 enrolled students and graduated 50 engineers and architects all of whom are engaged in honorable, responsible and lucrative pursuits. Nearly all these students and graduates came from Minnesota families; they have entered careers which from their very nature develop a broad catholicity of spirit and discipline for the highest usefulness in citizenship.

The State has passed the year of her majority. The trend of her development for the coming century is well defined. She will be in part an agricultural state; yet agriculture will be but one group of her industries. Manufactures, mining, transportation must occupy thousands of citizens. She is situated at the eastern frontier of an empire vast already in its area and its known resources. Along highways must be carried the traffic of all that region. The coal, the iron and the thousand and one articles of convenience and luxury must be brought in for the millions of consumers; the bread stuffs, the gold and the silver and many another product of the fields and mines must be sent out into other portions of the world in return. These highways must be made in perfect manner and kept in perfect condition; the power to move vast loads must be expended with the least loss and in the most effective manner. Engineers to build roads and railways, to construct and propel steamers and to conserve the uses of electricity will have here a constantly widening field of useful activity.

Minnesota has some of the most remarkable iron ore deposits in the world; they must be worked by trained miners and their ores must be treated by skilled metallurgists. Our neighboring States have silver, gold, copper, lead, iron, zinc and many other metals together with precious stones. The millions of dollars invested in these deposits and their exploitation must yield an income; to secure this income the highest ability, carefully trained, will be called into activity. Within nine years our commonwealth has sprung from obscurity to the fifth place among the iron producers of North America, to the third place in the production of hematite ores and to the first place in the purity and quality of her product. The supply appears inexhaustible, but the obtaining of it will command skill and science.

Manufacturing too is assuming vast importance. During the last year the three cities of Duluth, Minneapolis and Saint Paul turned out manufactured goods to the value of nearly $200,000,000; with the same ratio of production to population extended to all the cities of the State with 3,000 inhabitants Minnesota manufactures reach the value of over $250,000,000. This sum at current prices will buy the last wheat crop of the State almost ten times, or the Indian corn crop of that year, twenty-five times. Think what trained engineers can do in the manufacturing interests of the future even within our own State! The best guage of a nation's commercial and political rank among the nations of the world is found in the record of its mining and manufacturing achievements. The country that makes the largest use of its natural products is the leader in all that constitutes true national greatness.

There is probably no line of applied science which shows the relationship of the imaginative faculties and the trained business talent o men better than architecture. A few years ago architects were seldom seen, to-day no one thinks of building a residence, a factory or a business block without consulting an architect to secure symmetry and harmony and to guard against violations of the taste and artistic sense of the age. Minnesota constructs many million dollars worth of buildings yearly. Men must be had to design and superintend this construction of the present and to direct the vastly larger plans of the coming decade. We have among us already architects of national repute; should our State not lay the educational foundation for many more such men among her own sons?

Industrial chemistry is opening a wide field for skilful chemists. The technical applications of chemistry are many; the preparation of pure chemicals and drugs; the distillation of gases and liquids; the preparation of dyes, bleaching compounds and processes; the printing of cotton and woolen goods; the making of sugars and the preparation of scores of our modern foodstuffs all demand a skill and fitness for work of rare quality. When our cities are filled with chemical works and our prairies are dotted with sugar refineries, as now they are with creameries, there will be a large demand for the most skilful and ingenious chemists.

Let us for a moment take a broader view. We will not utter prophecy, for we should then produce results staggering to our senses; let us rather glance at the past. The census reports on our national progress during the past decade show some striking figures. For instance: The true valuation of property in the United States increased $20,000,000,000 in the ten years between 1880 and 1890; the capital invested in manufactures in 1890 was $4,600,000,000, one and two-thirds times as much as in 1880; $500,000,000 more was paid in wages to the employes of manufacturing establishments in 1890 than in 1880; more than four times as much steel was manufactured in 1890 as in 1880; the product of our mines and quarries for 1890 was almost twice that of 1880; thirty railroads which hauled 96,000,000 tons of freight in 1880, moved nearly 263,000,000 tons in 1890; the capital invested in railroads increased $5,000,000,000 in the same ten years; in 1880 electricity as applied for power was a new thing, to-day over $800,000,000 capital is invested in all branches of the industry. Hotels costing a million dollars each are almost too common to mention, as are manufacturing structures covering acres of ground. Truly "things are in the saddle and ride Mankind."

With such a glimpse of the past decade what may we reasonably look for in this? We have already outstripped all other nations in the magnitude of our industrial operations. What high engineering and technical skill has been required to secure this result! We can see in the future a demand for the highest and best. Some one must furnish this "best." Since Minnesota too demands it, let her prepare her own sons to become the best in effort and in skill and thus meet the home demand.

ELECTRICITY IN AGRICULTURE.

Geo. D. Shepardson, M. E., *Professor of Electrical Engineering.*

Within the past few years considerable attention has been paid to the uses of electricity in agriculture. This is simply a renewal of interest in a subject that has been studied for over a century and a half.

Electricity has been applied to agriculture in a number of different ways. Static electricity has been applied to seeds and to growing plants. Current electricity has been applied to the soil. Growing plants have been exposed to the action of electric lights. The heating effect of the current has been used in a number of processes more or less directly connected with agriculture. The mechanical effects of the current are being used to an enormous extent.

The opportunities for electricity in farming may be considered under two general classes. The first, in which it is applied directly to the growth of plants or animals; the second, in which it is applied in various ways to assist the farmer in his work. In the latter class, electricity is applicable in four ways: For furnishing power, light, heat and for the operation of telephones, signals, etc.

The recent discussions in the papers have generally centered about the use of electricity as a source of mechanical power for various purposes about the farm. This is a natural consequence of the great development of electrical transmission for other industries and presents comparatively few difficulties that are new. Probably the greatest difficulty is that of getting capital to meet the cost of experiments in designing the special machinery that would be necessary for some purposes such as plowing and harvesting; also to establish the power plants and obtain other equipment necessary. Some farms already have steam engines that could drive electrical generators for most purposes. Where farms are not large it would be best to have a central power station for supplying a district perhaps ten miles square. While the first cost of installing a plant for the use of electric power on farms and farm roads would involve the outlay of more capital than the farmers generally care or are able to expend; yet conservative estimates seem to show that the investment would be made profitable by the cheaper and safer handling of crops.

There are great opportunities for the electric motor in agricultural work. On the great prairies of the West is an immense area nearly level, free from stones, trees or other impediments to cultivation of the soil by power machines. The sole power now available is that of animals. Western farmers discard hand implements so far as possible and depend almost wholly upon animal power machines in the production of immense crops of grain. The use of animal power is subject to great limitations. There has been very great improvement in agricultural work in the present century due to the substitution of animal or steam power for manual labor. The application of animal power to the reaper performs the labor of thirty men. We harvest twenty acres of grain per day, thresh twenty acres and carry the grain five hundred miles per day to market, but plowing is limited to two acres per day. Western agriculture is held back by the inability of animal motors to plow wider, deeper and swifter furrows at less cost. There is now great interest in the matter of adapting mechanical power to this work. Several steam plows are being set up in Kansas, and at the last session of Congress, Senator Peffer introduced a bill to provide for the establishment of an agricultural power experiment station by the government, in which an effort might be made to determine the relative values of different motors. The problem is the dragging of a plow, harrow, seeding machine, reaper, or a wagon loaded with grain, half a mile in a straight line and dragging it back again until the whole field is worked. The electric motor seems particularly well adapted to such work since it is simple, has ample power, without excessive weight, carries neither fuel nor water and its few wearing parts can easily be protected from dust and weather. In 1879 electric motors were used in Europe for plowing fields near the beet sugar factory at Sermaize and did the work at the rate of two hundred square feet per minute.

It is not impossible that the electric motor may be successfully applied to the operation of plows, cultivators, mowers, reapers, rakes, etc.

Electric power is being applied very largely to moving vehicles and a considerable number of electric railways are being built through farming districts and connecting different towns. In some cases these roads are doing a considerable freight business and will be used more and more for hauling garden and other produce to the market. Enthusiasts have pictured railway lines on country roads and tramways connecting farm buildings to such an extent that the horse and wagon would be used only in a very limited way. The roads that have already been built to connect different towns are paying so good dividends that a large amount of capital is being invested in this direction. This has an important bearing in connection with the present agitation for good roads. Writers have vigorously advocated the advantages of electric railways running through the country as a means of bringing the farmer closer to the market, cheapening the cost of transportation, and rendering his entire life more comfortable. The same motors that would draw his produce to the market would also be useful in gathering the crops, since temporary tramways might be laid through the fields.

Stationary motors could easily be arranged for doing much of the work about the barn, which now is done by hand or horse power, such as running feed or ensilage cutters, corn shellers, cob grinders, threshing and seed-cleaning machinery, hay forks, horse cleaners, elevators, sorghum mills, cider presses, hay presses, and cotton gins. In the shop the motor could run saws for cutting stove wood, lumber and various other purposes; for running planer, lathe, grind stone, blower for forge and other machine shop tools.

A motor in the house could be used to advantage for driving washing machinery, churns, butter and cheese machines, coffee mills, and ventilating fans. Pumps could be used for forcing water into elevated tanks from which it could be used for fire purposes, drinking troughs, for washing vehicles or irrigation; also for the dairy house, laundry, kitchen, etc.

For many purposes a motor mounted on a truck would be of advantage, since the same motor could be moved about and used for different machines.

The electric light has already been used in agricultural operations to a considerable extent. The large sugar plantations in the Sandwich Islands are provided with electric light plants, which light up the fields at night, so that two sets of hands are employed and harvesting goes on day and night without interruption. A number of farms in the United States have adopted electricity for lighting buildings as a safer means than lanterns and oil lamps. The reduced fire risk more than compensates the increased cost. Strong argument may be advanced for the use of portable electric light plants in connection with threshing outfits. It is well known that enormous amounts of grain are lost because of the inability to get the grain under cover or get it threshed before it is ruined by exposure to the fall and winter weather. If threshing outfits were provided with electric light plants, they could operate two gangs of men and do nearly twice the amount of work in the short season available for threshing. Millions of money might be saved annually in Minnesota alone by the adoption of this improvement.

The heating effect of the current might be used to advantage in incubators, hot houses, and similar places that require a uniform temperature. Electric cooking utensils are rapidly coming into extended use. It is not impossible that trees may be cut and lumber sawed by means of wires heated to incandescence by electric currents. Indeed there has been considerable discussion of this subject during the past year.

The heating and chemical effect of high potential currents, such as are employed in arc lighting, has been used with great success in destroying undesirable vegetation. Near Memphis the experiment was made of connecting one terminal of a dynamo to the ground while the other was connected to a wire brush, which, dragged over the ground, successfully killed all the weeds and grass with which it came in contact. Doubtless a modification of this plan could be used, with success and economy, in destroying some of the pests of the farmer's life.

Electricity has been used in a number of cases for operating telephone and telegraph circuits in farming districts, in operating call bells, burglar alarms, electric heat regulators and for many similar purposes. In one farming community in Michigan a telegraph line, costing about $200, was run through the country connecting all the farms in a large district with a neighboring village. The farmers themselves or their children have learned how to operate the instruments, which have repaid their cost many times over by the saving effected, on account of better means of communication.

Electricity has been applied directly to the growth of plants and animals in a number of different ways. One has been that of turning night into day by placing electric lights in hot houses. Experiments show that periods of darkness are not necessary to growth and development of plants. The electric light promotes assimilation; it often hastens growth and maturity; it often intensifies colors of flowers and sometimes increases their productiveness. The experiments along this line have nevertheless led to conflicting and indefinite results. Under some conditions the presence of electric light at night is a great advantage, but in other cases it is detrimental. Considerable work has, however, been done in this promising field. A market gardener in England is reported to have increased his profits twenty-five per cent. by using the electric light for forcing lettuce.

Static electricity has been applied both to seeds and to growing plants. N. Spechnew soaked various kinds of seeds until they swelled and then subjected them to the influence of induced currents for one or two minutes. He found that electrified seeds developed in about half the usual time, and that the plants coming from them were better developed, having larger leaves and brighter colors. The final yield was not affected.

It is also thought that static electricity is a potent factor in the economy of nature, and has more to do with the growth and development of plants than is generally believed. That electricity has a marked effect upon the atmosphere, is shown by the well known effect of a thunder storm in clearing and purifying the air. It has long been known that objects in the air show a considerable difference of potential from the earth. It is commonly believed that the air itself is electrified, although some argue that the electrification comes simply by induction from the earth and that the air itself is not electrified. Abbe Nollet in 1746 found that the evaporation from plants and animals, liquids and solids, was much greater when they were electrified than when not. Abbe Bertholon in 1783 ran sharp metallic points into the air and connected them with other metallic points near plants. The plants treated in this manner were more vigorous than their neighbors. These experiments have been tried by many observers but with contradictory results. Cl. Bernard in 1877 studied the effect of putting wire cages around growing plants and making them grow in unelectrified soil and atmosphere. Plants treated thus were far less vigorous and productive than those growing in the free air and free soil. In 1878 Bernard studied the electrification of the atmosphere near trees and found that large trees of massive verdure act like a metallic cage. It is a matter of common observation, that plants do not flourish in the shade and it has been suggested by Sir William Thomson, Mascart, Bernard and others that trees act as electric screens, keeping the electricity of the air from plants underneath. Attention may be called to the fact that pine and cedar trees are especially effective in collecting atmospheric electricity, on account of the great number of sharp points presented by their needles. It is observed that most plants grow more rapidly at night than in the day time; also that at night the plant is generally covered with dew which is a good conductor and affords atmospheric electricity an easy path to the earth.

In view of these observations, the writer suggests that there is an increased evaporation from plants growing in an electrified atmosphere. which causes an increased circulation of the sap and so materially assists the growth of the plant.

That atmospheric electricity has much to do with the economy of plant life is also indicated by the fact, which seems to have been overlooked by former writers on this subject, that the leaves or other parts of nearly all forms of vegetation are provided with numerous sharp points such as spines, thorns, hairs, etc., which assist greatly in collecting or discharging atmospheric electricity. May it not be that in some cases this is their principal function? In this direction certainly lies a wide field that will abundantly repay careful study by competent investigators, who have at their service the best modern equipment, both of theory and apparatus.

Currents of electricity have been applied to the soil by many different experimenters and in several different ways. Lagrange had a number of pointed conductors in the air, connected with a network of galvanized wires buried in the soil. Others have buried plates of zinc and copper which were connected by an insulated wire so as to form an earth battery. Still others have buried in the ground plates or wires which were connected with a battery or other source of current. By each of these three methods current is sent through the soil and affects the vegetation in the vicinity, apparently by electrolytic action upon the soil.

The results obtained with current electricity applied in these ways have been as contradictory as those obtained from the application of static electricity or the electric light. Under some circumstances the current through the soil seems to exert a beneficial effect. In other cases it seems to be detrimental. Lagrange draws the conclusion that current electricity is better for plant cultivation, while static electricity is better for roots. A number of experimenters conclude that the use of electricity is good for some plants and bad for others. Still others claim it is not only useless but harmful in every case. The action of electricity upon plants seems like that of certain chemicals upon the human body. In small doses they are helpful stimulants; in large doses they are fatal poisons; or in still larger doses they are less harmful.

The present state of the art is practically this: Electricity doubtless has an action upon vegetation but the conditions for its best application are but little better understood to-day than they were years ago. It is noticable in the report of all the experiments that no quantitative measurements have been made of the electrical energy developed or applied. In view of the contradictory results obtained from experiments, we are sometimes forced to the conclusion that for most, if not all, plants a certain amount of static and current electricity is valuable. For knowledge further than this, there must be carried on, with greater care and intelligence, experiments in which a study may be made of the best methods of applying electricity and the proper amount to be applied under various conditions. It is the intention to carry on some such experiments at the University farm during the coming summer.

THE PRACTICE OBSERVATORY.

The need of an observatory for the use of the students in astronomy in the College of Science, Literature and the Arts and the students in Geodesy in the College of Engineering, Metallurgy and the Mechanic Arts has long been felt.

While much thought has been given to the question of a University Observatory the weight of opinion has been in favor of a practice observatory located on the University grounds and equipped with a full complement of small first-class instruments to be used by every student in astronomy and geodesy as the best suited to our needs. It was not until 1891 that the plan of such a students' working observatory was carried out.

The observatory building is constructed of blue limestone and is situated on the highest ground of the campus.

The instrumental outfit now consists of a transit circle, made by Segmuller of Washington (Fauth & Co.), having a telescope of three-inch aperture, to the axis of which is attached a graduated circle of eight-inch radius. A delicate spirit level is kept suspended from the axis. To the eye end of the telescope is attached a micrometer, which contains a reticle of spider lines, arranged parallel to the meridian, and a single thread parallel to the horizon, movable by means of a fine screw so graduated that fractions of a revolution of the screw may be read. The circle is graduated to five minutes. The single minutes and seconds are read by means of two microscopes attached to the pier. These microscopes are furnished with micrometers, by means of which the distance of the division mark from the zero point is directly measured to tenths of seconds. The spider lines of the reticle are illuminated by the light from an electric lamp which, entering through the hollow axis of the telescope, strikes against a metallic reflector and is thence directed to the eye end of the telescope. The whole telescope can be reversed on its bearings, so that the eastern and western pivots change places.

INTERIOR VIEW OF THE PRACTICE OBSERVATORY.

This instrument is fundamentally a meridian circle, but may be converted into a zenith telescope, by attaching a level to the telescope, parallel to the meridian. When used as a meridian circle, an observation consists of two separate measurements: first, the exact time a star crosses the meridian is determined; second, the altitude of the star when on the meridian is measured. The first is effected by observing the time when the star is on each wire of the reticle and taking the mean. The telescope should be so adjusted that the wire which corresponds to this mean will be in the meridian. The second measurement is obtained by taking the difference of readings of the circle when the movable wire is on the horizon and on the star. When the error of the clock and the latitude of the observatory are known, an observation gives the right ascension and declination of the star. When the right ascension and declination of a star are known, the observation gives the clock error and the latitude of the observatory.

A first-class chronograph, by the same maker, is used in connection with an astronomical clock, made by Howard of Boston.

A ship chronometer regulated to sidereal time is used with the transit for the *eye* and *ear* method.

Plans are already under consideration for the addition of a suitable equatorial telescope, which will soon be made and will furnish all that could be desired in that direction in a practical students' observatory.

FIRE-PROOF BUILDINGS.

J. E. WADSWORTH, C. E., *Assistant Professor of Civil Engineering.*

The losses from fire in this country are enormous, and well nigh appall one by their amount when expressed in dollars and cents. Edward Atkinson, in an article in the *Engineering Magazine* of May, 1892, on "fire risks on tall office buildings," makes this statement: "The fire tax of this country now costs—in fire losses, in the cost of sustaining the insurance companies and the excessive cost of the fire departments combined—a sum equal to the entire cost of sustaining the United States government in all its parts, together with the interest on the national debt." It will be seen from this, that from a financial standpoint alone, neglecting the danger to human life, which cannot be estimated in dollars and cents, there is a great need of making our buildings, if not absolutely fire-proof, at least more nearly so than now.

The subject of fire-proofing is not one of recent origin, although it has made great advances in these later years on account of the growth of our large cities, and the consequent congestion of the business of these cities in limited areas, necessitating large and expensive buildings in close proximity to each other, subjected to great fire risks. It is supposed that the stone vaulted roof of the cathedral is a result of an attempt to prevent the destruction of the roof by fire, as we read that, "after a fire in 1141 Bishop Alexander replaced the wooden roof of the nave (of the London Cathedral) with a stone vault." Mr. Hatfield (*Proc. Am. Inst. Architects, 1878*), gives as an example of the durability of a brick building with a vaulted roof, the Pantheon, which for "nineteen centuries of its existence, has been subjected to repeated conflagrations." Iron, in the shape of cast iron, began to be used in buildings about sixty years ago. In 1836, Dr. Fox of Bristol, proposed a system of flooring, composed of inverted tees of cast iron spaced about eighteen inches apart; on the flanges of these tees, spanning the opening between them, wooden strips one inch square were placed, leaving about one-half inch openings between the strips. Upon these wooden strips a thin coating of coarse mortar was spread, enough passing through the half-inch spaces to form a support for the ceiling plastering, which covered the bottom of the beam to the depth of one-half inch. On top of the mortar a layer of concrete was spread and leveled off for the floor. Thus the iron was covered and protected from the direct action of the heat.

Let us ask, what is meant by the term *fire-proof*, as it is used to-day? The *Encyclopædia Britannica* states: "It is now conceded that the thorough fire proofing of any building is almost impracticable." In a paper read by F. Schumann, C. E., before the American Institute of Architects in 1878, the statement is made that "warehouses for the storage of miscellaneous merchandise cannot, with our present knowledge, be constructed absolutely fire-proof; we can only apply devices that diminish the danger, by confining and localizing the conflagration." Mr. C. J. H. Woodbury, in a lecture before the Franklin Institute Jan. 23, 1891, says that "a fire-proof building is a commercial impossibility, because if one could be constructed so as to withstand the destruction of its contents, it would be good for little else and the cost would be prohibitive." It is said that when William A. Green was chief of the Boston fire department, he received a letter from an official in Berlin, asking for a description of the fire-proof buildings of Boston. He replied, that they had but one, the Beacon Hill reservoir, and they did not always feel quite sure of that. Since these opinions were given, the subject has been studied and advances made, so that buildings are being put up today which, it is claimed, can withstand the heat caused by the total combustion of the contents, and the buildings remain intact.

We have had, fortunately, a good test of one of these buildings during the past year. On October 31, 1892, the nine-story, fire-proof building of the Chicago Athletic Club caught fire and all the contents burned. The building was in course of erection and was therefore open from top to bottom, which would not have been the case had it been completed. If the amount of combustible material in the building was as great as it would have been were the building completed, it would afford a very complete test. The owners of the building employed the well-known engineers, Gen. Wm. Sooysmith and Mr. Isham Randolph, to make an examination of the building after the fire. The following are extracts from their report: "It is evident that the building, when it took fire, contained a very large quantity of combustible material. If the building had been completed, it would never have contained combustible material enough to have produced heat enough to do any considerable injury to the building by burning. This building furnishes an assurance that was lacking before; that the metal parts of a building, if thoroughly protected by fire-proofing properly put on, will safely withstand any ordinary conflagration. The integrity of the building does not seem to be impaired, and it may be made as good as new by replacing the parts injured."

The following is from the published statement of an official of the insurance patrol: "As far as we can see, the building itself is all right. The big girders are not bent, the brick floors are not broken and nothing has gone up but the slight brick partitions and brick casings around the pillars. A few unimportant scantlings that held the partitions in place are bent and must be replaced. On the whole the building is as sound as ever."

There has recently been another practical test: On April 2, 1893, the contents of about 30 or 40 offices in the Temple Court building in New York were destroyed by fire. The *Engineering News*, of April 6, makes this statement of the fire: "The framework of the building itself was practically uninjured. The fire stripped every particle of plastering from the walls and ceilings in some rooms, but the tile covering protecting the floor beams was not materially broken away in a single instance, and in only half a dozen or so places were patches of the fire-proofing removed from the columns." From these statements it will be seen that the definition of a fire-proof building is gradually changing, and we are coming nearer and nearer the ideal. It can be said now that a building can be built so as to withstand the complete combustion of its contents and remain intact; it only remains so to construct a building that the contents of one room may be destroyed and the rest of the building and its contents remain intact and we shall have a perfect fire-proof building.

Fire-proof Buildings. 17

In the construction of fire-proof buildings two precautions are to be taken, viz.: The precaution to prevent the fire from spreading from one building to another, and the precaution to prevent the spread of fire from one part of a building to another.

Buildings can also be divided into two general classes:

I. Buildings containing combustible material.

II. Buildings containing no combustible material.

Bearing in mind the precautions just given, it is seen that the treatment of class II. is comparatively simple. If the building is isolated from other buildings, or surrounded by buildings of the same class, it needs no protection save that the building be composed of non-combustible materials. This condition is generally met by making the building entirely of brick, iron and glass. But if the building is in close proximity to other buildings, which are combustible, it must have its sides and roof covered with materials that can resist the intense heat given off by the other building during its combustion. This will necessitate having brick walls, fire-proof shutters for the windows and a terra-cotta roof or a wooden roof, tinned. Precaution must be taken that all iron shall be protected by some nonconducting material, so that it shall not become heated enough to cause undue expansion. In the Boston fire of Nov. 28, 1889, the Ames building is said to have failed by reason of the expansion of the roof trusses, pushing the walls out and thus destroying the building before the fire had begun its work.

It is under class I. that most buildings are found. This class may be subdivided into sub-classes: (a) Warehouses, made as nearly fire-proof within and without as possible, by protecting all iron work and window openings, and by limitation of space. (b) Public buildings and dwellings, containing some combustible material in their structure but having all vital parts composed of fire-proof materials. The conditions of class (a) may be met by using fire-proof floors, keeping each floor separate from all others and providing all outside openings with fire-proof doors or shutters.

For fire-proof floors it has been found that porous terra-cotta arches with steel I beams make the best floors. A very complete set of tests was made on different flooring materials in Denver, Colorado, for the Equitable building in that city in 1891, and it was shown that porous terra-cotta was farsuperior to any of the others. (See *Engineering News*, Aug, 29, 1891, p. 180.) Terra-cotta is also a very good non-conducting material. W. W. Boyington, an architect of Chicago, claims to have heated a porous terra-cotta tile to a white heat at one end and held the other end in his hand.

Class (*b*) is, perhaps, the most common class met with and also the most difficult to deal with. It is to this class that the Chicago Athletic Club and the Temple Court building of New York would belong, and it is seen from the results of these tests that as far as the building itself is concerned the problem may be solved. The main point in this class of buildings, as in all the others, is completely to isolate different parts of the building, and under no consideration to allow two floors to be connected. This may be accomplished by having all flues, such as elevator shafts, stairways, openings for gas, steam and water pipes, electric wiring and transmission of power, enclosed in fire-proof shafts and connected with each floor, when necessary, by openings provided with fire-proof doors. The floors must be fire and water-proof, all iron work must be protected by some non-conducting material as terra-cotta, well fastened to the iron work. In the Chicago Athletic Club building the fire-proofing was attached to the columns by wooden strips, which burned out and let part of the fire-proofing fall. All unnecessary wood-work should be avoided and especially varnished wood-work. All windows should be protected with shutters. All shutters and fire-proof doors should be attached directly to the wall and not to any wood-work or any material likely to be destroyed by fire. This seems simple enough to understand, yet there are numerous instances where it has not been done. A single thickness iron door is almost valueless as a fire protection. A door which has safely withstood seven fire tests, is a door composed of two layers of boards, laid at right angles to each other to prevent warping, and covered with tin.

One great cause of fires is due to the fact that a wall which may be a secure protection against a high temperature intermittent in character becomes dangerous when the source of heat becomes continuous. There are many examples of fires arising from this cause, *e. g.*, a wall back of a grate, when the heat is made continuous either by a continuous fire or by a bank of ashes, has been known to set wood-work afire on the other side, while the same wall was perfectly safe for much greater heat, if intermittent. Numerous cases are on record where fires have arisen from seemingly no cause, but were found to have been caused by continuously heated steam pipes, the temperature of the pipes at no time being considered dangerous. There is one other reason for completely isolating different portions of a building, viz., the damage caused by smoke. In a great many cases the damage from smoke is as great or greater than from the fire itself, and in many cases it has caused a loss of life.

There is a great deal said these days about *slow-burning buildings*, and all buildings that cannot be classed as fire-proof are called *slow-burning*. A properly constructed slow-burning building is practically fire-proof, and is probably much more so than many so called fire-proof buildings. The most important feature of the modern slow-burning building is the use of continuous floors, thus avoiding as much as possible all holes, and the avoidance of any concealed spaces which may act as flues for a fire. Therefore the ceiling must follow the floor joists around to the floor above, not enclosing a concealed space between floor joists. In slow-burning buildings all walls are composed of some fire-proof material as stone or brick. The roof is composed of planking on which is laid tin or slate. All stairways, elevators, power or belt shafts, etc., are enclosed in fire-proof flues, and there is no communication with any floor except through openings provided with shutters. The floor is composed of two layers, the lower one of three inch planking. The floor joists are rather large timbers spaced far apart. All columns and joists are designed to sustain their loads when 1½ inches have been burned from all sides. In places especially dangerous, the wooden beams are covered with plastering.

CHARACTERISTICS OF THE TRUE ENGINEER.

Abstract of lecture delivered before the Society of Engineers, by WM. R. HOAG, C. E., Professor of Civil Engineering.

The following illustration was given of the nature of the fields of engineering: A wise landlord once upon a time owned a vast estate which was divided naturally into many fields and each field was by nature devoted to a particular line of work. In one field were rocks which could be quarried for use over the entire estate; in another were ores and minerals to be dug out of the earth; there were streams along the course of which were waterfalls by which all sorts of machiney could be run and electric light be furnished to all parts. In the most beautiful spot on the whole estate the mansion of the owner and the houses of the workmen were built. Men were hired and given perfect freedom of choice to enter any line of work the estate could furnish, and an accurate account of each man's work was kept, not of what he wanted to do or believed he could do, but of what he actually did. Every man felt that no injustice was done, but when his work was finished his reward followed.

Engineer John T. Howard, of England, was quoted in the statement that a university training must in the future play a large part in the education of an engineer. This fact is fast gaining recognition before the industrial world. The true engineer is not determined by the number of inventions patented. If this were so, we should have but one engineer,—Edison. He is not determined by the number of books written; as Rankine, or Weisbach or Thurston would then be the only engineers; he is not determined by the size of his income, or some of our railroad engineers, whose income exceeds that of the President of the United States, would be the only true engineers; nor is he the only engineer, who has given to the world one great thing. But he is truly an engineer who follows out a successful and thus useful career along some line of work which he enjoys and into which he can throw the enthusiasm of his whole nature. The man who calculates and tests and adjusts accurately every bolt and rivet and brace to the strain which is to come, is as true an engineer as he who superintends the construction of the bridge. The man carrying out perfectly a certain process, calculated to extricate a metal from its ore, and doing it carefully and thoroughly, is just as truly an engineer as he who originates the process.

With this ideal of what constitutes a true engineer we see that his acquirements must be partly technical. Several branches of study necessary in this technical training were noted. There are now 60 schools in the United States where technical education is given to fit for special duties. The graduates who obtain the best positions with corporations and in government work are those who devote themselves assiduously to the preparation these technical schools afford.

The practical side of the engineer's training was then dwelt upon. The necessity of noting the progress of discovery in the natural sciences as well as in inventions, and the application of those discoveries to the needs of daily life were emphasized.

Engineering in all its branches is now rapidly passing into a science without losing any of its elements as an art. To-day, scientific methods are employed without regard to precedent. The result is a wonderful advance along all lines of construction, and in valuable discoveries, from careful and well directed investigation.

DESIGN OF A HIGH-SPEED CORLISS EXPERIMENTAL ENGINE.

J. H. GILL, B. M. E.

This engine was designed to meet the needs of the department of mechanical engineering of the University for an experimental engine. The conditions imposed were: 1st. A compound condensing engine of ,100 horse-power. 2d. The steam to be taken from the University heating plant at a pressure of 75 pounds per square inch. 3d. The steam to be expanded at least 10 times between boiler and condenser, when working at the rated horse-power. 4th. The cylinders and receiver to be completely steam jacketed, and the piping so arranged that the steam may be cut off from the jackets completely or used in any one jacket as desired. 5th. Each cylinder to be provided with a separate flywheel, governor, etc., and thus be a complete engine by itself. The two engines were to be connected as one compound by a flange coupling on the crank shaft. The angle of the cranks was to be changed by shifting the coupling one or more holes. It was thought desirable to limit the size, so that most of the engine could be built in the University shops and thus furnish valuable shop practice for the students.

The condition of size necessitated a high speed to produce the required power. A four-valve engine was desirable on account of the variety of adjustments possible. As a high speed is impracticable with the ordinary Corliss dash-pot and releasing gear, it was decided to connect the valves direct to the eccentrics and use a shaft governor with a shifting cut-off eccentric; thus there could be obtained a positive valve motion. The speed selected was 200 revolutions per minute and stroke 18 inches. With 75 pounds gauge pressure and 10 expansions, these conditions required a low pressure cylinder 17 inches in diameter. To do equal work in each cylinder, the steam would have to expand about five times in the high pressure cylinder and twice in the low pressure cylinder. As it was thought more desirable to have nearly an equal number of expansions in each cylinder, a compromise was made between equal work and equal expansion, giving 3.5 expansions in the high pressure cylinder and 2.86 expansions in the low pressure cylinder.

The steam distribution was effected by means of modified Corliss valves, the modification consisting in using a double valve to admit the steam to the cylinder. This double valve is, in reality, two cylindrical Corliss valves, one inside the other. The outside or main steam valves and the exhaust valves are connected to a wrist-plate and this wrist-plate to the fixed eccentric in the ordinary way. The inside or cut-off valves are connected through an independent wrist-plate to the shifting eccentric and are regulated by the governor. They work in connection with the main steam valves in exactly the same manner as the double valve of the Buckeye engine. By means of this valve gear, the cut-off may be varied from zero to seven-eighths of the stroke, and at all but the extremely late cut-offs, the part is closed by the main and cut-off valves moving in opposite directions, thus giving a very rapid cut-off. The main steam valves control the admission, the exhaust valves the exhaust and compression, and the cut-off valves, working in their movable seats, the cut-off. All are independently adjustable, giving the greatest possible variety of adjustment for experimental purposes.

The governor is one of special design in which the principle of inertia is used in connection with that of the ordinary centrifugal governor. This combination gives a governor which will respond instantly to a change of load. It is readily seen that with a governor depending on centrifugal force alone, the speed must change perceptibly before the governor can act. But inertia supplies us with a force which will be called into action at the instant of any change in load. This will prevent, in a great measure, the instantaneous variation in speed, which is so objectionable in electric work.

The frame of the engine is modeled after that of the new experimental engine recently built for Sibley College by E. P. Allis & Co., of Milwaukee, and since adopted by them as their standard type of frame. The bearings are made large and the engine is designed safely to stand 150 pounds working steam pressure.

STEAM BOILER EXPLOSIONS.

H. E. SMITH, M. E., Asst. Professor of Mechanical Engineering.

The cause of boiler explosions is a subject which has often attracted the attention of engineers and scientists, and many theories have been advanced as to the causes of the terrible effects that usually accompany them.

The importance of having an accurate knowledge of these causes, from the standpoint of theoretical science as well as of practical engineering, would justify a much greater outlay of time and money than has hitherto been expended.

Among the many theories advanced, as to the causes of boiler explosions, the following may be mentioned:

1. *Excessive steam pressure.*—There can be no doubt that in rare cases this may be a cause, but in itself would not be sufficient to produce the enormous force exhibited in the majority of explosions.

2. *Electrical action.*—Audrand maintained that the evaporation of water in a boiler generated electricity and under such conditions as to cause an explosion. He argued that the electricity might collect on the brass fittings or tubes, in much the same manner as on a Leyden jar, and when discharged, cause the rupture of the boiler.

This theory apparently overlooks the fact, that the electricity which may possibly become free, collects on the surface of the boiler which is never insulated, by reason of the connecting steam pipes, and would thus be led off from the boiler. Besides, it is not proved that electricity itself explodes.

3. *Explosive gases.*—Some years ago, Perkins asserted that explosions were caused by the decomposition of water into its constituent gases; that by the reunion of these gases or by the hydrogen combining with any available air, an explosion would take place in the same manner as in the oxy-hydrogen gun.

Steam Boiler Explosions.   25

Another explanation was, that the oil which is returned to the boiler from the engine is decomposed, giving off hydrogen which mixes with the air in the feed water, and the compound fired by means of electric sparks obtained as previously stated.

These theories were overthrown by the investigations of the Commission, appointed by the Franklin Institute of Pennsylvania which demonstrated that water contained in a clean boiler heated red hot, is not decomposed; also by Shafhalt who has proved that one volume of oxy-hydrogen mixed with seven-tenths of one volume of steam will not explode.

The collection of carbonic oxide gas in the boiler flues, when the fires are banked, and its explosion when mixed with air has been advanced by some as a theory, but it is very improbable.

4. *Spheroidal state.*—A white hot metal ball, when plunged into soapy water, becomes immediately surrounded by vapor and produces no hissing sound; but when this vapor disappears, by reason of the cooling of the ball, a sudden evolution of steam takes place.

A condition similar to the above is claimed by some to be the cause of boiler explosions, and the lifting of the water from the heated surfaces, allowing them to become more highly heated has been known to be the indirect cause, at least, of some explosions.

Besides the above, may be mentioned: the super-heated water theory; the shock theory; the wedge theory and the Donny theory. The last contends that water, when deprived of air by any means, can be heated above its natural boiling point and has thereby an explosive tendency. This has been proved to be a very unlikely cause.

The last theory to be mentioned is one which is generally accepted as the most probable; namely, that of the stored energy in the steam and water due to high pressure, which is released when the pressure is reduced. If this be brought about quickly, as by the sudden opening of the steam valve or any collapse of fittings or flues, by which the compressive force upon the water in the boiler is diminished, the stored heat due to previous pressure and temperature is immediately available as energy. It was found by experiment, that a locomotive boiler at a pressure of 58.2 lbs., gauge pressure, discharged one-eighth of its contained water by continuous vaporization when, the fire being

drawn, the pressure was reduced to that of the atmosphere. If the steam so formed should be made to do work, being expanded down to atmospheric pressure, it has been found by calculation that the work done can be favorably compared with that of exploding gun powder, and that the destructive energy of one cubic foot of water, at a temperature which produces a pressure of 60 pounds per square inch, is equal to that of one pound of gun powder.

It is seen that if the pressure is quickly relieved from a boiler, through any cause, there is an enormous amount of energy available, which is ample to explain most, if not all, of the exhibitions of force in boiler explosions. Note the following: In the case of a plain cylindrical boiler, say of 10 horse-power at 100 pounds gauge pressure, and under ordinary conditions, the amount of stored energy is not far from 47,281,900 foot pounds or an amount sufficient to project the boiler over four miles. Again, a plain tubular boiler rated at 60 horsepower and carrying 75 lbs. pressure, under ordinary conditions, has 51,031,500 foot pounds of stored energy or enough to project the boiler one mile.

An instance of the explosion of a boiler of the first type is on record, where the boiler was first driven through a 16 inch wall, then several hundred feet through the air, cutting off an elm tree nine inches in diameter in its flight and partly destroying a house. Still other accidents attended by loss of life and great destruction of property might be cited. The only true preventive is good design and intelligent management.

ANGULAR VALUE OF BUBBLE-TUBES UNDER DIFFERENT TEMPERATURES.

J. JAY HANKENSON, B. C. E.

As is generally known all bubble-tubes have a slight curvature of the inner surface of the tube, and upon the degree of this curvature depends the angular value of one division of the tube. Any change in the degree of this curvature from whatever cause will change the value of a sensitive vial thereby rendering it useless for accurate work.

In connection with the U. S. C. and G. Survey work of Professor Hoag, it was found that the vial of the striding level on Level no. 3 did not have the tabulated angular value given by the official at Washington, and upon further investigation it was found that it varied with the temperature of the tube. The question then was: Does temperature change the value of all vials, or, is it due only to the mounting? As it was a government instrument we could not examine the mounting of the vial.

The tube was sent to Washington for examination. It was returned with "cork mounting," i.e., vial placed in "Y" and held in place by a piece of cork pressed by a light spring directly on the tube over the Y. This gives the freest mounting obtainable. The tube is tabulated to have different values at the extremes of temperature.

While this was going on a series of observations was made on a tube from the University transit circle. This is a very senitive bubble-tube and is very uniform in its curvature. It has the lightest of mounting, and no external cause can affect its curvature; any change must come from the glass itself. The observations were taken at temperatures ranging from 19° F. to 85° F. The time ranged through four months and observations were taken under as many conditions of detail as possible without altering the accuracy of results.

The level-trier used was one made by the writer, having a value of 3.882 seconds of arc for every 1-1000 of an inch of elevation of the micrometer screw.

drawn, the pressure was reduced to that of the atmosphere. If the steam so formed should be made to do work, being expanded down to atmospheric pressure, it has been found by calculation that the work done can be favorably compared with that of exploding gun powder, and that the destructive energy of one cubic foot of water, at a temperature which produces a pressure of 60 pounds per square inch, is equal to that of one pound of gun powder.

It is seen that if the pressure is quickly relieved from a boiler, through any cause, there is an enormous amount of energy available, which is ample to explain most, if not all, of the exhibitions of force in boiler explosions. Note the following: In the case of a plain cylindrical boiler, say of 10 horse-power at 100 pounds gauge pressure, and under ordinary conditions, the amount of stored energy is not far from 47,281,900 foot pounds or an amount sufficient to project the boiler over four miles. Again, a plain tubular boiler rated at 60 horsepower and carrying 75 lbs. pressure, under ordinary conditions, has 51,031,500 foot pounds of stored energy or enough to project the boiler one mile.

An instance of the explosion of a boiler of the first type is on record, where the boiler was first driven through a 16 inch wall, then several hundred feet through the air, cutting off an elm tree nine inches in diameter in its flight and partly destroying a house. Still other accidents attended by loss of life and great destruction of property might be cited. The only true preventive is good design and intelligent management.

ANGULAR VALUE OF BUBBLE-TUBES UNDER DIFFERENT TEMPERATURES.

J. JAY HANKENSON, B. C. E.

As is generally known all bubble-tubes have a slight curvature of the inner surface of the tube, and upon the degree of this curvature depends the angular value of one division of the tube. Any change in the degree of this curvature from whatever cause will change the value of a sensitive vial thereby rendering it useless for accurate work.

In connection with the U. S. C. and G. Survey work of Professor Hoag, it was found that the vial of the striding level on Level no. 3 did not have the tabulated angular value given by the official at Washington, and upon further investigation it was found that it varied with the temperature of the tube. The question then was: Does temperature change the value of all vials, or, is it due only to the mounting? As it was a government instrument we could not examine the mounting of the vial.

The tube was sent to Washington for examination. It was returned with "cork mounting," i. e., vial placed in "Y" and held in place by a piece of cork pressed by a light spring directly on the tube over the Y. This gives the freest mounting obtainable. The tube is tabulated to have different values at the extremes of temperature.

While this was going on a series of observations was made on a tube from the University transit circle. This is a very senitive bubble-tube and is very uniform in its curvature. It has the lightest of mounting, and no external cause can affect its curvature; any change must come from the glass itself. The observations were taken at temperatures ranging from 19° F. to 85° F. The time ranged through four months and observations were taken under as many conditions of detail as possible without altering the accuracy of results.

The level-trier used was one made by the writer, having a value of 3.882 seconds of arc for every 1-1000 of an inch of elevation of the micrometer screw.

drawn, the pressure was reduced to that of the atmosphere. If the steam so formed should be made to do work, being expanded down to atmospheric pressure, it has been found by calculation that the work done can be favorably compared with that of exploding gun powder, and that the destructive energy of one cubic foot of water, at a temperature which produces a pressure of 60 pounds per square inch, is equal to that of one pound of gun powder.

It is seen that if the pressure is quickly relieved from a boiler, through any cause, there is an enormous amount of energy available, which is ample to explain most, if not all, of the exhibitions of force in boiler explosions. Note the following: In the case of a plain cylindrical boiler, say of 10 horse-power at 100 pounds gauge pressure, and under ordinary conditions, the amount of stored energy is not far from 47,281,900 foot pounds or an amount sufficient to project the boiler over four miles. Again, a plain tubular boiler rated at 60 horsepower and carrying 75 lbs. pressure, under ordinary conditions, has 51,031,500 foot pounds of stored energy or enough to project the boiler one mile.

An instance of the explosion of a boiler of the first type is on record, where the boiler was first driven through a 16 inch wall, then several hundred feet through the air, cutting off an elm tree nine inches in diameter in its flight and partly destroying a house. Still other accidents attended by loss of life and great destruction of property might be cited. The only true preventive is good design and intelligent management.

ANGULAR VALUE OF BUBBLE-TUBES UNDER DIFFERENT TEMPERATURES.

J. JAY HANKENSON, B. C. E.

As is generally known all bubble-tubes have a slight curvature of the inner surface of the tube, and upon the degree of this curvature depends the angular value of one division of the tube. Any change in the degree of this curvature from whatever cause will change the value of a sensitive vial thereby rendering it useless for accurate work.

In connection with the U. S. C. and G. Survey work of Professor Hoag, it was found that the vial of the striding level on Level no. 3 did not have the tabulated angular value given by the official at Washington, and upon further investigation it was found that it varied with the temperature of the tube. The question then was: Does temperature change the value of all vials, or, is it due only to the mounting? As it was a government instrument we could not examine the mounting of the vial.

The tube was sent to Washington for examination. It was returned with "cork mounting," i. e., vial placed in "Y" and held in place by a piece of cork pressed by a light spring directly on the tube over the Y. This gives the freest mounting obtainable. The tube is tabulated to have different values at the extremes of temperature.

While this was going on a series of observations was made on a tube from the University transit circle. This is a very senitive bubble-tube and is very uniform in its curvature. It has the lightest of mounting, and no external cause can affect its curvature; any change must come from the glass itself. The observations were taken at temperatures ranging from 19° F. to 85° F. The time ranged through four months and observations were taken under as many conditions of detail as possible without altering the accuracy of results.

The level-trier used was one made by the writer, having a value of 3.882 seconds of arc for every 1-1000 of an inch of elevation of the micrometer screw.

drawn, the pressure was reduced to that of the atmosphere. If the steam so formed should be made to do work, being expanded down to atmospheric pressure, it has been found by calculation that the work done can be favorably compared with that of exploding gun powder, and that the destructive energy of one cubic foot of water, at a temperature which produces a pressure of 60 pounds per square inch, is equal to that of one pound of gun powder.

It is seen that if the pressure is quickly relieved from a boiler, through any cause, there is an enormous amount of energy available, which is ample to explain most, if not all, of the exhibitions of force in boiler explosions. Note the following: In the case of a plain cylindrical boiler, say of 10 horse-power at 100 pounds gauge pressure, and under ordinary conditions, the amount of stored energy is not far from 47,281,900 foot pounds or an amount sufficient to project the boiler over four miles. Again, a plain tubular boiler rated at 60 horsepower and carrying 75 lbs. pressure, under ordinary conditions, has 51,031,500 foot pounds of stored energy or enough to project the boiler one mile.

An instance of the explosion of a boiler of the first type is on record, where the boiler was first driven through a 16 inch wall, then several hundred feet through the air, cutting off an elm tree nine inches in diameter in its flight and partly destroying a house. Still other accidents attended by loss of life and great destruction of property might be cited. The only true preventive is good design and intelligent management.

ANGULAR VALUE OF BUBBLE-TUBES UNDER DIFFERENT TEMPERATURES.

J. Jay Hankenson, B. C. E.

As is generally known all bubble-tubes have a slight curvature of the inner surface of the tube, and upon the degree of this curvature depends the angular value of one division of the tube. Any change in the degree of this curvature from whatever cause will change the value of a sensitive vial thereby rendering it useless for accurate work.

In connection with the U. S. C. and G. Survey work of Professor Hoag, it was found that the vial of the striding level on Level no. 3 did not have the tabulated angular value given by the official at Washington, and upon further investigation it was found that it varied with the temperature of the tube. The question then was: Does temperature change the value of all vials, or, is it due only to the mounting? As it was a government instrument we could not examine the mounting of the vial.

The tube was sent to Washington for examination. It was returned with "cork mounting," i. e., vial placed in "Y" and held in place by a piece of cork pressed by a light spring directly on the tube over the Y. This gives the freest mounting obtainable. The tube is tabulated to have different values at the extremes of temperature.

While this was going on a series of observations was made on a tube from the University transit circle. This is a very senitive bubble-tube and is very uniform in its curvature. It has the lightest of mounting, and no external cause can affect its curvature; any change must come from the glass itself. The observations were taken at temperatures ranging from 19° F. to 85° F. The time ranged through four months and observations were taken under as many conditions of detail as possible without altering the accuracy of results.

The level-trier used was one made by the writer, having a value of 3.882 seconds of arc for every 1-1000 of an inch of elevation of the micrometer screw.

drawn, the pressure was reduced to that of the atmosphere. If the steam so formed should be made to do work, being expanded down to atmospheric pressure, it has been found by calculation that the work done can be favorably compared with that of exploding gun powder, and that the destructive energy of one cubic foot of water, at a temperature which produces a pressure of 60 pounds per square inch, is equal to that of one pound of gun powder.

It is seen that if the pressure is quickly relieved from a boiler, through any cause, there is an enormous amount of energy available, which is ample to explain most, if not all, of the exhibitions of force in boiler explosions. Note the following: In the case of a plain cylindrical boiler, say of 10 horse-power at 100 pounds gauge pressure, and under ordinary conditions, the amount of stored energy is not far from 47,281,900 foot pounds or an amount sufficient to project the boiler over four miles. Again, a plain tubular boiler rated at 60 horse-power and carrying 75 lbs. pressure, under ordinary conditions, has 51,031,500 foot pounds of stored energy or enough to project the boiler one mile.

An instance of the explosion of a boiler of the first type is on record, where the boiler was first driven through a 16 inch wall, then several hundred feet through the air, cutting off an elm tree nine inches in diameter in its flight and partly destroying a house. Still other accidents attended by loss of life and great destruction of property might be cited. The only true preventive is good design and intelligent management.

ANGULAR VALUE OF BUBBLE-TUBES UNDER DIFFERENT TEMPERATURES.

J. JAY HANKENSON, B. C. E.

As is generally known all bubble-tubes have a slight curvature of the inner surface of the tube, and upon the degree of this curvature depends the angular value of one division of the tube. Any change in the degree of this curvature from whatever cause will change the value of a sensitive vial thereby rendering it useless for accurate work.

In connection with the U. S. C. and G. Survey work of Professor Hoag, it was found that the vial of the striding level on Level no. 3 did not have the tabulated angular value given by the official at Washington, and upon further investigation it was found that it varied with the temperature of the tube. The question then was: Does temperature change the value of all vials, or, is it due only to the mounting? As it was a government instrument we could not examine the mounting of the vial.

The tube was sent to Washington for examination. It was returned with "cork mounting," i. e., vial placed in "Y" and held in place by a piece of cork pressed by a light spring directly on the tube over the Y. This gives the freest mounting obtainable. The tube is tabulated to have different values at the extremes of temperature.

While this was going on a series of observations was made on a tube from the University transit circle. This is a very senitive bubble-tube and is very uniform in its curvature. It has the lightest of mounting, and no external cause can affect its curvature; any change must come from the glass itself. The observations were taken at temperatures ranging from 19° F. to 85° F. The time ranged through four months and observations were taken under as many conditions of detail as possible without altering the accuracy of results.

The level-trier used was one made by the writer, having a value of 3.882 seconds of arc for every 1-1000 of an inch of elevation of the micrometer screw.

drawn, the pressure was reduced to that of the atmosphere. If the steam so formed should be made to do work, being expanded down to atmospheric pressure, it has been found by calculation that the work done can be favorably compared with that of exploding gun powder, and that the destructive energy of one cubic foot of water, at a temperature which produces a pressure of 60 pounds per square inch, is equal to that of one pound of gun powder.

It is seen that if the pressure is quickly relieved from a boiler, through any cause, there is an enormous amount of energy available, which is ample to explain most, if not all, of the exhibitions of force in boiler explosions. Note the following: In the case of a plain cylindrical boiler, say of 10 horse-power at 100 pounds gauge pressure, and under ordinary conditions, the amount of stored energy is not far from 47,281,900 foot pounds or an amount sufficient to project the boiler over four miles. Again, a plain tubular boiler rated at 60 horse-power and carrying 75 lbs. pressure, under ordinary conditions, has 51,031,500 foot pounds of stored energy or enough to project the boiler one mile.

An instance of the explosion of a boiler of the first type is on record, where the boiler was first driven through a 16 inch wall, then several hundred feet through the air, cutting off an elm tree nine inches in diameter in its flight and partly destroying a house. Still other accidents attended by loss of life and great destruction of property might be cited. The only true preventive is good design and intelligent management.

ANGULAR VALUE OF BUBBLE-TUBES UNDER DIFFERENT TEMPERATURES.

J. JAY HANKENSON, B. C. E.

As is generally known all bubble-tubes have a slight curvature of the inner surface of the tube, and upon the degree of this curvature depends the angular value of one division of the tube. Any change in the degree of this curvature from whatever cause will change the value of a sensitive vial thereby rendering it useless for accurate work.

In connection with the U. S. C. and G. Survey work of Professor Hoag, it was found that the vial of the striding level on Level no. 3 did not have the tabulated angular value given by the official at Washington, and upon further investigation it was found that it varied with the temperature of the tube. The question then was: Does temperature change the value of all vials, or, is it due only to the mounting? As it was a government instrument we could not examine the mounting of the vial.

The tube was sent to Washington for examination. It was returned with "cork mounting," i.e., vial placed in "Y" and held in place by a piece of cork pressed by a light spring directly on the tube over the Y. This gives the freest mounting obtainable. The tube is tabulated to have different values at the extremes of temperature.

While this was going on a series of observations was made on a tube from the University transit circle. This is a very senitive bubble-tube and is very uniform in its curvature. It has the lightest of mounting, and no external cause can affect its curvature; any change must come from the glass itself. The observations were taken at temperatures ranging from 19° F. to 85° F. The time ranged through four months and observations were taken under as many conditions of detail as possible without altering the accuracy of results.

The level-trier used was one made by the writer, having a value of 3.882 seconds of arc for every 1-1000 of an inch of elevation of the micrometer screw.

drawn, the pressure was reduced to that of the atmosphere. If the steam so formed should be made to do work, being expanded down to atmospheric pressure, it has been found by calculation that the work done can be favorably compared with that of exploding gun powder, and that the destructive energy of one cubic foot of water, at a temperature which produces a pressure of 60 pounds per square inch, is equal to that of one pound of gun powder.

It is seen that if the pressure is quickly relieved from a boiler, through any cause, there is an enormous amount of energy available, which is ample to explain most, if not all, of the exhibitions of force in boiler explosions. Note the following: In the case of a plain cylindrical boiler, say of 10 horse-power at 100 pounds gauge pressure, and under ordinary conditions, the amount of stored energy is not far from 47,281,900 foot pounds or an amount sufficient to project the boiler over four miles. Again, a plain tubular boiler rated at 60 horse-power and carrying 75 lbs. pressure, under ordinary conditions, has 51,031,500 foot pounds of stored energy or enough to project the boiler one mile.

An instance of the explosion of a boiler of the first type is on record, where the boiler was first driven through a 16 inch wall, then several hundred feet through the air, cutting off an elm tree nine inches in diameter in its flight and partly destroying a house. Still other accidents attended by loss of life and great destruction of property might be cited. The only true preventive is good design and intelligent management.

ANGULAR VALUE OF BUBBLE-TUBES UNDER DIFFERENT TEMPERATURES.

J. JAY HANKENSON, B. C. E.

As is generally known all bubble-tubes have a slight curvature of the inner surface of the tube, and upon the degree of this curvature depends the angular value of one division of the tube. Any change in the degree of this curvature from whatever cause will change the value of a sensitive vial thereby rendering it useless for accurate work.

In connection with the U. S. C. and G. Survey work of Professor Hoag, it was found that the vial of the striding level on Level no. 3 did not have the tabulated angular value given by the official at Washington, and upon further investigation it was found that it varied with the temperature of the tube. The question then was: Does temperature change the value of all vials, or, is it due only to the mounting? As it was a government instrument we could not examine the mounting of the vial.

The tube was sent to Washington for examination. It was returned with "cork mounting," i. e., vial placed in "Y" and held in place by a piece of cork pressed by a light spring directly on the tube over the Y. This gives the freest mounting obtainable. The tube is tabulated to have different values at the extremes of temperature.

While this was going on a series of observations was made on a tube from the University transit circle. This is a very senitive bubble-tube and is very uniform in its curvature. It has the lightest of mounting, and no external cause can affect its curvature; any change must come from the glass itself. The observations were taken at temperatures ranging from 19° F. to 85° F. The time ranged through four months and observations were taken under as many conditions of detail as possible without altering the accuracy of results.

The level-trier used was one made by the writer, having a value of 3.882 seconds of arc for every 1-1000 of an inch of elevation of the micrometer screw.

28   *J. Jay Hankenson.*

The values given in the following table are the results of these observations, 254 in number. The first column gives the temperature at which the observations were taken, the second the number of observations, and the third the average number of divisions through which the bubble moved for every 1-1000 inch elevation, or 3.882 seconds of arc that the trier moved. The mean of all these values gives 1.7991 seconds of arc per division of tube, as the value of the vial. But from an investigation of these results it is found that the residuals are both + and − for high or low temperatures and vary as much in value in either direction. Therefore, there is no indication that temperature affects the curvature of the vial. Hence we conclude that changes of curvature due to temperature must come from direct pressure on the vial from improper mounting and not from the glass itself.

AVERAGE OF ALL THE READINGS.

Temperature in degrees F.	Number of readings.	Average value.	Residual.
19	8	2.167	−.0069
20	11	2.1009	+.0595
21	3	2.1886	−.0280
36.5	4	2.235	−.0744
48	16	2.1815	−.0209
49	31	2.11271	+.0479
49.5	8	2.1716	−.0110
50	28	2.1880	−.0274
62	68	2.08926	+.0704
69	27	2.1872	−.0266
69.5	18	2.13865	+.0220
70	11	2.2639	−.1033
70.5	16	2.08145	+.0782
71	12	2.08877	+.0718
81	1	2.2250	−.0644
84.5	2	2.2195	−.0589

PIPING FOR STEAM ENGINES.

WM. A. PIKE, B. S., *Lecturer in Mechanical Engineering.*

Some recent experience of the writer has led him to believe, that in the matter of the piping between engines and boilers there is a field for investigation and experiment which has not been fully occupied. The conclusion that he has been forced to is, that, while every precaution is often taken to provide the best form of engine and boiler for any given plant, very little attention is paid to the piping connecting them, an omission which he believes has often rendered an otherwise well designed plant comparatively inefficient.

To illustrate this point an example will be taken of a plant consisting of ten Westinghouse compound engines and seven Stirling boilers. The engines were 16x27x16 inches, making 250 revolutions per minute and of nominal horse-power at a gauge pressure of 140 pounds. The steam from the boilers first passed to a drum 28 inches in diameter and 45 feet long. From this drum were 8-inch pipes averaging, perhaps, 76 feet long with 5 rightangle turns in each pipe, one pipe for each engine.

The result was that there was found to be a *drop* of pressure between boilers and engines of from eight to twenty and averaging fourteen pounds. To obtain some information on the cause of this drop, a Thompson indicator with an 80 pound spring was placed on the steam pipe close to the throttle valve of one of the engines. The motion given to the drum was coincident with that of the engine. Diagrams, of which figures *1* and *2* are copies, were obtained.

On studying these diagrams it was seen that at the point *a*, corresponding to the beginning of the stroke of the engine, the pressure was about the same as in the boiler; that the pressure fell off very rapidly as the engine called for steam, as at *a* and *b*, and remained nearly constant, averaging about fourteen pounds below boiler pressure until cut-off occurred at *c*. From this point the pressure rose.

Fig. 1.

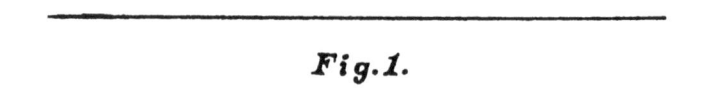

Fig. 2.

gradually, until at the end of the stroke *d* it was considerably above boiler pressure. From this point the pressure fell until it was again at about boiler pressure at *a*, the end of the return stroke. If the engine was not heavily loaded the return line would cross the out stroke line as in figure *1*, or if heavily loaded would make an open figure as in figure *2*. It is evident that when the valve opened at *a* there was a sudden demand for steam, in excess of the capacity of the pipe to supply. The steam surged back and forth in the pipes, because it was used, perhaps only one fourth of the time and the demand was then so very great. The result naturally would be that the steam, being in rapid motion, was suddenly stopped when cut-off occurred, causing it to rebound and tend to fly back in the pipe.

Now, it is just as impossible for steam to vibrate back and forth in a pipe, without expenditure of energy, as it is for any solid body to do the same thing. Evidently then, a certain amount of the energy of the steam must be used up in doing this work, besides the amount necessary to overcome the friction of the steam on the pipe surfaces.

It will also be noted, that as the engine only calls for steam one quarter of the time (if the cut-off is one-half), the demand on the pipe during that time is four times as great as the *average* demand during a whole revolution.

Two remedies suggested themselves, viz:—either largely to increase the size of the pipe, thus reducing the losses, or in some way to make the draft on the pipe nearly uniform. The latter plan was thought to be the better and an experiment made to ascertain if the idea were correct. The plan followed was to place a receiver, 36 inches in diameter and 5 feet long, between the 8-inch pipe and one of the engines; this was also connected through an intermediate receiver of 14-inch pipe with an engine on either side of the one above referred to. It will be seen that the object was to feed one, two or three engines, through the receiver with one eight-inch pipe. The theory was that the receiver, close to the engines, would act as a reservoir and cause the flow through the 8-inch pipe to be practically uniform.

The result was even better than was expected. It was found that with one engine drawing from the receiver, the *drop* was one pound,

Fig. 1.

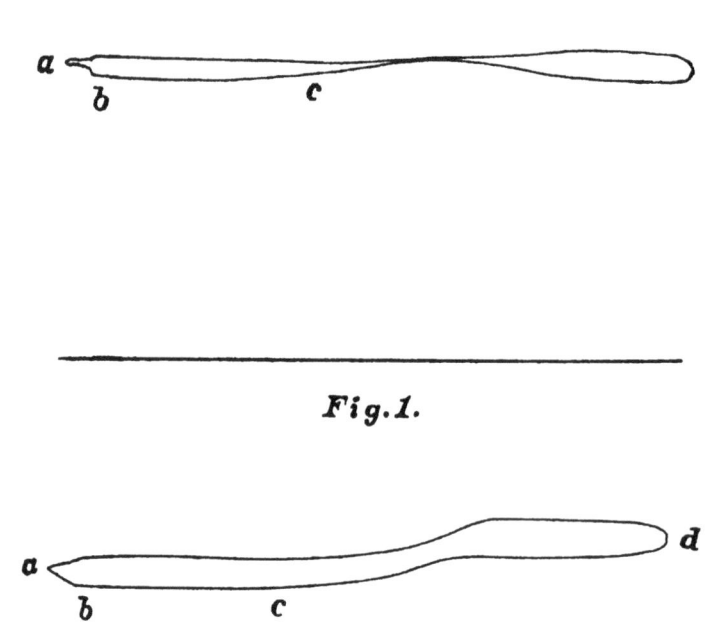

Fig. 2.

gradually, until at the end of the stroke *d* it was considerably above boiler pressure. From this point the pressure fell until it was again at about boiler pressure at *a*, the end of the return stroke. If the engine was not heavily loaded the return line would cross the out stroke line as in figure *1*, or if heavily loaded would make an open figure as in figure *2*. It is evident that when the valve opened at *a* there was a sudden demand for steam, in excess of the capacity of the pipe to supply. The steam surged back and forth in the pipes, because it was used, perhaps only one fourth of the time and the demand was then so very great. The result naturally would be that the steam, being in rapid motion, was suddenly stopped when cut-off occurred, causing it to rebound and tend to fly back in the pipe.

Now, it is just as impossible for steam to vibrate back and forth in a pipe, without expenditure of energy, as it is for any solid body to do the same thing. Evidently then, a certain amount of the energy of the steam must be used up in doing this work, besides the amount necessary to overcome the friction of the steam on the pipe surfaces.

It will also be noted, that as the engine only calls for steam one quarter of the time (if the cut-off is one-half), the demand on the pipe during that time is four times as great as the *average* demand during a whole revolution.

Two remedies suggested themselves, viz:—either largely to increase the size of the pipe, thus reducing the losses, or in some way to make the draft on the pipe nearly uniform. The latter plan was thought to be the better and an experiment made to ascertain if the idea were correct. The plan followed was to place a receiver, 36 inches in diameter and 5 feet long, between the 8-inch pipe and one of the engines; this was also connected through an intermediate receiver of 14-inch pipe with an engine on either side of the one above referred to. It will be seen that the object was to feed one, two or three engines, through the receiver with one eight-inch pipe. The theory was that the receiver, close to the engines, would act as a reservoir and cause the flow through the 8-inch pipe to be practically uniform.

Th   was even better than was expected. It was found that
 ith   draw   receiver, the *drop* was one pound,

with two engines a trifle under two pounds and with three engines about five pounds. From this experiment it was determined that one 8-inch pipe would feed two engines, with not more *drop* than is considered good practice and that with three engines the *drop* would be but little over one-third of that obtained with one engine drawing directly on an 8-inch pipe; but that a receiver, of ample size, as close as possible to the engine is of the greatest importance. Experience with other plants both with and without reservoir space has shown similar results.

The writer is of the opinion that considerable more care, than usual, could be given to the method of piping engines with a perceptible effect on the expense of steam power.

It is perhaps of interest to note, that the plant above referred to is to be re-piped with a receiver close to each engine, the receivers being supplied by a pipe much smaller than would be considered advisable if no receivers were used.

AN INVESTIGATION OF THE CORLISS DROP CUT-OFF MOTION.

RALPH P. FELTON, B. M. E.

Abstract of thesis presented for graduation.

The object of the investigation was to determine, in the case of various Corliss engines, the actual time which elapsed during the fall of the dash-pot piston, and also the time actually consumed in cutting off the steam at the admission port.

The attainment of this object involved the use of a graphical method, capable of analyzing the action of the dash-pot piston, and also furnishing the means for ascertaining the following additional points of interest:

The relative velocity of the engine valve and dash-pot piston at point of cut-off.

The nature of the drop motion, (*i.e.*) whether or not it is uniformly accelerated.

The relative speed of the upward and downward movement of the dash-pot piston.

The actual loss of work due to not having a theoretically perfect cut-off.

By means of an ingenious device, the motions of the valve and dash-pot piston were recorded on the drum of a Thomson indicator.

It was found that the time required for the dash-pot piston to fall, varied from .08 to .12 of a second.

Velocity of dash-pot piston at point of cut-off was 7.2 to 8 feet per second.

The time required for closure of steam port from maximum opening varied from .03 to .04 of a second.

Velocity of valve at point of cut-off was 1.3 to 1.7 feet per second.

Relative velocity of valve and dash-pot piston was about 5.5.

The average time of ascent of dash-pot piston was .15 of a second.

Relative velocity of upward and downward movements was in the ratio of one to two.

The drop motion is uniformly accelerated.

Loss of work due to not having a theoretically perfect cut-off, varied from .3 to 1.2 of one per cent.

PRECISE LEVELING.

WM. R. HOAG, C. E., *Professor of Civil Engineering.*

The instruments and methods employed in determining horizontal distance and direction for ordinary land and railway surveys are inadequate for the solution of the same problems when conducted on a continental scale. So, too, in the determination of differences of altitude, the engineer's level is not suitable when an extended system of levels is to be established covering a large area. In the first class of problems, the surveyor's chain and magnetic compass, and the engineer's tape and solar compass are used each for the different grades of work, giving an accuracy of 1 in 300 to 1 in 3,000 with the chain and of 1 in 2,000 to 1 in 20,000 with the tape for linear measurement; of 5 to 15 minutes with the magnetic compass and 1 to 5 minutes with the solar compass for direction; and from 30 seconds to 3 minutes with the engineer's transit in simple angular measurement.

A continental survey demands in linear measurement a probable error ranging between 1 in 100,000 to 1 in 1,000,000, and angles known within 1 to 5 seconds of their true value; with astronomical latitude, longitude and azimuth from 1-10 to 1 second of true value. About this same range in desirable accuracy obtains with altitude determinations from the rudest hypsometrical measures to the most refined precise leveling. With ordinary leveling 0.05 feet distance in miles represents the probable error for good work, but in precise leveling 0.005 feet distance in miles is not uncommon. While in precise leveling the two principal sources of error are theoretically allowed full correction through instrumental manipulation, all experience goes to show that the sources of error are so many and varied that the greatest care as to every detail is necessary to make the results satisfactory.

It is not intended in this article to give a complete account of the method of running precise levels or even to discuss systematically the various sources of error to which precise leveling is exposed, but rather to touch upon some of these errors and point out possible remedies which have been suggested to the writer during the past two seasons while using a precise level, of the Kern type, belonging to the U. S. Coast and Geodetic Survey.

The precise level differs from the engineer's level in three essential details; first, the bubble is mounted in a striding level which rides the collars of the telescope and is readily detachable. Second, the barrel of the telescope is provided with stops to enable it to be turned in the wyes just a half circumference. Third, one of the wyes is provided with a vertical motion, the amount of such motion being measured by the actuating micrometer screw. Among the details of minor difference from the ordinary level, might be mentioned a broad stand with rigidly braced tripod legs; three leveling screws; a means of lifting the telescope free of its bearings on the wyes; a high power inverting telescope; a bubble having a radius of about 500 feet; a small mirror for reading the position of the bubble, and two screens, one to protect the instrument against the direct rays of the sun, and the other, against wind.

The level-rods, usually two in number, are provided with a watch level to aid in plumbing; with a thermometer to indicate the temperature of the metal scale of the rod, and with a foot-plate into which the shoe at the lower end of the rod fits while being sighted upon.

Employing a target-rod, the usual practice with the Coast and Geodetic Survey is to have the rodman fix the target as near as convenient to the horizontal wire with the bubble near the center, then determine the small angle of the elevation or depression which the line through the center of the telescope and the target makes with an horizontal line at the instrument. It is the manner of measuring this angle to which we wish to call attention. The method being to determine it in terms of the micrometer screw, by employing the latter in causing the telescope to move through the vertical angle as defined by the horizontal wire and center of target for one limit, and the horizontal plane or "horizon" as it is called, as indicated by bringing the bubble to center or to an adopted division for the other limit. By taking the telescope in erect and inverted positions and the striding level in direct and reversed positions, the ordinary collimation and level errors are eliminated.

An examination of the record of a month's work shows ten extremes of the two micrometer values for the horizon which should in theory agree, very rarely to exceed 9 seconds or about 4 millimeters in a hundred meters, while their corresponding extremes on target are but three seconds or a little over 1 millimeter per 100 meters.

It thus appears that we are taking the same number of pointings on two objects to fix their angle of divergence when, from some cause, the pointings on the one are exposed to errors three times as large as on the other. Economy would suggest that more time be devoted to determination of horizon and less time to target, or some improvement be made in the method of determining horizon. The plan of taking two or more readings for horizon to one for target has been tried by the author with varying degrees of success. The objection, in theory, being that the repetitions are not independent determinations and hence are likely to repeat all errors with the same signs. In practice it appears that no considerable increase in accuracy is insured by multiple horizons.

The following plan devised by the author and hitherto untried so far as known to him, it is believed will greatly reduce the unreliability of the horizon pointing. Instead of attempting to bring the bubble to center or adopted division,—the former is much more difficult than the latter and possesses no real advantage with ordinary care to maintaining a constant length of bubble—*allow the bubble to settle, with the last movement in an adopted direction, near the 'center' or 'adopted division'* and either record its position or with an equivalent scale along the bubble tube expressing divisions of micrometer screw, read off, and apply the correction to micrometer at the time. This addition to or substraction from the micrometer reading, to reduce it to micrometer bubble-center, would be aided instrumentally by replacing the ordinary index mark, employed in reading the micrometer head, by a scale, the zero of which should correspond with the index mark and should extend both ways around the graduated head of the micrometer screw.

If the position of the bubble be on the minus side of center 3.5

divisions, the micrometer screw would be read for that graduation coming opposite the 3.5 division on the minus side of the screw or index mark. The particular points of advantage being, that the bubble's last movement is to be an adopted direction, which on account of adhesion and friction between the glass and the æther is not to be neglected, and that the time usually consumed, in vainly trying to bring the bubble to center, can be given in allowing the bubble to seek its true static position, which with the proper reduction scales, gives the true micrometer reading for a static horizon.

Except for employing target sights considerably inclined, which expedites matters very little on uneven ground unless considerable micrometer screw be used and is not to be employed in work of high precision, it is believed that a bubble tube of sufficient length and regularity could be constructed wholly to displace the micrometer screw.

A bubble tube with a range of 20 divisions of 2 inches value each, not an unusual length for tubes of this precision, would furnish a capacity in vertical reach of 20 millimeters at a distance of 100 meters or 5 millimeters at 25 meters and little difficulty would be found in getting the target quickly fixed within these limits.

TURNOUTS FROM RAILWAY CURVES.

NOAH JOHNSON, '94.

It is the purpose of this paper to discuss and point out the important relations between the rate of curvature of a turnout from a curved track and the rate of curvature of the track itself.

In figures I., II. and III., let D C = g = gauge; D B = d = switch throw; angle D A B = S = switch angle, and angle G F M = F = frog angle.

Presuming that these are the same in all three cases, let us first find the value of the line B F, the distance from the end of the switched rail to the frog, in each case. In Fig. I., since F K and D H are perpendicular to F G, and E F is perpendicular to F M, C L F = E F K = G F M = F. In the right angled triangle B F C, B F = B C ÷ sin B F C. But B C = g − d and B F C = 90° − (90° − ½ B E F − S) = ½ B E F + S. But ½ B E F = ½ (F − S), therefore B F C = ½ (F + S). Substituting the values of B C and B F C, we have

$$BF = \frac{(g-d)}{\sin \frac{1}{2}(F+S)} \qquad (1)$$

In Fig. II., calling B K F = K and E F K = F, we have in the triangle B F C, $BF = \frac{BC \sin BCF}{\sin BFC}$. But B C = g − d and B C F = 90° + ½ K, or sin B C F = cos ½ K. Now B F K = K F C + B F C and F B K = K C F − B F C = K F C − B F C, then B F K − F B K = 2 B F C. But B F K − F B K = F + S, therefore 2 B F C = F + S, or B F C = ½ (F + S). Substituting these values we have,

$$BF = \frac{(g-d)\cos \frac{1}{2} K}{\sin \frac{1}{2}(F+S)}. \qquad (2)$$

In the same manner it can be shown that in Fig. III., $BF = \frac{(g-d)\cos \frac{1}{2} K}{\sin \frac{1}{2}(F+S)}$. (3). Since ½ K is always very small for all ordinary railroad curves, cos ½ K may be called unity, in which case B K will have the same value for any ordinary curve as it has for a straight line of track, that is to say, it is a constant as long as F, S, g and d remain the same. Now in Figs. II. and III., F H and O B, being very small, compared with B F and O H and the angle K being also very small, B F is but slightly greater than O H. But B F is also seen to

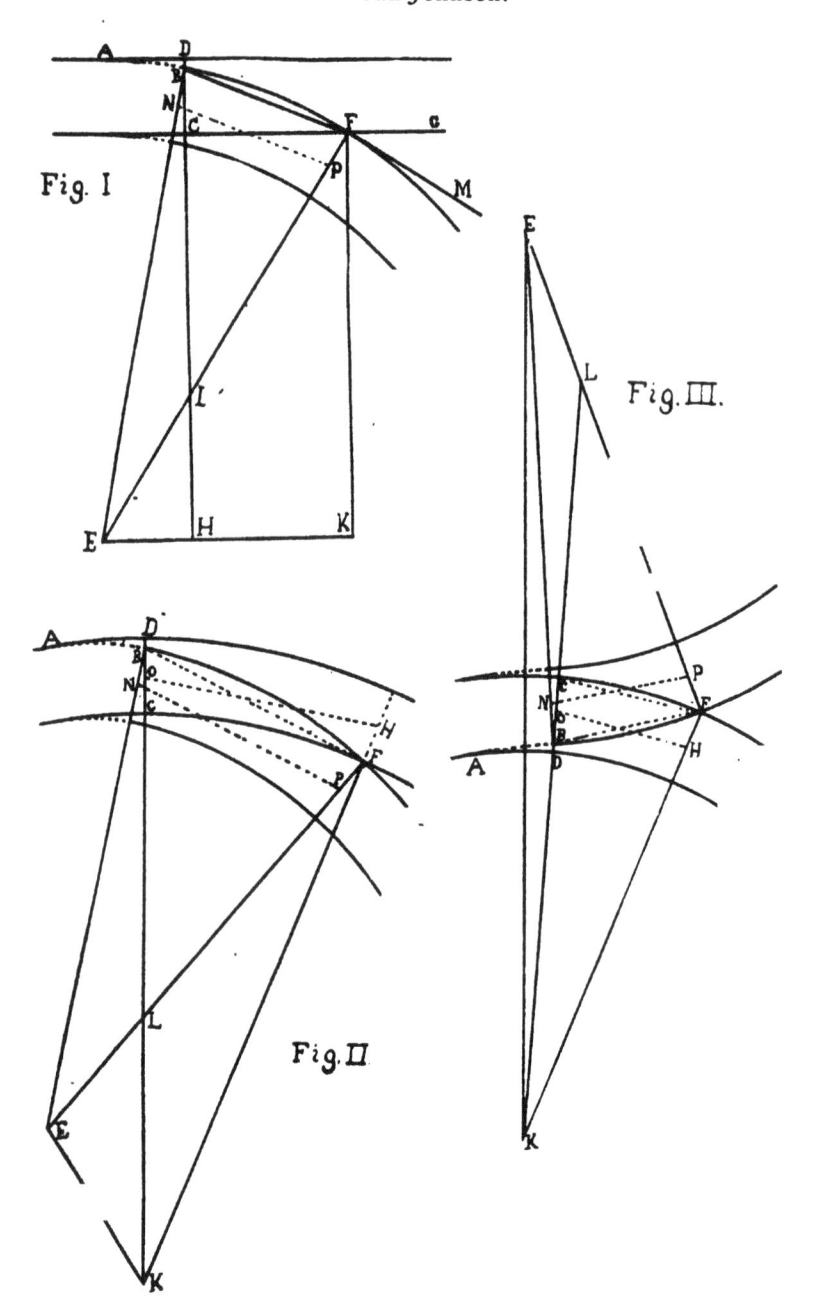

Fig. I

Fig. II

Fig. III.

be slightly greater than N P, therefore we assume that O H = N P, practically.

Now let $d = (F - S)\frac{100}{NP}$ = the degree of curvature of a turnout from a straight track, also let $d_2 = K\frac{100}{OH}$ = the degree of curvature of a main line of track from which a turnout is to be made.

Let $d_3 = BEF\frac{100}{NP} = [(F - S) + K]\frac{100}{NP}$ = the degree of curvature of the turnout in Fig. II.

Let $d_4 = BEF\frac{100}{NP} = [(F - S) - K]\frac{100}{NP}$ = the degree of curvature of the turnout in Fig. III.

Now $(F - S)\frac{100}{NP} + K\frac{100}{OH} = [(F - S) + K]\frac{100}{NP}$. (4)

and $(F - S)\frac{100}{NP} - K\frac{100}{OH} = [(F - S) - K]\frac{100}{NP}$. (5)

Therefore $d_3 = d_1 + d_2$ (6), and $d_4 = d_1 - d_2$. (7)

Thus equations (6) and (7) show the desired relations, viz.: That the degree of curvature, of a turnout on the inside of a curved track, is equal to the degree of curvature of a turnout from a straight track, *plus* the degree of curvature of the main track; also that the degree of curvature of a turnout on the outside of a curved track is equal to the degree of curvature of a turnout from a straight track, *minus* the degree of curvature of the main track. If this last quantity comes out *plus*, it is a curve of the opposite direction from the main track. If it comes out *minus*, the curve is in the same direction as the main track.

It is obvious that the same rule holds good in split-toe switches.

Now, from the nature of the assumptions which have been made it is evident that the less the rate of curvature of the main track, the more nearly will the results obtained correspond to those obtained by the strictly analytical methods. Taking therefore an extreme case, and solving for a turnout from a 15° main track curve by the two methods, letting $F = 7°$, $S = 1° 20'$, $g = 4.7$ ft., $d = .42$ ft., we find the degree of curvature to be as follows:

	Fig. II.	Fig. III.
Analytical method,	24° 46'	5° 14'
Short method,	24° 40'	5° 20'

Thus we see that for this extreme case there is a discrepancy of only 6' or less than .7% of the curvature of the main track. Remembering that the degree of curvature of most tracks is very much less than 15°, this short rule may be said to be perfectly applicable for all practical work, where only the ordinary degree of accuracy is called for.

THE COURSE IN CHEMISTRY IN THE COLLEGE OF ENGINEERING, METALLURGY AND THE MECHANIC ARTS.

JAMES A. DODGE, PH. D., *Professor of Chemistry*.

This course is planned and offered to enable students who expect to make chemistry a profession or business, to acquire the training in and knowledge of the subject which only several years of study, with good facilities, can give. Although numerous subjects besides chemistry are included—the higher mathematics for example—putting this course on a parallel with regular engineering courses, yet it is designed to require as its essential feature the detailed study of *applied chemistry*. This study must involve a large amount of time devoted to analytical chemistry, and to this are given about nine terms of work. Lectures on inorganic and organic industrial chemistry are given during several terms. The lectures are illustrated with specimens of crude materials and finished products. Opportunity is afforded for individual laboratory practice on these materials, and facilities are provided for the investigation of such special problems in applied chemistry as promise to lead to results of interest and value.

It is evident that the future managers of chemical works, the makers of acids, alkalies, phosphates, mordants, colors, etc.; the refiners of oils and sugars, etc.; the chemists of iron and steel works; the chemists of railway companies and mining companies; the chemists of public boards of health, can only be well prepared for their several callings and engagements by a thorough and systematic study of the several branches of chemistry successively to the most advanced. Such a study must necessarily extend over a considerable length of time, and be pursued at an institution able to furnish the facilities of a laboratory, a museum and a suitable library. The course in applied chemistry here offered, developed as it will be from year to year, will meet a national as well as a local demand for such facilities and opportunities.

TECHNICAL EDUCATION AND ITS RELATIONS TO THE MINING AND METALLURGICAL INTERESTS OF MINNESOTA.

WILLIAM R. APPLEBY, B. A., *Professor of Mining and Metallurgy.*

Note of an address delivered before the Society of Engineers, Nov. 19th, 1892.

The demand for an immediately practical course of training by the young men of the state was considered in the introductory words of the address. The possibility of satisfying such a demand was shown to lie in the proper establishment of technical courses that should be easy of access and fully equipped for all varieties of work. With the indecision concerning a profession that is so commonly found in young men beginning their education, there is special need that the differences between scientific courses and technical courses should be made apparent to them. While many men are led into the scientific courses by their belief that the word "science" is a synonym for practice, the result of such courses is not always what they had been led to expect and they find that their training has been in pure rather than in applied scientific work. For those that desire at once to become inducted into applied work the technical courses are provided. A scientific education is, then, defined as one that trains a man in the investigation of principles for the sake of knowledge, while technical education trains him in the investigation of these principles, but with the application of them held uppermost as the important end in view. A technical education is therefore both a science and an art education. While science teaches one to know, art teaches one to do, and the meaning of a technical education becomes plain.

Manual training is shown to be not necessarily a technical education for it trains only the mechanic while the true technical education trains the mechanical engineer. That the engineer is, in his proper comprehension, a man of ideas and not merely of things was then pointed out, with much care to make the distinction a clear one. The final definition of an engineer selected was as follows:—"An engineer is "a man who follows a vocation in which a professed knowledge of some "departments of science or learning is used by its practical application "either in advising, guiding or originating in the practice of an art "founded on it."

The place of the laboratory was then discussed and it was shown what, and how much importance must be laid to the instruction that one receives in the laboratory over and against that which one receives in the field.

After a brief exhibition of the important function of the engineer in the civilization of the day the foundation of the school of mining and metallurgy was indicated and its place in the development of the resources of the state. In closing a summary of the mining industries already established in Minnesota was given and the great importance of the iron deposits discussed somewhat at length. It was shown that Minnesota is already fourth or fifth in the order of iron producing states and the magnitude of business is indicated both by the tonnage of ore exported and the capitalization of the companies concerned in mining. To develop this most important property would demand the scientific attention of many expert men and the need for just such men was one of the reasons for the existence of the School of Mining and Metallurgy.

The address closed with a statement of the commercial position of Minneapolis as a center for smelting works and its relation with the properties west of it. The profession of the mining engineer was shown to be an important one for the state at this time. The hour demanded the men, and they must be furnished if possible from among the sons of the state itself rather than from abroad.

THE ELECTRICAL ENGINEER AND HIS WORK.

GEO. D. SHEPARDSON, M. E., *Professor of Electrical Engineering.*

The rapid growth of electrical industries has attracted the attention of a large number of men who look to it as a means of earning a living, and in many cases with a fond hope that it may prove a source of wealth. This is indicated by the fact that electrical manufacturers have on their books "for future reference," applications by the hundred from men and boys of all ages and qualifications. Indeed, so great is the supply of learners that it is hard for a beginner to secure a situation even for nominal wages.

It is also seen in the increase of the number of college students looking forward to electrical engineering as a life work. Five years ago the institutions offering courses in electrical engineering might be numbered on the fingers of one hand; now they are counted by scores. The growth of the work is illustrated by the fact that in 1884 there were 28 students in the electrical engineering course at Cornell university; while in 1892 the number had grown to 275. In the schools offering a course in electrical engineering a tendency exists for students to crowd into that course. In the University of Minnesota half the freshmen in the College of Engineering, Metallurgy and the Mechanic Arts are registered for the electrical engineering course.

There must come a reaction from this tendency, for there are equally great opportunities in other lines of engineering. Indeed, the work of most "electrical engineers" is largely something else with a little electrical work added.

The supply of beginners is greater than the demand. Many of these students who rush into the electrical course, are utterly unfitted to become engineers, having neither the ability, perseverance nor genius. Some have the idea that if they are too lazy to be farmers, too slow and dull to be successful in business life, or have not had and will not get education and culture enough to follow a so-called learned profession, they will become engineers. They may succeed in becoming threshing-machine engineers, but they would better save the time and money spent in trying to become what they were never intended to be. Engineering involves a long and thorough training. One must have a taste and a talent in that direction; must possess the ability to perform a great deal of hard and unromantic work, often for twenty-four or even sixty hours at a stretch; carry heavy responsibility; control men and manage work; do the work of three men at once when necessary, and be "up early and always at it."

The causes of this movement into electrical work are several. Perhaps foremost is the idea that electricity can do everything; therefore, the electrician can do everything and build up an enormous fortune.

It is true that some electricians have built up great fortunes and many are doing it to-day. Electricity is being used in nearly every line of industry and electrical processes or methods are rapidly displacing others on account of superior adaptability, efficiency and economy. The capitalist here finds safe and profitable investment, the inventor and engineer find ever-widening and attractive fields of labor. Many inventors in electrical lines as in others die poor. To be merely an electrician or merely a great inventor does not insure financial success. The men who grow rich in this field are of two classes; the sagacious ones with little or no electrical knowledge of their own, but with much knowledge of men and market, keen foresight and wide-awake business ability; and those who combine inventive genius and extensive knowledge of electricity in theory and practice, with more or less of the above endowments.

Such men would grow rich in almost any field of activity. Aside from dollars and cents there is a peculiar fascination in the study of electricity and a pleasure that comes merely from experimenting with the unseen but not always unfelt force. There is a certain satisfaction in being informed upon this wonderful form of energy. The great variety of phenomena offered is an unending source of interest. Many new phenomena are being discovered and new applications are continually being made. There is always open to the intelligent, inquisitive and reflective student the possibility of making new discoveries of greater or less importance. The latter, aside from financial possibilities, is indirectly a great attraction to the class of people whose highest object in life is to make themselves useful to others and to make this world a better place to live in. Which of these reasons is the most potent depends upon the individual.

Electricity is entering into modern life in so many different ways that an infinite variety of positions require more or less thorough knowledge of electricity. The variety is so great and shading from one class to another is so gradual that it is impossible to make any hard and fast classification. Comparatively few positions require simply electrical knowledge. Standardizing bureaus for calibrating instruments, testing apparatus; questions of theory, many lines of investigation, and some lines of invention might be classed as purely electrical, but as a general rule each one of these lines of work involves more or less thorough knowledge of chemistry, mechanical engineering and business methods.

One might divide electrical workers into two classes, the first embracing those who manufacture (this includes design, construction, testing and repair), and install or operate electrical apparatus. The second class includes a large number of people who have to do with electricity only in the abstract or indirectly, such as capitalists, promoters, patent attorneys, lawyers, editors, writers and teachers. The latter class is one of great importance. Electrical developments would be slow, indeed, without the help of capitalists. This assistance would be slow to come were it not for the useful work of the promoter or middleman. Much money has been wasted and injury done to the reputation of legitimate electrical enterprise by the wildcat speculation and cut-throat competition brought about by the zeal of unwise or dishonest schemers. It is a remarkable fact that the man who would not for a moment think of cutting for himself a $2.00 vest would not hesitate to rely upon his own judgment when considering the investment of thousands of dollars in some electrical scheme of which he knew nothing, except the representations and promises of an interested agent. In the early days there were no experienced engineers except those in the employ of the large companies. Many of these have graduated from the companies in the last few years, and are in business for themselves as disinterested and well-qualified consulting engineers. Capitalists are learning to make proper use of the experience of others, to the mutual advantage of themselves and the engineers.

The rapid growth and differentiation of electrical industries is accompanied by a corresponding development of the technical press. Many papers find a wide circulation, some of them devoting themselves to particular lines. Nearly all of the engineering papers have electrical departments, and the day seems not far distant when the great daily and weekly newspapers will employ electrical or engineering editors, or at least reporters who are intelligent on electrical subjects. A good demand exists for well-written articles treating of electrical matters in a popular style. In many cases it is true the men who are capable of writing correctly are too busy to write, and those who do write are only partially informed.

There seems to be a steady demand for good teachers in electrical engineering. Such a position requires considerable experience in various lines of commercial practice, combined with a thorough education and a willingness to work. It is an accepted fact that men who are competent to fill such positions obtain at least twice as large salaries when the same ability and energy are applied in commercial work. An editorial in a leading electrical paper says:—"It would seem as if "technical education of an engineering nature, owing to its importance "in the industrial world, demands the employment of the best men "obtainable, and if our leading universities expect to obtain these men "they must offer them such inducements that positions of this kind "will be chosen in preference to commercial work." As a matter of fact there were more than a dozen vacancies of this kind last September and some of them are still begging. The comparatively small salaries for this class of work are partially compensated by the opportunity for study and original investigation, the cultured society and other advantages that cluster about an educational center. Not the least is the satisfaction that comes from helping others. Electrical business is becoming so complicated that a growing number of lawyers are devoting their entire attention to obtaining electrical patents. For such a position an extensive knowledge of all sciences is desirable.

After completing a scientific or technical course and a law course, the preparation sought by many patent solicitors is to spend some years in the U. S. Patent Office as examiners, The government holds examinations at stated times and places for the purpose of examining candidates for these positions. The salaries are fair and the work is an excellent preparation for those who expect to become patent lawyers.

Closely allied to these are what might be called *electrical lawyers*, who make a specialty of electrical cases. It is a deplorable fact that a large part of the money made by electrical inventions has been used up in litigation between rival companies. The present tendency of the companies to consolidate makes it questionable whether this branch of the legal practice will continue to flourish as it has in years past.

The first class includes, further than were named, those who are generally called "electrical engineers" and the large number of people who call themselves "electricians." Under each of the sub-headings are all grades of work. It is assumed that the student is aiming at a high position, and we shall therefore consider the various lines of business of which he may become manager.

There is a good field for the electrical manufacturer. In electrical industries as in others there has been a tendency for manufacturers to combine and form what are essentially monopolies. In this way they obtain control of patents that would render competition impossible were it not that the courts do not generally render final decision until the patent has almost expired. The monopolies created by the control of patents have practically shut out competition in the telephone and storage battery business in this country. Now that the telephone patents are expiring, new telephone companies are springing up on all sides and some of them will do a profitable business. The General Electric Company controls an immense number of patents which, if sustained by the court, will give them a practical monopoly of electric railway work. In the early days of electric lighting when there were but few companies one "system" was used throughout each plant and the owners were usually compelled to buy all supplies from the parent or manufacturing company. The same was true in electric railway work. As new manufacturing companies were organized and began putting their goods on the market, it was found that certain parts of the apparatus of one company were better than the corresponding apparatus made by another company. Careful buyers soon began to use their own judgment, bought supplies wherever they could buy them best and whatever suited them best. This opened the field for manufacturers of a few special articles or lines of goods and to-day this is one of the most promising fields for manufacture. It requires so much capital to carry on a general electric manufacturing business by modern methods that most companies are contenting themselves with manufacturing a great many of a few things. If one could obtain control of the manufacture of some specialty for which there is or may be created a demand, there is a fair chance for it to succeed provided, of course, it is well managed.

It used to be the universal custom for manufacturers of electric lights and power apparatus to install plants themselves, but as the business became older, independent parties took hold of this business, buying their supplies wherever they pleased. This is true of other electrical lines except the telephone and perhaps a few others which were monopolies. The supply and contracting business has a large field and furnishes occupation for a great many.

The supply business is attractive to many and there is room for a supply house in almost every city. The character of the business will depend largely upon the size of the city and its nearness to great business centers. There are so many different lines of electrical work that a supply house must command an enormous capital in order to carry a stock of everything that may be needed. There is a growing tendency for supply houses to restrict their stock and handle a limited number of specialties.

Closely allied to the supply business is contracting. Most electrical contractors keep a certain amount of stock on hand and many of them are agents for manufacturers. The knowledge and training necessary for a contractor will vary with the nature and size of the work he is doing. If he is simply an "electrician" and "locksmith," he needs to know how to set up a sal-ammoniac cell and how to run annunciator wires, simply following the diagram given in the supply catalogue. Such men often think they can put in wiring for lighting houses and numerous sad tales might be told of their successes and the misfortunes of their patrons. The owners of a certain building in Minneapolis are said to have paid $11,000 for wiring which is entirely worthless on account of poor work. There is a steady business for contractors who wire buildings properly for lighting purposes; also for agents for various companies. This is probably one of the safest lines for establishing a steady electrical business.

In almost every section of the country there is a considerable amount of electrical repair work to be done. Electrical apparatus will wear out and meet with accidents. Much repair work is sent back to the factories, but in many cases the freight or express charges will amount to as much or more than the simple cost of the repairs. Large plants do their own repairing, but owners of small plants cannot afford to keep skilled help and therefore must have their repairs done elsewhere.

The consulting engineer who has established a reputation finds a wide demand for his services as a designer of new plants; for examining into new schemes or inventions that are seeking aid of capitalists; for testing plants in operation or in construction; for assisting in developing new inventions and in any similar cases that may be entrusted to him. He should be a man of broad education, of thorough technical training, wide experience in business and in the actual operation of electrical plants. The consulting engineer is often at the same time a contractor. The business of consulting engineer is often difficult, but is capable of being developed to a much greater extent than it is at present. There is an opportunity in this connection for excellent work to be done in educating capitalists who propose investing in electrical interests. Large sums of money that might have been turned to profit, if at the outset proper consultation had been made with competent engineers, have been wasted upon electrical schemes.

The third division of the first class, viz.: operators of electrical apparatus, includes a variety of positions, limited only by the number of purposes to which electricity is applied. The safe and economical operation of an electric concern of any magnitude requires intelligent supervision by a competent electrician. The demand for educated electricians is continually increasing. Technical students seem to think that the most promising, in fact, almost the only field for an educated electrical engineer is in electric lighting or transmission of power. These are, indeed, great fields and there is also a great rush of experienced engineers in these directions. One is inclined to repeat the proverb, "Where everybody goes, do not go." In other words, if one will strike out into new fields where he meets less competition, it is also likely that he will more easily make a success. There are other fields which are now open but in which little progress has been made simply because they have been left to inferior or poorly prepared men or have been given little attention. For example; inquiry was made at an electrical supply house on prices for electric gas lighting apparatus. The writer was shown a handsomely finished spark coil costing $80.00 and was assured that it was operated only by one cell of battery which would cost 50 cents. The probabilities are it would have been much more economical to use a cheaper coil and more battery. There seems to have been no investigation as to the best relation between the cost of coil and the cost of battery. There certainly is room for careful investigation and improved designing in this field.

Abundant opportunities for men who can combine electrical knowledge with good management are open in the superintendence of electric light and railway stations, telephone exchanges, telegraph lines, electro-plating and electro-metallurgical works and other establishments that apply electrical energy to commercial processes or operations. In each of these and in various other positions that might be mentioned, there is always need of intelligent supervision, a multitude of details to be looked after and new methods to be investigated. The financial failure or success of such establishment often depends upon the wisdom of the superintendent and the skill with which he manages details, which may be small individually, but, in the aggregate, determine whether there shall be dividends or assessments. A concrete example of this will be worth giving in some detail.

The writer is somewhat acquainted with the methods adopted by a large electric road in improving its service while reducing its repair account. It is thought that the result of this experience will be of interest and value to others who are struggling against heavy expenses in the repair shops. By careful management this road has accomplished wonders. The number of men employed in rewinding armatures has been reduced from sixty-five to less than twenty and corresponding improvements have been effected in other departments. This has been accomplished by the employment of trained electrical and mechanical engineers who are giving careful attention to details. The men in charge of the motors and cars are furnished with a book giving clear explanations of the various parts of the car and its equipment, illustrating and explaining so far as desirable the purpose of each part. They are taught how to take proper care of the motors, and how to make ordinary repairs in case of accident on the road. Specially trained inspectors are out on the line at all times looking after motors that are not working perfectly. These inspectors endeavor to see that everything is working properly. If anything is out of order they repair it at once or send it to the shop before a bad matter is made worse by neglect. Experience shows that these inspectors save their cost to the company many times over.

In the repair shops great attention has also been given to details. Careful observation of the scrap heap and of the "cripples" has pointed out a number of wrinkles which, small in themselves, have in the aggregate amounted to much. The adoption of taper fits on pinions and commutators has been the means of saving much time and much swearing in the shops. By altering the shape of the commutator bars and by pressing down the ends of the armature wires so as to form a sort of "goose-neck" where they are soldered to the commutator bars, broken armature leads have become a thing of the past. In all parts of the shops everything is made to gauge so as to be interchangeable. The repair shops are run on a manufacturing basis with great economy.

In one of the engine rooms a small change in the piping reduced the drop in pressure between the boilers and the engine from fifteen pounds to less than two. The saving of fuel from this single device will more than pay the salary of the chief engineer who made the experiment.

These are but a few of the improvements brought about as a result of the careful investigations and recommendations of trained engineers. It is a significant commentary on the claim sometimes made by ignorant or prejudiced men that educated engineers are not worth their salaries. It also suggests that many roads and other electrical industries that are running at a loss, barely paying expenses, or even those that are paying large dividends, would find it a valuable investment to secure the services of thoroughly trained engineers who can combine theory with practice.

The education and training that a person needs to fit him for his work will depend largely upon the sort of position he may occupy. In most cases the more he knows the better. The more knowledge he has of chemistry, mathematics, physics, mechanical engineering, business methods and human nature, the better prepared is he for his work. The field is so broad and the number of positions so widely different that it would be impracticable to discuss in detail the advantages of various ones or the qualifications desirable for each one. The quality and amount of electrical education that is desirable or necessary for the person depends very largely upon the work he expects to do. A university course is not necessary for certain lines of work. The demand for electrical engineers with a regular university training is comparatively limited, as is also the number of students ready to obtain such a preparation.

It is frequently suggested that men are making a great success in electrical work who have never been collegians and whose only schooling has been that of business and practical work. It is true that they make a success, but it is also true that the same men would have done much more if they had added a thorough training to their other attainments. The secret of their success without a college education is that they have an abundance of push, common sense and the faculty of using the results of other people's knowledge. It is an undeniable fact that a great majority of the men at the head of electrical progress are college men. It is notably the case in alternating current work, where there is hardly an exception. The men who boast that they never went to college and ridicule the green graduates who have not yet had time to tone theory with practice, are themselves reaping the benefits of the labors of trained scientific investigators and making their own living by it.

The young graduate has not finished his education. The technical course does not aim to fit a person for any one position, nor to develop him into a full-fledged engineer, but to give him a training in general theory and fundamental principles with a general idea of methods and purposes in commercial practice. He is given the tools for work, but he must learn the applied part by actual experience. For this purpose it is highly desirable that he spend some time in commercial work during his technical course. If two or three of the long summer vacations are spent in electric light or power stations, in repair shops or factories or in telephone or telegraph exchanges, he will learn to appreciate the practical bearings of what is taught in the class-room and laboratory, and will derive much more of value therefrom. Above all he should cultivate his common sense and reasoning powers. Students are likely to acquire the habit of accepting everything that is in the technical papers or text-books and thereby are often led into serious error. One should use one's common sense, train it, and if one finds one has but little, then electrical pursuits should be left aside.

TEST OF THE POWER PLANT IN THE ENGINEERING BUILDING.

JAMES H. GILL AND GEORGE B. COUPER.

Abstract of thesis for degree of B. M. E.

The plant consists of an 8 by 15 inch Buckeye engine running at 200 revolutions per minute; a Buckeye boiler 50 inches in diameter and 10 feet long, having forty-eight 3¼ inch return flues; a Wheeler surface condenser with air pump, etc. The engine was fitted with a special form of Prony brake.

The main objects of the test were, to determine the efficiency of the engine both condensing and non-condensing and to find the evaporative efficiency of the boiler.

The trials were of about two hours duration under each load, starting with 10 horse-power. The load was increased at each test until the maximum power of the engine was reached. Observations of pressures, temperatures, etc., were taken every ten minutes. A summary of the results is given in the accompanying tables:

ENGINE TRIALS.

\multicolumn{5}{c}{Non-condensing.}	\multicolumn{7}{c}{Condensing.}
Number.	I. H. P.	Water per h. p. hour.	Boiler pressure.	Ratio of expansion.	Number.	I. H. P.	Water per h. p. hour, for engine.	Water per h. p. hour, for engine and pump.	Boiler pressure.	Vacuum in inches of mercury.	Ratio of expansion.
1	13.40	46.50	66.4	5.92	1	11.76	38.85	56.78	52.5	24.38	9.53
2	17.61	37.60	63.5	4.10	2	14.45	28.20	35.20	57.0	23.91	....
3	18.01	39.40	58.8	4.11	3	17.24	31.21	38.65	59.9	23.80	6.96
4	20.27	39.08	54.3	3.07	4	22.32	29.94	37.15	57.5	23.02	4.79
5	24.25	33.70	63.0	3.31	5	24.95	27.82	29.67	60.6	23.10	4.37
6	26.67	34.86	59.4	2.60	6	26.71	30.30	38.67	57.8	22.67	3.13
7	28.33	35.56	58.4	2.40	7	27.08	31.31	40.26	55.2	22.00	2.78
8	29.03	33.60	61.3	2.43	8	27.15	32.16	40.65	56.3	22.24	2.97
9	31.51	35.30	63.7	2.33	9	33.75	30.99	36.25	59.3	21.00	2.33
					10	36.47	29.86	37.27	63.5	19.94	2.24
					11	39.90	29.08	30.46	66.5	18.30	2.06

BOILER TRIALS.

Number.	Total fuel burned.	Coal per square foot of grate area per hour.	Total water evaporated.	Water per lb. of fuel from and at 212° Fahr.	Water per square foot of heating surface from and at 212° Fahr.	Per cent. of moisture in steam.	Average boiler pressure.	Temperature of feed-water.
1	310.50	10.35	1845.0	6.110	1.350	1.93	59.1	195.5
2	420.05	14.60	2846.3	6.970	2.170	2.30	65.8	194.0
3	324.70	10.11	1889.5	6.400	1.380	1.73	52.5	125.6
4	385.00	16.74	2111.6	6.030	2.125	2.19	57.8	131.9
5	317.00	13.80	2217.4	6.520	2.240	1.88	56.3	159.3
6	395.50	13.68	2603.3	5.990	1.880	2.31	55.2	111.2
7	337.00	13.45	1890.0	6.176	1.554	2.07	60.6	124.2
8	400.00	17.40	2508.5	6.934	2.570	2.46	59.3	114.4
9	400.00	18.45	3037.6	8.350	2.860	3.11	63.5	117.1
10	383.00	16.65	2508.6	7.269	2.590	2.89	63.9	106.2

Referring to the foregoing table of engine trials, it is seen that when non-condensing the point of best efficiency is about 25 horsepower. The maximum power obtained was 31.5 horse-power.

When condensing the steam consumption is lowest at about 22 horse-power. Beyond this it increases, due probably to the condenser being too small. It was found impossible to maintain a good vacuum under the heavier loads, as is shown by the column headed *vacuum*.

The column headed *water per h. p. hour for engine and pump*, shows the true amount of steam used by the engine when condensing. This column shows the same increase as the preceding and probably from the same cause (the inefficiency of the condenser).

By comparing the results it is seen that the total steam used per horse power per hour, non-condensing, is less above 23.5 horse-power than when condensing.

THE OPERATION OF ELECTRIC STREET RAILWAY MOTORS.

ALBERT D. McNAIR, '94.

An eminent engineer has said that "the success of any electric road "is in the hands of the armature winders and motor men." In the early days of electric railroading as in the early stages of every new industry, many serious and costly mistakes were made. Electrical engineers themselves contributed their quota, and where leaders make mistakes what can we expect of followers?

Now that some years have passed since the successful introduction of electric motors in street-railway service, there is a fairly good supply of men who can operate and repair motors. In nearly all the larger cities this business is so extensive that the work has become definitely specialized.

Motor men do no repairing, except what may be done with hammer, pliers and screw-driver as emergency requires. If any accident prove too serious for such tools and the car cannot be moved, the motor man must wait and be pushed into the barn by the next car.

The repairers at the car barn can advantageously do only part of the repairing. Such things as burned out field coils, armatures and rheostats are removed bodily and sent to what is usually called the *armature room* to be repaired by a special class of workers. This room is called the armature room because the work of winding armatures constitutes the principal and most important part of the work done here.

It has long been common custom to compare the electric current to a stream of water. There are certainly many analogies between the so-called flow of the electric current, and the flow of a stream of water. Let us use one of those analogies in the present paper.

In many cases, turbine wheels or other water motors are turned by a current of water which is led from some source of supply through a

The Operation of Electric Street Railway Motors. 59

pipe to the point where the turbine is located. If now this pipe should break at some point or spring a leak by which an appreciable part of the water is wasted, the turbine wheel or other motor can no longer do its work properly.

It is the same with the electric current in passing through the electric motor and its accessories. When the insulation breaks and the current "springs a leak," the motor no longer does its work properly. Many readers of this paper are, no doubt, aware of this fact already, since it belongs to elementary instruction in electricity; but I am justified in referring to it because the subject of insulation is one of prime importance, and those who have not actually studied about electric motors do not realize how many are the places at which these leaks occur, and how necessary it is for the workman to be on the look-out for incipient breaks in the insulation. By tracing the current through the motors of a single electric car and its accessories we may see at how many points there is danger of such breaks.

In figure *1* is shown the path of the current from the place where it enters to that where it leaves the car. The diagram shows one of the arrangements of parts for the Thomson-Houston street railway motor, as it has been used in Minneapolis and other places. Many other arrangements are in use, but this is quite common and will serve to illustrate principles.

In nearly every car in Minneapolis there are two motors, each connected with its own car axle, and working independently of its neighbor. Figure *1* shows the course of the current from trolley to cut-out switch, thence to the fuse box, to rheostat, to field coils, to reversing switch, to armature, back to reversing switch, thence to ground and back to the dynamo, completing the entire circuit. The current which lights the car is taken off by a shunt circuit before it reaches the cut-out switch. This is not shown in the diagram.

The cut-out switch is used to open or close the circuit, and is always left open as indicated by the dotted lines, where the car is to stand still for any length of time. It is essentially a safety device. The fuse box contains what is called the fuse wire, which is supposed to melt and break the current when an excessive current passes through it. Next is shown the lightning arrester, through which no current passes under ordinary circumstances. A charge of lightning, however, is supposed to jump across the small air spaces, between the dots indicated in the figure, and go to the ground, thus saving the motor from injury. The fact of the case is, however, it does not protect against lightning half so well as is supposed.

FIGURE 1.

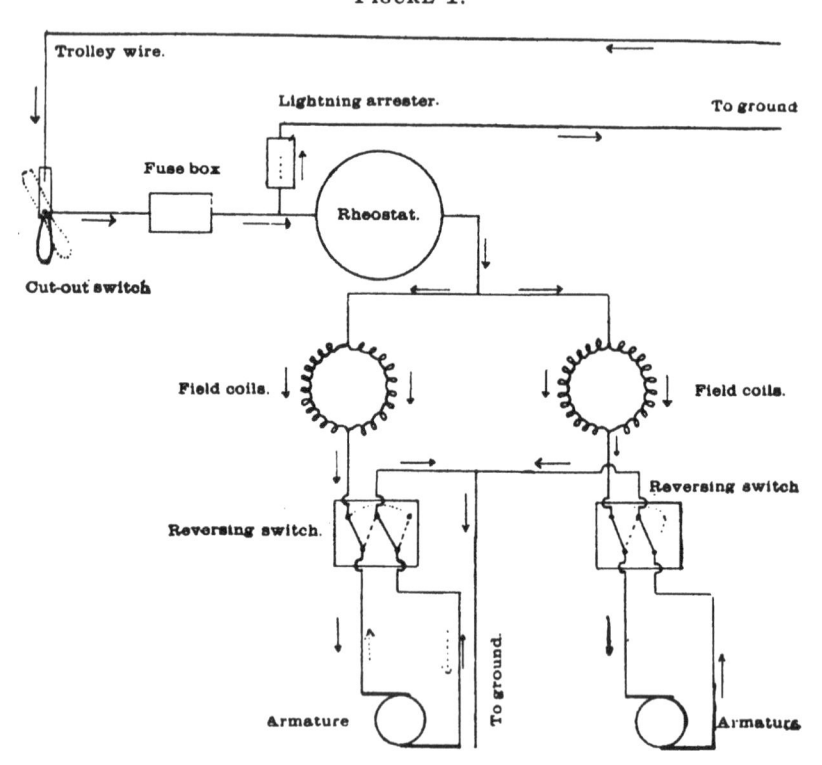

The next accessory is the rheostat. It is simply a resistance, usually made of a great number of thin iron plates, through which the current passes in series. The driver can vary this resistance by a turn of the handle at his side.

There is little trouble with insulation until the current reaches the rheostat. There it often leaves its narrow path. It breaks through the insulation, reaches the iron frame-work of the rheostat and from that it frequently reaches the ground; hence very little gets to the motor and the car will not go. Water is a great destroyer of insulation, and the constant cause of *short circuits* or *grounds*. Great heat produces similar effects. Where the rheostat is placed underneath the car, as has usually been the case, it will have water splashed on it on wet days in spite of all efforts to the contrary. It cannot be sealed up like a fruit can, because its parts must be readily accessible and air must circulate around it to keep it from getting overheated by the current. At the present time in Minneapolis, at least, many of the rheostats are placed on the front platform where the driver stands, and they last much better in consequence of being in a dryer place. The insulating material used on rheostats is principally mica, because it stands the heat well.

From the rheostat the current passes to the field coils which consist of several hundred turns of insulated copper wire, wound around an iron wire. These coils are usually well protected from water, but sometimes a little moisture finds its way in and induces either a short circuit or a ground. Oftener, however, the coils become very hot from a strong current of electricity passing through them. This heats the insulation to such an extent that it becomes charred or carbonized so that its resistance is materially lessened; therefore the current breaks through from one turn of wire to an adjacent turn, producing a flash of fire which burns the insulation on adjacent turns and often fuses several wires together, thus forming a short circuit.

The magnetizing force of the coil is lessened in practically the same ratio as the number of turns which are short circuited. This would decrease the counter-electro-motive force, which would increase the speed of the motor.

As before mentioned there are two motors under each car. If one of these motors has perfectly sound coils, and the other has coils in which many turns are short-circuited, the motor with the short circuited coils tends to run faster than the other. It is like having an energetic horse and a lazy one in the same team. The energetic horse not only has to pull all the load, but to drag the lazy horse along also.

Usually one field coil of one motor becomes short-circuited while the other coil is perfectly sound. This produces a distortion of the lines of magnetism, which in turn produces sparking, tending to destroy both brush and commutator.

If the current breaks through the insulation to the core of the field, the phenomenon of *bucking* occurs, the intensity of which depends upon when the current breaks through the coils. A *bucking motor* is quite as vigorous in its action as a bucking broncho. It is important, therefore, that the wire of the field coils be well insulated, well protected from water, and not subjected to an excessive current of electricity.

From the field coils the current passes through the reversing switch, each motor having its own switch. As will be seen from the diagram of figure *1*, each reversing switch has five posts, three on one side and two on the other. The outer posts, on the side containing three posts, are constantly connected with each other as shown by dotted line. The heavy lines, connecting posts on one side with posts on the other, show one position of the switch arms, while the short dotted lines show the reversed position. It will be seen at a glance, that reversing the switch will send the current through the armature in the opposite direction. Thus, if the switch arms should occupy the position of the dotted lines, the direction of the current through the armature would be indicated by the dotted arrows. The reversing switches need to be kept very dry and must be covered with canvas or other enclosing material, if placed underneath the car.

Coming next to the armature, we find that most of the accidents which occur in the field coils are also likely to occur in the armature. Here is the same danger of short circuits and grounds, and the same precautions are to be observed. A difficulty, which not infrequently occurs in armatures, is what is known as a *broken lead*. It is not such a serious difficulty as *bucking*, but it needs attention.

Let figure *2* represent a ring armature with an eight part commutator. The end elevation of commutator is shown, in which the black parts are the copper or brass bars, the blank spaces between representing the insulation. By a *lead* is meant the end of a coil which leads to a commutator bar. Suppose a certain lead to be broken, as shown at *D* on the right. Then the current, which enters at the brush *A* and

The Operation of Electric Street Railway Motors. 63

leaves at the brush B, cannot pass through the coils of the right half of the armature, but only through the coils of the left half. Thus, but half of the armature is working. Since in armatures commonly used, the leads are all covered over, we wish to know what are the external evidences of a broken lead.

FIGURE 2.

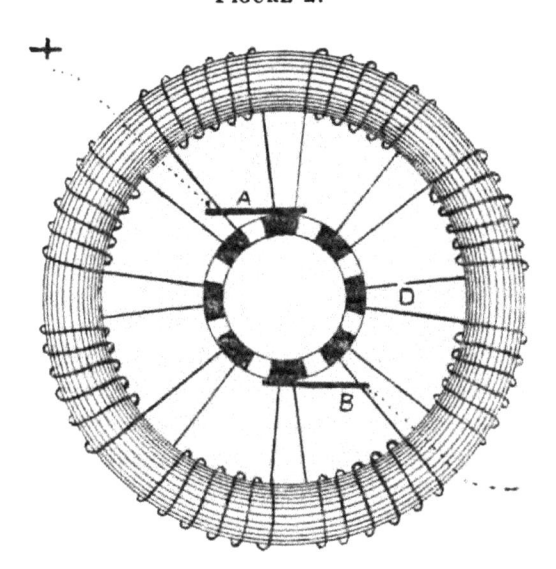

It will be seen from figure *2*, that as the armature with its commutator turns, the bar to which the broken lead is attached will pass under each brush once during every revolution. The current then will pass through the right or left half of the armature, according as the broken lead is on the left or right side. Thus the current changes from one side of the armature to the other every time the bar to which the broken lead is attached passes under a brush. A flash of fire, or as it is usually called, a *spit* of fire is seen every time this happens. Therefore when these spits of fire are seen, it is likely that there is a broken lead; but this does not necessarily follow. When the armature revolves rapidly, these spits of fire occur in such rapid succession that they appear continuous and resemble sparking due to other causes. To observe these spits of fire so as to distingush them from other forms of sparking it is necessary to watch the commutator at the moment of starting, or as the armature is running very slowly.

Another external evidence of the same difficulty is the burned condition of the edges of the commutator bar, to which the broken lead is attached, and also the adjacent bar; the insulation between is badly burned on the peripheral edge, which is the only part visible.

Sparking, from whatsoever cause, is bad for the commutator. It burns the surface of it, causing a poor contact between brush and commutator, producing much heat which is damaging to insulation.

The causes of sparking have been classified by Crocker and Wheeler in their work on the practical management of dynamos and motors, under the following heads: (1) armature carrying too much current; (2) brushes not set at neutral points; (3) commutator, rough, eccentric, or containing a high bar; (4) brushes making poor contact; (5) short circuited coil; (6) broken lead; (7) weak field magnetism; (8) high mica.

A point where the insulation needs to be especially good is at the brush-holders. Water sometimes splashes on the insulation at these holders, or the insulation becomes heated and the current breaks through and grounds on the holders. The inevitable result of a ground at this point is a vigorous buck, because it allows an enormous current to pass through the field coils, and produces a high counter-electromotive force and current, the energy of which is derived entirely from the momentum of the car. (For further details relating to bucking motors, see article by Professor Shepardson in *Electrical World* vol. xx, p. 160).

The mechanical details of electric motors are quite as important as the electrical details. A case in point is the armature bearings. The air space between the fixed pole pieces and the revolving armature is made very small; hence, if the armature bearings become worn, the whole armature settles down until, perhaps, its periphery touches the pole piece and the insulation is rubbed off the armature coils. Even if it does not settle enough to touch the pole piece, a distortion of field magnetism is produced which causes sparking and reduces the efficiency of the motor.

The matter of control of the motor is one of vital importance. It is not desirable to start the car with such suddenness as to make the passengers feel that they are being shot out of a catapult. Minneapolis citizens since the introduction of the electric railway have observed that many of the old cars started with a suddenness that is not known at present.

They can remember that even last summer (1892), the cars on the Interurban line started with a disagreeable jerk. Now the difference between starting gradually and starting by jerks is caused by applying a force accelerated gradually on the one hand and by sudden increments on the other. The constantly accelerated force, in the case of electric motors, is obtained by the use of the rheostat. But rheostats are costly to keep in repair, and they also transform much electricity into heat which is radiated and lost. Various substitutes have been tried for rheostats, to save the current which the rheostat wastes as heat. The old Sprague motors had the field coils made in separate parts, and a switch was so arranged that the current could be sent at will through some of the coils or all of the coils. The switch also provided for placing the coils in series or in parallels, as was desired. By means of these variations many different strengths of current could be obtained; but it was found that it was both difficult and expensive to keep the switch in order.

An arrangement similar in some respects was the series-multiple switch, used with the water-proof motors on the Interurban line last summer. In this arrangement there was a fixed resistance, which could be thrown out or in at will. There were five different points on which the controlling handle could be thrown. The arrangements for the different points were as follows: 1st, motors in series with the resistance in. 2nd, motors in series with the resistance out. 3d, motors in parallel with the resistance in. 4th, motors in parallel with the resistance out. 5th, same as the 4th, except that the loop was used. The loop has not been referred to before. A Thomson-Houston motor is said to be on the *loop*, when about half the field coils are cut out of the circuit. The loop is not a feature of the Edison or Sprague motors. The arrangement given above may be and has been modified to a great extent; but it serves to illustrate the methods by which different speeds may be attained.

THE STUDENT COURSE AT LYNN.

JOHN R. PITMAN, Ex—'93.

I have been asked to write an article on the General Electric Company's works at Lynn, Mass., touching more particularly upon the Expert, or Student course. Anything pertaining to the course is of interest to the average electrical student, who may have a vague idea of some day becoming an employe of this great corporation. To him this article is especially dedicated, with the hope that its few points may be of use in the future; for to write fully upon this subject would require more space than I am at liberty to use.

Lynn is a city of 60,000 inhabitants, situated about 10 miles to the northeast of Boston, and connected with it by two railroads, an electric line, and also by steamer. The surrounding country is very beautiful, and the numerous beaches in the vicinity afford great enjoyment to all lovers of old ocean. Lynn has always been widely and rightly known as the "city of shoes," but within the past few years a sturdy rival to the shoes has sprung up in the shape of electricity. How this occurred may be of interest to those who delight in viewing the wonderful strides made by this branch of science.

In the town of New Britain, Conn., was established by Professors Thomson and Houston twelve or thirteen years ago, a factory for the manufacture of arc light dynamos; the company was known as the American Electric company with a paid-up capital of $87,000. For two years they struggled along, overcoming many difficulties, but securing at the same time a goodly number of patents, and much valuable information. About this time the majority of the stock came into the possession of the president of the Brush Co.; and then it fell into the hands of several Lynn capitalists, and the firm known as the Thomson-Houston company. In the latter part of 1883 the factory was moved to Lynn, and since then the business has rapidly increased, different departments being added from time to time, until the floor space originally of 26,962 square feet has grown to be considerably over eight acres, and everything is badly crowded at that. They have just completed a number of fine buildings called the "River works," situated about one mile from the present factories, consisting of enormous iron and steel foundries, great machine, pattern, carpenter and blacksmith shops, some of the structures being 500 feet long and 120 feet wide. This will increase the floor space a number of acres, and the number of men employed, probably anywhere from one thousand to two thousand.

The new works located on the line of the Boston & Maine R. R., and the Saugus River have ample wharfage for coal and shipping. The buildings are supplied throughout with machines capable of handling the largest castings, one planer taking a piece 10x12 feet square and 25 feet long. The shops and yard are one network of tracks, on which both steam and electricity can be used. The whole plant is a credit to the company and is a much needed addition to the present factories. As the buildings have been increased, so have the employes, the forty-five men employed in 1884 had grown to 3,500 by January, 1892. At the present time, between four and five thousand are employed at the works, the weekly pay account running above $50,000. The stock has been correspondingly increased from time to time, until it reached the $18,400,000 mark in 1891. In the spring of 1892, a consolidation with the Edison people was effected, the companies becoming known as the General Electric, and the capitalization raised to $50,000,000. At this time a charter was obtained from the state of New York, which gave unlimited privileges. The combination now includes the following companies:—Thomson-Houston, Edison, Brush, Fort Wayne, and Schuyler & Bernstein.

I should be very glad to take the reader on a tour through the factories, but find that space will not permit it. So I shall refer my readers to an excellent article in the *Electrical Engineer* for June 29th, 1892, (Vol. XIII, No. 217) entitled "T-H" or "Among the Dynamo Builders of Lynn," by A. C. Shaw. This will personally conduct you through the works in a very satisfactory manner, and will leave me at liberty to devote my whole time to speaking of the Student course.

Let us assume, that sometime during the past year you sent in an application for entrance to the Expert department. Blanks are provided for this purpose, by the company, and you must fill one out; stating your name. age, residence, present occupation, experience, references, etc. You will find it to be of great assistance in gaining admission, if you will accompany your application with several letters from any prominent men you may happen to know. Upon reflecting that there are now over seven hundred applications on file, and that an average of four is received every day; also that the total number of students allowed in this department at one time, is only one hundred, you will realize that it is no small matter to gain admission; but do not let these figures discourage you. It is a rather curious fact that the applications are much more numerous in the fall than during any other season of the year; owing, probably, to the fact that many college men prefer a little rest after graduation before entering upon their life work. College men from all parts of the United States are gathered here, one might say from the whole world; as England, Germany, France, Russia, Spain, Norway, Japan, Australia, Mexico and several other countries are represented. Unless one is a college man, or has served on the Apprentice course, it is a very difficult thing to obtain admission, and it is becoming more and more so every year. Fully 85 per cent. of the men are college graduates. As one cannot become an expert if under age, the Apprentice course was started in August, 1892, as a sort of preparation school. Since then, however, the department has come under the same head as the Expert department, and is known as Student course, No. 2. One may enter this course if over seventeen. The work is somewhat mechanical in its nature, but there is enough electricity thrown in, however, to make it interesting. Two years spent at this kind of labor is considered good preparation for the Expert department.

The Expert course was inaugurated in 1886, and at that time there were enrolled about ten men, receiving in the neighborhood of $8.00 per week; the idea being to educate engineers to be of practical importance to the Thomson-Houston company. There was no scheduled course, and every man was guaranteed a position with the company.

At the present time the corporation does not promise to give the students a position upon their completing the course, but many of them get one, nevertheless, and to those that do not, the certificate received proves a valuable assistance at the doors of other companies.

A few days after sending in an application a communication will be received stating that the application has been placed on file, and will be given due attention. In time, if the application receive favorable action a letter will order the applicant to report for work at 6:30 a. m. on a specified date. It is a good plan to reach Lynn a few days before the date on which one is ordered to report. This gives time to get settled in convenient quarters. Valuable information about rooms, board, etc., may be obtained at the office of the Expert department.

The first morning, one must report to the person who hires all the men employed in the factories. He is given a number, and sent off in tow of a small boy, to the department to which he has been assigned, and which, in all probability, will be assembling railway motors, the dirtiest and most monotonous work in the whole factory. Let us now take a look at the work required in the course:

EXPERT COURSE AT LYNN FACTORY.

		Weeks.
1.	Shop Plant.	
	1. Wiring	4
	2. Shop motors	2
2.	Arc Department.	
	1. Arc lamp assembling	2
	2. Arc lamp testing	4
	3. Arc machine assembling and testing	5
3.	Incandescent Department, Direct.	
	1. Incandescent machine, assembling and testing	4
	2. Meters	2
	3. Winding armatures	4
5.	Stationary Motors and Generators.	
	1. Assembling and testing	4
	2. Railway and large generators	5
5.	Alternating System.	
	1. Machine assembling and testing	5
	2. Constructing transformers	2
	3. Testing transformers	2
	4. Testing mining drills and apparatus	2
6.	Railway motor testing	3
7.	Blacksmith shop	2
		52

In addition to the above, experts are required to do more or less work of an experimental nature, such as testing special machines, etc. In return for this labor one is reimbursed as follows: For the first three months, five cents per hour; for the second three months seven cents; for the succeeding three months, ten cents, and for the remainder of the year, twelve cents per hour. If one should remain on the course for more than a year he is paid at the rate of fifteen cents per hour.

The working day is nominally of ten hours, as follows: From 6:30–12, and from 1–6. This gives a total of ten and one-half hours per day, but by putting in this extra half hour every day the men are enabled to obtain Saturday afternoons off, which, as may well be imagined, is a great boon. The present factory hours are from 6:45 to 12 o'clock, and from 12:45 to 6 o'clock, the office men beginning at 7:45. Individual tools are supplied to the men and charged against their names. These must be returned upon leaving the factory or paid for. As regards a leave of absence, one can be obtained at almost any time. Plenty of special work of all kinds can be done, and many of the men put in quite a little time in the offices and draughting rooms.

Every few days there comes in a request from some company for men, and the most suitable ones are selected. They often go as far away as South America. Let us now watch a workman-student for a few minutes at his task. We shall find him on one end of Section C. Here the water-proof and other railway motors are assembled in pairs on stationary shafts, and geared together; they are then run on full load for ninety minutes. In this method one machine is run as a motor and drives the other as a generator. At the end of forty-five minutes their functions are reversed; the potential and current are noted, also the hot and cold resistances (the current for the latter being supplied by storage batteries), of the armature and field. Water rheostats are used for the load and railway rheostats for regulation.

The machines are then taken down, cleaned and sent to the car shop where they are painted and the finishing touches given; they are then sent to the shipping department.

About twenty-five or thirty of these w. p. motors, 30 to 50 h. p., that is, per pair, are turned out daily. The frame consists of two castings bolted together and forms a tight case with the exception of openings at the sides for the armature shaft and leads, these are afterwards closed with sheet iron covers. The pole pieces are cast with the frame and the single spool is bolted into the top frame. This field has three leads; loop, end and ground. When the loop is used some of the ampere turns are cut out and as the speed is dependent upon the strength of the field it is naturally increased, and the torque at the same time is diminished. Several different windings are in vogue for fulfilling different conditions. The armatures are iron-clad and of the drum type, being dressed at both ends and painted with water-proof paint. The motor is a series wound machine, and is single reduction, the gears being inclosed in gear cases partly filled with oil. The motor throughout is very well protected and the parts are compact and of good design.

When a student has completed his time on one subject, or it is thought desirable to shift him, he is given a card transferring him to another department. This card is generally given out Saturday morning, and the recipient reports to the new foreman the following Monday.

All except the office and draughting men must pass through a gate house on entering the factories in the morning and at noon, and pull a check, stamped with their number; this is deposited in a box. No man that pulls a check is allowed to work for that half day if more than an hour late.

Let us now enter the other end of Section C. We shall find a room where all transformers are tested, save the 30,000 watt. A number of boards are placed around the sides of the room, on each of which are four rows of terminals, the two upper being the positive and negative secondaries, the others being the primaries.

From the former, wires are run to banks of incandescent lamps overhead, the latter are connected with the dynamo, generally at 1,040, sometimes 2,080. One board will be for 600 watt transformers (12 lights) and each secondary will have that number of lamps on it, while another will be for 1800 watts, and will necessarily require more lamps per transformer, and so on. When the transformers are placed on trucks before the boards, the covers are removed and each one is tested with a magnet for grounds, they are then wired up to the boards and the switch thrown in. The transformers have two secondary coils which, if connected in multiple give 52 volts; if in series, 104. A run of five or six hours is made and then they are tested with a transformer, a variation with load of one volt being allowed, then the load is taken off by disconnecting the secondaries, and they are again tested, this time only a difference of half a volt being permitted. The machines are now flashed in connection with a standard. This shows whether the secondary coils are connected internally in a proper manner. The covers are then replaced, and the transformers sent to the shipper. When in use on 5,000 volts circuits, the cases are filled with oil to increase insulation, between the primary and ground, secondary and ground, and primary and secondary.

The wiring consists essentially of putting up new wires about the shops, replacing fuses and helping to construct switch-boards.

A great deal of the machinery is run by shop motors, as they are called, scattered all through the factories; some of them in almost inaccessible places, and a student is given a certain number of motors to look after. He arrives a few minutes before the whistle blows, starts them up, fills the oil cups, looks at the commutators, and sees that they are in good order generally. He must never allow them to stop if it can possibly be avoided. During the day he makes the rounds every half hour and takes readings on the voltmeter and ammeter, these he records and computes the horse power therefrom, the reports being handed in every evening.

Arc lamp assembling. This phrase explains itself. A little practice of this sort being very useful as one learns to understand the lamps thoroughly.

Arc machine assembling and testing. This is one of the most interesting and instructive subjects to be found in the whole course, as it requires a great deal of skill and care properly to adjust all the complicated levers and parts of the controlling device; also to set the commutator properly on the shaft and to adjust the brushes correctly. There is a great deal to learn about the machine and there is certainly no better way than this.

The assembling and testing of incandescent machines, railway and charge generators and alternators consists mainly, as the title implies, in putting the various parts into shape and testing the machines.

In the arc lamp testing room, the lamps are hung in long rows and the main things to learn are to adjust the carbons and to regulate the feed, both requiring a great deal of practice and care; for instance if the lamp hisses we know that the arc is too short, and so change the regulation, or if it should flame the arc is too long. The lamps are tested as a rule, for about six hours, three hours on each carbon. In adjusting the carbons, first one side is adjusted and then the other. The final test consists in pulling the carbons apart. If the flame does not go out it shows that there is something wrong with the lamp; as in case the carbons should give out or anything happen to the lamp, the current would not be short circuited through it and hence the other lamps on the circuit would go out. Finally the lamps are polished and sent to the shipping room.

Work on meters, winding of armatures and construction of transformers, consists mainly of standing by and watching other men labor. All that one is supposed to do is to pick up pointers, occasionally lending a hand.

Besides the regular course, which is very valuable, there is a vast amount of special work which is of exceeding value to the student.

The General Electric company offers a similar course in their factories located at Schenectady.

At Middleton, Conn., the same company has provided a special course in marine work, the number of students admitted at any one time being limited to twenty.

Upon completion of the prescribed course at Lynn, the student receives a certificate signed by the foreman and manager of the Expert department. This certifies that the bearer "has completed the Student course at Lynn factory in a satisfactory manner, and is deemed competent to install and operate the Thomson-Houston apparatus manufactured at the above named factory."

Certificates are also given at Schenectady and Middleton.

AN EXPERIMENT IN ELECTRIC RESONANCE.

Anthony Zeleny, B. S.

Two sheets of brass plate O and O^1, see figure 1, each 41 centimeters square were joined to opposite poles of a Holtz electric machine. To each pole of the Holtz machine were also attached five Leyden jars. To each of the brass plates was soldered a brass wire 2.1 millimeters in diameter and 32 centimeters long. These wires ended in the knobs a and b, which were made the terminals of the sparking gap A.

FIGURE 1.

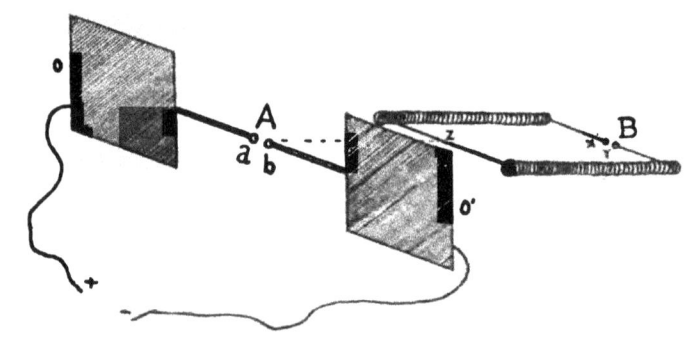

In this article let R = the resistance, in ohms, of the wires between the brass plates, plus the resistance of the sparking gap A at the moment the current is passing through it; L = coefficient of self-induction, in absolute electromagnetic units, of these same wires and of the sparking gap A at the moment the current is passing; C = capacity of the two plates in electrostatic measure; V = velocity of light or the velocity of the electro-magnetic ether waves, in centimeters per second; l = length, in centimetere, of the two wires between the plates plus the length of the air gap; d = diameter of the wires, in centimeters.

When a spark occurs at A, if $R < \sqrt{\dfrac{4\,L}{C}}$ the discharge is oscillatory, i. e., each visible spark is composed of many vibrations back and forth. In this experiment when the gap A is 1.3 centimeters long

$$L = 2\,l\,[\text{nap. log.}\ \frac{4\,l}{d} - .75] = 832.575.$$

$$C = \frac{.36 \times 41}{2} = 7.38$$ [The capacity of each square plate is .36 times the length of its side in centimeters and the capacity of two plates together, when placed in series as in the experiment, is equal to one half that amount.]

Since the velocity of light is 30 billion centimeters per second and since $T = \frac{2\,\Pi\,\sqrt{L\,C}}{V}$ the time of one complete vibration is .000,000,-016,417 of a second. The length of the electro-magnetic wave T V is in this case 4.92 meters.

A rectangle of brass wire having two opposite sides each 48 centimeters long and the other two sides made of loose coils of wire, as shown in figure 1, is placed near the brass plates. This rectangle can be increased in length by simply stretching these coils. There is a micrometer sparking gap at B. The capacity of this rectangle is constant or nearly so, but the coefficient of self-induction decreases as the rectangle is increased in length. Therefore the time of vibration in this rectangle, which is also equal to $\frac{2\Pi\sqrt{LC}}{V}$, decreases as the length of the rectangle increases.

When the rectangle is of just such a length that its time of vibration is equal to the time of vibration in the brass plates, the following resonance phenomenon occurs: When an oscillation at A goes from a to b, an oscillation is induced in the rectangle in the direction of $y\,z\,x$. Not being strong enough to jump across the gap B, this induced oscillation is reflected from x and when it reaches the point z, the oscillation in the brass plates is just returning from b to a. Another oscillation is induced in the rectangle, but this time in the direction of $x\,z\,y$. This oscillation is superimposed exactly upon the one already existing there and thus increases it in strength. This increased oscillation then proceeds to y and is again reflected, and when it reaches z is again increased as before. This goes on until the induced effect is strong enough to jump across the gap at B. Thus for every oscillatory spark at A, a spark occurs at B, which is far greater than one produced by simple induction. If the time of vibration in the rectangle, is not the same as that of the plates, the successively induced oscillations will be superimposed less and less as the difference increases and thus their combined effect will be weaker and the spark at B will diminish.

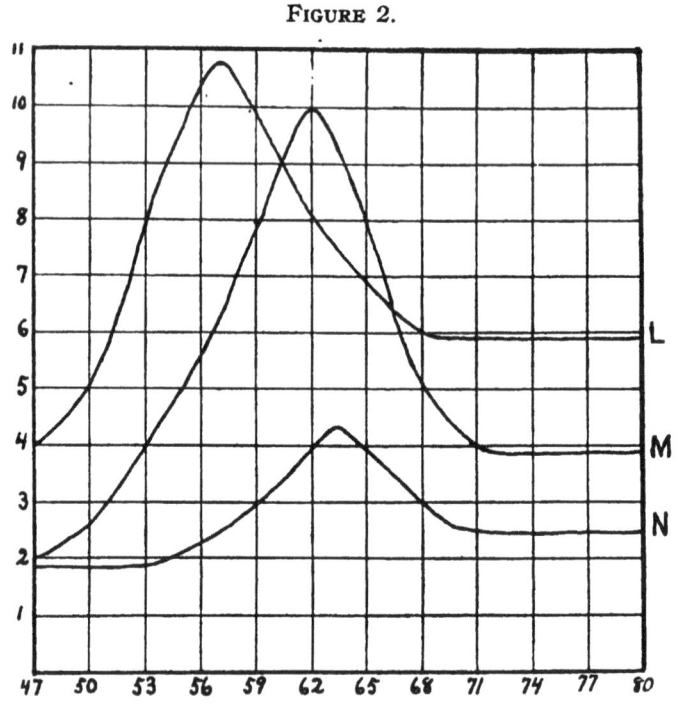

FIGURE 2.

The curves for the maximum sparking distance at B shown in figure 2, were obtained when the rectangle was placed 38 millimeters from the brass plates and its length continually increased by stretching the coils. The ordinates show the maximum length of spark in tenths of millimeters and the abscissae the length of the rectangle in centimeters. The curve L, was obtained when the sparking gap at A was 16 millimeters in length; M, when A was 13 millimeters; N, when A was 10 millimeters.

THE AUSTIN MASONRY DAM.

D. CUYLER WASHBURN, '93.

The new public works for the city of Austin, Texas, consist of a masonry dam across the Colorado river; a power house, with the many details of machinery always found in a large station; distribution mains for supplying the city with water; a large storage reservoir, or settling basin, and all necessary accompaniments.

This article will be confined to the dam, and a discussion of its stability and of the relative value of a square and a rounded top or spillway for such a structure.

Discussion for stability.—This particular dam under discussion is 60 ft. high above low water, with a rounded crest, as shown in the accompanying drawing, and a long sweeping curve at the toe which comes tangent to a horizontal line at zero point or low water. The center of gravity of the section of the dam can be found very accurately, if care be taken, by drawing the dam to scale on some hard homogeneous paper, cutting out the shape and balancing it on a pin point. We thus find the center of gravity to be 15 feet from front face of dam and 26 feet above base line or zero point. To get the force of gravity which acts through this point we take a section of the dam one foot in length, or multiply the sectional area by the weight of one cubic foot of masonry. We find the sectional area by cutting up the figure of the dam into triangles, then taking the sum of their areas which we find to be 1897.125 square feet. The dam is built of granite weighing over 150 pounds per cubic foot. Whence 1897.125x150=284,568 pounds as the weight or downward force acting through the center of gravity of the dam. Then 284,568x45 (distance from vertical line through center of gravity to toe of dam) =12,805,560, which is the moment of stability.

The dam is designed to stand a flood of 15 feet of water passing its crest, making 75 feet above the zero line which we consider the base of the dam. The force of the water tending to overturn the dam will be $62.5 \times \frac{(75)^2}{2} = 175{,}625$ pounds, acting at a point which is one-third

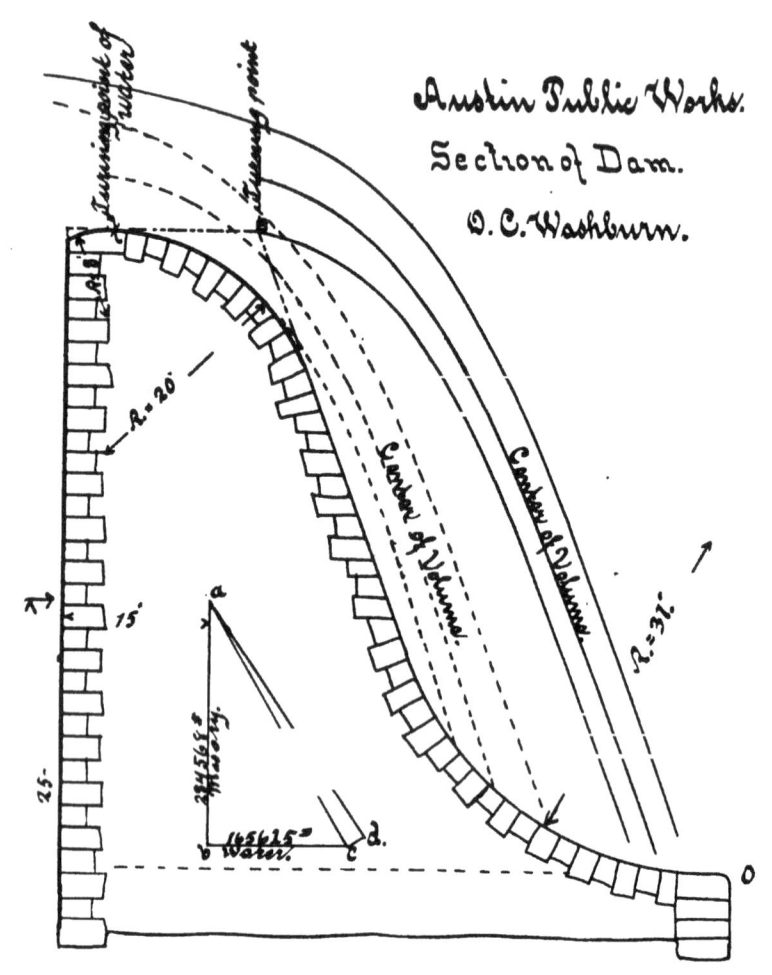

of 75 or 25 feet from the base of dam. Then the moment tending to overturn the dam is $175{,}625 \times 25 = 4{,}390{,}625$. Comparing this with the moment of stability we see that the latter is nearly three times as great, thus showing that we have a factor of safety of about three.

One of the conditions of safety for dams is that the resultant of water and masonry forces shall not pass outside of the middle third of the base. This may be solved graphically as shown in the drawing in figure $a\ b\ c$, which is seen to be within the required limit. For all conditions yet considered the dam is safe.

For possible sliding.—This investigation is rendered quite easy by consulting a table of coefficients of friction. Baker gives for this kind of masonry a co-efficient of .70, then weight of masonry 284,568x.70= 199,197 pounds resistance to sliding; but we know that the total water pressure is only 175,625 pounds, leaving a surplus of 23,572 pounds for safety, not considering the bonding effect of the masonry. This will be laid in the best hydraulic cement, which is capable of standing upwards of 1,000 pounds per square inch without crushing. The combination of all of these resistances gives a factor of safety of at least 6 to 8.

It will be noticed that the theoretical flow of water, as shown in the drawing, tends to leave the face of the dam. But the rushing of so large a volume of water over so small a space will produce a vacuum. The outside pressure of the air will bring the water against the wall, and thus prevent any disastrous pounding effect, which would be the case if the dam was built as first proposed and as shown by dotted lines in the drawing with outer corner of crest at e. The water tends to leave the dam about 7 or 8 feet. A force equal to one-fourth of a perfect vacuum is allowed to bring it back. Thus $\frac{15}{4}$x48x144= 25,920, overturning effect of the vacuum (the length of vacumm space being 48 feet). Laying off this force $c\ d$ to the scale of 50 tons to the inch at right angles to resultant, gives a new resultant $a\ d$ which is also within the required limit.

Flow of water over crest.—We now come to the best form of top or crest for dams. The square crest, as shown by dotted lines, was the form first proposed, and just above it is the position of a large volume of water flowing over. The curves of flow over the square and round crested dam are identically the same, and were obtained by plotting 31 curves of flow for different depths and an average taken, thus insuring very accurate results. The different positions of the curve of flow, relative to the dam, depend upon the form of dam, the water starting to fall on the round crested dam 13 feet farther up stream than in the case of the square crested dam. This 13 feet is utilized in bringing the jet of water nearer the face of the dam. All this is to prevent the dangerous effect of the pounding of the water on the dam. Since one practical test is worth much theory, it might be interesting to state what really did happen to this dam during a moderate flood in October, 1892. The water poured over the partly finished dam; then only 20 feet high, to a depth of 10 feet. The effect was to loosen large blocks of granite weighing six tons and tumble them into a heap about 300 or 400 yards down stream. Pieces of the shelving rock in the river bed, weighing over 30 tons, were torn up and moved 100 feet.

THE HISTORY OF PIG IRON.

Peter H. Christianson, *Instructor in Assaying.*

1. Genesis of the ores.—It is evident that at a very early period iron was quite uniformly disseminated throughout the rocks. A certain amount of it must have been held in solution by the oceanic waters, but this amount has not played a very important part in producing our economic supplies of iron.

At the present time all crystalline rocks contain iron in some form, and it is reasonable to suppose that the original primitive rocks all of which were crystalline contained as great a percentage of iron as those of the present day. Enormous quantities of these early crystallines have been disintegrated as proved by our enormous beds of clastic rocks, all of which must have been derived from crystalline rocks. Water is the principal agent in this disintegrating process; nearly all substances are soluble in it in a greater or less degree. Several years ago heat was supposed to be the great transforming agent; heat was made to account for all kinds of changes. In recent years microscopic investigations have given rise to the opinion that rocks were not violently altered by heat in a comparatively short time, but that rock alterations in a great measure are caused by the solvent action of water continuing throughout long periods of time. Although this action is slight in the case of many minerals, yet when we consider the vast ages through which it has been going on, there seems to be no difficulty in accounting for such vast transformations.

Aside from its chemical action, water contains a vast mechanical force. The results of this are more apparent than are those of its chemical action. The quartz of the crystalline rocks not being readily dissolved has been removed mechanically and deposited, forming quartzites or sandstones. These quartzite and sandstone beds may be taken to represent the free silica in the pre-existing crystalline rocks, but they do not carry all the iron which the earlier crystallines doubtless contained.

The world's supply of iron lies in the oxides and carbonates. The rocks associated with the latter are as a rule of later date than those associated with the former. The carbonates usually contain a less percentage of iron than do the oxides; yet occurring generally in the vicinity of coal fields and being less refractory than the oxides, their economic importance is quite considerable.

The oxides are as a rule associated with those older rocks which have been subjected to great alterations. Their mode of occurrence is exceedingly variable; sometimes we find them in irregular ore bodies, sometimes in beds and again in well defined veins. As no rule can be laid down touching their occurrence it is impossible to mark out any universal mode of origin. In some instances they may have been deposited as oxides from water entering a soluble stratum from which substances tending to precipitate the iron could be taken into solution. There are, however, other instances in which it has been proved beyond a reasonable doubt that the oxides have resulted from the oxidation and concentration of an original carbonate. It is quite probable that this method of concentration is much more universal than has been believed.

The originally disseminated and but slightly soluble forms of iron compounds were converted into the soluble carbonate through the agency of organic matter. The carbonate was dissolved and carried into some precipitating basin, as an inlet of the ocean. Here coming in contact with solutions of an alkaline nature, the iron was precipitated as an impure carbonate. In the course of time the ocean receded from these carbonate rocks leaving them subject to the action of percolating waters by which they have been oxidized and again concentrated.

This last named process of concentration has been thoroughly investigated by Prof. C. R. Van Hise, who has shown that the high grade hematite ores of northern Wisconsin were once impure cherty carbonates of iron. Since the Mesaba range in Minnesota is regarded as an equivalent of the iron range in northern Wisconsin, the same method of concentration has been attributed to its ores. There is a striking difference in the mode of occurrence of the iron bearing rocks of the two ranges. Those of northern Wisconsin are highly inclined while those of the Mesaba range dip but slightly. Again those of northern Wisconsin are cut nearly at right angles to the plane of sedimentation by dikes which are closely associated with the ore bodies and which have played an important part in their concentration. So far as known dikes are not associated with the Mesaba ores, only a few being known to exist on the entire range. It is quite probable, however, that the conditions for ore deposition are similar in the two localities, and that the level position of the iron bearing rocks of the Mesaba range has tended to produce results similar to those of the highly inclined formation cut by dikes. This view is substantiated by numerous thin sections from the Mesaba iron formation made and examined by the writer. These sections show that in certain localities carbonates are very abundant in the ore horizon. Thin sections made from the green schists, which at present are exposed in many places only a short distance north of the ore beds and which were once un doubtedly covered by the ore formation, prove that they contain a great amount of carbonate which is clearly seen to be of secondary origin, and which must have come from the once overlying ferruginous rocks. Again, sections from other localities show the ore horizon to be in a state of transition, the iron oxide filling the cleavage cracks of the carbonate. Considering these facts, in the light of what has been proved with reference to the iron ores of northern Wisconsin, there can be but little doubt that the genesis of the iron ores of the two districts is the same.

2. *Mining the ores.*—We have now the ores existing as oxides or carbonates in deposits of varying shapes. The plane of their greatest extent may be highly inclined to the surface. The position of such a deposit partakes of the nature of a vein, and the mining of it necessitates the employment of some method applicable to deep and extensive workings. Again, the plane of the greatest extent of the ore deposit may be more or less parallel with the surface, i. e., in the natural position of a bed. In the mining of such deposits the depth is usually not so important a factor.

Mining operations depend largely on the nature of the ore and its occurrence. Various methods are employed according to conditions and circumstances. These methods may be divided into two general classes, one of which is characterized by "open pits" and the other by "shafts and drifts." The open pit method is adapted to localities as the Mesaba range in Minnesota, where the ore lies near the surface. It consists in first removing the barren material from the surface of the ore. This is called "stripping." Any applicable method of excavation may be used, If the surface material is soft, like that on the Mesaba range, any of the common methods for handling sands, gravels and clays can be used. Often a considerable area is stripped before the ore is mined out. After the ore has been removed from this stripped area, the excavation thus made may be conveniently used to receive the strippings from the next area. If the ore is hard, blasting must be resorted to in order to loosen and break it up for future handling. Hand, steam or compressed air drills are employed for making the blast holes. Inclined tramways operated by stationary engines are usually employed for removing the ore and material from the pits. The ore is stored in stock-piles and ore-bins located on the surface.

The second method of shafts and drifts is the more universal. Shafts are rectangular, vertical, or inclined openings connecting the workings of the mine with the surface. Their walls are supported by heavy timbers. Through these shafts the ore is brought to the surface by means of steam or electric hoists, operated by engines in the hoisting-house, located near the top of the shaft. Where water is encountered the draining of the mine is accomplished by a more or less elaborate system of pumping or draining.

Leading from the shafts at various levels are drifts, i. e., horizontal rectangular openings, extending into or through the ore and dividing the ore body into numerous blocks. In a well regulated mine these drifts are so arranged and extended from time to time that a given amount of ore can be mined without materially increasing the rate of expenditure per ton.

Extending along the drifts from near the place where the ore is being taken out to the shaft, are tracks on which run cars carrying the ore to a cage or skip to which the hoisting cable is fastened.

At the mouth of the shaft the car containing the ore is removed from the skip and transferred along tracks to the ore-bins, stock-piles, concentrators or furnaces as circumstances require. When the transfer is to be made to ore-bins or stock-piles the tracks are generally supported on trestles twenty feet or more in height. This is for the purpose of letting gravity assist in loading the cars for further transportation.

3. Smelting the ores. The ore obtained from the mine passes into the hands of the smelter who extracts the metal from it. Smelting is carried on in a blast furnace. This is a huge, upright furnace, in which the ore, in connection with flux and fuel, is reduced and melted by the aid of a hot blast forced into the bottom of the furnace through pipes by means of fans or blowing engines.

The flux is the material with which many of the impurities of the ore will unite, forming a fusible slag which can be separated from the metal. All impurities cannot thus be removed; phosphorus and sulphur, the principal injurious impurities do not pass into the slag. Having great affinity for iron they can but partially be removed at the expense of considerable iron.

The fuel used may be either charcoal or coke. It should contain no sulphur nor phosphorus. Charcoal yields a superior pig iron, owing to its more suitable composition. It is, however, not adapted to large furnaces because of its friable nature. The fuel furnishes the reducing agent as well as the heat necessary, both for the reduction of the ores and the separation of the impurities.

Blast furnaces vary greatly in size and shape. The proportions of the various parts are more or less determined by the nature of the ore smelted and the fuel used. Nearly all blast furnaces are circular in section and vary in height from thirty-five to eighty-five feet. They are usually formed by lining an outer iron casing with several inches of refractory material. The lowest part of the furnace is called the hearth. This is where the greatest temperature is obtained. Into the lowest part of the hearth settles the melted iron and slag, which are drawn off at required intervals from the furnace through tap-holes. Above the tap-holes are the tuyeres. These are the pipes through which the air is forced into the furnace. They are usually made from wrought iron, having double walls between which water is continually circulating to keep them from melting in the enormous heat.

The widest part of the furnace, directly above the hearth, is called the boshes. Above the boshes is the stack, which is usually somewhat tapering towards the top to facilitate the collecting of the gases. The top of the furnace is closed by means of a specially constructed apparatus which prevents the gases from escaping and allows the charging of the furnace. As the furnace is charged from the top some contrivance in the form of a steam lift or inclined plane for elevating the charge is necessary. The gases are collected and used to heat the air for the blast in specially constructed heaters.

In the operation of the blast furnace the following steps may be noted: First, the calcination of the charge, i. e., the driving out of the moisture; second, the reduction of the oxide at a temperature below the melting point; third, the carbonization of the metal at a higher temperature, but yet below fusion; fourth, the fusion of the entire mass so as to enable the iron and the impurities to separate according to their specific gravities. These steps take place more or less in the order named as the charge descends to the bottom of the furnace. Here the slag and iron separate. They are usually tapped through separate tap-holes. The iron is run into moulds of varying construction, depending on the purpose for which the iron is to be used. Sand moulds allow it to cool slowly. This is desirable if the iron is for foundry purposes; but if it is to be converted into wrought iron, it should be cooled quickly in iron moulds.

The name pig-iron is derived from the manner of arranging the branch moulds around the feeder or main moulds. It is a term applied to the iron produced by the blast furnace process. The composition is quite variable. In general it is a combination of iron with carbon and silicon. It also usually contains small amounts of sulphur, phosphorus and manganese and sometimes traces of copper, arsenic, etc. Its specific gravity is from 7 to 7.5.

Silicon is one of the elements entering quite freely into the composition of pig-iron. It may be present in proportions varying from 0.1 to 5 per cent. Its presence is very necessary in pig-iron intended for the Bessemer process; but its presence in large quantities diminishes the strength of the metal for foundry work.

Aside from the iron the carbon is the most important element of pig-iron. Its percentage varies from 2 to 5.5. It may exist in the form of a chemical combination with the iron or in the graphitic state. Many of the physical characteristics of pig-iron, such as hardness, color and character of fracture, depend on the form in which its carbon exists. If the carbon is all graphitic the iron is tough, has a dark gray color and is best adapted to foundry purposes. Gray iron is produced by high heat in the blast furnace and cooling slowly when tapped. If nearly all the carbon is in the combined state, the pig-iron is more brittle, and has a white color and crystalline texture. Gray iron may be changed to white iron by rapid cooling.

A small amount of sulphur and phosphorus may be present in pig-iron, without injuring it for foundry purposes; but 0.03 per cent of sulphur or 0.3 per cent of phosphorus makes it unfit for coverting into wrought iron, and pig-iron containing over 0.08 per cent of phosphorus or sulphur is not suitable for making steel, except by the basic Bessemer process.

To summarize:

1. The iron was originally disseminated throughout the crystalline rocks.

2. The crystalline rocks have suffered enormous erosion and sub-aerial disintegration.

3. In the presence of organic matter the iron was taken into solution by the percolating waters and in a large number of cases was redeposited as a carbonate in some precipitating basin.

4. In case of the more recent rocks these carbonates have been but slightly changed while in the older rocks the carbonates have been converted into oxides and at the same time have passed through a process of concentration.

5. The ore is mined, i. e., it is loosened, loaded into cars or buckets, hoisted to the surface and prepared for further treatment.

6. It is transported to the smelter where it is reduced, carbonized, fused and separated from the impurities in a blast furnace.

7. The product is moulded into commercial pig-iron.

THE CHEMISTRY OF IRON.

J. Frank Corbett, '94.

In treating of the Chemistry of Iron a review of its most essential physical properties is necesssary, since the physical and chemical properties are so intimately related.

Iron has a peculiar gray color and strong metallic lustre, which are heightened by polishing. In ductility and malleability it is inferior to several metals, but exceeds them all in tenacity. At ordinary temperatures it is hard and unyielding. This hardness may be increased by heating and suddenly cooling, thus rendering it brittle. When heated to redness it is remarkably soft and pliable so that it may be beaten into any form or welded. Its texture is fibrous. Its specific gravity is 7.788, varying according to the degree to which it has been rolled, hammered or drawn, and is increased by fusion. It is attracted by the magnet and may itself be rendered magnetic. It retains this property however only within certain limits of temperature. Iron ceases to be attracted when heated to an orange red. This description applies to commercial wrought iron, which is nearly chemically pure. The difference in appearance and properties commonly seen in wrought iron, cast iron and steel is due to the rapidity of cooling, amount of rolling and presence of other chemical elements. These three divisions all merge into one another by such imperceptible degrees that no sharp line of distinction can be drawn.

The crude product obtained from the reduction of iron ores in the blast furnace is known as cast iron. It is not malleable, particularly when hot, but may be hardened by sudden cooling. Carbon and various impurities always exist in cast iron in small proportion to no great injury, but the quantity must be limited. Certain amounts of sulphur and phosphorus are especially injurious. Other metals, such as copper, manganese, chromium, nickel, titanium and cobalt, are generally found, but in such small quantities as not to affect the iron.

Wrought iron is the term applied to the more or less refined metal produced from pig iron or from the ores; it is malleable and ductile, both hot and cold, but cannot be tempered.

Steel forms an intermediate link between ordinary cast iron and wrough iron, uniting in a degree the properties of both; its disinguishing character is its capability of being hardened or softened by rapid or slow cooling. It cannot be said where steel begins and where it ends. It is a member of a series commencing with the most impure pig iron and ending with the softest and purest malleable iron. In ductility and malleability steel is far inferior to wrought iron, but exceeds it in hardness. Its texture is also more compact and it is susceptible of a higher polish. It sustains a full red heat without fusion, being more fusible than wrought iron, but less so than cast iron. Cast steel is the closest grained and most uniform variety. Cast iron and steel crystallize in the regular system while wrought iron is fibrous, a condition resulting mainly from mechanical causes. Pure iron is generally obtained as a powder, and as such is a grey uniform mass; cannot be made permanently magnetic; has a constant specific gravity and at ordinary temperatures will burn in open air.

Iron in its ordinary state has a strong affinity for oxygen. In a perfectly dry atmosphere it undergoes no change but when moisture is present its rusting is rapid. In the first part of the change ferrous carbonate is formed; but the carbonate rapidly passes into the hydrated sesquioxide and carbonic acid is evolved. Rust of iron always contains ammonia, a circumstance which indicates that the oxidation is accompanied by decomposition of water, the nitrogen in the air uniting with the hydrogen set free by the iron uniting with the oxygen of water. Chevallier has observed that ammonia is present in many of the iron ores.

Heated to redness in the open air, iron absorbs oxygen rapidly and is converted into the black oxide of iron. Chemically pure iron, prepared by reducing pure sesqui-oxide of iron by hydrogen, may be obtained at a heat below redness; but the iron, when thus reduced, on being exposed to the air takes fire spontaneously and burns to the black magnetic oxide. This property, which Magnus has noted in nickel and cobalt prepared in a similar manner, appears to depend upon the finely divided and expanded state of the metallic mass; for when the reduction is effected at a red heat, which enables the metal to acquire its natural degree of compactness, this phenomenon is not observed. Iron decomposes the vapour of water by uniting with its oxygen at all temperatures from a dull red to a white heat, a singular fact when it is considered that at the very same temperature the oxides of iron are reduced to the metallic state by hydrogen gas.

The action of nitric acid on iron is attended by a series of very remarkable phenomena, which were first observed by Prof. Schoenbein in 1830. He observed that nitric acid, sp. gr. 1.35, though capable of acting with great violence on ordinary iron, was perfectly inert on a portion of iron wire which had been made red hot previous to its introduction into the acid. He found that this indifference to nitric acid may be communicated: 1st, by mere contact from one iron wire to another; 2d, by submersion for a few moments in strong nitric acid; 3d, by making it the positive electrode of a galvanic current, the negative electrode having been introduced into nitric acid. Faraday, who also experimented along this line, found that the same property is communicated to iron by contact with platinum and that the same effect is not limited to nitric acid, but extends to various saline solutions which are usually acted upon by iron. Acid of less specific gravity than 1.299 does not produce passivity; but if an iron rod is dipped into concentrated nitric acid and the whole immersed in dilute acid, the entire rod is passive.

These facts are explained by Varrenne, by supposing a gaseous film to be deposited on the surface of the iron, which protects it and the gas to be absorbed in the dilute acid. In the case of the partially immersed rod the gas bubbles are removed from one part and adhere to another. Ramann considers the passivity due to a layer of oxide of iron, which is soluble in dilute but insoluble in strong nitric acid. This view is supported by the fact that iron is rendered passive by potassium chlorate, chromic anhydride and hydrogen peroxide, since these are all oxidizing agents and the passivity of iron is removed by rubbing or by heating in reducing gases.

Iron is distinctly a metallic element; it replaces the hydrogen of acids forming two series of compounds, represented by the formula

Fe X_2 and Fe X_3 respectively; X representing Cl, N O_3, $\frac{SO_4}{2}$, $\frac{PO_4}{3}$, etc. Those compounds of the first class are called ferrous compounds; of the latter, ferric compounds. The ferrous salts are readily oxidized to the ferric form. Many salts, both normal and basic of both series, have been isolated, and numerous double salts are also known. Ferric oxide forms compounds with several oxides more basic than itself; e. g. with potassium oxide, barium oxide and calcium oxide. These compounds may be regarded as ferrites derived from the hydrates. There are also some salts known as ferrates derived from the hypothetical ferric acid.

Iron readily combines with the halogens. It is interesting to note that the only compounds of iron that have been gasified are the ferrous and ferric chlorides. The latter has a molecular formula, Fe Cl_2 at 1300° to 1600°; but there is evidence of the existence of molecules of Fe Cl_4 at lower temperatures. The ferric chloride appears to exist as $Fe_2 Cl_6$ and Fe Cl_3. The haloid salts of iron are soluble in water, but are decomposed by much water. A peculiar property of iron is that at a high temperature it will absorb 46 per cent of its volume of hydrogen, indicating that a hydride may exist.

Three oxides of iron have been isolated Fe O, $Fe_3 O_4$ and $Fe_2 O_3$. The first and last unite with acids to form ferrous and ferric salts respectively. According to Lefort, a few salts corresponding to $Fe_3 O_4$ are known; e. g. $Fe_3 Cl_8$, $Fe_3 (SO_4)_4 + 2 SO_2 + 15 H_2 O$. Besides these oxides we have also, as the anhydride of the hypothetical acid of the ferrates, an oxide represented by FeO_3. The ferrous oxide is a black, generally, pyrophoric powder which burns to ferric or ferro-ferric oxide, according to completeness of combustion. The sesqui-oxide, $Fe_2 O_3$, is a red powder, to which almost all compounds of iron will change if exposed to the air. The ferro-ferric oxide, $Fe_3 O_4$, is a black substance most commonly seen as scales on iron. It does not seem entirely improbable that this so called oxide may be a ferrous salt of ferric acid.

Carbon and iron unite in many different proportions. Carbide of iron may be made by exposing iron and carbon, away from the air, to intense heat for a long period of time. This compound contains about 10 per cent of carbon and is found in some grades of cast iron, though a considerable per cent of carbon in other grades of cast iron is mechanically mixed. Upon the amount and condition of carbon in cast iron depends, to a large degree, the grade of iron—ranging from white to gray. The former is hard and brittle, sometimes breaking like glass from sudden change of temperature; while the latter is softer and much more tenacious. The white may be converted into the gray by exposure to strong and long continued heat, and cooling slowly. Conversely the gray may be changed into the white by being heated and rapidly cooled.

The oxy-salts of iron are included in two general classes—ferrous and ferric. The former are mostly normal and are soluble in water. They include a large number of double-salts and exhibit many isomorphous forms with the corresponding salts of cobalt, nickel, manganese, zinc and magnesium. The ferric salts include many double and basic salts, some of which are insoluble in water. They are generally analogous to the persalts of alumina, chromium, cobalt, nickel and manganese. The following is a list of important salts of iron: carbonates, nitrates, sulphates, antimoniates, chlorates, chromates, phosphates, titanates, tungstates, iodates, borates and arsenates.

The organic salts of iron, so far as numbers are concerned, occupy considerable space, but are rare and unimportant in nature. They find their chief use in medicine, and are generally basic and double salts. Two important salts are citrate of iron and tartrate of iron and potash.

Iron alloys with manganese, aluminum, lead, mercury, antimony, copper, chromium, tin and zinc.

Owing to certain similarities of chemical and physical properties, iron, cobalt, nickel, manganese and chromium are generally grouped together. Below is a brief table showing some of these similarities:

	Iron.	Nickel.	Cobalt.	Manganese
Atomic weight	55.9	58.6	58.6	55.0
Specific gravity	7.9	8.8	8.5	8.0
Specific heat	112.0	108.0	107.0	
Melting point	1,800.0	1,600.0	1,500.0	1,500.0

All are magnetic and crystallize in cubic form.

THE BESSEMER PROCESS.

HARRY E. WHITE, '93.

The discovery of large quantities of high grade Bessemer ore in Minnesota, together with the greatly increased use of Bessemer steel, makes the Bessemer process a subject of unusual interest.

Steel is a compound of iron and carbon generally containing from 0.3 to 2 per cent. of carbon. The temper depends upon the amount of carbon and the manner of its combination with the iron. Steel must be free from impurities; 0.1 per cent. of phosphorus renders it unworkable at ordinary temperatures, and 0.3 per cent. of sulphur renders it unworkable at a red heat. Steel holds in its chemical composition an intermediate position between wrought iron and cast iron; the point where wrought iron ends and cast iron begins cannot be precisely determined. Steel possesses the physical property of greater strength in proportion to its weight than wrought iron, and retains to some degree the malleability of the latter. This has led to the extended use of the milder varieties of steel as a material for construction. It is especially useful where lightness and strength united with durability are required.

The former methods for the production of steel, however, were too costly to make its use practicable in many lines of manufacturing, but the Bessemer process has in a large measure met the demand for a cheap material of high quality.

There are many other processes at present used in the manufacture of particular varieties of steel, or to work up impure ores, rather than for general production.

The essential difference between the Bessemer and other processes is, that in the former the iron is completely decarbonized and a definite amount of carbon again introduced by the addition of *spiegeleisen*, a compound of iron, carbon and manganese. In nearly all the other processes the iron is only partially decarbonized. In the older processes phosphorus is largely eliminated, while it is not sensibly affected by the ordinary Bessemer process. Phosphorus can be disposed of, however, by a modification of the ordinary or acid process, called the basic process.

The Bessemer converter is a round pear shaped vessel with a contracted opening from which the metal is poured. The peculiar shape of the opening prevents the escape of the metal during the blow. The converter consists of two parts, the outer or wrought iron shell made of plates, five-eights to seven-eights of an inch in thickness. The lining is a foot or more in thickness, and consists of ganister or other refractory material of an acid character. The bottom, which in nearly all American converters is removable, is covered with the same material as that of which the lining is composed. In the lower part of the bottom is the air space, which is connected with the body of the converter by means of tuyeres. The converter is set in a band or trunnion, having two arms, one of which is solid and is geared so as to rotate the vessel. The other arm is hollow and through it the compressed air for the blast reaches the air space. The converter, after having been properly lined, is heated to redness. It is then turned down to receive its charge of molten metal, the blast is turned on and the vessel brought back to a vertical position.

During the first three or four minutes of the blow a shower of sparks accompanied by a pale flame issues from the mouth of the converter; next an intense, unsteady yellow flame is seen for five or six minutes. The pressure of the blast is then lessened, and the *boiling* stage continues for eight or ten minutes, when the flame becomes a pale rose color for a moment and then disappears entirely, showing that the carbon, silicon and manganese are oxidized. At this point the converter is quickly turned down and the blast shut off; for if the process were continued further the iron would be oxidized. The desired amount of carbon is then added in the form of molten spiegel. The manganese serves the double purpose of preventing the oxidation of the iron, and after becoming oxidized of forming a silicate of maganese. It thus serves as a flux for the silicon.

The operation of the basic process does not differ materially from that of the acid process. In the basic process there is a basic lining to the converter, generally of dolomite instead of the acid ganister of the ordinary process. The iron for the basic process should be low in silicon, and it should contain at least 2 per cent. of phosphorus to generate the heat which in the acid process is produced by the oxidation of the larger amount of silicon. While as good a quality of steel can be produced from pig-iron high in phosphorus, the lining of the furnace is carried away more quickly by the basic than by the acid process. Owing to the large amount of ore in this country, low in phosphorus, but little use has been made of the basic process, which is used quite extensively in Germany and England.

The size and number of converters used in a plant depend on the nature of the output. If the production is small and for some special purpose, one 3 or 4 ton converter is used, but if large production is desired three 10 or 12 ton converters would be required to meet demands. Usually two converters are in constant use, one in blast, another discharging and receiving a charge, and the third undergoing repairs or ready to take the place of either of the others.

The iron in large plants is taken direct from the smelter. It is either run in by inclined spouts or carried by means of huge ladles. Owing to better arrangement and management, more steel can be produced per man in this country than in Europe.

In 1872 the United States produced 142,954 long tons of steel; in 1890, 4,277,071 long tons, while Great Britain produced only 3,679,043 long tons during the same year.

The Bessemer process has played an important part in the steel production of this country. In 1890 the United States produced 3,688,871 long tons of Bessemer steel, while only a little over 500,000 long tons were made by all other methods. Great Britain produced nearly 1,700,000 long tons by all other processes.

In all methods for producing steel, except the Bessemer, the great item of expense is labor. Great Britain with her cheap labor would have made it impossible for the United States to attain its present rank in the steel industry had it not been for the Bessemer process.

WIND MILLS FOR ELECTRIC LIGHTING.

GEORGE H. MORSE, '93.

With the introduction of any new departure from existing industrial methods the following questions must be satisfactorily answered: First, can this be done? Second, how may it best be done? Third, will the returns pay a proper interest on capital invested.

The first two of these questions, with reference to the generation and storage of electricity by the use of wind mills, have been well answered by the recent experiments of a number of able engineers.

Experiments of Charles F. Brush. Mr. Brush, electrician of the Brush Company, has in his private grounds at Cleveland, Ohio, a fine example of an electric wind plant. Neither time nor money has been spared in making this plant a thorough success. A very good description was given in the *Scientific American*, December 20, 1890, p. 389. The wind-wheel which is of the American pattern having a diameter of 56 feet, furnishes power to run the plant which contains 350 incandescent lamps, two arc lights and three electric motors. About 100 of the incandescent lamps are in every day use. The electricity is furnished by one 1200 watt Brush dynamo having a speed at full load of 500 revolutions per minute. The dynamo is automatically arranged to be thrown into circuit at 330 revolutions per minute, the electromotive force being limited to 90 volts. There are 408 secondary battery cells, each having a capacity of 100 ampere hours. These cells are charged and discharged in 12 batteries of 34 cells each.

Professor Blyth's experiments. A description of these experiments was given in the *London Electrical Review*, and an abstract of the same may be found in the *Electrical Engineer*, August 31, 1892. Professor J. Blyth devised a peculiar form of apparatus in order to do away with some of the inconveniences attendant upon the use of the more common form of windmill.

Quoting from the article mentioned above: "It consists of four hemispherical cups attached to four arms, and moving in a horizontal plane about a vertical axis. From the theory of this instrument (Robinson's anemometer) which is, however, only approximate, it appeared that whatever might be the speed of the wind, the speed of the cups attained a certain *terminal* value, such that the couple due to the wind pressure was exactly equal to that produced by the resistance to the motion through the air, and the friction on the bearings.

Last summer I erected a machine of this kind, which has been considerably improved within the past three months. The cups are replaced by semi-cylindrical boxes attached to four strong arms, each about 26 feet long—the opening of each box is 10 feet by 6 feet, and the vertical shaft is a long rod of iron 5 inches in diameter. At the lower end it carries a massive *pit* wheel which actuates a *train of gearing*, and drives a flywheel, 6 feet in diameter, with the requisite speed for driving a dynamo connected with it by a belt in the ordinary way. This machine worked most satisfactorily, and with a fair wind speed gave about 2 electrical horse-power. I also tested it in a strong gale, by allowing it to run with no load, and the result was perfectly satisfactory as a safe terminal speed was attained, and all racing avoided. Hence I think that electrical windmills, at least for small installations, are likely to assume this form, as there is no limit to the size and strength with which they may be constructed, and if necessary, several could be placed in any well exposed position, each having its own dynamo and set of accumulators. During the past few months I have increased the power of the machine by adding an auxiliary box to each arm with a gap between it and the previous one. This I find to be better than merely increasing the size of the previous box."

Experiments, using the Halladay windmill.—Two notable experiments upon this subject have been made, in each of which the Halladay wind-mill has been used with great success. This mill is of the American type and seems to be a favorite, owing to ease of regulation.

Messrs T. E. Carwardine & Co.'s wind plant.—We find in the *Engineer*, London, April 1st, 1892, the following: "An electrical installation has been made at Messrs T. E. Carwardine & Co.'s flour mill in the City Road, E. C., and seems to fulfil both the requirements indicated (i. e., storage accommodation and method of regulation).

One of Messrs Alfred Williams & Co.'s Halladay windmills erected on the roof of the building on substantial timber supports, hitherto used for making wheatmeal and working elevators, has been set to drive an Elwell-Parker dynamo capable of developing a current of about 30 amperes with 70 volts pressure. The windmill drives this at a rate which, taken with the use of the governor and cut-out employed, is sufficiently uniform to charge a battery of 28 E. P. S. accumulators. From this battery sufficient electricity is obtained for two 1,500 candle power arc-lamps and several incandescent lamps. The windmill consists of a sectional wheel with a vane at the back, the whole arrangement being mounted on a turntable. The vane acts as a rudder and keeps the wheel always facing the wind. The wheel, which is 30 feet in diameter, consists of a skeleton framework, into which a series of wooden sections are centered, and these are connected with counterbalance weights which act as governors, and cause the sections to open and shut according to the strength of the wind blowing, thus obtaining a comparatively uniform speed. By means of a sliding contact worked by a governor on the dynamo shaft, the charging circuit of the electrical apparatus is switched on when the speed is high enough, and switched off when it drops too low, and there is also an automatic switch which reduces the existing current when the speed is too high, and thus prevents too much current being forced into the cells at any time. In addition to this there is a resistance in the main circuit which aids the automatic excess switch in its action. The governor controls a lever which short-circuits and opens up resistances which are arranged in the shunt of the machine, and so regulates the electro-motive force according to speed."

The following, taken from the *London Electrician*, January 20, 1893, forms a sad sequel to the above: "The City Road windmill which has been used for the electric lighting of Messrs Carwardine & Co.'s flour mill, besides other work, has been attacked by the county council as a sky sign, and now the Ecclesiastical commissioners, as freeholders of the land have obtained an injunction restraining its further use on the ground that it is an erection contrary to the terms of the building agreement. A rather interesting experiment is thus brought to an end. We understand that the plant gave every satisfaction, the power obtained being sufficient to light the premises with 27 16 c. p. lamps and 3 arc lights. The storage cells are in excellent condition, and have given no trouble whatever. At the time the injunction was granted Messrs Carwardine were about to erect a motor, to be driven by the accumulators, for the purpose of hoisting sacks."

Experiments at Cape of the Hague (Translated from the French, *Annales Industrielles*, 1889, t. 1., colon 454).—Under the patronage of the duke of Feltre, experiments were made during 1886 and 1887 by the engineer Raoul de L'Angle Beaumanoir at the Cape of the Hague, upon the use of wind-power for generating electricity. The experiments were carried out in the most complete manner. The plant was installed in the North light-house at Havre, the primary object being a study of the application of electricity, thus obtained, to the economical lighting of the beacon lamps. During the progress of these experiments, numerous trips were taken upon the Continent and to London for the purpose of studying the wind and also the best form of wind-mill, dynamos and accumulators for this use.

A 40 foot Halladay mill was finally selected, a ball governor being substituted for the original centrifugal regulator. This mill was brought to France from the United States and was guaranteed by the French agent to give 18 *force de cheval* (17.8 horse-power), measured on the wind shaft with a wind of 33 feet per second, but was found in reality to give this power at a wind velocity of 23 feet.

Two dynamos were made by the Brush Company at London for the above plant. They were of the Victoria type each having a normal voltage of 75 volts, one ranging from 1 to 4 horse-power, the other from 4 to 16 horse-power. The out-put of these machines is variable and quite closely proportional to the number of revolutions. The larger machine while making from 250 to 650 revolutions per minute gave from 40 to 160 amperes. The smaller machine gave for a speed of 100 to 260, a corresponding current of 8 to 40 amperes. The dynamos were arranged to work alternately, according to the power furnished by the mill, the object in having two being to avoid the great loss of efficiency that would occur in any one machine working under such wide variations in speed. Between each dynamo and the series of accumulators were placed automatic circuit breakers.

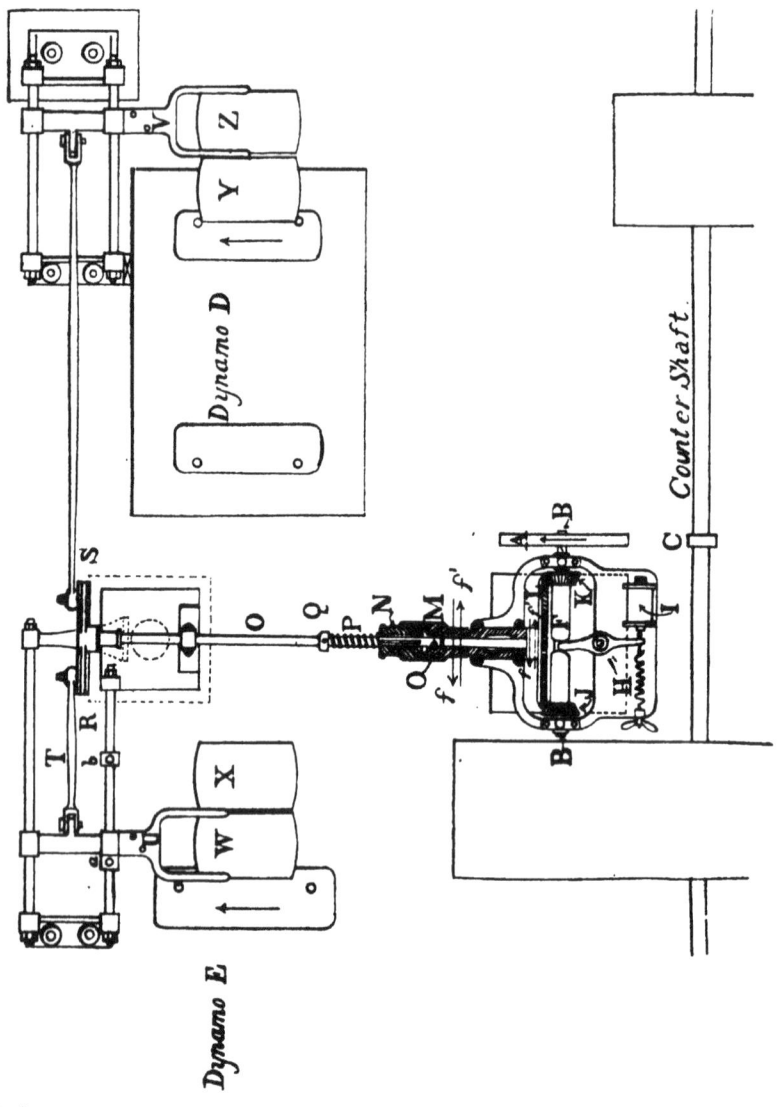

A most peculiar and original feature of this installation is the automatic arrangement employed to set either dynamo in motion according to which one is better adapted to the strength of wind blowing. Referring to the figure, A is a pulley keyed to a horizontal shaft B, which is constantly kept in motion (always in the same direction) by means of a belt running from a pulley C on the counter shaft. A reversing gear F, consisting of a hollow cylinder, slides on shaft B and revolves with it by means of a sliding feather. At its center the cylinder is turned to a smaller diameter and is embraced by two prongs of a fork, the latter forming the end of a horizontal lever oscillating about the point G. The other end is attached on one side to a spring H, the tension of which may be regulated, and on the other to the plunger I of a solenoid, around which passes a derivation from the circuit of each dynamo.

When the current reaches the maximum strength fixed for the smaller dynamo, the solenoid acts upon the lever, pushing the cylinder F to the left. When, however, the current sinks below this maximum, the spring acts so as to return the cylinder F to the right. On the shaft B are mounted two bevel-gears J and K, each being keyed, not to the shaft, but to one end of the reversing cylinder. If the solenoid prevails in placing the cylinder F, so that the gear J engages with L, it will turn the latter in the direction of the arrow f, being impelled by the shaft B. If, however, the spring prevails, K engaging with L will give motion to the latter in the direction of f.'

The wheel L is keyed to a horizontal shaft which carries at its other extremity one part of a clutch M, the other part of the clutch N is mounted on a horizontal shaft O, the extremity of which moves freely in the first half M. The part N is held against the part M by means of an adjustable spring P, pressing against the stop Q. When the two faces of the clutch M and N are pressing against each other, as indicated, the shaft O takes the motion of L, either in the direction of the arrow f or f' At the end opposite the clutch MN, the shaft O carries two arms, R and S, diametrically opposite each other. The central portion of these arms is cut away to form guides. In the arm R slides a pin attached to a connecting rod T, which moves a forked piece U, the latter traveling upon two fixed cylindrical guides. This fork embraces the approaching side of the belt which drives the large dynamo E. The arm S is attached in the same manner to another fork V, which shifts the belt of the smaller dynamo D. W is the fixed pulley of the dynamo E and X its free pulley. Y is the fixed pulley of the dynamo D and Z its free pulley. The arms R and S are so arranged that the dynamo E will be running when D is at rest, and *vice versa*.

An inclination is given to the teeth of the clutch MN, such that when the shaft O meets any considerable resistance, the half N will be forced back upon the spring P, thus freeing O from the motion of L. The resistance referred to is found whenever the forks U or V reach the end of travel, which is determined by the adjustable stops a and b.

The experiments carried out at the North light-house have proved very satisfactory, although from a financial point of view, the cost of electric energy thus obtained, appeared to be rather high owing to the excessive cost of accumulator cells.

In the accompanying table, average costs and performance of apparatus, under the peculiar conditions imposed, have been in all cases adhered to.

The dynamo and accumulator plant must have a capacity equal to at least twice the required average, to allow for strong and continuous winds.

Although 50 and 45 per cent. efficiency respectively appears low for the above apparatus; yet we should remember that the assumed power of the windmill is to be obtained only under a wide variation in speed.

TABLE SHOWING THE COST OF ELECTRIC LIGHTING BY WIND POWER.

Diameter of wind wheel in feet.	Cost of geared wind mill, shafting and tower.	Useful horse-power developed. Wind sixteen miles per hour.	Average number of hours this horse-power will be developed per day.	Expense of power, cents per hour.				Watts recovered from dynamo, 50% efficiency.	Cost of dynamo.	Expense of generating electricity.			Watt-hours recovered from accumulators per day. Efficiency 45%.	Number of lamps, sixteen candle-power, 110 volts.	Number of hours lamps will run per day.	Required capacity of accumulators in ampere hours.	Cost of 58 accumulator cells.	Cost of automatic battery regulator.	Expense of storing the electricity.			Total expense of obtaining electricity per hour.	Total cost of the eight hours daily storage.	Equivalent number of lamp hours.	Average cost of lamp. hour from wind power, in cents.
				Interest on first cost at 5% per annum.	Depreciation at 5%.	Attendance.	Oil.			Interest on first cost.	Depreciation at 5%.	Oil.							Interest on first cost.	Depreciation at 20%.	Attendance of plant.				
8½	$ 153	0.04	8	0.26	0.26	0.06	0.04	14.9	$ 15	0.03	0.03	0.05	53.7	1	0.84	4	$ 21	$20.00	0.07	0.28	0.24	1.32	10.56	0.84	12.57
10	164	0.12	8	0.28	0.28	0.06	0.04	44.8	20	0.03	0.03	0.05	161.1	1	2.52	4	21	20.00	0.07	0.28	0.24	1.36	10.88	2.52	4.32
12	177	0.21	8	0.30	0.30	0.06	0.04	78.3	25	0.04	0.04	0.06	279.6	1	4.38	5	27	20.00	0.08	0.32	0.24	1.48	11.84	4.38	2.70
13	181	0.25	8	0.31	0.31	0.06	0.07	93.2	30	0.05	0.05	0.06	335.7	1	5.26	6	32	20.00	0.09	0.36	0.24	1.60	12.80	5.26	2.43
14	227	0.28	8	0.39	0.39	0.06	0.07	1[illegible].4	30	0.05	0.05	0.06	376.0	1	5.89	7	37	20.00	0.10	0.40	0.24	1.81	14.48	5.89	2.46
16	301	0.41	8	0.51	0.51	0.06	0.07	1[illegible].9	40	0.07	0.07	0.10	550.5	2	4.31	10	54	20.00	0.12	0.48	0.24	2.23	17.84	8.62	2.08
18	350	0.61	8	0.60	0.60	0.06	0.07	227.5	50	0.08	0.08	0.10	1,433.4	5	4.49	26	124	20.00	0.25	1.00	0.24	3.08	24.64	22.45	1.10
20	373	0.79	8	0.64	0.64	0.06	0.10	279.7	60	0.10	0.10	0.30	1,761.9	6	4.60	32	153	20.00	0.30	1.20	0.24	3.68	29.44	27.60	1.07
22	544	1.23	8	0.93	0.93	0.06	0.10	48	1[illegible]	0 22	0 22	0.30	2,890.4	10	4 53	52	249	20.00	0.46	1.84	0.24	5.30	42.40	45.30	0.94
25	584	1.34	8	1.00	1.00	0.06	0.10	[illegible]	140	0.24	0.24	0.40	3,148.9	12	4.11	57	273	20.00	0.50	2.00	0.24	5.78	46.24	49.32	0.94
30	679	2 40	8	1.16	1.16	0.06	0.13	895.2	175	0.30	0.30	0.40	5,639.8	20	4.42	103	470	20.00	0.84	3.36	0.44	7.95	63.60	88.40	0.72
36	743	2.95	8	1.27	1.27	0.06	0.13	1100.3	2[illegible]	0.36	0.36	0.50	6,932.2	25	4.34	125	571	20.00	1.01	4.04	0.24	9.24	73.92	108.50	0.68
40	842	4.42	8	1.44	1.44	0.06	0.13	1648.7	270	0.46	0.46	0.50	10,386.6	38	4.30	190	868	20.00	1.52	6.08	0.24	12.33	98.64	163.40	0.60
50	1,592	6.88	8	2.73	2.73	0.06	0.16	2566.2	30	0.51	0.51	0.60	16,167.3	60	4.22	294	1278	20.00	2.22	8.88	0.24	18.64	149.10	253.20	0.59
60	1,902	10.00	8	3.26	3.26	0.06	0.16	3730.0	4	0.68	0.68	0.80	23,499.0	85	4.33	427	1857	20.00	3.21	12.84	0.24	25.19	201.50	368 00	0.55

104   O. J. Anderson.

ESTIMATION OF OVERHAUL BY PROFILE OF QUANTITIES.

O. J. ANDERSON.

The method of estimating overhaul by "profile of quantities" was first introduced into this country in 1872 by F. Reineker, then first assistant engineer of the Pennsylvania Railway Company, Pittsburg, Pa., for use in his department. About 1882 it came into extensive use on the South Pennsylvania Railway, and has since been adopted by many leading railway engineers.

This method is very flexible in its application to various kinds of earthworks, and can be used to advantage in making preliminary distribution of the material before the work begins. It is especially useful in adapting mountain grade lines to the best ground, where the side slopes are very steep and the actual profile of the center line does not show the real amount and location of the material to be moved.

The method is presented with an example of the work as executed. The subject treated involves the following steps:

(a) Compilation of data.
(b) Plotting the profile.
(c) Taking off the results.

(a) *Compilation of data:* The paper on which the data is recorded should be ruled in six vertical columns, as in the accompanying table.

Column *1* contains the number of the station. In columns *2* and *4* the volumes of earthwork, in embankment or excavation at the successive stations, are recorded. Embankments are designated + and excavations —.

In rock work or where mixed material is encountered care must be taken to allow for proper shrinkage of the several materials. The percentage of shrinkage is embodied in the summation of the quantities before the ordinates are plotted on the profile, and a column is

Estimation of Overhaul by Profile of Quantities. 105

entered for the corrected quantities of fill. Column *3* contains these corrected quantities. This column appears only where an allowance is made for shrinkage.

Columns *5* and *6* are obtained by the algebraic addition of *3* and *4*. The ordinate at any point of the profile of quantities is equal to the algebraic sum of the volumes as far as that point, and should be verified by this principle at convenient points as the summation proceeds.

STA.	Increments +	Increments −	3 Cor. for shrinkage	Ordinates +	Ordinates −	STA.	Increments +	Increments −	3 Cor. for shrinkage	Ordinates +	Ordinates −
1	2		3	4	5	6	1	2	3	4	5	6
1				316		316	26	425	467		2,421	
2				442		758	27			481	1,940	
3				603		1,361	28			486	1,454	
4				715		2,076	29			454	1,000	
5				799		2,875	30			650	350	
6				886		3,761	31			1,075		725
7				1,010		4,771	32			679		1,404
8				1,074		5,845	33			875		2,279
9				963		6,808	34			900		3,179
10				355		7,163	35			1,076		4,255
11	94	103				7,060	36			1,213		5,468
12				102		7,162	37			1,532		6,900
13				493		7,655	38	653	718				6,182
14	172	189				7,466	39	795	875				5,307
15	172	189				7,277	40	843	927				4,380
16	170	187				7,090	41	843	927				3,453
17	169	186				6,904	42	206	227				3,226
18	87	96				6,808	43	750	825				2,401
19	875	963				5,845	44	633	696				1,705
20	3,426	3,768				2,076	45	546	600				1,105
21	970	1,067				1,009	46	506	557				548
22	909	1,000				9	47	760	836		288	
23	760	836		827			48	1,570	1,727		2,015	
24	570	627		1,454			49			710	1,305	
25	455	500		1,954			50			1,398		93

(b) *Plotting the profile.*—Ordinary profile paper may be used. The stations are laid off horizontally along the lower edge of the paper the same as for ordinary profile. For plotting the quantities, instead of referring to a datum plane at the lower edge of the paper, the initial line or axis of abscissa is located somewhere on the profile, from which the ordinates are measured, + above and − below. Connecting the ends of these ordinates is a curve composed of elements between the successive stations. If the increment is − the corresponding element of the curve will incline downward, as shown by the full line of the curve, indicating an excavation; but if the increment is + this element of the curve will incline upward, as shown by the dotted part of the curve, indicating an embankment, so that cut or fill is always indicated by full or dotted lines respectively.

Balancing line.— Some points of this line are usually fixed by the conditions of the problem. Where there are no fixed points, the balancing line may be determined by investigating for points of forward and backward haul, which make the work a minimum. The balancing line is drawn through these points. In the accompanying figure the point *a* is the beginning of the work and hence fixes the balancing line through that point. The point of forward and backward haul of the cut *df* is *e*. The portion *fe* is hauled forward to build up the embankment *fg*, and *ed* is hauled backward for building up the portion represented by the dotted line *cd*. The balancing line is horizontal, since the ordinate of a given amount of fill plus its percentage of shrinkage corresponds to equal ordinates of cut, and a line joining these ordinates must be horizontal.

If the quantities are plotted, as is sometimes done, without making corrections for shrinkage, the balancing line will be inclined. This line is then located by taking the algebraic sum or difference (the sum if the curve points upward as *cde*, the difference if it points downward as *abc*) of its intersection with the full line of the curve, and the shrinkage of quantity of material equal to that represented by the portion of the curve included between the above intersection and the point where the curve changes from a full to a dotted line. This sum or difference will be the ordinate of the intersection of the dotted line of the curve with the balancing line; hence, two points being known, the line is determined. The point of intersection with the full line of the curve, if not known, can be found by the same process as for horizontal balancing lines. In the curve *jklm* let *j* be a fixed point of intersection with the balancing line; *r* its ordinate; *k*, the point where the curve changes from a full to a dotted line; *s*, the ordinate of *k*; *l*, the intersection of the balancing

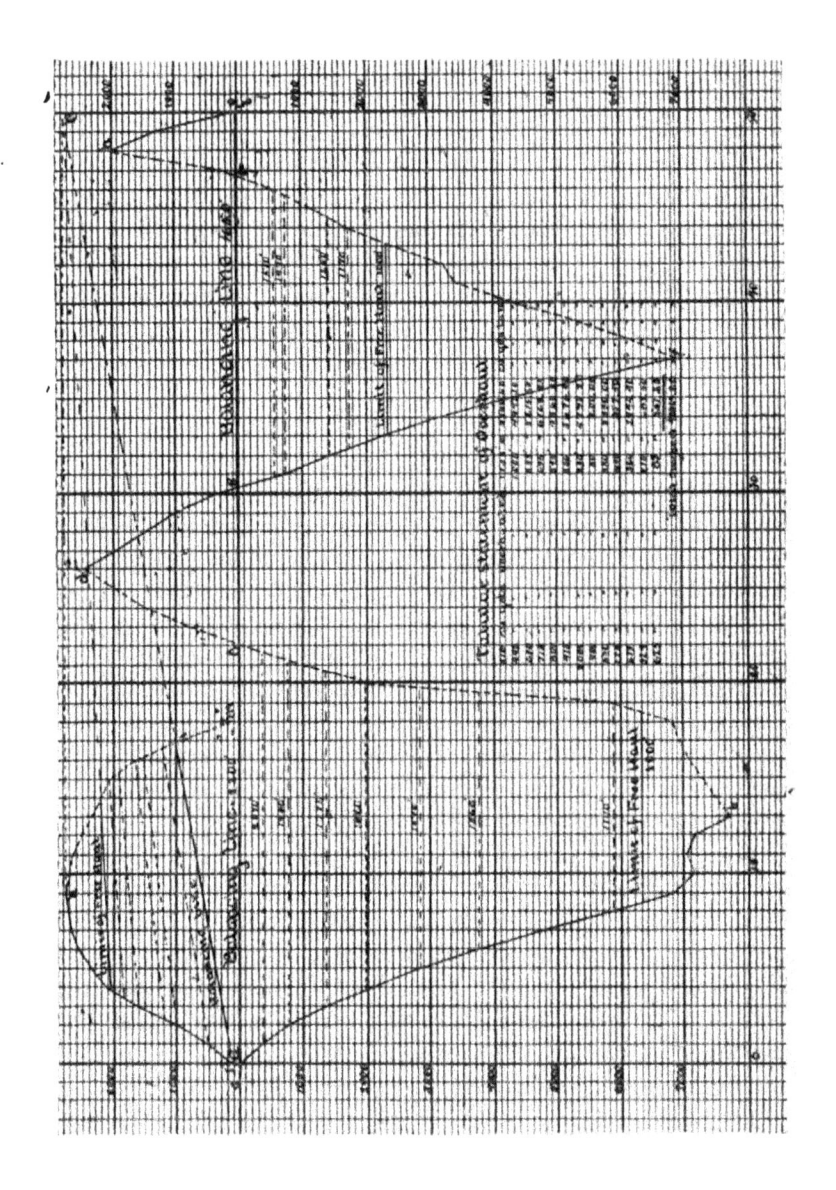

Estimation of Overhaul by Profile of Quantities. 107

line to be determined; t, the ordinate of l, and p the percentage of shrinkage; then $t=r+(s-r)p$. The point l is now located at the intersection of the dotted lines of the curve with the ordinate t.

Lines of haul. These are lines drawn to each point of flexure of the full and dotted lines. If the percentage of shrinkage has been embodied in the algebraic summation of the quantities before plotting, or when there is no shrinkage, the lines of haul are always parallel to the balancing line. If no correction for shrinkage has been made in tabulating the ordinates, we proceed as follows: Produce the portion of the balancing line which passes through the points j and l of the curve $jklm$ to the point P, its intersection with a horizontal line drawn through the ordinate of k (the point of division of cut and fill). The lines of haul are now drawn from this point, as a pole, to the points of flexure of the full and dotted lines of the curve.

(c) *Taking off the results.* The principle here involved is the same as in the ordinary *center of gravity* method. Each element is here treated separately. The center of gravity of any element is assumed to lie midway between the intersection of the curve with successive lines of haul, and hence the distance between the center of gravity of an element of excavation and that of a corresponding element of embankment is the mean of the two corresponding lines of haul. The distance between the centers of gravity of cut and corresponding fill, less limit of free haul times element of cut, is the overhaul for that element. The sum of the overhaul for all the elements is the total overhaul of the cut under consideration.

The only respect in which this method differs from the ordinary *center of gravity* method is that instead of dealing with the whole cut, we here divide it into its elements and solve for each one separately. Let u_1 be the overhaul of the first element, u_2 of the second, u_3 of the third, etc.; v_1, v_2, v_3, etc., the cuts of the corresponding elements; x_1, x_2, x_3, etc., the distances from the origin to the centers of gravity of these cuts; z_1, z_2, z_3, etc., the distances from the origin to the centers of gravity of the corresponding fills; l, the limit of free haul.

Then $u_1 = v_1 [x_1 - (z_1 + 1)]$ is the overhaul for the first element; $u_2 = v_2 [x_2 - (z_2 + 1)]$ is the overhaul for the second; $u_3 = v_3 [x_3 - (z_3 + 1)]$, that of the third, etc.

Estimation of Overhaul by Profile of Quantities. 109

Adding we have for the total cut under consideration $u_1 + u_2 + u_3$ + etc. $= v_1 x_1 + v_2 x_2 + v_3 x_3 +$ etc. $- (v_1 z_1 + v_2 z_2 + v_3 z_3 +$ etc. $) - 1 (v_1 + v_2 + v_3 +$ etc.$)$ or $U = VX - VZ - LV = V[X-(Z+L)]$, which corresponds to the equation used in the ordinary *center of gravity* method.

To illustrate the method here employed, let us consider the element between the balancing line and the first line of haul in the curve *abc*.

Element of cut = 316 cubic yards.

Balancing line = 2,200 feet.

First line of haul = 2,070 feet.

Limit of free haul = 1,000 feet.

$316 \times (\frac{2,200 + 2,070}{2} - 1,000) = 3,586.6$ cubic yards to 100 feet.

In finding the element of cut between two lines of haul we take the difference of ordinates from the profile, or if great accuracy is desired, we read the elements of cut directly from the data prepared for plotting the profile.

The length of the balancing line and the length of the lines of haul are found by taking the difference of the abscissa of their intersections with the right and left branch of the curve. When these quantities are determined, they are tabulated and the equations solved, as illustrated in the tabular statement of overhaul on the accompanying fac-simile profile sheet.

PEN AND INK RENDERING,
JUNIOR YEAR.

PEN AND INK RENDERING,
JUNIOR YEAR.

PEN AND INK RENDERING,
JUNIOR YEAR.

PEN AND INK RENDERING,
JUNIOR YEAR.

BOOKS ✦ ✦ ✦

ENGINEERS: We shall be pleased to answer any inquiries regarding engineering and scientific books and magazines. When you are buying get our prices. The fact that we patronize the Year Book entitles us to preference over competitors who do not advertise, other things being equal. Is that so?

The University Book Store.

H. G. BUSHNELL, President.
CYRIL MITCHELL, Vice-President.
C. F. MOFFETT, Secretary.
W. M. HORNER, Treasurer.

Draughtsmen's Instruments and Supplies at the University Firm,

MOFFETT, BUSHNELL & CO.,
323 HENNEPIN AVENUE,
MINNEAPOLIS, MINN.

15 Per Cent. Discount to University Students.

WILLIAM RICKETSON HOAG, C. E.
PROFESSOR OF CIVIL ENGINEERING OF THE UNIVERSITY OF MINNESOTA.
(See Biography Page 1.)

THE
YEAR BOOK
OF THE
Society of Engineers.

The Year Book is a scientific publication issued by the Society of Engineers in the College of Engineering, Metallurgy, and the Mechanic Arts in the University of Minnesota. It is essentially technical in its scope and contains articles contributed by active and honorary members of the society and lectures before the society by eminent engineers.

This is the second publication of the Year Book, and as in the previous year, is in charge of an editorial committee representing the departments of Civil, Mechanical, Electrical and Mining Engineering.

The price of the Year Book is fifty cents per copy, or by mail (post paid) sixty cents.

Address

The Year Book of the Society of Engineers,
University of Minnesota,
MINNEAPOLIS, MINN.

1894.
MINNEAPOLIS PRINTING CO.,
MINNEAPOLIS, MINN.

THE UNIVERSITY OF MINNESOTA
College of Engineering, Metallurgy
AND
The Mechanic Arts.

CYRUS NORTHROP, LL. D., PRESIDENT.

CHRISTOPHER W. HALL, M. A., DEAN.

THE COLLEGE OF ENGINEERING, METALLURGY AND THE MECHANIC ARTS of the University of Minnesota is intended to prepare students for the active practice of the professions of CIVIL, MECHANICAL, ELECTRICAL, CHEMICAL and MINING ENGINEERING and METALLURGY.

Applicants for admission must be at least fourteen years old and must show proficiency in the following studies: English Grammar and Composition with an Essay; Elementary and Higher Algebra; Plane and Solid Geometry; History of the United States and of Greece and Rome; Physiology; Natural Philosophy; Chemistry; Botany; Mechanical Drawing; German or French; English; while four year's work in Latin may be offered in lieu of these languages, students are urged to present their preparation in German or French.

The work of the freshman and sophomore years is intended to lay a foundation in mathematics, physical sciences and elementary engineering work for the more strictly technical work of the junior and senior years. The equipment of the college is very complete. The instruction is made thoroughly practical. For this end the situation of the institution is most favorable.

In one or the other of the Twin Cities some large work in charge of the best engineering skill is constantly going on. In the cities of the Northwest, notably Minneapolis, Saint Paul, Duluth and West Superior, are many manufacturing establishments, electric light and power stations, metallurgical works, ore docks and railway shops; in Minnesota and other neighboring states, easily accessible from the university, are situated some of the most interesting mining districts of North America. Visits and excursions are planned at convenient times for investigating the practical application, in a large way and under business methods, of the principles studied in the class room.

Upon completion of the four years course of study the appropriate bachelor's degree is conferred.

Tuition is free.

For further information, address the president or dean of the college, Minneapolis, Minn.

THE SOCIETY OF ENGINEERS

IN THE

UNIVERSITY OF MINNESOTA.

OFFICERS, 1893-1894.

NOAH JOHNSON, '94, *President*,
   HARRY L. TANNER, '95, *Vice President*,
     ALBERT C. WEAVER, '95, *Secretary*,
       PLINEY E. HOLT, '96, *Treasurer*,
         BURCHARD P. SHEPHERD, '95. *Business Manager*.

THE YEAR BOOK OF THE SOCIETY OF ENGINEERS.
1893-1894.

EDITORIAL COMMITTEE.

CHAS. H. CHALMERS, *Managing Editor*,
   ANDREW O. CUNNINGHAM, *Business Manager*,
     NOAH JOHNSON, *President of the Society*,
       BURCHARD P. SHEPHERD, *Assistant Business Manager*.

DEPARTMENT EDITORS.

GEO. E. BRAY,	*Mechanical Engineering.*
C. H. CHALMERS,	*Electrical Engineering.*
H. C. CUTLER,	*Mining Engineering.*
W. C. WEEKS,	*CivilEngineering.*

SOCIETY OF ENGINEERS

IN THE

UNIVERSITY OF MINNESOTA.
LIST OF MEMBERS
MAY, 1894.

HONORARY MEMBERS.

C. W. Hall, M. A.
Wm. R. Hoag, C. E.
J. E. Wadsworth, C. E.
Geo. D. Shephardson, M. E.
Wm. R. Appleby, B. A.

H. E. Smith, M. E.
W. H. Kirchner, B. S.
J. H. Gill, B. M. E.
B. E. Trask, C. E.
G. H. Morse, B. E. F

ACTIVE MEMBERS.

Abbott, Arthur L.
Adams, A. A.
Adams, Geo. F.
Ackerman, Christ.
Andrews, W. F.
Atkins, Clifford
Atty, Norman B.
Becker, George
Bell, Frank A.
Bestor, Frank C.
Beyer, Adam C.
Bird, C. E.
Bishman, Adam E.
Blake, Robert P.
Bohland, Jno. A.
Bray, Geo. E.
Brackenbury, Cyril
Bryan, Albert R.
Buck, Daniel
Burch, Albert M.
Burch, Frank E.

Burgner, Linneus P.
Burt, Austin
Byorum, Henry E.
Carswell, Robert E.
Casseday, Geo. A.
Chadburne, R. W.
Chalmers, Charles H.
Chapman, Leslie H.
Chesnut, Geo. L.
Childs, Hubert G.
Christianson, Peter
Coleman, Lee M.
Cross, Charles H.
Cunningham, Andrew O.
Cutler, Harry C.
Dinsmore, Louis L.
Donaldson, Ezra S.
Dustin, Fred G.
Ellis, Sydney A.
Erikson, Henry A.
Ford, Robert E.

Garland, Albert E.
Gilman, Jas. B.
Hamilton, Herbert C.
Hannay, Jno. R.
Hastings, Clive
Hewett, Frank
Hibbard, Truman
Hildreth, Jas.
Hill, Arthur L.
Hilferty, Charles D.
Hjardemaal, Edward
Hoffman, William L.
Holt, Pliny E.
Hughes, Thomas M.
Hugo, Victor
Huntington, Guy B.
Iverson, Lewis
Jaugensen, D. F.
Johnson, Noah
Jones, Cloyed P.
Joslin, Max A.

Joy, Samuel P.	Olson, Jacob S.	Swartz, Harry M.
Kane, Joseph P.	Pease, Levi B.	Swem, Daniel R.
Kernohan, Robert B.	Pfau, James	Tanner, Harry L.
Kinyon, Fayette C.	Phelps, Clyde E.	Tanner, Wallace N.
Lackor, Harry D.	Porter, E. A.	Tilderquist, Wm. M.
Laidlow, Chas. P,	Pratt, Edward E.	Towne, Burton A.
Lang, Jas. S.	Rhame, Geo. A.	Tunstad, B. E.
Latham, William	Rounds, Fred M.	Wales, Rowland T.
Lincoln, Robb E.	Rucker, Wm. C.	Walker, F. B.
Lindman, Eric F.	Savage, Edward S.	Weaver, Albert C.
Linton, Jas. H.	Savage, F. J.	Weeks, Wm. C.
Lonie, Jas. H.	von Schlegel, Fred.	Webber, Fred W.
Long, Fred P.	Scofield, E. H.	Wentworth, Romeyn W.
Magnusson, C. Edward	Shephard, Burchard P.	West, Wm. J.
Markhus, Olaf G.	Sherburne, Walter H.	Wheeler, Herbert M.
Maughan, Herbert C.	Sillman, Henry D.	Whittelsey, Henry
McCrea, Almeron W.	Simmons, Harry F.	Wilkinson, C. D.
McKinstry, Wm. R.	Smith, E. E.	Will, Otto
McNamara, Jno. A.	Smith, L. V.	Wing, Geo. F.
Mills, Eugine C.	Snoad, Geo. R.	Wood, Daniel B.
Mooney, Francis X.	Squires, R. W.	Wood, Frank D.
Murray, James	Stack, Wm. E.	Woodford, Geo. B.
Myers, Mortimer A.	Staughton, Neville D.	Woodman, Howard A.
Neil, Victor A.	Sterling, F. D.	Yale, Washington
Norris, Jno. M.	Strong, A. W.	York, M. A.
Nye, Carl M.	Sprague, C. B.	Zimmerman, Frank
O'Leary, Arthur F.	Stuart, Newton P.	Zintheo, Clarence J.

CORRESPONDING MEMBERS.

Anderson, Christian, B. C. E., '88	Portland, Ore.
Anderson, Ole J., B. C. E., '93	Nicollet, Minn.
Andrews, George C., B. M. E., '87	Minneapolis, Minn.
Aslakson, Baxter M., B. M. E., '91	Indianapolis, Ind.
Avery, Henry B., B. M. E., '93	Minneapolis, Minn.
Batchelder, Frank L., B. C. E., '93	St. Paul, Minn.
Barr, John H., B. M. E., '83; M. S., '88	Ithaca, N. Y.
Burch, Edward P., B. E. E., '92	Minneapolis, Minn.
Burt, John L., B. C. E., '90	Minneapolis, Minn.
Burtis, William H., B. E. E., '92	Minneapolis, Minn,
Bushnell, Charles S., B. M. E., '78	Minneapolis, Minn.
Bushnell, Elbert E., B. M. E., '85	New York. N. Y.

Carroll, James E., B. C. E., '91 — Minneapolis, Minn.
Chase, Arthur W., B. E. E., '93 — Oshkosh, Wis
Chowen, Walter A., B. C. E., '91 — Chisago, Ill.
Coe, Clarence S., B. C. E., '89 — Wenatschee, Wash.
Couper, Geo. B., B. M. E., '93 — Minneapolis, Minn.
Crane, Fremont, B. S., '86; B. C. E. '87 — Prescott, Ariz.

Dann, Wilbur W., B. C. E., '90 — Minneapolis, Minn.
Dawley, William S., B. C. E. '79 — Danville, Ill.
Dewey, William H., B. E. E., '93 — Minneapolis, Minn.
Douglas, Fred L., B. C. E., '91 — Trenton, N. J.

Erf, John W., B. C. E., '93 — Minneapolis, Minn.

Felton, Ralph P., B. M. E., '92 — Minneapolis, Minn.
Furber, Pierce P., B. C. E., '79 — St. Louis, Mo.

Gerry, Martin H., B. M. E., '90; B. E. E. '91 — Minneapolis, Minn.
Gilman, Fred H., B. C. E. '90 — Minneapolis, Minn.
Gill, James H., B. M. E. '92 — Minneapolis, Minn.
Gillette, Lewis S., B. S. '76; B. C. E. '76 — Minneapolis, Minn.
Goodkind, Leo, B. Arch., '92 — St. Paul, Minn.
Gray, William I., B. E. E., '92 — Minneapolis, Minn.
Greenwood, Williston W., B. C. E., '90 — New York, N. Y.
Guthrie, John D., B. E E., '93 — Minneapolis, Minn.

Hankenson, John J., B. C. E., '92 — Ithaca, N. Y.
Hayden, John F., B. C. E., '90 — Minneapolis, Minn.
Hendrickson, Eugene A., B. S., '76; B. C. E. '76 — St. Paul, Minn.
Higgins, Elvin L., B. C. E., '92 — Hutchinson, Minn.
Higgins, John T., B. C. E., '90 — Minneapolis, Minn.
Hoag, William R., B. C. E, '84; C. E. '88 — Minneapolis, Minn.
Howard, Monroe S., B. E. E., '92 — Minneapolis, Minn.
Hoyt, Hiram P., B. C. E., '93 — Minneapolis, Minn.
Hoyt, William H., B. C. E., '90 — Duluth, Minn.
Huhn, George P., B. E. E. '91 — Minneapolis, Minn.

Leonard, Henry C., B. C. E., '75; B. S. '78 — Minneapolis, Minn.
Loe, Erie H., B. M. E., '88 — Minneapolis, Minn.
Loy, George J., B. C. E., '84 — Spokane Falls, Wash.

Mann, Fred M., B. C. E., '92 — Boston, Mass.
Mathews, Irving W., B. C. E., '84 — Waterville, Wash.
Morris, John, B. M. E., '88 — Minneapolis, Minn.
Morse, George H., B. E. E., '93 — Minneapolis, Minn.

Nilson, Thorwald E., B. M. E., '90 - - Minneapolis, Minn.

Pardee, Walter S., B. Arch, '77 - - Minneapolis, Minn.
Peters, William G., B. C. E., '83   Tacoma, Wash.
Plowman, George T., B. Arch., '92 - - Minneapolis, Minn.

Rank, Samuel A., B. S., '75; B. C. E., '75   Central City, Col.
Reed, Albert I., B. C. E., '85 -   Hastings, Minn.
Reidhead, Frank E., B. E. E , '93 - - Minneapolis, Minn.

Smith, Louis O., B. C. E., '83   LeSueur, Minn.
Smith, William C., B. C. C., '90   St. Cloud, Minn.
Springer, Frank W., B. E. E., '93 - - Minneapolis, Minn.
Stewart, Clark, B. S., '75; B. C. E., '75 - Minneapolis, Minn.

Thayer, Charles E., B. C. E., 76 - - Minneapolis, Minn.
Trask, Birney E., B. C. E., '90 - - Minneapolis, Minn.

Washburn, Delos C., B. Arch., '93 - - Minneapolis, Minn.
Woodmansee, Charles C., B. Arch., '86 - - St. Paul, Minn.
Woodward, Herbert M., B. M. E., '90 - - Milwaukee, Wis.

THE YEAR BOOK

OF THE

SOCIETY OF ENGINEERS.

MAY, 1894.

PROFESSOR HOAG.

William Ricketson Hoag, C. E. was born in 1859 of parents who had previously emigrated from the state of New York and settled on a farm in Fillmore County, Minnesota. His education began in the common school of his neighborhood, but as soon as he was old enough to help on the farm his attendance was confined to the winter school. The habits of hard work and of enduring exposure so formed were however not the least valuable part of his training. At an early age a taste for mathematical studies appeared and a fondness for mechanical invention. Of the latter tendency several inventions in toys and machines bear testimony.

It was his fortune when grown up to pursue his high school course in Rochester, Minnesota, under good instruction. This course was pleasantly interrupted by a visit to the Centennial Exposition in 1876, and it was this experience which determined his future profession. To fit himself for that he proceeded after a year spent in teaching a common school, to the University of Minnesota and was enrolled as a student in the course in Civil Engineering in September, 1878. Mr. Hoag very wisely extended the usual time for under graduate study by accepting employment in the engineering service of the Northern Pacific Railroad then in course of construction For three years his time was about equally divided between professional study and practice. The year after his graduation as B. C. E. in 1884, was devoted to professional work which gave him unusual opportunities to study the best American practice in railway maintenance and operation. It was not without reluctance that he gave up work in the field and office to accept an instructorship in civil engineering in his Alma Mater, but his love for mathematical pursuits and a desire to advance his studies in applied mechanics prevailed. These studies occupied his leisure while passing through the successive stages of instructor and assistant professor to that of professor of civil engineering, and enabled him to obtain the degree of C. E. which at his university is a master's degree. A brief course of special studies in Cornell University formed part of the preparation for this degree.

In 1887 Professor Hoag was made acting "assistant" of the United States Coast and Geodetic Survey, and placed in charge of its operations in Minnesota, and he has since devoted the long vacations of each year to base line measurements and primary triangulations now embracing several counties along the Mississippi River.

In 1892 Professor Hoag was designated by the board of regents of the University of Minnesota having charge of the Geological and Natural History Survey of the state, as State Topographer in charge of the topographical operations and mapping contemplated by the authorizing law.

It is the intention, under an amicable arrangement, to do much of this work in connection with the U. S. geodetic operations and thus save the cost and trouble of measuring base lines, erecting stations, and generally of duplicating the work of that survey.

Professor Hoag has not merely become an expert engineer and topographer, but he has been deeply interested in the pedagogy of his own and allied specialties. He has visited many of the leading polytechnic institutions to study their organization and methods.

Impressed with the importance of consultation and co-operation among persons engaged in the various branches of engineering, he was the first to suggest the movement which culminated in the section on engineering education of the congress of engineers which convened in Chicago in the memorable summer of 1893. So successful was the undertaking that it was resolved to make the organization a permanent one, and it continues as the Society for the Promotion of Engineering Education, with Professor Hoag as a director.

In the same summer he was complimented with an appointment as a member of the board of judges of the World's Columbian Exposition, in the department of liberal arts. Naturally his assignments were germane to his profession, and they brought him into close acquaintance with some of the most distinguished men in his line. For some weeks along with them, he was engaged in examining the most complete collection of instruments of precision ever assembled. The writer is able to testify to the high value of the services thus rendered and to the consideration in which the professor was held by his colleagues. It was his privilege to purchase a large number of admirable instruments for use in the University and in the surveys under his charge.

The attainments of a long and eventful course of studies has only whetted the appetite of the subject of this sketch for still higher attempts. He still expects to study systematically, both in this country and abroad. For that reason he has postponed publication, contenting himself with the official reports expected of him.

It might be expected that such a scholar would be an able and earnest teacher, and so his pupils find him, and delight in the style and manner of his instruction, as well as in the information imparted. His devotion to his Alma Mater has been proven by the fact that offers of more lucrative employment elsewhere have not overcome his loyal attachment to her.

WILLIAM W. FOLWELL.

(See Frontispiece.)

THE TRUE FOUNDATION OF THE STATE UNIVERSITY.

CHRISTOPHER W. HALL, M. A., *Dean of the College of Engineering, Metallurgy and the Mechanic Arts.*

To those who reflect upon the purpose of the public school system the conviction of the close adaptation of this system to the needs of the American youth is irresistible. Leaving for the moment the mistakes and shortcomings of the system on the one side and the special diversions toward special purposes on the other, we see the general trend of the system to be in the direct line of preparing the young men and young women of America for citizenship. The European idea of familiarity with the three R's is left far behind and the advanced position has been taken that the highest type of citizenship demands the highest type of preparation for its duties. Indeed the western states have launched boldly out into the support of the state university as a grade of the public school system. A moment's reflection convinces that this position is correct and tenable. As the nation trains the officers of its army and navy, so the state must train those who are to hold positions of trust in its every community. As a major general is worth more than a private so is the well-trained citizen worth more than he who is ignorant and indifferent.

If the university be recognized as a grade of the public school system, should not the additional fact be recognized that it must be as useful in its place as any other grade below it. It must be in close touch with the sentiment of the commonwealth. If that commonwealth be largely an agricultural community so agriculture must be fostered to train the very best agriculturists to do its foremost work. In the present day and generation competition enters largely into all the affairs of life and in America competion is already becoming the most dominant factor in affairs. How to meet this condition successfully is one of the most practical questions of the day. It should press upon college and university faculties as well as those who have in charge other and more material lines of business. The young man to be successful in this intellectual business which he proposes to enter must have the best preparation and the best capacity, both physical and intellectual, for meeting all the exhausting demands and exigencies to which he may come. American industries are becoming vast. A recent census bulletin points out these interesting facts: There are over 7,500 mechanical and manufacturing establishments in the state of Minnesota alone; over $127,000,000 is invested in capital to carry on the business of these establishments; over 88,000 persons are employed at an annual wage of over $4,000,000. The Commissioner of statistics just says that the crops of Minnesota in 1893 covered over 7,000,000 acres of ground, which acreage must require another 88,000 laborers to till. Here then along the two lines indicated lie the greatest physical activities of the commonwealth. The great effort of modern civilization is to reduce physical effort by the application of intellectual activities, and with an imperfect knowledge of the laws of nature, a still imperfect control over these laws, the intellectual is brought to bear upon the physical in a constantly increasing ratio.

As commonly understood the grammar grades and high schools prepare for the common duties of life. All their work is thoroughly practical. It develops a man's intelligence in an everyday way for everyday use. Any workman will acknowledge the intimate relation between the education he receives and even the commonest duties and occupations of life. The man in more elevated occupations who comes into more constant contact with keen and pushing men in his business relations sees the necessity of a broader view and more catholic spirit than is called for in casting machines, bringing out ores or raising crops. As the highest positions for agriculture are advancing to a stage requiring the keenest intellectual ability and the clearest knowledge; as the mechanic arts are advancing from the stage of trained artisans to that of broad and positive engineers; as the education of business and professsional men is advancing even beyond the mere college and into the sphere of graduate studies and research, so the preparation commensurate with the advancing demand upon technical professional life must be advanced into fresh and even unexplored fields.

In a country like ours the foundations must first be laid in the hard sense and every day comfort of the people instead of in the lines of luxury and superfluity. It is requiring a constantly higher degree of technical skill to overcome the obstacles of nature environing mankind and adapting the laws of nature to his use. It is the first duty of the Commonwealth to educate for these very industries which are fundamental. A physician is a necessity but statistics show that the demand for physicians increases with the complexities of an advancing civilization; lawyers are a necessity yet lawyers are not useful until society is enlightened enough to quarrel without killing. The first Napoleon was at least half right in abolishing all higher education save the technical and professional schools because it was so bad. In other words, it was not in touch with the needs of the times. But good roads are always an advantage; better tools and machines are always devised; laws of nature are discovered and brought into use; the many ores of the metals are sought out by the most economical method. These lines so useful in all grades of human communities and so essential to all progress towards intelligence and a broad utilization of those amenities which add every enjoyment to living should be cultivated constantly. As the applications of natural forces multiply and their uses become more and more refined, the field at once broadens from the mechanic arts into engineering refinement and activity.

It will be said that the American college has accomplished a noble work. That is granted. The college was created for a specific purpose in the furtherance of certain ideas and dogmas with which the state has only incidentally to do. In the institutions founded by the state where "any person can find instruction in any study" the practical aspects of learning stand out sharp and clear against the minds of the people and for the common weal. They cannot come into competition with the specific purposes which the college fulfills, and each institution is strengthened in moral and intellectual force by the success of the others in its own sphere. There is enthusiasm in numbers; there is strength in united action; there is momentum in the high intelligence of a commonwealth. The *Renaissance* of the country consists in laying a broad and firm foundation of scientific knowledge and practice beneath all the great activities of modern life. State Universities are a natural outgrowth of this *Renaissance*. Their work in developing the impulse to our material activities is but begun.

THE BESSEMERIZING OF COPPER MATTE.

H. C. CUTLER, B. E. M. '94.

The development of the process of Bessemerizing copper matte is comparatively modern. It was first experimented upon by Lemereckow, a Russian, in 1867. Several experiments were also made about this time by other Russian engineers, but were only partially successful. These attempts were followed by those of two Swedish engineers. In 1879 Mr. P. Holloway announced that he had made the process successful, took out patents, but after trying for some time finally gave it up.

The problem to be solved was different from that in the treatment of iron where only from 9 to 10 per cent.of the weight of the iron was to be oxidized; since in the copper matte the percentage of weight to be oxidized ranged from 40 to 80.

In the Bessemer process the iron, sulphur and carbon develop from 7,800 to 8,000 calories while the sulphur and iron in copper matte only produce from 1,500 to 2,200, although the matte requires less than the iron.

It was finally left to M. P. Manhes, of Lyons, France successfully to solve this process. He took out patents in his own country in 1883, and in the United States in 1892.

The brief was as follows:

"A process for converting copper matte of low grade into white metal, and the conversion of this product into pig copper." The fundamental claims were these: "Charging the matte in a molten condition into a converter, forcing jets of air through the molten mass and maintaining it in a molten condition and at the proper temperature by the combustion of the sulphur and the iron in the matte, continuing the operation until the sulphur and iron have been separated from the metallic copper. A converter,for reducing pig copper from copper matte, having a wind belt encircling the converter above the bottom, a series of tuyers extending through the lining of the converter and communicating at the outer ends with the wind belt. Removable stoppers located in the outer wall of the wind belt and in alignment with each of said tuyers whereby a drift bar may be inserted successfully through said tuyers to remove obstacles."

An experimental plant was erected by pupils of Manches at Butte, Mont. Here copper matte containing 72 per cent. CU. was converted in 20 minutes in a single operation into black copper—98.9 per cent. CU.

The appliances for the process are in general those of the ordinary Bessemer plants, differing somewhat in details. The converter is about the same except that the tuyers are horizontal and a short distance above the bottom (instead of being vertical and at the bottom.) It was found that the blast blowing through verticle tuyers into the melted copper would chill it causing copper to be projected out. It would also lower the temperature, hindering combustion. In the horizontal elevated tuyers the blast enters the matte and melted slag above the copper.

M. Auerbach, superintendent of the works at Bogoslowski, Russia, places the tuyers at an angle of 45 degrees, and so calculates the quantity of the charge that when the white metal or black copper is formed, the surface of the separation of the matte, copper and slag is almost at the outlet of the tuyers. In this way he does not blow through the slag nor does he oxidize the product rich in copper which sinks to the bottom.

The process now used at the Parrot works in Butte is as follows for poor ores.

(1) Melting ores without any previous roasting to produce a matte containing 25 to 30 per cent CU. and a fluid scoria.

(2) Fusion of this matte in a cupola furnace.

(3) Treatment of the matte in a converter and bringing it up to 72 per. cent CU.

(4) Fusion of this matte in cupola.

(5) Treatment in the converter for black copper.

(6) Fining and refining in a reverberatory furnace.

For rich ores the operations with a sufficient number of converters requires only process 1-5 and 6 as the mattes can be run directly from the shaft furnaces into the converters; the cooling of the mattes after they have been made is entirely unnecesssary.

The ore is first melted in a shaft furnace as in any of the other processes. The liquid matte is then run through a launder lined with refractory material or carried by means of large wells on trucks directly into an inverted converter.

The converter is made in three sections of wrought iron boiler plate. The upper contracted part is called the hood. Inside of this are wrought iron hooks to hold the lining. The lower section is surrounded by the wind or blast chest. Through this $\frac{5}{8}$ in. holes are made into the converter. The outer holes are closed by means of removable plugs. Through these holes the tuyers are kept open.

The number of tuyers for the best work is from 18 to 20.

The mouth of the hood is about $\frac{1}{4}$ the diameter of the body. When the converter is in position the mouth is in connection with a flue which conveys off the gases. The whole converter swings on a trunion.

As the charge is run into the converter a light blast is turned on to keep up the heat and, when full charge is in, the convertor is swung up into position. At Butte a charge is enough to produce about a ton of metal, while at the new works of the Boston and Montana Co. at Great Falls, five tons are produced at a time.

At first sulphurous acid and other gases pour out in a dense white cloud. From this time on no fuel should be used unless at the end of the operation the temperature should fall too much when a stick of wood must be thrown in to again raise it. The burning of the sulphur and iron should produce sufficient heat. The reactions that now take place are as follows:

$$2 FES + 3O_2 = 2 FEO + 2SO_2$$
and $FEO + SIO_2 = FESIO_3$

This stage is characterized by dense white clouds tinged with rose and green. The rose dissapears first and the white gradually dimminishes and green becomes more constant. Finally the close of this stage is indicated by both white and green changing to pale blue. When this color becomes constant it shows that the FEO has all combined with the SIO_2 If zinc and lead are present in the matte the white cloud is very dense, but only two or three minutes are required to drive off any foreign metal.

After all of the FEO has dissapeared the blast is turned off and the convertor turned down and slag run off into slag pots. Then the convertor is turned back and blast again turned on. A rose color gradually replaces the blue until finally the blue dissapears. The rose deepens to red, then to reddish brown and gradually diminishes. The precise moment when all the sulphur has disapeared is hard to tell, and considerable experience is required.

At the proper moment the convertor is turned and the copper poured out into moulds.

The time occupied varies greatly with the grade of matte, initial temperature and force of blast.

It usually takes about twenty-five minutes to go from 40 per cent to 73 per cent matte and 30 minutes from 73 per cent matte to 99 per cent black CU.

The slags contain from five to six per cent of copper and are returned for fluxes to the blast furnace.

They are usually basis sub silicates of iron, and sometimes highly magnetic. An analysis of some of the matte at Butte was as follows:

C U=77.61   Single operation
S=20.65   CU=70
                        S=24
FE=1.22   FE - 3.9
Insolubles—.38   AS - .8

Analysis of black CU:
CU - 98.5 to 98.8
S - .9 to .8
FE = .6 to .4

The copper is absolutely free from arsenic and antimony. The high temperature of the blast and the intense oxydizing action carries them off, as they dissappear before they have time to act on the copper.

The zinc, tin and lead are separated without difficulty. Some of the cobalt is scorified but some remains in the copper. The nickle and bismuth seem also to concentrate. The Bessemer process seems to be no better for them than other processes, but the elimination of the arsenic and antimony is the most important result.

Sometimes with a low grade matte, two fillings are made before pouring there by saving heat.

It has been found that it is not economical to produce mattes of 72 per cent in the shaft furnace. That it is not economical to go in one operation from matte to black copper unless the matte contains 72 per cent CU. So that the latest works using low grade ore have abandoned the idea of one operation.

The process is applicable to any ores containing sufficient sulphur and copper to form a matte. And the best results are obtained from ores having a regular composition. Then the number of operations does not have to be changed for each new installment of ore. The amount of fuel consumed is about 90 per cent of the weight of matte.

A plant of three converters is needed to keep operations running. When one of the converters is running, the second is cooling and the third is being relined, dryed and heated up. The lining of the converter is made of pounded quartz mixed with enough fire clay to make it plastic. This is made up into balls and stamped into the bottom of the converter first, the hood having been removed. Then a barrel about the size of the hole to be left is put into the body and the side lining stamped around it. The hood is then relined and put on and the converter is dried, usually by building a fire in it. It takes about twenty-four hours to dry one and they should never be used until thoroughly dry.

As to the life of a lining accounts differ.

Originally they would only last about seven or eight operations, especially when the mattes were poor in copper.

In later years mechanical devises have been added for the introduction of SIO_2 in the form of a fine powder into the converters. The FEO unites more readily with this than with the SIO_2 in the lining, thus preserving the lining longer. In this way a lining has been made to stand from eighteen to twenty-four operations. At the same time too much SIO_2 introduced tends to chill the slags so that there is a limit to the amount that can be used.

After every twenty to twenty-five operations the lining becomes too much worn to be used longer. The converter must then be cooled. This is best done by introducing water in the form of fine spray into the interior. This becomes vaporized and escapes in the form of steam. Finally the whole converter is filled with water. The water is then poured out, the loose lining dropping with it. A man then enters and the lining is patched. By this patching it will often be made to last for several months. But when too badly worn a new lining must be stamped in. It generally requires about eight hours and often more on account of the lining being baked so hard that it becomes very difficult to separate the parts.

A. W. Stickney has suggested a new converter made in three parts which are held together by clamps and catch pins. The parts are all relined separately and graphite or sand introduced between joints so that they separate easily, even after long usage. The bottom is removable and the lining can be dumped while hot, thus saving the time that would be necessary to cool·

A recent form employed by M. P. Manhes consists of a horizontal cylindrical vessel of plate iron lined internally with fire clay and provided on one side only with a row of tuyers. The depth of these below the surface of the molten charge is regulated by the partial rotation of the cylinder. It is charged and emptied in the same manner as the Bessemer converter.

The use of this system at the Parrott works in Butte has been attended with great success. The ores there are sulphuret and quartzore with a general average of twelve per cent CU. Two operations in the converter are used. A matte of twenty per cent CU is obtained from the Reverberatory furnace. A remelting furnace is used and the matte run from this into the converter. The first operation is to eliminate the iron and to obtain a rich white metal of about sixty-four per cent CU. This operation lasts about twenty-five or thirty minutes.

In the second operation the white metal is remelted in the furnace and then treated in the converter. This operation lasts from one hour and thirty minutes to two hours. The result is black copper ninety-five per cent CU. About seventeen men are required for a set of three converters.

The cost has been about $13.55 per ton of copper or about $\frac{2}{3}$ cents per pound.

The amount of coal used for every thing is about $\frac{2}{3}$ less than the English method.

For works producing 100 tons of copper per month, the number of workmen required for everything is about 70 men. The entire cost is about $\frac{1}{3}$ of that by the English method. The total loss in copper does not exceed one per cent. This process is now used at the new works of the Boston and Montana Company at Great Falls, Mont. They have modified the Parrot Company's process somewhat, in that they have done away with the smelting process, running the matte direct from the shaft furnaces into the converters. In this way an immense amount of fuel is saved.

The same method is used by M. Auerbach at his works in Bogoslowsk, Russia, and seems to give entire satisfaction, thus proving, contrary to the opinion of A. W. Stickney, that it can be done.

Finally, we have to sum up all the advantages:

1st. Simple apparatus which can be constructed rapidly and cheaply.

2nd. The works occupy but little space.

3rd. The operations can be easily learned.

4th. The work is not difficult and can be carried out by persons of ordinary intelligence.

5th. The losses are not large.

6th. The time gained is so great that the increased output for the capital required makes it comparatively cheaper than the old method. Taking it for granted that the ores can be concentrated, which may not always be the case, two operations in the converter will be necessary. Whether this can be done will depend first, on the price of copper, second, on the capital available for the erection of a double plant.

7th. If the ores contain the precious metals the rapidity with which they can be concentrated into a matte or black copper seems to be a very great advantage.

THE UNIVERSITY HEATING PLANT, WITH THE RESULTS OF A RECENT TEST.

H. E. SMITH, M. E., Asst. Professor of Mechanical Engineering.

With the rapid growth of our University and the increase in the number of buildings on the Campus, the economic and efficient system of heating these buildings, has been and is, an important financial question.

Previous to the year 1889, each building had its own system of heating, or drew its supply from two horizontal tubular steam boilers located in the basement of University Hall.

With the addition of the Chemical and Physical Laboratory building and Pillsbury Hall and in view of the addition of several other buildings, then planned and since built, it was made apparent that the old system could no longer adequately supply the new demand. It was then decided to establish a central heating plant and supply each building from this source. To this end a substantial brick structure of ample size for future needs was built in the summer of 1889 and equipped with three horizontal return tubular boilers and the other accessories of a modern heating plant. The steam from this plant was conducted to the several buildings, through mains laid in trenches, walled with masonry and filled with mineral wool to prevent radiation of heat.

The connection to University Hall was, however, made through a larger trench or tunnel of sufficient size to admit of passage through it by a workman in case repairs became necessary.

Pipes were also laid in these trenches for the return of the water of condensation from the buildings to the boilers. The increased demand made upon the plant by the erection of the Medical College building in 1892, made it necessary to add two more boilers to those already in place.

The connection to this building was made through a much larger and more convenient tunnel than that connecting University Hall with the plant. The first winter proved that the system was not all that could be desired as to economy and steps were taken to ascertain what changes could be made that would tend to raise the efficiency or at least indicate the most economic policy of the future.

With the increased number of furnaces another serious question arose as how to best abate or at least modify the nuisance arising from the smoke that passed from the chimney of the plant.

This question was made more imperative by the decision of the Board of Regents, adopting a very light stone for the new Library Building about to be erected near the heating plant. Smoke burning devices are being investigated by Mr. J. H. Gill an instructor and post graduate in the Mechanical Engineering Department of the University, and from the results of several experiments that have been made, the solution seems possible, though with what economy, if any, is yet to be determined.

It was deemed advisable, as a first step in the general question, to determine just what the conditions were, under which the boilers were working, and the evaporative efficiency of the boilers and settings.

To this end a test was made by a number of students of the Engineering College under direction of the Mechanical Department, on the second of December, 1893. The plant was working under normal conditions as near as was possible to make them.

The following data and results were obtained from the trial:

The plant consists of 5, 60-inch, 15 ft. horizontal tubular boilers, each containing 42, 4-inch tubes, set in brick work and provided with the usual safety valves, gauges, feed-pumps, etc. A steam "header" connects all the boilers and acts as a reservoir from which the steam is distributed to the buildings.

The gases are discharged through an iron drum or flue connecting all the furnaces to the base of the brick stack.

Arrangements were made whereby all water, whether the returns from the buildings or water from the city mains, could be weighed and its temperature taken before being delivered to the boiler.

Each of the three steam pipes leading to the Medical College Building, University Hall, and the Chemical Building and Pillsbury Hall, were tapped near the "header" and fitted with sampling tubes for the attachment of calorimeters for determining the quality of the steam. For this purpose a "separating calorimeter" was placed on each of the above first and last named mains and a "throttling calorimeter" on the University Hall main.

Eight experiments with each instrument at regular intervals through the trial showed that the average moisture in the steam was 1.3 per cent., a very good showing, but due, no doubt, to some extent to the "header" which probably acted somewhat as a separator.

The temperature of the flue gases, feed water and outside air were taken at fifteen minute intervals throughout the trial, as were the readings of the steam pressure and draught guages.

Two samples of the escaping gases were taken during the trial, and an analysis made of them.

All coal was weighed before being delivered to the furnace, and at the end of the trial the ash and clinker found in the ashpit or hauled from the fires, weighed up.

At the beginning of the trial, the water was brought to the normal level in the boilers and the height marked on the water glasses, the ash pit cleaned, the condition of the fires noted and the height of water in the supply tank measured. At the end of the 10-hour run, during which the steam pressure and other conditions were controlled as in every day practice, the water level in the boilers and the condition of the fires were brought to correspond as near as possible to those of the start.

All sources of leak or waste water and steam was carefully looked to at intervals and corrections made for them when necessary.

The following is a summary of the data and results of the trial:

Date of trial... ....December 2nd, 1893
Duration of trial..10 hours.
Grate surface, 5 furnaces each 4ft. long, 4ft. wide................80 sq. ft.
Heating surface, total...3,850 " "
Ratio of heating to grate surface...............................48 to 1
Coal used, Youghiogheny, bituminous.

AVERAGE PRESSURES.

Steam Pressure, by guage......................................67 lbs.
Force of chimney draught (in base of stack)............ .706 in. of water.
" " "   " (far end of flue).............. .201 " " "

AVERAGE TEMPERATURES.

Outside air... 7.7 deg. Fahr.
Of. Steam...312.4 " "
" escaping gases.....................................564.1 " "

The University Heating Plant.

"feed water..167.96 " "

FUEL.
Total amount of coal consumed............................16,000 lbs.
Total amount of ash and clinker.... .... 1,925.5 "
Percentage of ash...12 per cent.
Percentage of combustible.............. .. ..88 " "
Coal burned per hour......................................1,600 lbs.
Combustible burned per hour.......................... ......1,407.45 "
Coal burned per sq. ft. of grate surface per hour.... .... 20 lbs.
Coal burned per sq. ft. of heating surface per hour...............415 lbs

CALORIMETER TESTS (AVERAGE.)
Quality of the steam............ :....,..............1.3 per cent moisture.

WATER.
Total weight of water pumped into boilers and apparently
  evaporated...109,009.1 lbs.
Water actually evaporated, corrected for quality of steam....107,592 "
Equivalent evaporation, *from and at* 212 deg. Fahr..........115,715.2 "
Water evaporated per pound of coal from actual pressure and
  temperature of feed water............................. 6.72 "
Equivalent water evaporated per pound of coal from and at 212
  deg. Fahr.. 7.23 "
Equivalent water evaporated per pound of combustible from
  and at 212 deg. Fahr.................................. 8.22 '
Efficiency of furnace..54.7 per cent.
Rated H. P. of each boiler...................................60
H. P. of each boiler as shown by test, taking the Centennial
  Commissioner's unit as standard.............. 68
Actual over rated H. P.............. ....13 per cent.

FLUE GASES.
Carbonic acid	(CO_2)	2 per cent.
Carbonic monoxide	(CO)	0.4 per cent.
Oxygen	(O)	16.5 per cent.
Nitrogen	(N)	81.1 per cent.

The analysis of the flue gases seems to show that far more air was admitted either through the ash pit or through leaks in the setting than should have been for best economy. The effect of too much air is to dilute the furnace gases and carry off heat to the stack that otherwise might be absorbed by the boiler. The above results, do not show as good economy for furnaces and boilers as could be expected and pointed out some needed changes. They also demonstrate that the present capacity of the plant is not adequate to supply the demand made upon it, without undue forcing and consequent low economy. It will be readily understood what even a small percentage in increased economy means, in a plant of this size by considering the cost of fuel per year for heating our buildings.

Between the first day of October, 1892, and the first day of March 1893, 4,080,000 lbs. of coal were burned being an average of 13.6 tons per day at a total cost of over $9,000.

The use of crude petroleum as a fuel is being investigated, with a view of possible adaptation, if all the conditions warrant such a change.

It is planned during the next year to ascertain the proportion of steam taken to heat each of the buildings and the condensation of the mains, besides other interesting points.

TRIGONOMETRIC LEVELING AND ABNORMAL REFRACTION.

*W. R. HOAG, C. E., *Professor of Civil Engineering.*

In all topographical surveys the sea elevations of a few stations are determined by precise levels, this being the most accurate method known. Between these stations certain important lines of levels are run with the engineer's Y level establishing the elevation of all important stations along the route.

These comparatively few stations, whose elevations are thus well known, furnish the hypsometric control of the whole survey.

In this earlier and more accurate part of the work, on account of the instruments and field methods employed, atmospheric refraction plays an unimportant part in the work, but in establishing a sufficiently large number of secondary elevations from these and of carrying elevations through a chain of triangles from one primary bench mark to another, the trigonometrical method is found to be quite satisfactory as to accuracy and offers a very ready means of giving elevation to a large number of the secondary and tertiary triangulation stations for the safe control of the survey.

The trigonometrical method requires a knowledge of the atmospheric refraction present at the instant of observation, to attain the best results. This can be inferred somewhat from local conditions as to temperature and barometric pressure.

When these vertical measures can be made at each end of a line upon the opposite stations, called reciprocal observations, we are enabled to eliminate the effect of refraction wholly if the two observations are made simultaneously. If not, the error due from refraction can be corrected so far as it is normal or can be determined from attendant conditions.

Since some of the primary, and most of the secondary lines cannot be ob-

---
*Acknowledgements are due the Superintendent of the U. S. Coast and Geodetic Survey for all the observations and results furnishing the basis of this article. W. R. H.

served reciprocally it becomes important to know, as nearly as possible, what the coefficient of refraction is under ordinary atmospheric conditions—this is usually given as about .072—expressing the ratio of the amount of curvature in the refracted ray to double the curvature in the geodetic line connecting the two stations—and shows that a ray of light passing about parallel to the surface of the earth is refracted into a curve of a radius about seven times that of the earth. If this path of a ray of light were the arc of a circle the coefficient would be independent of the length of the line and a constant, except for the very small change due to the radius of the earth for different latitudes and azimuths. For distances over 25 miles it appears from work in this state that for most localities the average value of the coefficient of refraction is about .07 and that for distances less than this it is much less constant and of much lower value.

The average value on 12 lines ranging from 5 to 12 miles long being .023; while 13 lines from 12 to 18 miles long have an average coefficient of .046, and those above 20 miles give .067.

Each set of reciprocal observations made to determine the difference of height of the two stations furnishes incidentally the means of computing the atmospheric refraction present at the time these observations were made, allowing that it be the same at the two different times. From the close agreement of the coefficient under different atmospheric conditions it is hoped to discover from these reciprocal observations good values for the coefficient for use in the state survey.

Some half dozen lines passing over bodies of water, such as our lakes, the Mississippi river or Lake Pepin, exhibit such abnormal refraction that its persistency must be established by observations through a cycle of ordinary atmospheric changes or all but reciprocal measures must be regarded of but little value and to nothing but simultaneous observations can be given the weight ordinarily due reciprocal measures.

Three of these lines give for the refraction coefficient about .01, two give .003, while two others indicate a negative refraction of .018 or show that the ray of light instead of being concave to the earth is really convex.

The presence of a body of water has long been regarded as highly objectionable where vertical measures are to be made and in the case of conducting a line of precise levels across a river so unreliable is the refraction element that nothing but simultaneous observations are taken and these are repeated many times to eliminate possible erratic values. Since the phenomenon of refraction is due primarily to the ray of light passing through media of different densities arranged in horizontal strata according to the law of pressure and consequent density, we readily see why, in the case of a line of levels passing necessarily within a few feet of the surface of a body of water furnishing conditions wholly different as to heat and moisture from the ordinary land surface, we may reasonably expect to find an abnormal coefficient of refraction.

The lines above noted pass from 450 to 700 feet above the surface of the body of water whose width does not exceed as a rule one tenth the length of the line, yet the results show unquestionably that even upon lines passing at these heights above a water surface, affecting directly only a small part of the line, the pressure of the latter renders the use of the normal refraction coefficient highly questionable, if not positively erroneous. Sufficient observations have not yet been taken on the lines exhibiting abnormal refraction to warrant offering a new coefficient, yet it seems highly probable from the observations already made and discussed, that it must be reduced from $\frac{1}{2}$ to $\frac{2}{3}$ of its ordinary value to suit the local conditions present on these particular lines in the Minnesota State Survey.

EARTH CURRENTS AND THE EARTH DYNAMO.

JAMES S. LANG, '95.

With the advent of modern electrical transmission, there has come to inventors and scientists in general, a partial realization at least of the tremendous waste of energy continually going on around us. Those more sanguine and less practical would harness the tides which roll against our coasts and compel the moon herself as she swings on in her path, a corpse among worlds, to drive our factories and light our streets.

And yet such visions are not wholly illusions, but are capable of being developed to a certain degree, by careful study of the conditions, with proper discrimination between the practical and the impractical.

One class of phenomena worthy of more attention than it has received is that relating to earth currents. This is not to be wondered at, as investigation flows in well defined streams, and the attention of investigators has been turned to subjects of more immediate pecuniary value.

The theory of their cause which I present here, and which is a result of the action of what I designate as the Earth Dynamo, appears to be borne out by the facts and phenomena observed.

It has been known for many years that there was an intimate connection between the sun spots, the magnetic needle and earth currents. This connection is so well marked and its action is so certain, that it would indicate conclusively that magnetic polarity is possessed to an intense degree by the sun. That the magnetic poles of the sun coincide very nearly with its geographical poles is shown by the fact that in the absence of spots, no changes in the terrestrial magnetic state occur, which have a periodicity equal to the time of solar rotation. The sun, then, we are forced to conclude, may be considered as a magnet, in the same way that we look upon the earth as one, and as such possesses its lines of force, which after having passed from pole to pole inside the sun, envelope and pass the earth at a distance of 91,000,000 miles on their return course. Far beyond our farthest planet these lines

Earth Currents and Earth Dynamo. 23

spread out in gigantic ellipses, bringing planet and asteroid alike into magnetic touch with the seething monitor of our solar system. (See Fig. 1.) And our planetary system is again and again permeated with weaker lines of force having their origin in other suns than ours, and our sun himself as he whirls ceaselessly onward feels with his magnetic nerves the influence of the siderial system of which he is a single atom. The milky way stretching

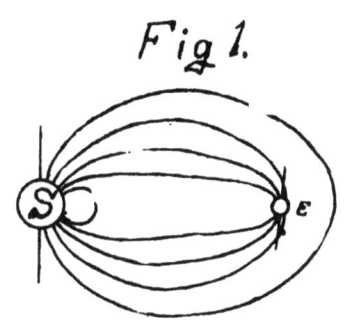

Fig 1.

across our sky points to our minds the direction of lines of force which have hurled together suns, stars and systems. Multiply the range of our telescopes by infinity and you would see other galactic belts, around, beyond and parallel to our own stretching away into endless space—indices of lines of force—the handiwork of an infinite creator, the harmony of whose productions exceeds our wildest flights of imagination.

Thus we see that one factor of the magnetism of the earth is due to the lines of force passing through the sun, which penetrate the earth on account of its superior permeability.

That the magnetic poles of the earth are not stationary is shown by the constant diurnal variation of the magnetic needle. Their virtual traces on the earth's surface, revolve in small circles around the traces of the main pole, and the angle between the two tangents to these circles drawn to a magnetic needle, gives the amplitude of diurnal variation of that needle under normal circumstances.

In Fig. 2 the angle e between the tangents W n an E n is the amplitude of diurnal variation of the needle at n. Then again these lines of force passing through the earth at an angle with the line of mean stationary poles, may be resolved into its two components, one of which is parallel to the direction of mean poles, the other perpendicular to it—the former stationary, the latter revolving, as regards the earth, once daily, pointing always in the direction of the sun. Let F L represent, in intensity and direction, the earth's magnetic polarity the trace of which revolves in small circle W E. The two components into which F L may be resolved will be F m and F E.

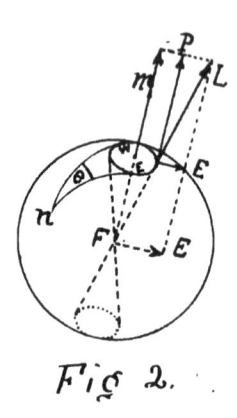

Fig 2.

Thus it is plain that if a wire be suitably placed* on the surface of the earth, it cuts this revolving component E F of the earths magnetism and we have earth currents as a result. We may consider the earth as a typical spherical armature, familiar to all, in the conductors of which, at times when the equatorial component grows large, we have high voltages.

We see then, that earth currents may be due to the revolution of this earth armature in a magnetic field of which the sun is the immediate inducer. We are then led farther—why not wrap our earth with a suitable system of conductors and wrest from our mother earth an amount of power equal to that wasting at every tide on every sea.

It is easy to determine its feasiblily, for we know that the intensity of the revolving equatorial component is equal to the sine of half the angle of the diurnal variation of the compass, times the intensity of the earth's magne-

---

*Ganot's Physics, Sec. 894.
Faraday's Researches in Electricity, Sec's 187 and 192.
Mascart & Joubert, Elect. and Mag. Vol I, Sec. 531.

Earth Currents and Earth Dynamo. 25

tism. A difference of potential of one volt being produced in every conductor which cuts 100,000,000 lines of force per second, if we surround the earth with a coil of one hundred wires, it may be easily shown that a difference of potential of upwards of sixty thousand volts would be produced under normal conditions.

For wires extending both above and below the equator with turns connected in series, the following method of calculation will apply. Selecting such a position on the earth's surface that data applicable to that place will be approximately a mean for the whole surface, we may say that the actual intensity of the earth's magnetism is equal to the horizontal component times the secant of the angle of inclination of the dipping needle.

Taking horizontal component equal to .20 of a line per sq. cm. and inclination equal to $60°$, we have for the intensity of earth's magnetism .40 of a line per sq. c. m.

The revolving equatorial component of this will be equal to .40 times sine $7'$, when diurnal variation is $14'$, which equals .0008144 of a line per sq. cm. The area swept over by a single wire placed as designated will be 2x48,375,360,000 sq. ft. or 2x44,940,709,440,000 sq. c. m.

Therefore the voltage induced will be $\frac{44940709440000 \times .0008144}{10^8}$ which equals 742. This multiplied by 100, the number of turns of wire used, equals 74,200 volts.

If the difference of potential was 60,000 volts in round numbers, and number 0000 wire used, one eleventh of an ampere would flow and the total power produced would be 5454 watts or less than eight horse power. Although the amount would rise into hundreds of horse power at those times when earth currents become abnormal, the enormous cost of the plant would forbid its practical application.

A ROTARY PUMP.

Geo. B. Couper, *B. M. E.* '93.

The following is a description of a rotary pump, the patents for which are controlled by the Cooper-Hampton Electric Co., Minneapolis, Minn.

The exterior of the pump consists of a casing of cast iron, T Fig 1, which completely surrounds the working parts. The middle piece of the casing has a standard, not shown in the drawing, by which the pump is supported. The drum A is keyed to the shaft S and contains the pistons Q. The arrangement of the pistons in the drum is shown in Fig. 2.

The piston rods are connected to a disc M by the universal joints R. The disc M is connected to the shaft S by the universal joint L. There is another disc O, shaped like a basin, in which is placed the disc M. The disc O is fastened to the casing T by trunnions not shown in the drawing. These trunnions allow it to be rocked back and forth on a horizontal axis but prevent it from turning with the shaft 3.

N is a babbit metal ring placed between M and O to give a good bearing surface. The disc M is held into the disc O by the nut P which is screwed to a part of M projecting through O. B is a brass sleeve screwed into the drum A. C is a cap screwed into the sleeve B and fastened with a set screw. It might be well to state here that the drum A, although keyed to the shaft S is lose enough on the shaft and key to allow it to be moved lengthwise of the shaft. C is connected, by means of the bolt G, to the diaphram H. The space I behind the diaphram is open to the pressure side of the pump connections so that the pressure, under which the pump is working, acts upon the diaphram and tends to force it outward. This pressure is thus transmitted through the bolt G, cap C and sleeve B to the drum A and serves to hold it up to its seat which is the end of the casing. J and K are ports for the intake and discharge of the pump.

Fig. 2 is a diagram to show the action of the pump. Here we have simply the drum A; the pistons and their rods represented by the lines a b, a 'b'; the disc M and the stationary disc O.

A Rotary Pump.

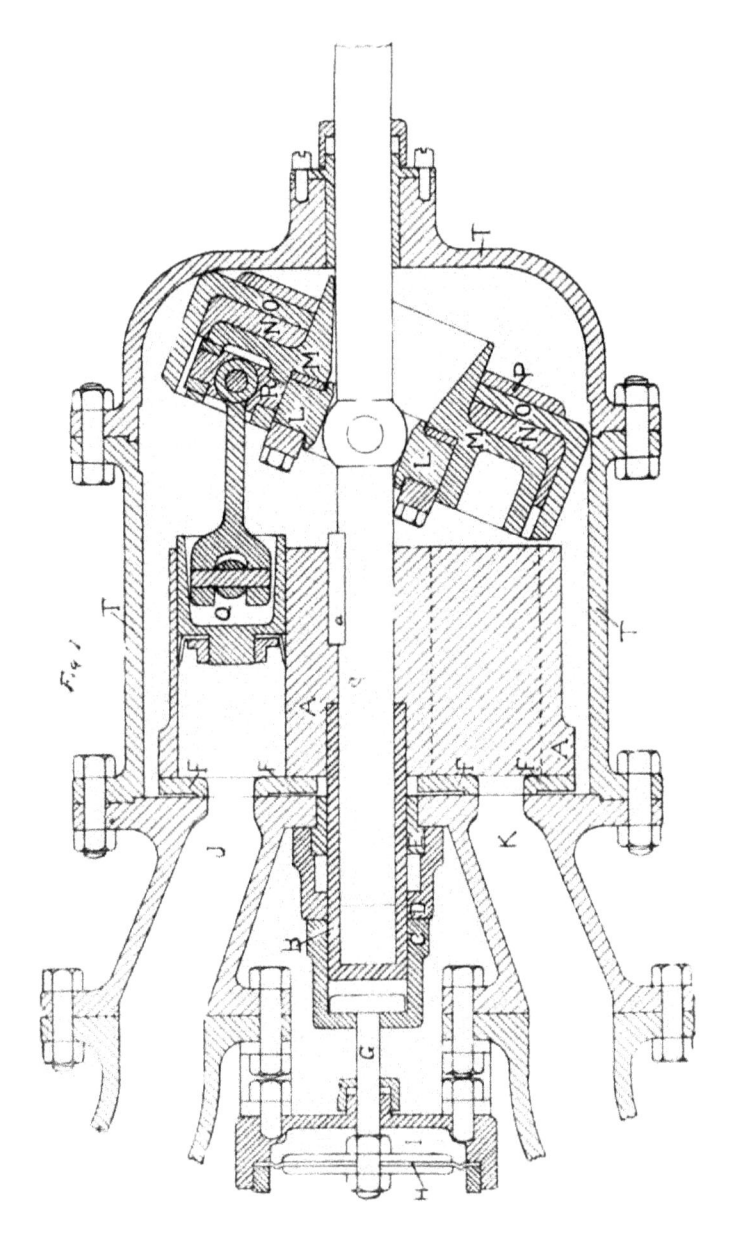

The discs O and M are tipped backward as shown by the line a a Fig 3. If, as in Fig. 2, the disc M is rotated upon O, in the direction of the arrows any point as (a) will be carried around in a path (a c a'). In doing this the point (a) travels forward a distance (a'e) Fig 3. As the disc M continues to rotate the point a will be carried forward until it reaches its fartherest point which is on the axis (x y). After it passes this point it is carried backward in the direction of the arrow (d). (In following this motion one must bear in mind the fact that the drum A, containing the pistons, and the disc M, to

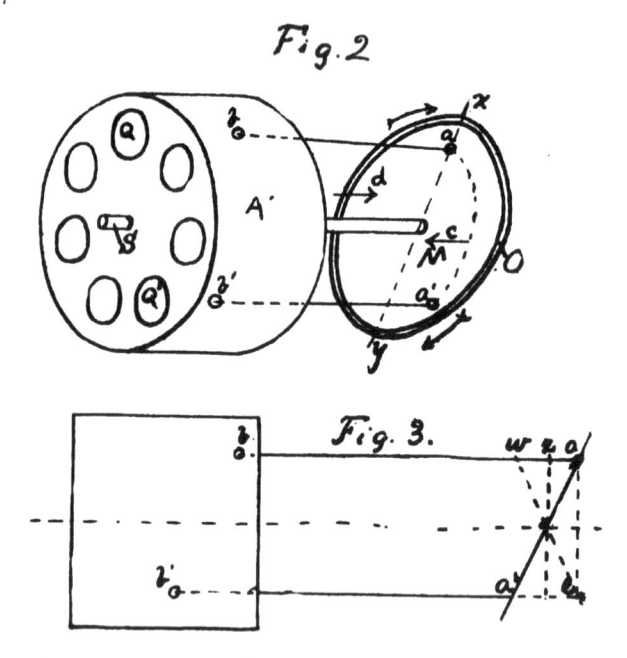

Fig. 2

Fig. 3.

which the piston rods are attached, are both connected to the shaft and S hence must move together in the plane of rotation.) Thus it is seen that the rotation of M gives the pistons a reciprocal motion similar to the motion of an ordinary piston pump.

In order to direct the flow of the liquid there are two ports so placed that one is open to the pistons while they are on their backward travel, thus allowing them to fill, and the other is open to the pistons during their forward travel, thus allowing them to discharge.

The two parts are separated by bridges placed at the points where the pistons reverse their horizontal motion. In Fig 1, the parts are not shown in their proper position with reference to the tip of the discs M and O but are rotated 90° in order to better show their connecction with the drum. As stated above the disc O is supported on trunnions which allow it to be tilted forward and backward. If it is brought to a verticle position as (Z) Fig. 3, it will be readily seen that in rotating the disc M, any point on it will have only the one motion, that of rotation in a plane perpendicular to the shaft. Consequently the pistons which are attached to M will simply rotate about the shaft and have no horizontal motion. With the disc in this position there is no discharge through the ports although the pump is in continuous rotation.

To carry the action of the pump still further suppose the discs to be tilted beyond the verticle as at (w) Fig. 3.

If the action in Fig. 2 is carried through again with the discs in this new position it will be found that any point moving from (a) to (a') will have exactly the opposite horizontal motion given to it. This brings about a complete reversal of the flow through the ports and constitutes one of the main features of the pump.

The trunnions which hold the disc O project through the casing and are controlled by a worm acting on a crank arm. The disc can thus be regulated while the pump is in motion and the discharge changed from its maximum gradually up through the point of its discharge and finally over to the maximum of reversed action.

As the pump is designated to be run direct connected to an electric motor the ability to change the amount of discharge through such a wide range, without changing the speed of rotation, becomes of great importance. To give some idea of the capacity of these pumps some figures are given below.

A pump with pistons 3 inches and a stroke of 3 in., making 400 revolutions per minute will discharge about 250 gallons per minute. And working under 180 pounds pressure will consume about 25 mechanical horse power.

BRIDGE VIBRATION.

W. C. Weeks, B. C. E., '94.

From time to time short articles on the subject of bridge vibration have appeared in current engineering literature, the majority of which gave evidence that the matter was of considerable importance, and as yet only partially understood and appreciated. Considering this a good field for investigation, it was selected as the subject of graduating thesis by J. B. Gillman and the author. Prior to 1875, there seems to have been nothing done, toward recording the actual motion of a bridge under passing trains, and the first attempts made about that time resulted in crude diagrams devoid of information. Bridge deflections, being easy of measurement have often been taken by means of a pole resting on the ground or bed beneath midspan, and a pencil attached to the pole and a card to the bridge, or vice versa, it is immaterial which, the length of the plotted line being the measure of deflection, but of course gave no indication of the intermediate bridge motion. Deflections have also been measured by means of a level and rod—the level being placed upon the bank and the rod held on the bridge span, the deflection being read directly by the leveler during the passage of a train.

It soon became apparent, however, that by moving the card or pencil in a plane parallel to the bridge, much clearer diagrams could be obtained, the lateral motion preventing the confusion of the vertical motion of the bridge. Several such diagrams were published by the Commissioner of Railways of Ohio in 1881 and were the first diagrams of bridge vibration ever published. These were obtained by Mr. S. W. Robinson, M. Am. Soc. C. E., from stations erected at mid-span, and about this time it became evident that in order to obtain intelligible diagrams from which deductions could be made an instrument which would give a uniform speed to the recording paper must be employed, and in the same railway report for 1881, the general character of such an instrument was fully set fourth. Two years later, a Mr. Biadego used a similar instrument on a three span continuous girder bridge, notice of his experiments being given in a foreign journal.

In the latter part of 1884, Mr. Robinson constructed an instrument and obtained vibrations from thirteen different bridges, resulting in 193 diagrams which, when taken with the accompanying exhaustive analytical calculations forms the most complete investigation which the subject has ever received. His instrument in brief consisted of two drums, upon one of which was wound the recording paper, the second cylinder winding the paper from the first at the time it received the markings of the recording pencils, the speed of the drums being controlled by a centrifugal governor. The paper was speeded about fifteen inches per minute—this being the only attempt to record the passage of time. Recently a prominent engineering journal described an instrument known as Frankels Deflection Meter for measuring bridge vibration, which has the feature of being attached to the bridge member to be examined the static condition of pencils being secured by means of wires, fastened to a weight resting on the ground or bed beneath mid span. This feature will be referred to later.

The instrument used by the author during the period of field observation, as well as its workings, may be well understood from the cut. In brief it consisted of two cylinders, upon the smaller of which was wound the recording paper; to the larger of the cylinders mounted on a spindle the rotating force was applied, this force being supplied by the descending weight. On the upper part of the spindle was fastened a cog wheel, meshing into a series of cogs, terminating in a rotary governor, the speed of which was controlled by the impact of the weights free to respond to centrifugal motion. By regulating the point of impact, the speed of the system could be readily controlled. At a distance of about fourteen inches from the recording drum, was placed the pivot for the recording arm carrying the pencil, this arm having at a distance equal to three-fourths of its length an appliance for fastening a light iron rod. This light rod was in turn fastened to the main wire, which was attached to the bridge, and was kept taut by the steel spring at a tension of from two to four hundred pounds. Thus it will be seen that the vibrations are increased one-fourth in amplitude, but the cuts have been reduced this amount and are therefore actual size. Directly beneath the recording pencil was placed another pencil, subject to the influence of an electro magnet. This pencil ordinarily drew a continuous line, but at certain intervals an electric circuit was completed by means of a special clock work device, and a break marked in the line. Thus the passage of time was recorded at the same time as the vibrations. The pencils were held against the paper by means of springs. The whole was mounted upon a tripod, the legs of which were generally frozen into the ice to insure greater stability. A wind break was usually erected to protect the instru-

ments from tremors. At no time was the recording wire affected by wind, even when, at one time one hundred and thirty feet were exposed to a high wind. At first, an attempt was made to record the passage of wheels over the panel point, but the high speed of so many of the trains observed made this a matter of great difficulty, and the plan was not carried out. The general working of the indicator was very satisfactory, considering the varying conditions under which it was used.

No attempt was made to obtain a record of the lateral vibrations of the bridges. This, in most cases, would have necessitated the erection of a station at mid-span, entirely free from the bridge, involving much expense and the results would not be of sufficient value to warrant such a proceeding. Lateral vibration does undoubtedly occur from the wandering of trucks, crooks in the rails and wind action, but they are probably not of the cumulative type.

Bridges may be considered as elastic bodies, and to vibrate in accordance with the laws of such bodies. With this understanding, bridges have certain rates of vibration depending upon the weight, span, and moment of inertia of the structure. They may be set in a state of vibration by any of the following conditions, viz.: a sudden blow, a sudden release of the bridge from a condition of strain, or by a succession of light blows regularly repeated at time intervals corresponding to the time of vibration of that particular bridge. At high speed, the weight of a locomotive and train is applied to bridges in a very brief interval of time. This sudden application of load is believed to produce strains greatly in excess of those produced by the same load when not suddenly applied. This would cause a bridge deflection in excess of the normal; the bridge would then spring back a little past its normal position; in returning the normal would again be passed; in short the bridge would be in a state of vibration. Now if this motion were supplemented by a succession of light blows from any cause, as unbalanced locomotive drivers, eccentric wheels, etc., delivered at intervals of complete vibration of the bridge, a cumulative vibration would be induced which under extreme cases, might endanger the stability of the structure. That vibration is severe, causing overstraining of the metal and excessive wearing at connections may be understood from the remarks of Mr. Morison, M. Am. Soc. C. E., who has built some of the largest steel bridges in America: "There are two sources of motion in a bridge under strain, one perfectly legitimate and one which should be avoided in every possible way. The legitimate source of motion is the change of length of metal due to strain. The illegitimate source is the lost motion—and I believe more harm is done to bridges by vibration due to this lost motion than comes from every other source.

It is possible to design a structure in which no metal under any ordinary condition will be overstrained, and yet without such overstrain vibration can exist which would be entirely inadmissable." Mr. Thompson, M. Am. Soc. C. E., also an engineeer of wide experience, says: "I think some of us could design a structure that would meet current requirements but that would be so full of vibration that it would be merely a question of a decade or two as to its endurance. I know a bridge of 150 foot span that has been visibly overstrained in one year and another of 125 foot span that did not withstand it so long. It would seem to me that if these structures have much of motion, it does not make much difference whether the iron is sustaining 6,000 or 8,000 or 10,000 lbs. per square inch, they will not endure."

The first cause of bridge vibration previously mentioned, that of suddenly applied load, has in no case been found to cause cumulative vibration, the very nature of the cause tending to a lessening amplitude. On all diagrams which show cumulative vibration a certain time interval elapsed before the attainment of their maximum amplitude, showing that the suddenly applied load was not the cause. Bridge camber doubtless tends to modify vibrations from this source, but it would appear that the usual method of allowing only one half of the working strains per unit area for suddenly applied load would give an excess of metal not required. The rate of vibration of a stringer, floor beam, or even a structure like a bridge is very rapid, in the latter case ranging from two to five per second. When double area is allowed for the effect of this sudden application of full load, the theoretical condition of instantaneous full loading is assumed which in the case of a bridge is never realized. To obtain this, the load would have to be applied in less than the time of one-half of one complete vibration, or in from one-fourth to one-tenth of a second. At a speed of 50 miles per hour which is probably as fast as most trains run over bridges, the time for an engine to arrive on the middle of a 150-foot span is about one second, or four times the time required for sudden application. On stringers, floor beams and in some cases posts, the load arrives almost instantly at its ·full intensity but on the trusses the loading is comparatively slowly applied. This may well be inferred from study of diagrams. If sudden application pro_ duced double stress, then double deflection would result in accordance with Hooke's law. Diagrams fail to show any difference in deflection for speeds ranging from only a few feet per second to forty and fifty feet. Often additional deflection was caused by vibration at the higher speed but the excess was due to cumulative vibration as shown by the record made each side of the point of maximum deflection.

This superadded deflection caused by vibration would indicate the increased stress in the trusses over that which would be caused were the bridge prevented from vibrating. The maximum amplitude of vibration observed on the whole number of diagrams taken would indicate an increase of about 25 per cent. in the chord strains. Were this the greatest maximum which would ever occur, the reduction of the unit stresses for the dead load 25 per cent. would suffice for the live load working stresses. However, it is possible that at rare intervals bridges suffer cumulative vibrations greatly in excess of anything yet observed and until this matter is cleared up, we are dealing with uncertainties. Were it shown by many observations, that bridges do not suffer an excess of strain by vibration greater than 25 per cent. of that due to the simple load, it would appear that the only distinction needed between dead and live load stresses is the reduction of the unit strains due to the latter by 25 per cent. and thus a considerable saving of material.

Vibration due to the second cause, the sudden release from strain, has not been found to exist. On none of the bridges examined has release from strain been found to induce appreciable vibration.

But the third cause, that of repeated impulse, has been found to be a potent cause for cumulative bridge vibration. These impulses may come from external influences, such as unbalanced locomotive drivers, eccentric or flat car wheels, low or worn rail joints, vibrating of cars and engines on their springs, or from conditions within the structure itself, such as equality of driver circumference and panel length, yielding stringers, and lost motion in details. Plainly, then, if these are the causes of vibration, the vibrations themselves may be greatly modified and perhaps entirely corrected by presenting the most unfavorable conditions.

As long as locomotive drivers are built with counterweights there will be an unbalanced vertical component, tending to produce vibration. When the excess of non-balance is up, the weight on the bridge is decreased, and when it is down, the weight is increased and under proper train speed, cumulative vibrations result. Eccentric car-wheels probably do not cause excessive cumulative vibration, though in some cases exceeding that caused by the engine for that particular bridge and train speed. Diagram C_1 and P_2 show vibrations due to train, the engine in each case having passed entirely off the bridge; both diagrams showing the vibration to be of the slightly cumulative type. The important effect of speed is shown by reference to diagrams C and C_1. Both engines were of the same weight and driver circumference, but the difference in speed of 7.5 feet per second, was the most probable cause of considerable vibration in one case and none in the other, due to engine load. At the same time we see how changes in load intensity interfere with vibrations, the successive application of heavy and light cars upon a bridge changing the rate of vibration of the system.

Low or worn rail joints are not often allowed to exist on a bridge, and it is not probable that the impulses from that source are synchronous. The vibrations of cars and engines on their springs are modified by so many conditions, such as kind of springs, position, eccentricity and rigidity of car loading, that it is quite probable they almost wholly neutralize each other. The equality of driver circumference and panel length becomes particularly liable to induce cumulative vibration when the excess of non balance is down at mid panel, the greater weight being applied to the deflecting stringers and the lesser weight at time of passing floor beams. A recent diagram obtained by the author shows the deflection of the floor beams relative to chord to be almost double the ordinary when the excess of non-balance is down at mid panel. If the panel length permit of the passage of car trucks over floor beams at the same time and the yielding of stringers be appreciable, the successive regular rise and fall under proper train speed will cause cumulative vibration. On bridges having panel length differing considerably from driver circumference and half car length, no vibration due to trains was observed and in no case was there found to exist vibration due to passenger cars, probably on account of their varying length. Also locomotive vibration was very light, being scarcely distinguishable from the tremors of the train. Double track bridges are singularly free from vibrations; on the three examined, the maximum observed amplitude did not exceed one-sixteenth of an inch, in most cases being only about one-thirtysecond of an inch. Considerable depth of truss would tend to diminish vibration as well as increase the stiffness of the structure. Diagrams B and B_1, taken from a double track bridge of panel length 30 ft. 9 inches and depth 42 ft. are selected from many to show these points. It will be observed that the deflection is small and the vibrations insignificant. Loose details by the freedom given to members would in extreme cases permit of blows induced by vibrations, upon the connections and thus at once introduce undue strains and excessive wear.

The only plate girder bridge examined had no appreciable vibration and though of seventy-five feet span the maximum deflection observed did not much exceed three-eighths of an inch. Further investigation would probably show plate girder spans free from the evil of vibration. See diagrams E and E_1. One highway bridge was examined for vibration; the steel arch bridge across the Mississippi from Nicollet Island to the West Side. To the pedestrian, the vibrations at times appear excessive but when they are measured they prove very small. The maximum amplitude observed did not exceed one-eighth of an inch with about the same amount of quiescent deflection for the passage of a street car.

In the design of a bridge the avoidance of conditions favorable to vibration should be observed. The vertical component of non balance in locomotive drivers should be as small as possible. Eccentric car wheels and low or worn joints should not be permitted on bridges. Equality of driver circumference and panel length should not exist and no multiple of panel length should equal extreme car length. Stringers should be deep to obtain rigidity. Car loading should be eccentric and the train loading variable throughout its length. Speed of trains has been found a very important cause of excessive bridge vibration. Diagrams showing considerable cumulative vibration have been taken from bridges that under similar conditions with the exception of train speed, have shown almost no vibration whatever. Details should be rigid and tight fitting. Diagrams from wooden bridges show them to be in a state of excessive tremor during the passage of freight trains at considerable speed; to be only slightly affected by passenger trains, and to be frequently subject to light cumulative vibration which sometimes becomes excessive. Some of these conclusions were arrived at by Mr. Robinson together with others of lesser importance, and the whole subject of bridge vibration is in accord with simple and generally well understood laws.

The future will probably see the bridge indicator in much more general use. The great number of bridges built and yet to build demands some efficient method of inspection and this the indicator offers. There is certainly a relation between inherent weakness, deflection and vibration and further investigation will doubtless make it of practical use. Favorable diagrams could also well be made a condition of bridge acceptance. Every test of a new bridge should include diagrams of bridge motion under various conditions. We have seen how a bridge could be designed to meet general requirements and yet be so subject to vibration as to last only a few years. The diagrams from such a bridge would show at once its weakness and cause its rejection. But before this can be attained, an indicator capable of being operated on the bridge regardless of the condition of affairs beneath midspan must be constructed. Relative stability could be secured by means of a wire fastened to a weight resting on the ground or bed, and attached to the bridge with the interposition of a spring to eliminate bridge motion. The indicator itself could well be made in two parts on separate base boards, one to furnish the necessary rotary motion, the other to carry only the two cylinders and the recording pencils. The object of this is to obtain complete isolation from bridge motion for the part supplying the rotary motion, that the speed regulator may not be affected, and to give extreme lightness to the recording part, thus reducing its inertia and permitting its response to the finer bridge tremors. The instrument if made as a whole, could be almost wholly isolated by mounting it upon the middle of a long flexible board, supported at the ends, and the middle held by a wire, weighted to the ground below, at some deflection greater than that to which the bridge would be subject. The length of bridge was taken by the author as a base, by which to measure the speed of trains, the time of passage being recorded by a stopwatch and observer. Electricity could in the future be advantageously employed to this end, and its more extended use would allow more time for observation of conditions by the operator. Experience shows that the indicator should be made to do as much as possible. It is to be hoped that in the future, the taking of bridge indicator diagrams will be much more common and our knowledge of the stresses and strains, to which these important structures are subject thereby increased.

Center span of M. St. P. & S. Ste. M. Ry. Bridge, a single track wooden Howe truss of 150 feet span with trestle approaches. Panel length 10 feet, height 24 feet, 1 inch, in. to in. width 14 feet, 6 inches, in. to in. 19 feet, 10 inches out to out. Span has floor beams near panel points. Ties about three inches apart, rails out to out: 22 feet 1 inch. All track joints opposite. Bridge is near Bruce, Wis. Center span in use 5 years. All track in good line.

A MACHINE FOR EXPERIMENTING ON FRICTION.

GEO. E. BRAY, '94.

The following is a description of a machine which has been designed for the purpose of experimenting on friction, and by means of which the relative values of metals to be used for bearings may be determined.

The principles of the machine were obtained from an article in the Engineering News of May 18th and 25th, 1893, giving a description of a similar one designed and constructed at Cornell university.

This machine has the advantage over others constructed for a like purpose in that the different materials may be tested at the same time and under the same conditions, thus eliminating the error which is apt to exist when tests are made upon the materials separately.

Discs of different materials such as steel, wrought iron and cast iron are keyed to a shaft as represented in Fig. 1, and the blocks of copper, babbitt metal and different alloys used for lining journal bearings are pressed against the discs. The shaft is then rotated and the temperature of the bearing surface can be directly observed and the coefficient of friction computed. (The method to be given later).

The machine is designed so that it can be run at the following speeds, 134, 400, and 1,187 revolutions per minute when connected by an open belt to the counter shaft which is to make 400 revolutions per minute.

The speed cone pulley on the counter shaft is to be of the same dimensions as the cone on the machine; the same belt may then be shifted from one set of pulleys to another.

A A Fig 1 represents the bed of the machine. S the shaft resting on a support at each end, also supported between the discs. The disc D and the block B are the two metals to be experimented with. There are three discs keyed to the shaft S, 11 inches apart.

The pressure between the block B and disc D may be determined by means of the lever L L[1], having one end supported by a knife edge resting on a pair of scales, and the other end connected to block B by means of links which act on the knife edges. The pressure between the block and disc may be increased by turning the hand wheel W. By observing the pressure on the scales K the total pressure between the disc and block can be computed by taking moments about the point O, Fig 1. For example let the pressure on the scales be 500 lbs. and let X = pressure betwen the block and disc, then 500xL^{1}O=XxO L or 500x21 =Xx3.5 or X=3,000 lbs. To this weight must be added the dead weight consisting of block, links, knife edges, etc.

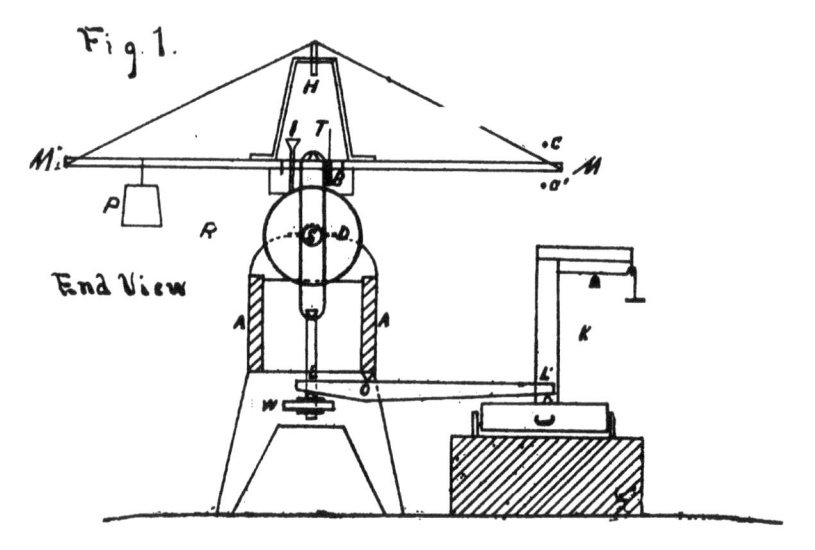

Fig. 1.

End View

The upper part of Fig. 1 represents that portion of the machine by means of which we are enabled to determine the length of lever arm required by a given weight, to overcome the action of friction between block and disc, as the shaft is revolved. The block B is securely screwed to the lever arm MM^1. The tendency of the block to turn, as the shaft revolves, is prevented by pins C and C^1 fastened to a frame not represented. The weight W can be moved along the lever arm M or M^1 until an equilibrium is obtained.

The parts of the machine are so arranged that the shaft may be turned in either direction. If then we balance the block with the weight on one side and then reverse the direction of the shaft and balance with the weight on the other, we can by taking the average of the two lever arms eliminate any error which may exist from the knife edges not being exactly in line with the centre of the shaft.

From the formula $fPr=RW$, we are able to obtain values for f, for different metals and at different speeds: where f=coefficient of friction, P=pressure of block on disc, r=radius of disc, W=weight, having lever arm R. See Fig. 1.

The arms MM^1 are prevented from bending by a wire fastened to the ends and stretched over a screw H which may be raised or lowered. The bearing is to be lubricated by the oil well I, and the temperature is taken by means of a thermometer, T, placed in a mercury well as shown at S.

Although this machine was designed for experimenting on friction, it may also be used for determining the durability of lubricants or the most economical rate of feed for a given lubricant under any load.

. This machine is exceedingly simple and easily operated, and is destined to be of great value in the experimental labratory.

INSTRUMENT EQUIPMENT OF THE DEPARTMENT OF CIVIL ENGINEERING.

The higher professional work of the civil engineer has in abstract designing little to do with engineering instruments, as, for in the design of a bridge, a system of sewerage or of water supply, yet in the final execution of the design a very important part falls to the locating engineer, and the final success of the undertaking is measured in no small degree by the accuracy with which the design is executed, and can be done most efficiently by the skillful use of the modern engineering instrument.

All projects for railroads, canals, water storage, land drainage, land subdivision etc., depend in a much larger sense upon the field instrument while the work of the topographer and geodesist is rendered possible only by instruments of the highest precision which the mechanician's skill can furnish.

From these considerations it has been the constant aim of the department to furnish instruments with which the student, after a critical study of their mechanism and adjustments, executes all problems and projects which are calculated to illustrate theoretical discussions. This plan at the same time gives the student a fair amount of skill and handicraft in the use of the various instruments.

Only the best grade of instruments have been purchased and nearly every American maker of field instruments is represented. This is to acquaint the student with the different types he is sure to meet in professional work as well as to guide him to the intelligent selection of an equipment upon entering private practice. To allow each student the fullest opportunity to become skilled in the use of instruments argues as small squads as the nature of the work will allow; whenever possible, as in reconnaissance and in approximate topographical surveys, individual work is insisted upon. With the present size of classes in Civil Engineering this calls for a large number of instruments which have been generously supplied.

Instrumental Equipment of the Department of Civil Engineering in 1894.

Besides the instruments shown in the cut, the department, through the courtesy of the Superintendent of the Coast and Geodetic Survey, enjoys the use of certain instruments adapted to special work in triangulation and magnetics, which are officially in charge of Prof. Hoag who is an Acting Assistant of the Coast and Geodetic Survey in charge of the triangulation of Minnesota.

The college also has a transit house fully equipped with a transit circle with circles reading seconds and with registering chronograph and clock for advanced work in astronomical geodesy.

Instrumental Equipment of the Department of Civil Engineering in 1894.

DESIGN OF A STEEL ARCH BRIDGE.

A. O. CUNNINGHAM, C. E, '94·

The object of the structure is to span the Great Northern R. R. tracks which pass through the campus of Minnesota University. The bridge is 110 feet from centre to centre of end pins and 32 feet wide.

It is composed of four steel ribs placed nine feet four inches apart, hinged at the ends and parabolic in form. Each rib is three feet deep, being made up of four $5x3\frac{1}{2}x\frac{7}{8}$ inch angles with coverplates, and a three-eighths inch web plate. Ornamental metal plates are fastened along the outside of the web of the two outside girders.

As aesthetic design was an important factor, it was desirable to make the rise of the ribs small. While it was necessary for the arch to be of such a height as to admit the passage of trains under any part of it, the topography of the ground required the structure to be placed at as low a position as possible, otherwise high embankments for approaches would be necessary, thus spoiling the aesthetic effect. For these reasons the rise was made only ten feet.

The roadway rests upon the crown of the arch, being supported along the ribs by posts eleven feet apart. As the structure will be used by pedestrians chiefly, the sidewalks were made eight feet wide, thus leaving a carriage way of sixteen feet.

The floor beams are eighteen inches deep, and made continuous over four supports nine feet four inches apart, their ends overhanging two [feet. The posts, which support these beams, are made of angles, the outside ones being covered up by ornamental imitation posts. The joists are nine inch; twenty-seven pound I beams placed thirty-seven inches apart and connected to the web of the floor beam.

The flooring is composed of two layers of 3x12 inch pine plank upon which are placed cedar blocks. The side walks are of plank, being slightly raised above the roadway.

Design of a Steel Arch Bridge.   47

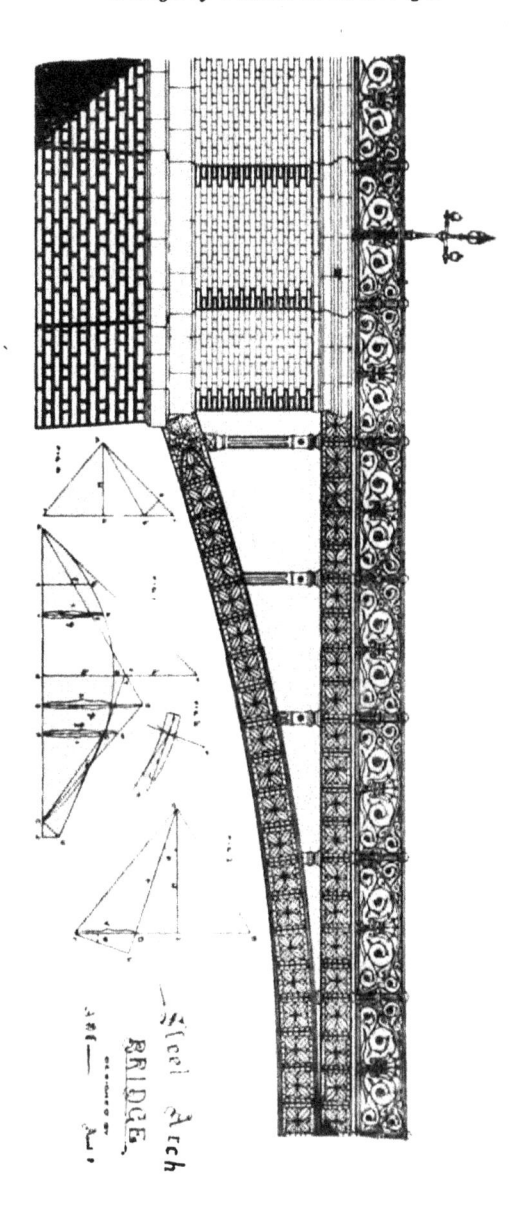

The lower part of the abutments is of limestone faced with grey sandstone, the thrust and weight of the steel ribs being borne by the skewbacks—granite blocks set normal to the thrust of the arches and backed by granite and concrete. The upper part of the abutments is made of common brick faced with grey colored pressed granite brick, the cornice and all trimmings being made of grey sandstone. The abutments have rectangular projections on each side on which to place ornamental lamps. All iron work is painted black, so that the soot, which will naturally accumulate from the smoke of passing trains, will not discolor it, while at the same time this color will make a decided contrast with the grey abutments.

The stresses in the arch ribs were found both analytically and graphically, the results of the two methods checking very closely. On account of lack of space, however, only the former method will be presented, even this being necessarily abbreviated.

The methods pursued in proving the following formulae are in most cases similar to those employed by Professor Greene in his work on "Trusses and Arches."

In Fig. 1, let AGC represent the steel rib, loaded at a point directly under B. In Fig. a, supposing H, the horizontal thrust of the rib, to be known, and cR and Rb the vertical reactions at the points C and A respectively, then abc will represent the force polygon, and ABC the funicular polygon. Calling the vertical reaction at $A = P_1$, and that at $C = P_2$, and supposing an imaginary section to be cut in the rib at any point G, then since $\Sigma M = 0$, G being the centre of moments,

(Eq. 1) $M = P_2 \cdot EC - \Sigma Wl - H \cdot EG$.

ΣWl meaning the summation of the loads into their respective lever arms. Again, taking moments about the point F in the funicular polygon, we have

(Eq. 2.) $P_2 \cdot EC - \Sigma Wl - H \cdot EF = 0$,

$\therefore H \cdot EF = P_2 \cdot EC - \Sigma Wl$.

Substituting this value in equation (1), it becomes,

$M = H \cdot EF - H \cdot EG = H \cdot FG$;

which means that the bending moment at any section of an arch rib, due to any system of vertical loading, equals the product of the vertical ordinate from that point to the proper funicular polygon multiplied by the horizontal force of the force polygon.

Design of a Steel Arch Bridge.

Referring again to Fig. 1, and taking any point on the rib G, the bending moment in this section causes an elongation of the fibres on one side of the neutral axis and a compression on the other, producing an exceedingly small angle which is designated a, and shown more clearly in Fig. b. Since a varies as the bending moment H. F G it must equal (H. F G) k in which k is some constant. Now in Fig. 1, if the point C were not fixed, the bending moment at G would cause it to take some other position such as U (greatly exaggerated in the figure). The position taken would depend not only upon a, but also upon the distance G from C. Since a is very small, U C = G C. a ∴ U C = G C (H. F G) k, and J C = U C $\frac{EG}{GC}$. Substituting the value of U C in this last equation, it becomes J C = (H. F G) k. E G. and taking all points in the rib into consideration, the whole horizontal displacement Σ J C = H. k. Σ F G. E G. But since C is fixed, no displacement can take place ∴ Σ J C = 0. Now neither H nor k can equal 0

∴ Σ F G. E G = 0

which means that the summations of the products F G. E G. for every point where the funicular polygon lies on one side of the rib, must equal the summation of the products for every point where the funicular polygon lies on the other side. There can therefore be but one polygon which will satisfy this condition. In order to obtain this proper polygon it will be necessary to derive the value of H, the horizontal reaction, but before a value of H can be derived, a value for the maximum ordinate to the funicular polygon must be obtained. To get this value let the maximum ordinate (Fig. 1) B E = Y_0, let the half span = m, the height or rise = h, the distance from the middle of the span to the position where the load is placed = n, and x = distance from left abutment to any ordinate E G. Then A E = m+n and E C = m-n.

Now since the ordinates to a parabola from the line A C are proportional to the products of the segments into which they divide the span, then

E G = $\frac{h}{m^2}$(2 h x—x^2), also E F = $\frac{y_0}{m+n}$x

The required condition is that Σ F G. E G = 0, or Σ(E G−E F) E G = 0
(Eq 1) ∴ Σ E G^2 = E G. E F.

Now substituting the values of E G and E F in equation 1, we get for Σ E G^2 the value $\int_0^{2m} \frac{h^2}{m^4}$ (2 m x—x^2)^2 dx, this reduced = $\frac{16}{15}$ h^2 m, and for

$\mathbf{\Sigma}$ E G. E F between the limits A and E we get the value
$$\int_0^{m+n} \frac{y_0}{m+n}x \cdot \frac{h}{m^2}(2mx-x^2)\,dx, \text{ this reduced} = \frac{hy_0}{m^2}\left[\tfrac{2}{3}m(m+n)^2 - \tfrac{1}{4}(m+n)^3\right].$$

For the portion between E and C if we write m-n for m+n and reckon x from C to the left, EF will equal $\frac{y_0}{m-n}x$ while E. G will be unchanged, so that the integration of the right hand member of equation 1, between the limits $x=0$ and $x=(m-n)$ will give the same result as between the limits A and E, if -n be placed instead of +n, that is, $\mathbf{\Sigma}$EG.EF between the limits A and E $=\frac{hy_0}{m_2}[\tfrac{2}{3}m(m-n)^2-\tfrac{1}{4}(m-n)^3]$

These two values of the right hand member of equation 1, being added give the value $\frac{hy_0}{m^2}(\tfrac{8}{3}m^3-\tfrac{1}{6}mn^2)$.

Equating this value with the value found for $\mathbf{\Sigma}$ E G^2, we get
$$y_0 = \frac{32}{5}h\frac{m^2}{5m^2-n^2} = \frac{32}{5(5-f^2)}h$$

if $n=fm$, where $f=$ a fraction of the half span.

To obtain the value of H, we have by taking moments about C (Fig. 1), and then about A,
$$P_1 = W\frac{m-n}{2m} \quad \text{and} \quad P_2 = W\frac{m+n}{2m}$$

P_1 and P_2 being the reactions at A and C respectively caused by a load W. Also by similar triangles in Fig. 1, and Fig. a
$$\frac{H}{P_1} = \frac{m+n}{y_0},$$

whence, substituting the values of P_1 and y_0 in the last equation, it becomes
$$H = W\frac{m}{h}\left\{\frac{1-f^2}{2} \cdot \frac{32}{5(5-f^2)}\right\}$$

The values y_0, P_1, P_2, and H being expressed in terms of known quantities, we can now, by first finding the values of z (the ordinate to the parabola from the line AC) obtain the stresses in any part of the arch ribs due to any system of vertical loading.

Besides, however, the stresses caused by the dead weight and the ordinary live load, there are stresses exerted on the arch ribs by temperature, wind and change of length from thrust.

To obtain a formula for thrust from temperature, it must be remembered

A Design for a Steel Arch Bridge.

that the amount of flexure varies inversely as the modulus of elasticity and the moment of inertia, and directly as the bending moment. So if we put E for the modulus of elasticity, I for moment of inertia at the crown of the arch and H_t for temperature thrust, we must write $\frac{Ht \cdot \Sigma F G \cdot E G}{E.I}$ to obtain a quantity equal to the change of inclination. But we wish to introduce a quantity to represent the change of span that would take place on account of temperature were it not checked. This quantity must be added to or subtracted from the above fraction according as the temperature is above or below that at which the arch was designed. If $t =$ no. of degrees, $e =$ coef of elasticity of material, then this quantity will be ± 2 tem. As there can be no bending moment at either abutment, the line of thrust must be in the line joining the two hinges, and the bending moment at any point will equal H_t times the ordinate to the rib. The expression $\Sigma F G \cdot E G$ then becomes $\Sigma E G^2$, and now writing the whole expression we have $\frac{Ht \cdot \Sigma E G^2}{E.I} \pm 2 \text{ tem} = 0 \therefore Ht = \frac{1}{8} \cdot \frac{t. e. I. E}{h^2}$ since $E G^2$ in deriving the value of y_0 was shown to be equal to $\frac{1}{8} h^2 m$. Having now all the necessary equations we shall commence on the design of the bridge.

Looking at the structure it is seen that the rib is divided into ten equal parts, and therefore the load on the rib between centre to centre of pins is applied at nine points. These points are placed 11 ft. apart. Calling the first point to the right of the pin 1, and the others in numerical order, the centre point, the one at the crown of the arch will be 5.

To get coefficients for any parabolic arch for the values of y_0 and H we have already found.

$$y_0 = \frac{32}{5(5-f^2)} h, \text{ and } H = \left\{ \frac{(1-f^2)}{2} \div \frac{32}{5(5-f^2)} \right\} \frac{m}{h} W$$

Points along rib   1   2   3   4   5

$\frac{32}{5(5-f^2)}$ 1.4679   1.3793   1.3223   1.2903   1.280.

Multiply these factors by h and we get value of y_0.

$\frac{(1-f^2)}{2} \div \frac{32}{5(5-f^2)}$ 0.1226   0.2320   0.3176   0.3720   0.3906

Multiply these factors by $\frac{m}{h} W$ and we get value of H. To compute bending moments, let y = ordinates from A C to the inclined line of the funicular polygon, and let z = ordinates to the parabola, then

$$M = H(Y - Z).$$

A table of bending moments can now be made to get values of (y-z) at different points along the arch.

The ordinates z are proportional to the product of the segments into which they divide the span, or $z = (1+f) m (1-f) m \dfrac{h}{m^2} \therefore z = (1-f^2)h$. It will be remembered that the value of $(1-f^2)$ has previously been found.

Since the span is divided into ten parts and suppose a weight placed at point 6, then y_0 at this point has been found to equal $1.2903h$ and y will be successively on one side $\tfrac{1}{6}$, $\tfrac{2}{6}$, $\tfrac{3}{6}$, etc. of y_0, and on the other $y = \tfrac{1}{4}$, $\tfrac{2}{4}$ and $\tfrac{3}{4}$ of y_0. Putting this down in the following manner, we can get coefficients which may be used for any parabolic arch.

Values for coefficient of M (load placed at point 6 or point 4.)

Points on Arch	1	2	3	4	5	6	7	8	9	
$y = \dfrac{f}{6} y_0 =$	.2151	.4302	.6453	.8604	1.0755	1.2151	.9678	.6452	.3226	$= \dfrac{f}{4 y_0}$
z =	.36	.64	.84	.96	1.0000	.96	.84	.64	.36	h
y−z =	−.1449	−.2098	−.1947	−.0996	+.0755	+.2551	+.1278	+.0052	−.0374	

Multiply these factors by H=0.372.

| M = | −.054 | −.078 | −.073 | −.037 | +.028 | +.123 | +.047 | +.002 | −.014 | mW |

This computation has been carried out for a load at each joint successively, and is tabulated below. Since the moments are the same at similar points on each side of the centre of the arch, the table need only be carried out as far as the fifth point.

M=cmW. Values of c.

Load placed on pts	1	2	3	4	5	6	7	8	9	H
Mom at 1	+.136	+.076	+.025	−.014	−.041	−.054	−.054	−.044	−.024	.123 $\tfrac{m}{h}$ W
" " 2	+.082	+.171	+.076	+.002	−.050	−.078	−.083	−.068	−.039	.232 "
" " 3	+.037	+.085	+.153	+.047	−.028	−.073	−.087	−.075	−.043	.318 "
" " 4	+.002	+.017	+.055	+.123	+.025	−.037	−.065	−.063	−.038	.372 "
" " 5	−.023	−.032	−.018	+.028	+.109	+.028	−.018	−.032	−.023	.391 "

To obtain the shear on a vertical section, it must be remembered that one of the properties of the parabola is that a tangent at the springing of the arch will intersect the middle ordinate at a distance h above the crown, equal to the rise of the arch. If then in fig. a, we draw a i parallel to the

Design of a Steel Arch Bridge. 53

tangent A S at A and call the distance R i $=$ Z, then from similar triangles
$$Z : H :: 2h : m \therefore Z = \frac{2h}{m} H.$$
The vertical shear at the abutment on the left will then be
$$V = P_1 - Z_1 = \frac{m \div n}{2m} W - \frac{2h}{m} H.$$
Since the max shears will occur midway between the points of division of the arch, a table of vertical shears has been made at these points. Placing the load W at point 6, as was done in making the table for the moments, then
$$P_1 = 0.4 W; P_2 = 0.6 W; H = .372 \frac{m}{h} W; Z = .744 W.$$
Z will diminish at a constant rate, and at the middle of the first space $=$ $(.744 - \frac{744}{10})$ W $= 670$ W; for each succeeding ordinate it diminishes $\frac{744}{10}$ W.

As an illustration of how to make a table of vertical shears the following is given.

Values of coefficients of V, load applied at point 6, or point 4 if signs are changed.

Pt. along arch	½	1½	2½	3½	4½	5½	6½	7½	8½	9½
P	.4	.4	.4	.4	.4	.4	—.6	—.6	—.6	—.6
Z	.670	.520	.373	.223	.075	—.075	—.223	—.367	—.521	—.669
P—Z	—.270	—.121	$\div$.027	$\div$.177	$\div$.325	$\div$.475	—.377	—.228	—.079	$\div$.069

By the preceding process the following table of coefficients of vertical shears has been computed.

$$V = kW \quad \text{values of } k.$$

Load Placed on Points.	1	2	3	4	5	6	7	8	9
V at ½	+.678	+.382	+.128	—.000	—.204	—.270	—.272	—.218	—.121
" " 1½	—.272	+.475	+.255	+.079	—.047	—.121	—.145	—.125	—.072
" " 2½	—.223	—.432	+.382	+.228	+.109	+.028	—.018	—.032	—.023
" " 3½	—.173	—.339	—.491	+.377	+.265	+.177	+.109	+.061	+.026
" " 4½	—.125	—.247	—.364	—.475	+.422	+.325	+.236	+.153	+.075

It is not necessary to continue this table further since the values on the other side of the centre will be the same only with the signs changed.

From these values of verticular shear the tangential thrust and the normal shear can be computed. For in Fig. 2, let H equal horizontal thrust, then O L the line parallel to the tangent at the point will equal tangential thrust, or calling this T, we have $T = \frac{H}{\cos \theta} + V \sin \theta$, the angle θ being the angle made with H by the tangent, and V=DC. The normal shear which we will call S=V cos e.

Since the tangential thrust must be combined with the thrust caused by moments, it will not be easy to determine a general expression for the amount of load to produce a maximum stress in either flange. So in the working out of this design the stress in each part of the flanges was determined by placing a single load at each point of division. The stresses were tabulated in place of bending moments, those of like sign being added together.

The loading for which the bridge was designed was the same as that designated class A in "Coopers' Specifications for Highway Bridges." All parts were designed according to the above specifications, and since these required different formulae for compression members, depending on the kind of loads, the maximum stresses were tabulated separately. The amount of metal required for each was then found and the sum of material taken.

Cooper's specifications also require that the maximum stresses in all members subjected to both kinds shall be increased eight-tenths of the smaller. The following table gives the maximum stresses as found in this design with such increase wherever necessary.

+ —Compression — Tension	Live	Temp. $\pm$ 75 deg.	Dead.	Wind.	Section required.
Stress in Upper Flange at 1	+ 81455	+ 58945	+84931	+1713	4—5x$3\frac{1}{2}$x$\frac{5}{8}$ angles
" " Lower " " 1	+129534	+ 58945	+82349	+1713	and 2—5-16x$10\frac{3}{8}$ plates.
" " Upper " " 2	+111480	+103881	+84817	+3045	4—5x$3\frac{1}{2}$x$\frac{5}{8}$ angles
" " Lower " " 2	+146038	+102881	+79245	+3045	and 4—5-16x$10\frac{3}{8}$ plates.
" " Upper " " 3	+ 92121	+134266	+84545	+3997	4—5x$3\frac{1}{2}$x$\frac{5}{8}$ angles
" " Lower " " 3	+127750	+134266	+77657	+3997	and 4—5-16x$10\frac{3}{8}$ plates.
" " Upper " " 4	+ 95861	+153096	+81448	+4568	4—5x$3\frac{1}{2}$x$\frac{5}{8}$ angles
" " Lower " " 4	+ 92349	+153096	+70267	+4568	and 4—5-16x$10\frac{3}{8}$ plates.
" " Upper " " 5	+ 76855	+159374	+84104	+4758	4—5x$3\frac{1}{2}$x$\frac{5}{8}$ angles
" " Lower " " 5	+ 77108	+159374	+76023	+4758	and 4—5-16x$10\frac{3}{8}$ plates.

The wind load upon the ribs was applied at the points where the vertical loads come upon the arch, the floor beam at the crown and the pins at the abutments resisting the tendency to overturn. The stresses were then found as if the ribs were straightened out forming three trusses.

Since the floor beams are continuous girders, Clapyeron's theory of three moments was used in finding the stresses in them. A road roller was found to be the heaviest load that could come upon them. The position of the wheels causing a maximum stress was found, concentrated loads were placed at these points and stresses found from this loading. The dead load was considered as evenly distributed and stresses due to it found separately.

Finally the following equation was applied to the ribs to ascertain how much camber in the arch would be necessary, so that no stresses might be produced by change in form of the arch-ribs when loaded.

Let v = distance between the springings in order to compensate for the amount of compression which would arise from loads, then

$$v = \tfrac{8}{15} \cdot \frac{H}{A.E.} \cdot \frac{3m^2 + 2h^2}{m}$$

in which A represents the area of cross section at the crown, the other letters representing the same quantities as before.

Applying this formula, v = .08, a value practically equal to zero, showing that no camber is necessary.

The amount of metal used in the bridge is about 180,000 lbs, and since the arch ribs are only 110 ft. long and with a small rise, they can be built in the shops thus reducing the cost. All the sections and connections have been made as simple as possible, and it is estimated that such a bridge can be built for $10,000. This of course does not include the cost of the abutments.

DIFFERENT FORMS OF THE TELEPHONE.

ROBERT E. FORD, 95.　　　　　　　　　　　　　F. B. WALKER, 96.

Since the telephone interests have for the past decade been in the hands of one company which controls most of the fundamental patents, many inventions of merit have been kept in comparative obscurity and are unfamiliar to the general public. It is the object of this paper to mention a number of such inventions, together with several forms of the telephone which are in successful operation in foreign countries.

The earliest and simplest form of telephone is the Bell receiver, the details of which are familiar to all. The receiver of the Swiss Telephone Administration is probable the most efficient and carefully designed of this type.

The compound magnet consists of four laminae and is four and one-half inches long; the diaphragm is two and one-fourth inches in diameter, one hundredth of an inch ($\frac{1}{100}$in) thick; the coil is wound with 2500 turns of number 40 B. & S. copper wire.

There is another single pole telephone known as the "Crown" telephone, in which intense magnetization is secured by using several horse-shoe magnets with like poles at the center, the other poles being attached to the circumference of the diaphragm.

An odd form of telephone has a long conical tube of magnetized steel with a coil around the smaller end. An iron or steel tube outside completes the magnetic circuit from the larger end of the magnet to the edge of the diaphragm, making it similar to the "Crown" form except that the sound travels through the hole in the magnet.

The crown telephone may make use of a steel cup instead of the horse-shoe magnets. An instrument of this form constructed by the writers, took the form of a cylindrical box with an iron ring for sides and steel plate magnetized radially, for the bottom. A quarter inch iron core in the axis of the cylinder carried a coil of the usual dimensions. The diaphragm was fastened to the ring, and a brass ear-piece outside protected it from injury.

This form is very compact but is difficult of construction and not capable of high magnetization. The latter fault might be remedied by laminating the steel plate.

In order to increase the number of effective lines of force, horse-shoe magnets are used, carrying coils on the poles, both of which are presented to the active portion of the diaphragm. One of Prof. Bell's first telephones was constructed this way, and the Siemens, Gower, Kotyra and other forms make use of the same idea.

The useful portion of the wire in the two coils seems to be that section between the poles, the rest being useless resistance. To make all of the wire active, D'Arsonval uses a cylindrical pole-piece carrying the coil, the other pole entirely surrounding it.

The Ader telephone, an instrument of the simple two-pole form, has one striking peculiarity; directly above the diaphragm, fastened to the mouthpiece is an iron ring, called by the inventor the "over-excite" (sur-excitateur.) DuMoncel found that the more nearly equal are the masses of the magnet and the armature, the stronger is the reciprocal induction. The iron ring therefore tends to increase the effect of the magnet. This effect might be obtained by increasing the mass of the diaphragm itself, but the result would be secured at the expense of its vibrating capacity. The Ader instrument is extremely sensitive to the delicate inflections which constitute the timbre of the human voice.

If several receivers are connected in series on a line carrying a telephonic current, the sound in each is but little less than if there were but one. In order to combine the effect of several receivers in one instrument. Goloubitzky designed a telephone in which the effect of four magnet poles was concentrated upon one diaphragm. The C shaped magnets were set at right angles to each other. The four coils were connected up with like pole coils in parallel and the two pairs of coils in series. We have no data as to the increase in effect, but since the amplitude of vibration is increased but little with an increase of the pull, it is not likely that the increased efficiency would warrant the additional expense of construction.

The next type of instrument which might be considered is a modification of the two pole type, having two active poles but differing from it in that the diaphragm is between the pole faces. An instrument of this variety consists of a horse shoe magnet having iron cores, which carry the coils, projecting from the interior pole faces. One of these cores is hollow, providing an opening for the sound waves to travel from the mouth-piece to the center of the diaphragm. This portion of the diaphragm is the most active and is also under the maximum intensity of magnetization, a combination which gives the maximum effect.

A radical departure from the usual construction of telephones and an apparently new principle is brought out in the Field iron-clad receiver and transmitter.

As in the ordinary Bell receiver, a core carrying a coil of wire is attached to the pole of a bar magnet and close to the end of the core is a diaphragm of soft iron. In addition to this is a second pole piece projecting towards the center of the diaphragm and connected with the same pole of the steel magnet by an iron box which encloses the coils and diaphragm and to which the latter is fastened. The coils are so connected as to induce like poles on the ends toward the diaphragm. The diaphragm is therefore acted upon by two equal poles of like polarity and is under no strain whatever; the magnetic lines of force take the path of least reluctance, which is around through the box; and when there is no current in the coils, the number of lines through the diaphragm is practically zero. When a current traverses the coils, a part of the lines of force are diverted through the cores and act upon the diaphragm, thus inducing sound waves which traverse a hole in the outer pole-piece. There is also a two pole receiver of similar construction.

A telephone company of Chicago has also brought out an instrument involving the principle of a diaphragm under no strain and for which is claimed many advantages in the matter of a louder and more distinct reproduction of speech.

The magneto telephones, to which class all the foregoing instruments belong, are limited in their field of successful operation to short lines of a comparatively low resistance. If two telephones of very low resistance were connected over a line of no resistance, the efficiency of transmission would approach 100 per cent.; but when there is introduced into each instrument a resistance of 100 ohms or so and several hundred more into a line between them, the loss amounts to a considerable per cent. of the total difference of potential, and leaves very little effective current for use in the

Different Forms of the Telephone. 59

receiver. To compensate for this loss, Edison introduced a battery into the circuit of the receiver, and devised a transmitter in which the sound waves operated, a variable resistance causing the changes in the line current; and to increase the potential of the line currents, the battery circuit is connected with the primary circuit of an induction coil with the secondary on the line.

The commonest and the most effective method of varying the resistance in the transmitter is by employing carbon, which has the property of changing its resistance when pressure is applied to it. Familiar examples of carbon transmitters are the Edison, Blake, Humming, Berliner, etc.

The Edison carbon transmitter was the first of its kind. It depends in its action upon the difference of pressure on a carbon plate between two metallic plates, one of which vibrates in unison with the diaphragm.

The Blake transmitter which has been the commercial instrument for short distance work, consists of a small carbon button in contact with a platinum point; the latter rests directly upon the center of the diaphragm. The pressure between the electrodes is regulated by varying the tension of the springs which support them.

In the Marche transmitter the metallic point is replaced by a carbon button attached to the center of the diaphragm. The Berliner transmitter is of a similar construction, except that the movable electrode is held in position by its own weight. Lighter or heavier carbon pencils are used according to the resistance in the circuit. For long distance work a similar instrument is employed having three contacts. The Christensen transmitter has a horizontal carbonized diaphragm on which rests by its own weight, a soft carbon button with a flat surface. This contact is capable of carrying a rather high current without heating and hence was used on lines of high resistance.

In a French microphone transmitter, the variable resistance is obtained by placing in the circuit, six balls of hard carbon in a glass tube. The variations in pressure affect all the contacts of the balls equally, thus multiplying the effect.

Reference has been made earlier in this paper to the Hunning transmitter. This instrument stands at the head of the class known as granular transmitters from the form in which the carbon is utilized and combines many features of originality and efficiency. It consists of a vibrating platinum foil diaphragm, behind which, at about a distance of about one-eighth inch is a fixed metallic or carbon plate. The granulated carbon is loosely packed between them. Several important improvements and modifications have been brought out, among which may be mentioned one in which the platinum diaphragm has been replaced by a thin carbon disc. The Hunning solid back form used on the New York—Chicago line, employs a mica diaphragm to which is fastened a copper plate in contact with the carbon granules.

The Berliner Universal transmitter is very similar to the Hunning, but uses a horizontal carbon diaphragm as one electrode.

To avoid the packing of the carbon granules which is so detrimental in other forms, an instrument has been patented which eliminates compression of the carbon by having fixed electrodes. Attached to or supported against a mica diaphragm is a small hard rubber box nearly full of granulated carbon and in which are two vertical plates set perpendicular to the diaphragm, which serve as electrodes. The vibration of the carbon changes the resistance and at the same time keeps it in a loose state.

Another method of varying resistance is by changing the area of contact of the electrodes, either by plumbago blocks sliding upon each other as in a later Edison form, or by employing liquids at the electrodes. Elisha Gray proposed a platinum wire dipping in water and some instruments employ a jelly like liquid instead of carbon between electrodes. Brequet has perfected a mercury telephone depending upon the principle that the level of contact of mercury and another liquid in a capillary tube is dependent upon the potential difference at these surfaces.

Next might be mentioned a few phones which are more noticeable on account of peculiarities of construction or theory than upon their commermercial value. The first to be considered is a receiver which Edison evolved soon after the Bell form appeared. The essential feature of this invention is a chalk cylinder saturated with some saline compound, as caustic potash, revolving at a constant speed, and upon the periphery of which rubs a platinum strip soldered to the diaphragm. The current in passing from the cylinder to the platinum strip decomposes the salt which increases or decreases the friction at this point, causing an intermittent pull on the diaphragm.

Different Forms of the Telephone.

It has been found that the length of a magnet depends to a greater or less extent upon the degree of its magnetization. An instrument based upon this principle has a permanent magnet fixed at one end, carrying at its other extremity one of a pair of carbon electrodes. An iron diaphragm close to the fixed end of the magnet, when vibrating under the influence of the voice, induces magnetic changes in the magnet which vary the pressure upon the carbons.

Ever since the telephone was introduced electric currents exclusively, have been used to reproduce speech at a distance. (So called mechanical telephones which merely substitute a more elastic medium for the air, are not considered.) Of late, however, an exception to this rule has been brought out which is entirely magnetic in its action. It consists of a single magnetized wire between the two stations, and an iron diaphragm at each end. The vibrations of one diaphragm affect the magnetism in the wire causing similar vibrations at the other end. This system is not affected by induction from parallel lines carrying currents.

DETERMINATION OF MANGANESE IN IRON ORES.

C. D. WILKINSON, '95.

In the analysis of manganiferous iron ores it is difficult to find methods that are quick and accurate, and still adapted to the ordinary laboratory.

The so called "acetate method" was formerly most used.

It depends on the precipitation of all the materials except Mn from a neutral solution by sodic acetate. If Co, Ni, or Cu are present, they must be precipitated by H_2S.

The method is as follows:

The ore is dissolved in HNO_3, evaporated and redissolved in HCl The solution is then neutralized by Na_2CO_3. Sodic acetate is then added and the solution diluted to about 700 c. c. with boiling water is boiled for ten minutes.

After settling, the clear solution is decanted and the precipitate thrown into a filter and washed, dissolved in HCl and after being reprecipitated, is washed again. This is repeated three times and the filtrate saved.

The filtrate is then concentrated to about 300 c c., and the MnO_2 is precipitated with Bromine water.

After heating to expel the excess of Br the precipitate is filtered out, washed and dissolved in SO_2 water.

The excess of SO_2 is then driven off by boiling and 5 to 10 c. c. of a solution of microcosmic salt is added.

The solution is heated and NH_4OH added, drop by drop, until a precipitate begins to form. It must be constantly stirred so as to obtain a silky, crystallic precipitate.

Add a slight excess of ammonia and cool the solution.

Filter, wash, ignite and weigh the precipitate as $Mn_2P_2O_7$ containing 38.74% Mn.

This method is accurate if great care is taken, first, in the neutralization by Na_2CO_3 and precipitation by sodic acetate, and second, in the precipitation with microcosmic salt and NH_4OH. It is, however, very evident that the operation is long and tedious.

FORD'S METHOD.

Ford's method is based on the action of $KClO_3$ in presence of HNO_3.

The solution is made in HCl evaporated, diluted and the silica filtered out. After evaporating again, 100 c. c. of HNO_3 and 5 gms. $KClO_3$, are added.

After boiling for 15 minutes, 50 c. c. of HNO_3 and 5 gms. more of $KClO_3$ are added, and the solution heated until yellowish fumes, from the decomposition of $KClO_3$, are no longer given off.

The solution is cooled as rapidly as possible, and filtered with an asbestos filter. Pump and wash two or three times with HNO_3 free from nitrous fumes.

The precipitate and asbestos are transferred to the beaker in which the precipitation was made and 10 to 40 c. c. of So_2 water poured in.

As soon as the precipitate is dissolved 2 to 5 c. c. of HCl are added and the solution filtered from the abestos.

The solution is heated to drive off the excess of SO_2 and the MnO_2 precipitated with bromine water.

Boil off the excess of Br, add ammonia to alkaline reaction, boil and filter.

The ferric hydrate is then dissolved with hot dil. HCl, and the filtrate allowed to run into the beaker in which the precipitation was made.

If Cu, Co, or Ni are present they will now be precipitated with H_2S gas.

The Mn is then precipitated from the solution as MnO by microcosmic salt as in the "acetate method."

This method, though shorter than the acetate method, is yet long and has the same troublesome precipitation with microcosmic salt.

BARIUM CARBONATE METHOD.

A commonly used method is one in which the precipitation is made with $BaCo_3$.

Five gms. of the ore are dissolved in 40 c. c. con. HCl and 2 c. c. HNO_3.

The solution is evaporated, diluted, and the silica filtered out and the filtrate made up to 250 c. c.

Sodium carbonate is added to the filtrate from a burette until the acid is neutralized.

Then milk of barium carbonate is added to precipitate the Fe and Al as hydrates. After this has stood two or three hours the hydrates of Fe and Al are filtered out.

After HCl is added to the filtrate, it is boiled. Then the Ba is precipitated from the hot solution with hot $H_2 SO_4$ (dil.) This is allowed to stand 24 hours and is then filtered.

NH_4 OH is added to the filtrate in a flask, to alkaline reaction. Then $(NH_4)_2$ S is added to precipitate the Mn as MnS. The flask is corked and allowed to stand 24 hours.

Filter and wash with water containing $(NH_4)_2$ S. The precipitate is then dissolved in HCl and heated to boiling and $Na_2 CO_3$ added to alkaline reaction, giving a precipitate of $Mn CO_3$.

Filter, wash, ignite and weigh in a porcelain crucible as $Mn_3 O_4$ and calculate the MnO_2.

This method is quite accurate but two of the precipitations require 24 hours, so it necessarily takes some time. It could be somewhat shortened by weighing as the sulphide. If a number of determinations are to be carried on at the same time, and a quick analysis is not required, this does not take much actual work and is hence a good method.

VOLUMETRIC METHODS.

1. Ferrous Sulphate Method.

This method depends upon the oxidizing power of the $Mn O_2$ in the ore.

Weigh out two portions of ferrous sulphate or ammonia-ferrous sulphate, one of two gms. and the other of 4 to 8 gms.

Put second amount in a flask with one gm. of ore, and run in CO_2 gas. Then add about 30 c. c. of water and 10 c. c. $H_2 SO_4$, and continue the stream of CO_2 to exclude the air.

Now while the ore is being acted on by the acid, dissolve the 2 gm. portion of $Fe So_4$ and titrate with a standard solution of potassium permanganate.

Then cool the other solution and make up to 300 to 400 c. c. and titrate with the permanganate.

The difference between the two titrations is the amount of Fe SO_4 oxidized by the MnO_2.

This is a very easy, accurate and rapid method, when all the Mn exists as MnO_2.

If the potassium permanganate has been standardized with Fe SO_4 and the Zn used in reducing was not pure, it is well to put into each solution three gms of Zn, dissolved in H_2SO_4.

2. Oxalic Acid Method.

This is identical with the ferrous sulphate method except the use of oxalic acid instead of the ferrous sulphate. The acid, in this case evolves CO_2, hence that apparatus may be dispensed with.

3. Peroxide of Hydrogen Method.

In this method the ore is put in a flask arranged to save and measure the gas evolved.

To a gram of the ore 20 c. c. of peroxide of hydrogen is added, and the gas evolved is measured, corrections having been made for temperature and pressure. The gas is oxygen, half from the MnO_2 and half from H_2O_2.

The method is not very satisfactory if ordinary methods for collecting the gas are used; but if a volumeter can be used it is very accurate.

4. Volhard's Method.

The solution is made in 2 c. c. of HCl, 4 c. c. of HNO_3 and 6 c. c. of dil. H_2SO_4 and boiled until fumes of sulphuric anhydride appear. The solution is then neutralized with ZnO, precipitating the iron. It is then titrated with a potassium permanganate solution. The titration is complete when a faint pink shows on the edge of the solution. To obtain the value for the Mn multiply the value of the permanganate for iron by 0.2946.

The reaction is $3MnO + Mn_2O_7 = 5MnO_2$.

Colorimetric Method.

One gram of material is dissolved in HNO_3 and boiled.

Then 0.4 gram of pure peroxide of lead is added. First add a little of the lead and when the violent ebullition has ceased the rest is added and boiled for exactly two and a half minutes longer, then removed from heat, and cooled in water away from light, and allowed to settle for an hour. The clear solution is then poured off into a graduated tube to match, by dilution, a standard solution containing .0001 grams of Mn as a permanganate in each c. c. of solution.

The method is accurate to .02 per cent Mn for material with .15 to 1.5 per cent. Mn.

Great precautions have to be taken, however, against *any* impurities.

Crookes in his "Select Methods," says that on account of the fact that manganese is used for the oxygen it gives up, a method for the determination of the "available oxygen" is the best for use in a technical laboratory.

References to the methods may be found as follows,—

ACETATE METHOD.

Blair's "Chemical Analysis of Iron," pp. 103-8.
Thorpe's "Dict. of Chem." under Manganese.

FERROUS SULPHATE METHOD.

Pages 109-10 of Blair.
" 130-131, Hartley's "Quantitive Analysis.

FORD'S METHOD.

"Transactions of American Institute of Mining Engineers," Vol. IX, p. 397.
Furmans "Manual of Practical Assaying," p. 194.
Blair's "Chemical Analysis of Iron, p. 109."

COLORIMETRIC METHOD.

Crooker Select Methods—Trans. Am. Inst. of Mining Engs., Vol. XV, p. 104.
Furmans "Manual of Practical Assaying," p. 200.

VOLHARD'S METHOD.

Furmans "Manual," p. 199.

In the analysis of an ore containing manganese, if the manganese itself is desired, Fords method would probably be the best to use.

But if the oxidizing power of the MnO_2 is desired, some one of the methods for the determination of the "available oxygen" would be the best

In running an ore for a complete determination the Ba CO$_3$ method is useful.

The following extract from the "School of Mines Quarterly: Nov. '93" published by the Columbia School of Mines, has a bearing on several of the processes.

"Carnot—Compt. Rend. CXVI, 1375

"Ignition of Mn. oxide when perfectly pure yields Mn$_3$O$_4$, but the presence of even minute quantities of other substances alters the result. Evaporation of a Mn solution in HNO$_3$ does not afford all of the Mn as MnO$_2$ unless it is repeated two or three times.

The KClO$_3$ precipitation in Con. HNO$_3$ requires repetition also in order to obtain pure MnO$_2$. The most ready method for obtaining a higher Mn oxide of a constant composition, consists in adding H$_2$O$_2$ to the Mn sol. and then ammonia in excess and boiling. The precipitate is Mn$_6$O$_{11}$. By determining the "available oxygen" in this precipitate the Mn may be determined.

Resolution and reprecipitation of the Mn$_6$O$_{11}$ is necessary if Cu, Zn, Ni or Co are present. Co requires five precipitations, the others three. MnO$_2$ if precipitated by Br in ammonia solution requires prolonged washing.

In comparing these methods I have made analyses by Ford's method, Ba Co$_3$ method, Ferrous Sulphate method and Hydrogen Peroxide method. The results are as follows:

Ba CO$_3$	6.81
Ford's Method	6.23—Best.
Ferrous Sulphate	6.59

This last result gave very much trouble. As it stands it is probably a better result than the other two, but I do not think the method is to be depended on.

The hydrogen peroxide gave very poor results, as the amount of gas evolved was small, and the utmost care must be taken, or in heating, heated air will be driven over instead of gas.

Probably for highly manganiferous compounds these volumetric methods may be very good, but for an ore where there is liable to be some other oxidizing agent, they are not very useful.

Results with the oxalic acid method in all cases ran below 6 per cent.

On the whole I think the methods very unsatisfactory, the older and more reliable methods being very slow, while the new ones are not accurate.

THE ELECTRIC TELEPHONE.

MORGAN BROOKS, of the Elect. Eng. and Supply Co., Minneapolis.

A Lecture before the Society of Engineers.

Now that the fundamental patents covering the art of telephony have expired we may look for a greatly increased use of the telephone. The most important patent, that dated January 30th, 1877, has recently expired. This patent of Alex. Graham Bell covers the receiver in general use throughout the country, and it also covers a very good form of transmitter. A very satisfactory conversation can be carried on by means of a pair of such magnetic telephones alone, one at each end of a short line.

As this is the simplest form of telephone, it may be well to describe what actually takes place. The sound vibrations of the voice acting upon the diaphragm of the sending telephone cause it to vibrate precisely, as does ones ear drum. The diaphragm, being made of sheet iron, has a strong inductive reaction upon the permanent magnet close behind it, and produces a vibratory current of electricity in the helix of copper wire at the end of the telephone magnet. This is according to the law discovered by the great physicist Faraday. The currents produced by the diaphragm by induction are believed to be similar in form to the sound waves produced by the voice, so that they truly represent the sounds spoken against the diaphragm. Being carried over the line wire to a precisely similar telephone at the far end of the line, this current produces a corresponding reaction on the receiving telephone magnet which attracts and repels the diaphragm of that telephone, reproducing not only tones of the voice, but also spoken words. The reproduction is beautifully clear, but it seems incredible that the induced currents caused by the vibration of the diaphragm should have sufficient power to transmit speech without the use of any battery whatever.

The rapidity of transmission is that of electricity, which is practically instantaneous. Were the *sound* vibrations transmitted through the wire as in the mechanical or string telephone, it is easy to see that conversation could not be carried on, even over a comparatively short line. Since sound travels at the maximum rate of 10,000 feet per second through copper wire, it is easy to calculate that it would take nine minutes for a single syllable to be transmitted from New York to Chicago.

On long circuits, conversation by means of magneto telephones alone, is too faint for commercial use. In order to overcome this difficulty, many forms of microphone transmitter have been invented. The microphone transmitter is essentially a different instrument from the magneto transmitter, since it depends for its power, upon the variation of current, obtained from a battery. This variation of current is produced by the vibration of the microphone diaphragm, which causes a variation, either in the pressure upon or in the surface of contact of a poor conductor, such as carbon included in the battery circuit. The action depends upon the variation of the current thus produced, reproducing sound waves in the receiving telephone, which is generally the regular Bell magneto receiver. At first these variable resistance transmitters were placed directly in circuit with the receiving telephone, but finding the amount of battery required for long lines to be excessive, it was afterwards arranged in local circuit with the primary of a small induction coil, the secondary being connected through the receiver to the line. This is the arrangement in universal use to-day.

The Blake transmitter, which is the one in most general use, consists of a single carbon button, with a platinum pin head resting against this, and controlled directly by the vibration of the diaphragm.

The long distance transmitter is similar in principle, but instead of having a single contact point of variable resistance, has a small chamber filled with granulated carbon, giving a great many points of contract, and permitting the use of a stronger battery.

In the ordinary subscriber's telephone set, is the calling device, generally consisting of a magneto generator, which is really an alternating current dynamo of simple construction, the armature being turned by a crank. The alternations produced by this machine cause a polarized bell to ring even at great distances.

The switch connections in the magneto bell are somewhat complicated, but need not be described in detail, as they are not a fundamental part of the telephone system.

In order to make the telephone available, some form of switch board or central exchange device becomes necessary. The early exchanges used a form of switch board designed upon the same principle as the telegraph switch boards then in use. Improvements of construction were rapidly introduced, until now the largest exchanges, such as at Broad st., New York, are exceedingly intricate in their arrangements. This latest exchange is very similar to the one on exhibition at the World's Fair, and is skillfully arranged for the most rapid connecting of lines. The principle is practically the same, whether the lines are single grounded circuits or double metallic circuits, and the connecting device consists generally of a single or double cord, with plug at the end to be inserted in the sockets representing the lines to be connected. In any but the smallest exchanges it is necessary to have the sockets or "jacks" representing the several lines placed within the reach of every operator, making what is known as the multiple board. By this repetition of sockets, any operator is able to make connections with any lines in the exchange without assistance from any other operator, thereby facilitating rapid communication.

An ingenious device is arranged to notify the operator when the line is already engaged or "busy." Modern practice allows only one subscriber on a line, but in the early days, where several were connected upon one line, much interference and delay were caused when two or more wished to talk at the same time.

In the matter of line building, copper has now nearly superceded iron for telephone lines. The advantages of copper for telephones are very great, and were realized by the telephone companies earlier than by the telegraph companies. The present standard of line construction, adopted by the long distance telephone companies is very much better than the standard of even the Western Union Telegraph Co.; one result of which is shown in the case of a severe sleet storm, when the telephone lines are the last to yield.

Metallic or return circuits are required for all long lines, and should be used everywhere. When properly put up, a metallic circuit, even where subject to the disturbances of parallel electric light or street railway lines, will be perfectly quiet. This is only accomplished by a careful system of construction, whereby the times are transposed at frequent intervals.

For crossing rivers and for under-ground work in cities, cables are required, which are a decided disadvantage to telephones, as compared with over-head lines, one mile of cable, even of the best quality, having a greater retarding influence on telephone conversation than ten miles of over-head wires. Nevertheless, in nearly all large cities, telephone wires are entirely under-ground, and the service given is admirable.

The possibilities of sub-marine conversation seem to be limited to about 100 miles. While it may not be impossible that conversation will be carried on across the Atlantic, it certainly can not be done over cables similar to those now in use for the telegraph.

* * * * * * &

The American Bell Telephone Company has been so successful in holding a monopoly of the telephone interests, that it has succeeded in educating the public to the belief that Alexander Graham Bell was the original inventor of the telephone. It is not generally known that Dr. S. D. Cushman, now living at Chicago at the age of seventy five, invented, or perhaps more properly, discovered, a magneto speaking telephone in the year 1851.

At that time Mr. Cushman was engaged in building a telegraph line for the Erie & Michigan Telegraph Company, in the neighborhood of Racine, Wis. Thunder storms were frequent during the summer of 1851 when he was thus engaged, and he devised a form of lightning arrester to protect the telegraph instruments. In order to ascertain whether his lightning arrester performed its work, he placed an electro magnet in the circuit between the line wire and the ground and put a piece of sheet iron over this magnet, not quite touching it, and so arranged that if the current passed through the electro magnet, the piece of sheet iron would be attracted and held in place by a permanent magnet arranged in connection with the apparatus. This was practically a magneto telephone, although not devised for that purpose. There were two of these instruments, one at the Racine office and one at the end of the line in a swamp, some ten or fifteen miles distant. Mr. Cushman heard peculiar noises coming from his instrument at Racine which sounded like the croaking of frogs. On going out one day to the distant end of the line, signals were exchanged by tapping on the sheet iron with a lead pencil. Afterwards various sounds were transmitted, and even spoken words.

This was considered a great curiosity by Mr. Cushman, but he did not realize its value and little was done to develop its use.

It is by no means extraordinary that Mr. Cushman did not push his telephone invention, since few supposed it would have any particular value; indeed we all know that the telephone and phonograph were both classed as equally worthless toys in the early days of the Bell telephone.

That the telephone was not considered a merely useless toy by all, is shown by the following translation from M. Chas. Boursule, printed in Vol. XXIV, of L'Illustration, Paris, Aug. 26th, 1854,

"We have gone still further. By the employment of the same principle, and by means of a mechanism rather complicated, it has been possible to reach a result which at first would seem to be almost a miracle. Handwriting itself is produced at a distance, and not only handwriting, but any line or any curve; so that, being in Paris, you can draw a profile by ordinary means there, and the same profile draws itself at the same time at Frankfort. Attempts of this sort have succeeded. The apparatus has been exhibited at the London Exhibition. Some details, however, remain to be perfected. It would seem impossible to go beyond this in the region of the marvelous. Let us try nevertheless to go a few steps further. I have asked myself, for example, if the spoken word itself could not be transmitted by electricity, in a word. if what was spoken in Vienna may not be heard in Paris? The thing is practicable in this way:—

"We know that sounds are made by vibrations, and are made sensible to the ear by the same vibrations, which are reproduced by the intervening medium. But the intensity of the vibrations diminishes very rapidly with the distance; so that even with the aid of speaking tubes and trumpets, it is impossible to exceed somewhat narrow limits. Suppose that a man speaks near a movable disk, sufficiently flexible to lose none of the vibrations of the voice; that this disc alternately makes and breaks the connection with a battery; you may have at a distance another disk which will simultaneously execute the same vibrations." * * * * *

However this may be, it is certain that in a more or less distant future, speech will be transmitted by electricity."

As early as 1861, Philipp Reis, a German, made a great many experiments with his telephone, and endeavored to transmit speech, but unsuccessfully. He succeeded very well, however, in transmitting musical and vocal sounds. These interesting experiments of Reis were published in Germany and were familiar to Prof. Bell.

Between 1870 and 1874 Daniel Drawbaugh, an obscure mechanic in

Pennsylvania, made sundry experiments with telephones, using pulverized carbon as a transmitter, very similar in design to the one patented by Berliner.

It is somewhat curious that Elisha Gray, of Chicago, filed a caveat for a telephone on the same day, Feb. 14, 1876, that Bell filed his first application. Indeed it is by no means certain that Gray's caveat was not filed earlier in the day than Bell's application.

Gray in his caveat claimed articulate speech in the following words: "It is the object of my invention to transmit the tones of the human voice through a telegraphic circuit, and reproduce them at the receiving end of the line, so that actual conversations can be carried on by persons at long distances apart." *In Bell's first patent, no mention is made of articulate speech.* Furthermore, in the opinion of the United States Supreme Court delivered by Chief Justice Waite in the famous telephone cases, occurs the following:—"It is quite true that when Bell applied for his patent, he had never actually transmitted telegraphically spoken words so that they could be distinctly heard and understood at the receiving end of his line."

In view of the above mentioned facts, Prof. Bell cannot be considered the original inventor of the telephone, although he is entitled to credit for pushing it into successful use.

The American Bell Telephone Company has, with admirable management, hitherto suppressed all competition. The number of Bell instruments in use in the United States has reached the enormous number of 566,000. The exchanges in his country give employment to about 10,000 persons.

The use of the telephone however, has not nearly reached its limit, since with the removal of excessive royalty charges, it will spread rapidly in factories, hotels, office buildings and residences. Also many small towns can now afford to have a telephone exchange, since they will not have to pay the royalty hitherto demanded.

MAGNETIC DECLINATION AND INSTRUMENTS USED FOR ITS MEASUREMENT.

NOAH JOHNSON, C. E. '94.

It is well known among engineers and navigators that the deviation of the magnetic needle from true north is not only different for different localities but varies from time to time for the same place. Ever since the discovery of the existence of these changes, which dates as far back as 1635, scientists have been at work investigating and establishing as far as possible their laws. In this country the work has been carried on chiefly by the Coast and Geodetic Survey to which we are indebted for most of the data and all the systematic compilation and published results.

From a scientific point of view the result of these investigations, together with those of intensity and dip, are of great interest since they contribute towards further and more important discoveries in regard to much magnetic and electrical phenomena as yet unexplainable. From a practical side they are of importance not only to the navigator who is to a great extent dependent upon the needle for guidance, but also to the land surveyor who is frequently called upon to retrace old lines originally established by the magnetic compass.

To do the latter intelligently it is essential to know not only the present declination, but also the amount and direction of annual change so as to be able to compute what it was at the time of the original survey.

It is obvious that to determine the rate of change for a certain locality it is necessary to make two or more series of observations with intervals of time between, and even then the law of change will have to be modified each time a new series of observations is made, for the secular variation may be compared to the oscilations of a pendulum which comes to rest momentarily at its extreme positions and moves fastest midway between these points, thus the secular change makes a complete oscilation in about three centuries in this country. So, while a great deal has been done already towards the unfolding of these important and universal laws, yet owing to the comparatively brief time over which the observations extend our laws are as yet very imperfectly known and are subject to such modifications and corrections as further observations can furnish.

In selecting a point for the observation of declination it is of first importance to have it free from all local attractions such as masses of iron, water pipes, electric wires, etc. The observer must be careful that he has nothing about his person such as jack-knives, keys, etc., which will attract the needle.

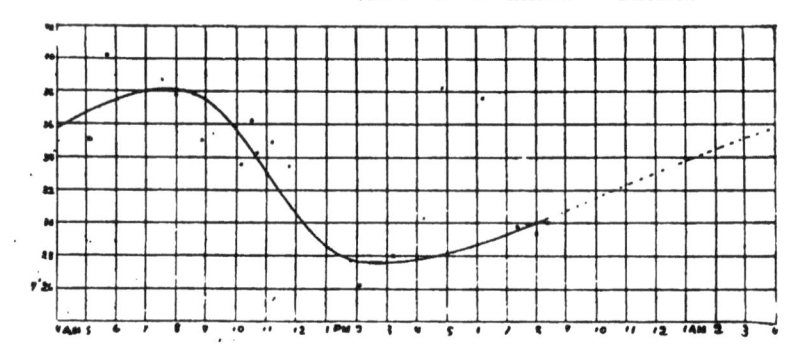

DIAGRAM SHOWING DIURNAL VARIATION OF MAGNETIC DECLINATION.

Observations taken at Woodbury June 20, 1893 with Compass Declinometer No. 744.

From the diagram it is seen that the declination does not remain the same throughout the day, but reaches a maximum eastern elongation at 8 o'clock in the morning and a minimum at about 2 o'clock in the afternoon. This curve is the result of observations made at Woodbury triangulation point June 20th, 1893, with the compass declinometer, and while this instrument is not one of extreme accuracy, yet it exhibits here the typical diurnal curve.

It is customary to observe only at the maximum and minimum periods of the day and take the average, which is generally within half a minute of the mean of the whole day.

In September, 1877, the U. S. Coast and Geodetic Survey conducted observations for declination at the University azimuth station, and it was found to be 10° 13.4′ E. In August, 1891, it was again observed and found to be 9° 5′, showing a decrease of 4.886 per year between 1888 and 1891.

In April, 1894, a third series was made. The University station being supposed to be affected by local attraction, a point was chosen on the university farm having a longitude of 3' 20'' less than the old station. The results here showed a declination of 8° 53.4'. Adding 3.3, to this to reduce it to the old station since declination increases about one minute for each minute increase of longitude in this locality, we have a declination of 8° 56.7'. This shows an annual decrease of 3.32' for this period of time. and seems to indicate that the declination is changing much slower now than during the first period. It is not altogether improbable, however, that the observations of 1891 were effected by local attraction.

For the purpose of carrying on these investigations of the magnetic declination as well as dip and intensity, different forms of instruments have been contrived, of which the best and only independent one is the magnetometer. This instrument is the one adopted by the Coast Survey and is so constructed that the index errors can be eliminated and constants determined by independent means. The magnet bar is suspended by a silk fiber and is therefore not subject to the friction and irregularities of a pivot and socket, but will continue to oscilate for a long time after being disturbed, and in fact is never found to be stationary, hence the two extreme positions are read and averaged for the center. Very close-work can be done with this instrument provided all the torsion of the fiber has been removed.

Any instrument having a magnetic needle may be used for determining declination after the index error, which all such instruments have, has been ascertained by comparisons with the magnetometer. The reliability and accuracy of the results will vary with the grade of instrument and delicacy of suspension of the needle.

In this connection an investigation was made of all the important instruments of this class belonging to the department of civil engineering, with a view of ascertaining the probable error of the settling of the needle for each, and their relative values for this kind of work. Among the instruments examined there were two which were especially designed for this class of work,—the compass declinometer, the property of the U. S. Coast and Geodetic Survey, and the Bamberg instrument imported from Germany in the fall of '93 by the university.

The following table gives the number of observations made with each, the probable error in minutes of one settling, and the value of each instrument, calling the value of the Bamberg unity.

Instrument.	Number of observations.	Probable error of one observation.	Value.
No. 1. Bamberg.	135	.45	1.0
2. Compass Declinometer.	24	2.8	.026
3. Young & Sons transit.	20	1.15	.15
4. Plane table needle.	20	.74	.4
5. Gradientor Needle.	20	1.21	.14
6. Heller and Brightly transit.	23	1.2	.14

From the way the declinometer needle behaves at times it is supposed to have some irregularity about its cavity, and is rendered almost unfit for service on this account.

In conclusion it may be said that sufficient accuracy can be easily gained in retracing magnetic lines with such instruments as number three or six of the table.

To furnish such reliable declination throughout the state is the object of the systematic magnetic survey now being prosecuted by Professor Hoag, which survey is maintained by the joint support of the U. S. Coast and Geodetic Survey and the Geological and Natural History Survey which is under the control of the Regents of the University of Minnesota.

THE MINING OF NATIVE COPPER AND ITS MANUFACTURE IN THE LAKE SUPERIOR REGION.

ARTHUR EDWIN HAYNES.
Asst. Prof. of Mathematics.

(Abstract.)

The author emphasized that he did not write as a mining engineer but as one who, through the kindly and highly appreciated courtesy of engineers, mine superintendents, captains, etc., of certain mines in the copper region of Lake Superior, had seen much to interest him in the work of mining and manufacturing copper.

Copper occurs as native copper sometimes carrying silver. It is bedded in the rock in three ways:

1. In veins filling fissures, as in the Phoenix and Cliff mines.
2. In "amygdules" or masses, as in the Atlantic and the Quincy mines.
3. In the cement uniting pieces of rock, as in the Calumet and Hecla and Tamarack mines where the rock is called "Calumet conglomerate."

The dressing of the copper bearing rock usually requires the use of a stamp mill. Exceptions are where the masses are large and can be easily freed from the attached rock in which case they are sent directly to the smelting furnaces. Masses of copper of great size are sometimes found. In the Quincy mine they have reached the size of 100 tons. Some years ago in the Minnesota mine in Ontanagon county a single mass was discovered containing 500 tons of copper. It was so large that it had to be cut with "cape chisels," a process taking nearly a year's time. Several theories have been advanced to explain the deposition of copper in the three ways named. The one preferred by the writer is that of Director Wadsworth of the Michigan Mining School. The following quotation taken from the writings of Director Wadsworth states his theory concisely: "That the copper was deposited from water, with or without electro-chemical action, is shown by the fact of its being found inclosed entirely in minerals known to be formed by water only; also by its enclosing such minerals; by its being found in disconnected or isolated masses in the lavas and elsewhere and by its greater abundance where there are to be seen the most signs of water action. Had the copper been deposited by igneous agencies, it would have had a channel or line of passage, and been continuous along that line of passage while all the different masses would have been connected together downward, unless separated by fractures or faults."

Three things are especially noticeable in this mining region: First, the magnitude of the mining operations; second, the amount of annual product, and third, the great depth the mines have reached. As an example the writer instances first the engines of Calumet and Hecla Mining Company aggregate 40,000 H.P. one engine alone, the Superior, being 4700 H.P.

The great pumping engine, the Michigan, at the Stamp mills on Torch Lake has a capacity of 60,000,000 gallons daily. The Company uses 50,000 pounds of dynamite per month and 1,000,000 tons of rock are stamped annually. This Company produced over 30,000 pounds of refined copper in the year closing April 30th, 1893, its total product to date is nearly 400,000 tons and it has paid in dividends about $40,000,000 in the last twenty-five years. The Calumet and Hecla mine has reached a depth of 3750 feet on a slope of 37½ per cent. A shaft now being sunk will reach 4500 feet, it being already over 3,000 feet in depth. The Tamarack mine has two vertical shafts of over 3,000 feet depth each.

The method of dressing the *copper bearing rock* is then discussed. Being brought to the surface it is crushed into pieces, ranging from one to four inches in diameter, and sent to the stamp mill. Here it is delivered to large bins from an elevated trestle from which it is fed automatically under the foot of the stamp. The stamp weighs about five tons, moves twenty-two feet per second and strikes one hundred times per minute. A stream of water of about three inches in diameter runs under the stamp and carries away the stamped rock and copper to sloping inclines called jiggers where by constant shaking the copper is separated by gravity from the rock. The fine copper held in suspension is finally carried to revolving slime tables where it is collected. The material *too low* in value for treatment is elevated by immense sand wheels, fifty feet in diameter, and poured into a chute and thus carried into the lake. To give an idea of the extent of the mining operations conducted by the Calumet and Hecla Mining Companies it was noted that this waste material of the two mines in operation in 301 days would cover 100 acres of ground to the depth of five feet.

The waste in the operation of extracting the copper from the rock is only four-tenths of one per cent. The stamped copper, together with the mass copper extracted from the mine, is smelted. During the processes of rabbling and poling the copper is refined. When these processes are complete the copper is ladled into moulds of various forms such as ingots or cakes in which forms it is shipped. A large amount of the copper of this region is manufactured into wire and sheet copper in the near vicinity, Dollar bay, on Portage lake. Shaft No. 1 of the Tamarack mine was begun February, 1882. June 20, 1885, at the depth of 2770 feet the Calumet lode was struck. This shaft cost $500,000 before copper was found in paying quantities. At the present time about $1,500,000 in dividends has been paid. The shaft is now over 3,000 feet deep. It has passed entirely through the Calumet lode since that lode slopes at an angle of $37\frac{1}{2}$ per cent. The lode is reached on various levels by cross cuts.

The writer emphasized the remarkable accuracy of the underground surveying of the mining engineers who have had these mines in charge. He also called attention to some instances of remarkable mechanical engineering exhibited in the mining operations.

The descent into a mine was then described. Attention was called to an ingenious device for getting into and out of a mine seen at the Quincy and called a "man-engine," a piece of machinery which can not well be described. An aneroid barometer carried in the hand during the descent and ascent would show the progress made by its rapidly moving needle. Even the ear drums, as they are compressed and released from pressure, painfully emphasize to the visitor the rapid progress of his journey. The cars run up and down this shaft at a surprising speed; only 45 seconds is sometimes used in making the descent of three-fifths of a mile, a rate only about one-third that of a stone falling down the shaft.

The temperature in the mines does not follow the law of increase sometimes seen, for it remains nearly constant at about 70° Fahr.

The timbering of mines was next described. Pine and hemlock are the two woods used. Trees from 2 to 4 feet in diameter are selected, cut to the length that will fit between the "hanging" and "foot" walls at right angles to their surfaces; thus the hanging wall is kept from crushing in and closing the mines. Timbering mines is necessary in working the conglomerate rock. The timbering of the Tamarack mine consists of rough logs while that of the Calumet and Hecla is of logs squared by sawing. An idea of the extent of timbering in the Tamarack mine was given by citing that about 5,000 logs from 18 to 27 feet in length were used per month.

Mines in the amygdaloid rock do not require timbering since the rock of the hanging wall is sufficiently tenacious to hold its weight.

Machine drills and dynamite are employed to loosen the copper bearing rock. Compressed air is the motive power for the drills and also assists in working the ventilating machinery. The diamond drill is used for exploring in vertical and horizontal directions. The core of rock obtained from this machine furnishes accurate information with regard to the material through which the drill is passing. At noon the blasts are generally fired and the concussion is so great that often lights are extinguished at long distances away. The aneroid barometer shows easily the pressure on the air caused by the heavy blasts. 20 to 50 tons of rock are often brought down by these blasts and great damage is done to the timbers by flying pieces of rock. Accidents are very rarely traced to the handling of dynamite although 30,000 pounds per month of this explosive is used in the working of one of the mines. According to the report of the mine inspector of Houghton county in 1892 over 7500 men were employed in the copper mines and less than one in a thousand were killed. In all but one of the cases on record carelessness in the use of dynamite was the cause of death.

A STUDY OF ARC LAMPS

GEO. D. SHEPARDSON, *M. E.*, *Professor of Electrical Engineering.*

The arc light furnishes a most interesting field for investigation. Considering the arc from the physical standpoint there are many problems well worth months or years of study. The same thing may be said of the action of the lamp considered as a mechanism. The theory of the arc lamp has been developed and published to some extent, but it must be said that a vast amount of information has been obtained by manufacturers, central station operators aud wiremen who keep the results of valuable experience to themselves or communicate it only to their fellow workmen. Many points in connection with the operation of lamps under different circumstances are well known in a general way, but detailed information is difficult to obtain. Much of the data desired could easily be obtained by the judicious use of an ammeter, voltmeter and watch. The scarcity of curves or tables giving accurate and detailed information about the action of arc lamps is undoubtedly due to the lack of interest or knowledge on the part of those who have the necessary facilities and the lack of opportunity with those who have the interest. A variety of investigations have been carried on during the year in the electrical laboratories of the University of Minnesota, some of which suggest fruitful fields for future theses or other special studies. In order to arouse further interest on the part of others and to present to the public some information which seems difficult to obtain from technical literature up to the present time, it has been thought desirable to present some of the results obtained from a series of experiments by two advanced special students in this department.

The curves presented are taken from experiments upon a number of arc lamps by Messrs. Scofield and Lackore. The lamps under investigation include the Thomson-Rice, Thomson ribbon feed, Waterhouse & Gamble, Wood, Kester and Ward lamps. The Ward alternating current lamp, the Brush, Clarke and Standard were not included in these experiments, since they were added at a later period. From these tests the curves of three lamps were selected as illustrating in a marked way some peculiarities of lamps under different conditions. The two differential lamps represent a large group of lamps in which the series coil carrying the total current tends to separate the carbon points while the shunt coil tends to approximate them. The shunt lamp is of a well known type that is extensively used. In this lamp the regulation is effected entirely by means of the shunt coil, the series coil being cut out of the circuit as soon as the lamp begins to operate. Each of the three lamps selected is of well known make and is extensively used. Each lamp had been previously used to a considerable extent by the students in the laboratory, and may not have been in the best possible condition or adjustment. It is safe to presume, however, that they were in at least as good condition as the average lamps upon the commercial or street circuits.

The curves in Plate I are selected from those of a number of lamps operated in series on the circuit of the Thomson-Houston arc machine which maintained the current very constant. Each lamp was adjusted so as to keep the voltage as constant as possible with a current of 6.8 amperes. In order to take readings from the different lamps as quickly as possible, the terminals of each lamp were connected to a multiple-point double pole switch so that by moving the switch from point to point the voltmeter was connected with each lamp successively. One observer would operate the switch and read the voltmeter, while the other recorded the readings. In this way the readings were taken at intervals of less than five seconds, so that the variation of voltage of each of the six lamps could be read at intervals of one-half minute. The curves in Plate I are plotted from these half minute readings on three of the lamps, the ordinates representing volts and the abscissae representing time. The curves, therefore, simply connect the half minute readings and do not strictly show what would be given by a continuously recording voltmeter. The readings are close enough together, however, to represent the actual operation of the lamp within a close degree of accuracy. The curves on Plate II are plotted from observations taken at intervals of only ten seconds so that the greater fluctuations in Plate II are partly due to the larger scale of the time ordinates. Readings on each lamp were taken at half minute intervals for thirty minutes while the current was maintained constant at 5.5 amperes.

A Study of Arc Lamps.

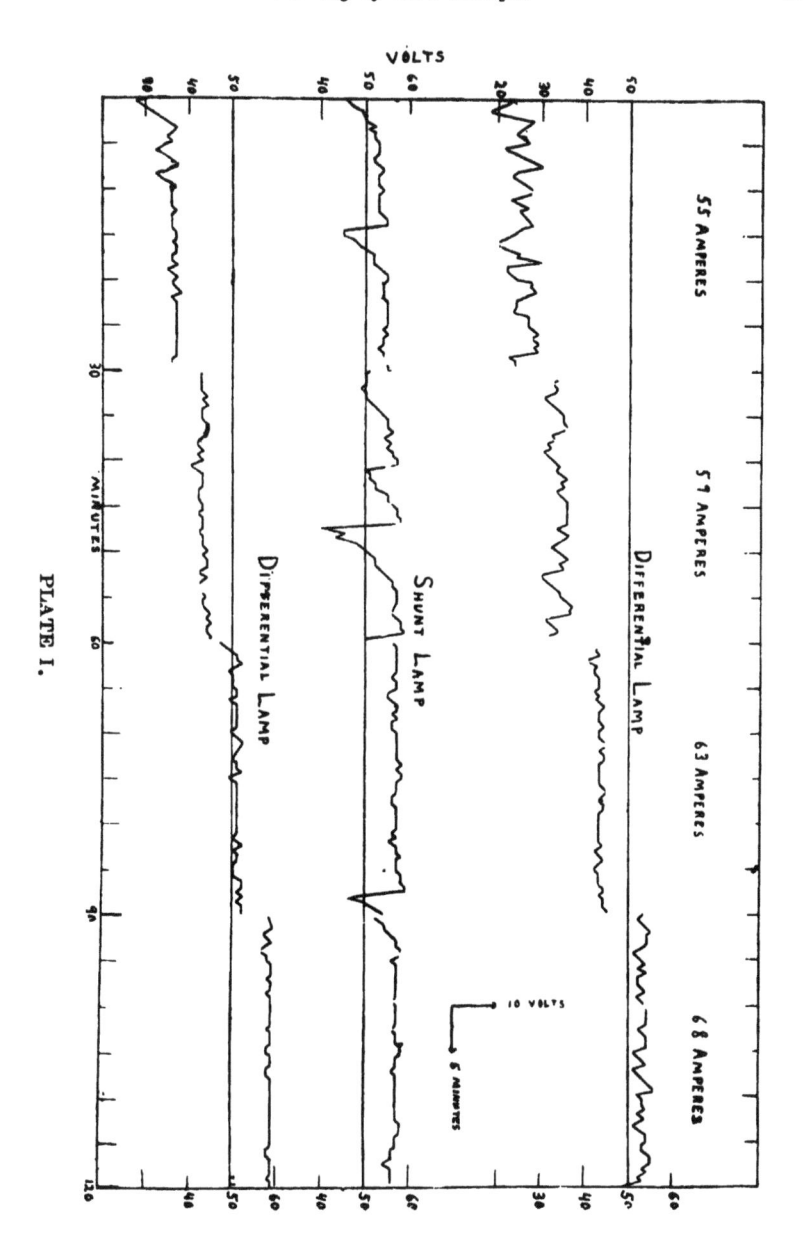

PLATE II.

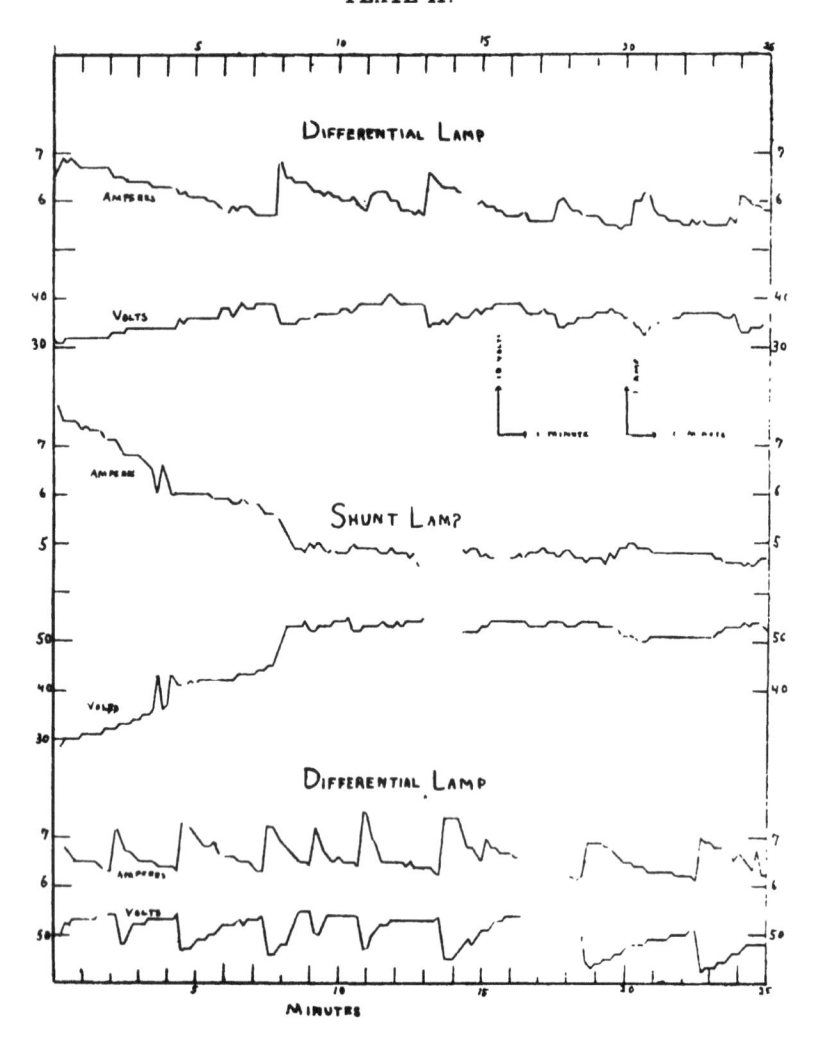

The dynamo was then regulated for 5.9 amperes and a similar set of readings taken over a period of thirty minutes. The current was next increased to 6.3 and then to 6.8 amperes. The current was afterwards increased to 7.3 amperes, but the readings were taken before the lamp coils and the carbons had become heated to their normal temperature; consequently that portion of the curves is not strictly comparable with the others and is therefore omitted. In the Plate the successive portions of the curves are left detached in order to avoid confusion.

In the curves for the differential lamps it is interesting to note how the voltage rises with each increase of the current, this being due to the fact that the series coil carrying the total current tends to separate the carbons. The increased force of the series coils and the consequent tendency to lengthen the arc is partially compensated by the resulting increase of the difference of potential around the arc which causes a greater current to flow through the shunt coil. The increased strength of the shunt coil tends to draw the carbons together, and to some extent balance the greater pull of the series coils. This compensating action is further complicated by the fact that with the carbons separated a definite distance and the arc of definite length, the resistance of the arc and the consequent difference of potential around it are inversely proportional to the area of the arc, which again increases directly with the strength of the current.

It is also interesting to note how much more closely the lamp regulates with the higher currents. This may be due to the fact that the force necessary to operate the mechanism of the lamp is comparatively constant while the operating forces, viz.: the electro-magnets, depend upon the strength of the currents in the series and shunt coils. We note, therefore, that with small currents the action of the lamp is much more irregular and the system unbalanced, while with the higher current and voltage, the feeding is more uniform.

In the case of the shunt lamp the series coil is absent and therefore does not introduce a variable factor on account of the changes in the strength of the main current. The voltage is consequently much more uniform, as shown by the curve for the shunt lamp in Plate I. The carbons used in the shunt lamp happened to be somewhat shorter than those used in the other lamps. The carbons also burned out faster on account of the higher potential of the shunt lamp during most of the run, consequently the weight of the carbons decrease in a faster ratio on the shunt lamp than on the differential. This would account for part of the gradual rise in the potential. In the shunt lamp the average voltage varies from 53.7 volts with 5.5 amperes to 54.5 volts with 6.8 amperes while in the differential lamp, shown in the lower curve, the voltage varies from 35.8 volts with 5.5 amperes to 58.7 volts with 6.8 amperes. The shunt lamp therefore is much more independent of changes in the current strength than the differential lamp. It naturally follows that when shunt lamps are used on arc circuits, close regulation of the dynamo is not so essential as with the differential lamps.

Plate II. gives the curves from simultaneous readings of the current and potential on the same three lamps on a constant potential circuit. Each lamp was connected in series with an adjustable resistance and connected across the terminals of the compound wound Edison dynamo which maintained constant potential upon the line. The lamps were taken as they had been left by previous experimenters and were not especially adjusted for the experiments. No great care was taken in determining the proper resistance necessary for the best operation of the lamp, so that too much weight must not be given to the curves in Plate II as comparing the relative advantages of shunt and differential lamps. The principal value of the curves is for showing the intimate relationship between the current and potential taken by lamps in series with a dead resistance on constant potential circuits.

The fact also appears that the curves of the amperes and volts are nearly exact counterparts of each other, the current decreasing as the potential increases and *vice versa*. This naturally follows from Ohm's Law. The readings for the upper two pairs of curves in Plate II were evidently taken as soon as the lamps were lit. It is interesting to note how the voltage gradually rises as the carbons and magnets become heated.

It is to be noted also that the fluctuations are more violent on the differential lamps than on the shunt. This is to be expected from reasoning similar to that applied to lamps on "constant current" circuits. On the other hand one would expect that the differential lamp would maintain the arc of more constant length than the shunt lamp for the reason that as the carbons burn away, the strength of the shunt coil becomes greater and the series coil weaker, both tending to make the carbons feed together again. The curves for the differential lamp show that they do work together and vigorously. It would appear therefore that for a single lamp across a constant potential circuit, a shunt lamp would regulate more closely and give more satisfactory light than the differential lamp.

It would be valuable to compare the regulation of a shunt lamp and series lamp, each in series with a dead resistance and connected to a constant potential circuit. Another interesting study would be the action of two or more lamps of the same kind in series across a constant potential circuit. In this case simultaneous readings of the potential at each lamp and the current through the group would give interesting curves. Other problems would be to determine the precise effect of varying the potential on the main line, of changing the dead resistance in series with the lamps; also of changing the adjustment of the springs or levers in the lamps themselves. These are only a few of the many problems that have suggested themselves. That the presentation and discussion of these curves may serve to arouse a deeper interest in the precise study of arc lamps and to induce others to communicate the results of their own investigations is the earnest wish of the writer.

ELECTRIC WELDING.

Harry D. Lackore.

Electric Welding has already been shown to be of great commercial value in every branch of manufacture to which welding is applicable, and is fast superseding the old method.

Wilde discovered this important application of electricity over a quarter of a century ago, it being expressly stated in his English patent, number 1412 in 1865, concerning his dynamo machine and its application to welding. He does not however seem to have put it to practical use.

To Prof. Elihu Thomson belongs the distinction of producing electric welding apparatus for commercial use. He discovered the process independently of Wilde, while lecturing before the Franklin Institute several years ago, by the accidental melting together of the large wires of an induction coil.

In the simple process of electric welding brought out by Professor Thomson, the materials to be welded are held in suitable clamps to which the terminals of the generator are attached; the surfaces to be welded are brought into contact, and a current of large volume is passed through the pieces. As the resistance of the joint is comparatively high, heating takes place here and is further increased, because the resistance of the heated metal is greater than before. The current is gradually increased until a temperature is reached at which the metal softens; the current is then shut off and the clamps are pressed together by a lever, hydraulic press or electro-magnet, to make a good metallic union; the article is withdrawn from the clamps and the burr reduced as in ordinary welding.

The principle upon which the process is based is as follows: The total amount of mechanical work done by current (C) in passing through a resistance (R) during the time (T) is $= RC^2T$ and the equivalent amount of heat (H) obtained by dividing the expression by the mechanical equivalent of heat (J) thus

$$H = \frac{RC^2 T}{J}$$

Amoug the many advantages of the electric weld is the simple mechanical process, requiring very little skill of the operator in comparison with that of the hand and eye for ordinary fire welding; the operator need only understand the proper color of the welding heat of the metal he is working, which is readily learned. The work is in full view and is not moved during the operation, except to remove and reduce the burr at the weld, this being done with the same heat as the weld.

Enormous currents are used but with so small a potential ($\frac{1}{2}$ volt) that there is no possible danger to the operator, and it is not capable of giving any sensation to a person.

There is no unnecessary waste of fuel as the heating is local and does not extend far from the weld. This has been illustrated as follows: Cotton covered wire one-fourth inch in diameter can be welded without searing the insulation more than three-quarters of an inch from the weld. Professor Thomson has passed a piece of iron through a hole in a flat piece of wood, and then raised a burr on the iron on each side of the wood without charring the wood.

The time required to make a weld varies according to size of work, from a fraction of a second to about two minutes. It is not necessary to have engine capacity equal to the maximum power required, for the time is so short that the momentum of a fly wheel will give surplus energy the same as in a drop press.

The first commercial use of welding by this process was in July, 1888, when the Thomson Electric Welding Company installed a plant for John A. Roebling's Sons Co., at Trenton, N. J.

Professor Thomson has successfully used current from continuous, alternating and unipolar machines and secondary batteries. Alternating currents may be economically produced in large volume at low E. M. F., being easily controlled and allowing of being distributed at a high pressure and small current wherever needed. They have another beneficial effect which is of importance in all welds of large cross section. With a continuous current the center will become overheated before the outside comes to a welding heat; this is due to the outer part being exposed to radiation which keeps down the temperature. On the other hand it is true that the heating of the metal increases the resistance and thus tends to equalize the temperature, but not enough in all cases. When the alternating current is used the effect of self-induction is very marked and has a tendency to concentrate the heat at the surface.

Prof. Thomson has two methods or form of welders which he calls the Direct and Indirect Systems.

In the Direct System the dynamo armature contains two windings, one of fine wire in series with the field magnet coils, the other being simply a bar of copper in the form of a letter U nearly surrounding the armature core. This bar, having a very low resistance, safely carries a current sufficient for welding purposes, the terminals being attached directly to the clamps of the welder. It is impracticable and undesirable to change the welding current suddenly or by switches. This is effected by a set of resistance coils placed in the circuit which passes around the field magnets. By putting in more or less of the resistance coils the strength of the magnets is varied and the welding current altered accordingly.

In the Indirect System which is more convenient for large work or where a number of welders are operated by one dynamo, the welding current is produced by the conversion of a comparatively small current of high tension to a large current of low tension by means of an inverted induction coil or transformer. The dynamo current is conducted through many turns of fine wire around a soft iron ring and upon this same ring is a single turn of a large copper bar in which the welding current is produced by induction.

These currents receive from 4,000 to 15,000 alternations per minute; and the amount of the secondary or welding current is controlled by varying the current in the primary coil by means of a variable resistance or choke coil.

There are two other processes which are in every day practical use. The first is that of the C. L. Coffin, who conceived the idea of placing carbon pencils on the clamps of the machine above and below the material to be welded, and making an electric arc between them. In the base of the clamps are located electro-magnets, which (when the material has reached the proper welding heat and the current cut off from the carbons) are thrown into the circuit and cause the pieces of material to be pressed together steadily. At the same time the carbons recede and when the pressure has been applied enough, the switch is moved one more point and a small anvil moves forward under the work and an electric hammer is put in motion to reduce the burr.

Mr. Coffin has several modifications of this method of heating the point of contact or heating the surfaces to be welded, by application of an electric arc formed between two converging carbon pencils. To this machine he has attached an ingenious device for directing the arc upon the work. He found by placing a magnet near the arc that the arc would be repelled, and the magnet would act virtually as a blowpipe to a flame. Applying this to the welder with converging carbons he was able to throw the arc down upon the work. He claims also that the magnet increases the electrical resistance of the joint.

Another point discovered by Mr. Coffin is that, during the operation of welding, if the pressure is stopped and a slight tension applied with care not to rupture the weld, that the electrical resistance and consequently the heating effect is increased. This force is applied only a short time and then the weld is finished in the usual manner.

In the Bernardos System of welding the positive pole of the circuit is connected to the material to be welded and the negative pole attached to a handle provided with a slot to hold a carbon pencil and having a guard covered with asbestos to protect the hand. To start the arc the workman touches the metal to be welded with the carbon pencil while the current is turned on, and then draws the pencil back producing the arc and thus heating the part to be welded.

For pipe welding a very interesting machine has been built. A vertical shaft arranged with a feather key and depressing lever carries a carbon holder at the lower end, which may be set at any desired eccentricity. This shaft is mounted on a carriage which is moved back and forth by means of a crank and connecting rod. The carbons describe a circle about $2\frac{1}{2}''$ in diameter making about 400 rev., per minute, while the carriage travels about $4''$ making 30 strokes per minute. The motion is produced by a small motor mounted on the machine and controlled by a switch.

In the operation of this machine the tube is placed on a mandrel having a fire brick block below it; the carbon is brought down to start the arc and then raised about three inches and set spinning. The end of the arc is at first confined one point, but as the metal becomes heated it soon spreads out and after a few seconds describes a larger circle than the carbon; the current used is about 250 to 300 amperes.

Pieces of steel are laid on the seam, and when working heat is reached the tube is shifted a foot or so on the mandrel, and the weld completed as usual.

To give some idea of the metals which may be welded by the electrical process it may be said that it will weld nearly ever known metal or alloy from the most refractory metal to the alloy which fuses at 162° F. It will join any two metals when the fusion point of one metal is not too far in excess of that of the other.

Below is given a partial list of the metals which may be welded:

METALS.

Various Grades of Tool Steel,
Various Grades of Mild Steel,
Steel Castings,
Chrome Steel,
Mushet Steel,
Stubbs Steel,
Crescent Steel,
Bessemer Steel,
Cast Brass,
Gun Metal,
Brass Composition,

Fuse Metal,
Type Metal,
Solder Metal,
German Silver,
Aluminum alloyed with Iron,
Aluminum Brass,
Aluminum Bronze,
Phosphor Bronze,
Silicon Bronze,
Coin Silver,
Various grades of Gold.

COMBINATIONS.

Copper to Brass,
Copper to Wr't Iron,
Copper to German Silver,
Copper to Gold,
Copper to Silver,
Brass to Wr't Iron,
Gold to German Silver,
Gold to Silver,
Gold to Platinum,
Silver to Platinum,
Wr't Iron to Mushet Steel,
Wr't Iron to Stubbs Steel,
Brass to Cast Iron,

Tin to Zinc,
Tin to Brass,
Brass to German Silver,
Brass to Tin,
Brass to Wild Steel,
Wr't Iron to Cast Iron,
Wr't Iron to Cast Steel,
Wr't Iron to Mild Steel,
Wr't Iron to Crescent Steel,
Wr't Iron to Cast Brass,
Wr't Iron to German Silver,
Wr't Iron to Nickel,
Tin to Lead.

The process is coming into use in all departments of manufacture for welding, from the smallest wire used for electrical conductors, to large steel rails. It is also used for forging, bending, tempering, etc. The hardest steel is successfully worked, as in welding band saws and broken teeth to saws.

The U. S. Government finds this process especially profitable to use in the Army and Navy, in making projectiles which were originally bored out of solid steel. They now make them by welding a steel tube of the proper thickness to the head and then to the butt of the shell, thus accomplishing in a few moments what formerly required hours of costly labor.

In the new wire wound guns of the Army and Navy, made by winding wire over a steel tube and shrinking rings over the whole, the welding of the wire is done by electricity. The wire is $\frac{1}{10}$" square with breaking limit of 180,000 pounds, which gave a good idea of the strain an electric weld will stand.

The process is also of great use on ship-board where it can be used for repairing broken parts without altering either length or shape, such as piston rods, valve stems, etc., for heating, tempering, upsetting, pipe bending, etc.

It has been found possible to weld wire cables up to $1\frac{1}{4}$' in diameter, the weld being of peculiar construction and distinguished by high tensile. strength, great flexibility and admirable wearing qualities. The cable was made of hard drawn, low carbon steel of great uniformity. The welding of the wires separately is accomplished by shrinking a sleeve over the end about $\frac{1}{16}$" from the end and then cutting circular grooves between the layers of wire, (of which there were four layers and a center wire). This allows for the upsetting of the ends of the separate wires without welding them into a solid mass, which would not allow of the strain being evenly distributed when the cable joint ran over a sheave.

In testing, the unannealed cable gave a maximum tensile strength of 67,-700 lbs. per sq. inch and the annealed fell to 50,500 lbs., the test of the annealed was made that the cable might be in the same state as at the weld.

The tensile strength of welded cable gave most perfect results, averaging over 50,000 lbs. some going as high as 53,000 and 54,000 lbs. which is over 88% of the maximum load of the unannealed cable.

One might go on almost indefinitely with regards to many special uses to which electrical welding may be applied, but we will close with a few notes from practical work in order to give an idea of the time, power expended and cost of electric welding.

(1.) Round bar tool steel $\frac{7}{8}''$ dia., 11'' long (between clamps) was raised to forging heat in one minute by expenditure of 32 H. P.

(2. Flat bar machine steel $\frac{1}{4}''$x1'' and 17'' long was heated for forging in one minute by 34 H P.

(3.) Balling stock balls, tool steel $\frac{1}{4}''$ dia , 5'' long, heated in one half minute by $27\frac{1}{2}$ H. P.

(4.) A bar of machine steel 1'' square and 12'' long, brought to white heat in two and one-half minutes by 36 H. P.

These were all evenly heated the entire length. Taking the last case as an example and 2 pounds of coal used per hour per H. P. consumed, wehave seventy-two pounds of coal used per hour for 36 H. P. which is a consumption of about 3 pounds of coal for $2\frac{1}{2}$ minutes.

In order to compare the cost of electrical and best methods of fire welding, below is given a week's work taken at random by Messr's Clark, Chapman & Co., Gateshead on Tyne, England. This firm makes electric and other machines and uses one electric welder and also does the same work by fire method.

The figures show very well the profit in using the electric weld. The firm states that the weld is superior to the fire weld, it decreases the number of smiths and they are able to weld in forms unattainable before.

The following figures are given for the week ending March 18th, 1891:

Cost of operating Electric Welding plant.	£	s.	d.	Cost of manufacture of the same product by present method.	£	s.	d.
1 Smith 53h.@ 33sh.	1	13	0	180 Winch stags @ 2d.	1	10	0
1 Lad " @ 14sh.		14	0	80 Connecting rods @ $4\frac{1}{2}$d.	1	10	0
1 Engineer 53h. @ 23sh.	1	3	0	41 Brake straps @ 8d.	1	7	4
Cleaning Boiler		1	6	212 Winch handles @ $2\frac{1}{2}$d.	1	19	9
Coal 36 cwts		17	0	168 Eccentric rods @ 2d.	1	8	0
Wear and tear and deprecation	12	0		168 Piston rods @ 2d.	1	8	0
Interest on £1000		19	0	Total	9	3	1
Glaze wheel gringing			9	Deduct Expenses of El.weld.	6	1	0
Total	6	1	0	Saving	3	2	1

Deducting expense of Electrical welding from that of the fire welding we see there is a saving of £3 2sh. 1d. or over 34%.

THE REGULATION OF TURBINES.

WM. M. THDERQUIST, '95.

Owing to the facility with which power can be transmitted by electricity, more attention has been given, by engineers and manufacturers, during recent years, to the utilization of water power. Many valuable waterfalls all over the country have been allowed to run to waste, not because the cost of building the necessary dams and constructing the power plant have been too great compared to the power derived, but because the nature of the locations have been very unfavorable to the establishment of any factories of any kind near the falls, and because no suitable method of transmitting the power to any greater distance has been known. Now since electricity solves this question of transmission as well as distribution of power, we can readily utilize almost any water fall and with it produce heat, light or motion several miles away.

According to the census of 1880, not more than five per cent of the total available water power in U. S. was then utilzed. To-day about 700,000 wheels are used, out of which 68,000 are scattered among small country places, while the remaining 2,000 are located at the larger power plants such as are in operation at Niagara, Holyoke, Lowell, Minneapolis and Spokane.

The turbine water wheel which up to this time has been used almost exclusively at all water power plants has now reached such a state of perfection that very much improvement, which will increase the efficiency, can not be made. The average effciency is about eighty or eighty-five per cent., though in special cases as high as ninety per cent. has been reached. Most of the energy lost is due to friction of the water flowing through the pipes, and through the wheel itself, to axle friction, and to residual energy of the water when leaving the wheel, and as these losses often amount to ten per cent. or more, the efficiency can not be increased very much by any improved design of the wheel alone. But there is a great need of improvement in the methods of regulating the speed. As long as the load remains nearly constant or fluctuates very slowly, the present methods of regulation may do fairly well, but for electrical work where the load may vary several per cent. almost instantaneously, or at least within a few seconds, the present methods are found very defective and the turbine gives unsatisfactory work.

The methods to be used in any particular case will depend on the nature of the wheel; whether it be a reaction or an impulse turbine. In reaction turbines the construction is such that the space between any two consecutive vanes must be completely filled with water and the flow of water must be continuous if the best results shall be gained; while in an impulse turbine the shutes are not completely filled, air being allowed to fill the remaining space. In both classes the flow may be either in a radial or axial direction, i. e., flow parallel to the radius either inward or outward, or parallel to the axis of rotation. In reaction turbines it may perform work by going both in a radial and in an axial direction, and they are then called mixed flow turbines. Most American turbines are of this class. In the reaction turbine the water after striking the vane must follow the vane, while in an impulse turbine it may take its natural course. All reaction and most impulse turbines have guide vanes to direct the course of the water before entering the wheel. These vanes are set so that in impulse turbines the water shall strike the wheel in a tangential direction, or nearly so, while in reaction turbines the water strikes the wheels at a considerably larger angle.

All turbines now made belong to these two classes, hence, in dealing with methods of regulation we have but two classes to consider.

The reaction turbines work to best advantage only when all the passages or buckets are full, and in any method of regulation where the wheel does not work full, the efficiency is reduced, and the method is not a perfect one. A very common method is to vary the quantity admitted to the wheel. This can be done in several ways. Let us first consider the method of cutting off the water supply by a head-race gate or sluice. Suppose that the opening to the flume carrying the water to the wheel is closed by the sluice sufficiently to admit only half the water with which the wheel works at its full power. The velocity of the water through the wheel will remain the same, but the quantity is only one-half and the work done can be only one-half of that at full power, and hence the speed of the wheel is only one-half. To get the maximum efficiency of a wheel, its speed should be about half the speed which it will obtain when revolving under no load. But in ordinary practice the speed should be constant. Then in this case the wheel will have to revolve with the speed it obtains under no load, and its work will be zero, and, consequently, the efficiency is equal to zero. Though this case may be an extreme one, it nevertheless shows that with this method of regulation the efficiency varies between wide limits and the method is consequently very defective.

A similar method is to put the sluice in the tail-race instead of in the head-race. This is mostly done when suction tubes are used. A suction tube, or what is also known as draft tube, is an air tight tube connected below the turbine and reaching below the surface of the water in the tail-race. The water enters this tube on leaving the turbine and as it falls it tends to produce a partial vacuum and thus the pressure against which the water issues from the wheel is decreased and its velocity is increased by a proportional amount. By using this tube the turbine can be placed within a certain limit (in practice about 28 ft.) above the tail-race and give the same power as if placed at the lowest point in the tail-race. The use of such a tube often makes the access to the wheel more convenient. Suppose, now, that the sluice placed into the lower end of such a suction tube be closed until only half the quantity of water used at full power be allowed to pass through the wheel. Since the flume is full of water and the cross section of the wheel the same as before, but the quantity of water reduced to one-half, the velocity of the water must also be reduced to one half of that at full power. But since the work done is proportional to the square of the velocity, the work will be equal to one-fourth of that at full power, and this method is worse than the previous one.

Beside this rapid fall of efficiency the sluice in this latter case must be moved through a great height in order to reduce the velocity of flow any appreciable amount, as illustrated by the following figures taken from experiments made at Mulhouse, England;

Open height of gate	Cu. ft. per min.	Efficiency.
.426 ft.	296.1	71.8%
.176 "	271.7	63.1 "
.095 "	253.9	34.9 "

The decrease in velocity of flow is comparatively small compared to the distance the gate has been moved, and as the pressure on the gate is quite large for high heads, two much time and work is wasted.

A throttle valve may be used instead of the sluice, and placed either above or below the wheel, but the results will be the same as with the sluice.

Another method is to partially close the guide passages. This violates the theory on which the wheel was designed viz:, that it shall be full of water, and there is consequently a fall in efficiency, but it is not as rapid as in the previous methods. This method is therefore more generally used for reaction turbines. The mechanical devices for effecting this regulation are numerous, but among the best may be mentioned: the use of a separate slide for each guide passage, movable guide vanes, and a circular sluice or ring placed between the guide vanes and the wheel, and which can be turned around the axis of the wheel as a center, thereby partially closing the guide passages.

All the above methods are objectionable because they reduce the quantity of water, while the area of the cross section of the wheel remains the same, and the wheel does not work under the best conditions. A more perfect method then, would be to reduce the quantity of water by decreasing the areas of the guide passages and of the cross section of the wheel similtaneously. This would practically amount to the same as if it were a smaller wheel. With the head of water and the velocity of the water the same, the efficiency would remain constant, or as nearly so, as it is possible to get it in practice. This method has been tried in different ways but has been found impracticable on account of too complicated mechanical construction. Something similar has been done by constructing the wheel and guide apparatus with partitions so that each division acts as a seperate turbine. Any one division can then be closed or opened by slides or sluices without effecting the remainder of the wheel, and the efficiency will remain very nearly constant. This is one of the best methods in use.

While reaction turbines have an efficiency slightly higher than impulse turbines, the latter are much more easily regulated. The use of the reaction turbines must necessarily be limited to larger falls as the diameter of the wheel must be made to suit the quantity of water available. Not so with the impulse turbines. They can be constructed with any diameter no matter what the volume of water may be, and they are therefore more suitable for high falls with small volume than the former. The simplest and most practicable method of regulatng them is to vary the number of guide passages or nozzles, and this is about the only method used.

The efficiency remains the same, as the quantity of water is changed while the velocity remains constant, and the condition, that the wheel passages shall not be full of water, is not violated. This regulation can be made with the same devices as used for reaction turbines regulated by partially closing the guide passages.

In all these methods where regulation depends on changing the quantity of water passing through the wheel another factor comes in, which, though it may seem small, is of quite great importance where the load is changing rapidly and the regulation must be rapid. When the quantity of water passing through the wheel is reduced the velocity of the column of water above the wheel must also necessarily be reduced, and this surplus kinetic energy is used up in increasing the pressure and consequently the velocity through the parts of the wheel yet remaining open. When again the quantity of water is increased gravity must overcome the inertia of the water and increase its velocity. All this takes time while the change of the load may be almost instantaneous. This can be remedied only by constructing a gate in the flume at the same height with the wheel and as the wheel is partially closed, this gate is opened allowing the surplus water to take this course and thus the column of water retains its velocity. But this water passing through the gate is of course wasted.

A method of regulation which has been introduced into electric plants is to keep the turbine running at full power, but to keep the load constant by cutting in and out resistance in the circuit. This has been done automatically by an ordinary ball governor.

For automatic regulation of the gates in all these methods ordinary ball governors are mostly used. The governor is connected to an intermediate machine, run by the turbine, and is used simply to put this machine in and out of gear and to reverse it. This machine is then connected to the gates, or regulating apparatus.

A machine has recently been put upon the market in which the governor is used to work a switch in an electric circuit. In the circuit are two electro-magnets which when magnetized attract two counter-balanced pawls. The pawls are fastened to the end of a lever arm, which receives motion sote, dead oil or other wood preserving compounds. This pavement is not in great favor in this country; Minneapolis has not used it at all, but St. Paul has some of it and is not satisfied with it.

This class of pavement is used a great deal in Europe, particularly London and Paris, and according to late reports, some of the most luxurious pavements have been put down in this manner in Paris, viz.:

Foundation, by rolling and compression of ground, concrete foundation proper, upon which is laid sawed blocks of selected sapless wood with joints filled with composition, as the wood is imported, selected and chemically treated, the cost of such a pavement is about four dollars per yard, and the covering is supposed to be replaced after four to six years service.

Similar pavement has been put down in Chicago around the Grand Central Passenger Station, in the following manner:

Eight inches of broken stone foundation, in this was bedded two inch by eight inch stringers, on these laid three inch surfaced white pine plank, and on the latter, the blocks made of two inch by five inch by five inch were laid, every fourth row nailed.

The price of this pavement was two dollars and eighty-five cents. Trials have been made in Berlin, Germany, and Columbus, Ohio, with oak blocks put down as above, but proved a failure in both places, by becoming very slippery and dry-rotting.

CEDAR BLOCK.

Cedar blocks on a sand or earth foundation is rarely laid at present, and if laid is only fit for light traffic, it has not been laid in the northwest at all. The most common method of constructing a cedar block pavement is, to use a plank foundation in the following manner: Stringers, two inch by twelve inch planks, were put down embedded in sand about every six feet, upon these stringers rest two by twelve inch plank and directly upon these the blocks.

The joints between the blocks are filled with gravel, tamped in properly, and then covered with one inch thick layer of clean gravel, generally the joints, after being rammed, are also filled with hot paving composition. A few years ago the cedar pavements were put down, as described,here in Minneapolis, but it was thought that nothing particularly was gained by the tar filling and its use was discontinued.

I am however, of the opinion that tar should be used and have so recommended.

It assists materially in a sanitary measure by making the joints water tight, and prevents the filling of the joints with foreign matter,—particularly offal from horses where they are permitted to stand hitched along the curb lines as for instance on Nicollet Avenue. It would also prevent water from getting under the blocks during heavy rain storms and float the pavement. Tarring was discontinued after 1887 and since that time damages as mentioned have occurred, but not before 1887.

Cedar pavement costs us here about eighty-five cents per yard, and bids were received March 9th, as low as seventyfive and seven-tenths cents per yard, the lowest figure probably made in the United States (of course without tar filling.) Some discussion has of late arisen, with reference to so called River cedar against Track cedar. With River cedar is understood cedar that is cut, driven and logged in common with other pine logs, and accordingly lies in the water a long time before being sawed. Track cedar is cedar that has been cut along railroads and is shipped by cars directly to the the mills; it appears that in Minneapolis the track cedar is to be preferred; any defects in the blocks are however more readily seen in a River cedar block, than in a Track cedar block.

The best method of putting down a cedar paving is to use a concrete foundation six inches thick, it adds to first cost, and would cost in Minneapolis one dollar and fifty cents per yard, but will keep up better, and when worn out only the blocks need to be replaced. The trouble with cedar is principally in the sappy ring that surrounds every block from one-quarter of an inch to one inch in thickness; this sap rots fast and wears down and makes the pavement rough after a few years wear. In a few instances this sap has been removed, for instance, in Grand Rapids, Mich., and the pavement is claimed to be a success. The extra cost per yard is about thirty cents. The sap is removed in the woods by a machine with circular knives that trim the blocks after they are sawed. Cedar is still laid a great deal, but in Cleveland and Cincinnatti none is laid, and in Kansas City and Omaha its use is being very much opposed of late.

Stone pavements can properly be divided into cobble stone and dressed stone paving. The cobble stones are simply boulders of a convenient size, placed together on a sand or earth bed, and rammed together with hand rammers.

Such pavements are extensively used in Philadelphia and also on the steep grades in Cincinnati, being there preferred by the teamsters to anything else. They make a cheap, but very rough roadway and cannot be recommended at all.

Dressed stone paving is made of various kind of stones, such as limestone sandstone and granite.

Limestone is not advisable to be used, but some very good pavements are made of sandstone, viz.: the so-called Medina sandstone, an Eastern stone very closely grained, and as I understand, a Colorado sandstone has been used a great deal in Denver and Omaha.

I am familiar with the Medina stone pavement, and find that it makes a very good pavement, extensively used in Cleveland. Granite is extensively used but compared with some of the southern granites, so called Georgia granite used in Cincinnati and Columbus, our St. Cloud granite is far superior.

The ordinary granite pavements are generally made in the following manner, the blocks are cut from four to six inches in width, six inches to seven inches in depth, and about ten inches to fourteen inches long. The top surface is rather rough. The blocks are set in sand as close as possible with joints broken. The joints are filled with sand, then the blocks are rammed with a sixty pound iron shod rammer, joints again swept full of sand and then again rammed, and the surface is then covered with small gravel.

In many cities, a six inch concrete foundation is put down first upon which the blocks rest, and the joints are filled with paving composition. In Cleveland and here in Minneapolis, it has not been found necessary to use a concrete base, due to the sandy sub-soil.

A Granite pavement on concrete costs, throughout the east, about three dollars and fifty cents to four dollars per yard, but we get our paving for this season for one dollar and fifty-seven cents without concrete of course. Granite is unpleasant to ride over and very noisy, but it will stand more wear and tear without repair than any other pavement.

The so-called Belgian block pavement consists of blocks as above described, a little smaller and with the top surface practically cut smooth.

For future use in Minneapolis, I have recommended granite blocks from three to four inches wide with top practically smooth; the blocks not this way would add about thirty cents per yard to the expense.

VITRIFIED BRICK PAVEMENTS.

At the present writing, I suppose there are fifty cities in the United States that have introduced brick pavements, but in most of these places, these pavements have not been down more than four and five years and real facts as to the merits of brick for pavements have not yet been collected.

Brick was first introduced in Charleston, West Virginia, some twenty-three years ago, and next in Bloomington, Ill. The brick first used, was not a vitrified brick however, but a brick made of fire clay. Paving bricks are manufactured a great deal in Ohio, Illinois, and of late in Iowa, and a great number of varieties are now made, viz:

The Hallwood block,
The Clayton block,
The Hayden block,
The Porter fire clay brick,
The Canton,
The Galesburg brick,
The Iowa brick etc.—

The principal features of these various kinds are—

The Hallwood, Clayton and Hayden blocks are burned especially for street paving purposes. The other kinds are burned for paving purposes certainly, but what cannot be used for this purpose, is used for building purposes, and hence as much care is not taken, as the manufactures have two chanches to get rid of the brick; first the tendency is to get as many bricks in a lot accepted for paving purposes, the price being higher, and the rejected brick are sold for other purposes. These bricks are of the ordinary size. The Hallwood and Clayton bricks are of the following sizes: nine by three by four inches deep, burned solid. The Hayden block is ten and one-half by five and one-half by five and one-half inches and is made hollow for a depth of about three inches, with walls of one inch in thickness left and is grooved on top. The claim is made that this block can be burned easier than the larger solid blocks.

In Columbus, Ohio, where more brick has been used than in any other city, the authorities have adopted for use only the following kind—Hallwood, Clayton and the Porter fire clay brick.

A first class brick pavement should be laid in the following manner: First - The ground first to be rolled, then to be covered with a six inch bed of concrete well tamped, upon this a cushion of sand from one to two inches thick, upon this selected brick, with as close joints as possible in straight rows, the brick to be either rammed or rolled, and finally the joints should be filled with distilled tar composition. The main trouble is in selecting the good from the poor brick, a very difficult task. The bricks should first, before being put down, be closely inspected, then after being laid be subjected to another overhauling.

In some cities, brick pavement after being laid is flooded with water so as to make the final inspection by judging the soft brick on account of the color due to absorption.

In my opinion, the best paving brick is manufactured in Ohio with Illinois a close second. Owing to the great distance from these markets, the freight expenses add so much to the cost of the brick pavements that it will be very questionable if brick can be adopted in Minneapolis in competition with cedar and granite. It appears, however, that the proper clay for paving brick is to be found near our city. Samples of the Coon Creek clay have been sent to Ohio and burned there with excellent results. Clay from Monticello, has also been sent to Ohio for burning and examination in general, I have not yet received the results, however, but expect them to be favorable.

The following analyses show the chemical character of Columbus, O., Galesburg, Ill., and the Coon Creek, Minnesota, clay:

	Silica.	Alumina.	Perox of Iron.	Lime.	Magnesia.	Potash & Soda.	Water.	Undet'rmin'd	Total.
*Columbus, O.	66.30	18.62	9.78	0.04	0.84	1.89	....	2.17	100
*Galesburg,	68.69	17.95	7.25	0.76	1.47	2.83	....	1.05	100
Coon Creek,	60.31	23.77	7.96	0.00	0.00	2.41	5.55	....	100

*Character due to chemical constitution:

Pure alumina will resist the highest temperature of the blast furnace, in which crystalline quartz (silica) will be only slightly effected, both being practically infusable. Alumina shrinks, warps and cracks greatly in drying, but gives plasticity and adhesiveness to the clay and strength to the product.

Silica prevents cracking and distorting, the more silica being present, the less shrinkage.

But the more silicia, the less plasticity and adhesiveness of the clay and the less strength and greater brittleness. Lime and magnesia while infusable in themselves or with alumina, fuse in the presence of an excess of silicia, as do also several other common ingredients of clay and form a vitrified brick.

It is found that potash has the most active fluxing effect on clay, after which follows soda, lime, magnesia and iron in the order named. To "vitrify" a clay should contain at least three per cent potash, or three and one-half per cent. of soda, or five per cent. of lime or magnesia, or eight per cent iron, or a compound proportion of any or all of these fluxes equal to these amounts.

It will be seen from the above that the Coon Creek clay will make proper brick. Tests on absorption from Coon Creek and Monticello bricks, give remarkably good results; only about one per cent after four days in water. A first class paving brick should not absorb over one and three-quarters per cent in ten days and salt glazed brick should not be used. To build a proper brick-making plant that can turn out seventy-five thousand brick per day, will cost at least one hundred thousand dollars and I am informed that steps are being taken to establish such a plant in case the projectors of the enterprise convince themselves that brick pavements will be introduced in Minneapolis.

The cost of an imported Hallwood brick pavement would be about two dollars and seventy cents per square yard. The cost of a Galesburg brick pavement would be about two dollars and fifteen cents per square yard.

These rates are based upon an all rail rate for freight. Basing estimates upon a fair price for Coon Creek brick, the cost of pavement made of them would be as follows:

    Hallwood size of brick....................$1.90 per square yard.
    Galesburg size of brick.................... 1.80 "   "   "

*Abstract from a paper read before the International Engineering Congress of the World's Columbian Exposition, by Daniel W. Mead, Assoc., M. Am. Soc., C. E.

Asphaltum Pavements.

A few months ago, Mr. J. W. Howard, B. L. C. E. of New York read a very interesting paper upon Natural Asphaltum and its compounds at a meeting of the St. Paul Society of Civil Engineers, and a short time ago, this paper has been prepared and printed for the use of the Professors and Students of the Reusselaer Polytechnic Institute and others at the request of the Reusselaer Society of Engineers.

As this paper is a most exhaustive one on the subject, I will refer you to it for full information on the subject and give here only some extracts of same.

The importance of Asphaltum can readily be judged by the volume it occupies in the industry of Asphalt paving viz: in seventy-two American and fifty-five European cities with an investment of at least $55,000,000 of Municipal funds, and a working capital of at least $15,000,000.

The growth has been very rapid during the last forty years in Europe and twenty years in America.

Definition of Asphaltum—Asphaltum, called generally in English Asphalt, is the Latin for the Greek word (asphaltos).

It is a natural mineral pitch and is practically unchangeable at ordinary temperatures. When pure, or nearly pure, it is dark brown or black. When struck with a hammer at ordinary temperatures it is brittle. Its fracture is dull and conchoidal. It is sectile. When rubbed or freshly broken it emits a bituminous odor. Its specific gravity in its natural state is from .96 to 1.68, due to impurities and porosity. It is not soluble in water. It is soluble in carbon disulphide. It begins to melt at 190° F. and flows at about 200° F. It is partly soluble in alcohol, turpentine, ether, naptha, and petroleum. Its hardness at 70° F. is $2\frac{1}{2}$ to three Dana Scale.

In America, the public has been led into the error of supposing that the by-product of the gas works, namely, gas-tar and coal-tar, were asphalts. This is due to the efforts of either unprincipled or ignorant persons, who have desired to sell coal tar preperations, which, while they have valuable uses, should not be sold or used where asphaltum is needed. They are the products of distillation. The volatile oils contained in them, if exposed to the air or sun, soon disappear and black powder remains. Real asphaltum retains its fixed oils and cementing qualities, hence its lasting properties in its natural and artificial compounds.

Refined Asphaltum is the product resulting from the refining by heat, at a temperature below 400° F. of crude asphaltum. The water and vegetable matter are removed. Commercially refined asphalt used in engineering, architecture, etc., still contains some earthy matter. It is not necessary to fully remove this, because in the mixture subsequently used it would be replaced when mixed with sand, limestone, dust, etc., as occasion demands.

Asphaltic cement is a combination of refined asphalt, and residuum oil of petroleum of a standard specific gravity at a given temperature. The proportions used vary with the objects for which the asphaltic cement is prepared. From fifteen to eighteen pounds of residuum oil is used to every one-hundred pounds of refined asphalt. They are mixed hot and agitated by air or mechanical means.

Asphaltic Limestone.—Asphaltic limestone is a natural compound of asphalt and limestone. Maltha, and other bituminous substances, are sometimes present. Sulphur, sand and a few other foreign substances are generally found in it. Its fracture is irregular, and its color is from brown to chocolate black and black when freshly broken. The color is darker in the specimens containing the larger amounts of asphaltum. The percentage of asphaltum permeating the limestone varies in different deposits, and in different parts of the same mine, from less than one per cent. to above twenty per cent. The grain of asphaltic limestone is very fine. Under the microscope the smallest particle is coated with asphaltum. If cut by a sharp blow of an axe, it appears grayish white along the cut, due to forcing out of the asphaltum and leaving particles of limestone exposed. If a piece contains above ten per cent of asphalt, it can be warmed and broken in the hands. A piece heated over the fire falls apart. If heated in a pan and held for an hour at a high temperature, the asphaltum is driven off and a gray powdered limestone remains. A bituminous odor is perceptible in all samples when freshly broken if they contain above four per cent. of asphaltum.

Asphaltic limestone is sometimes called "rock asphalt" and "bituminous limestone." In France while often referred to erroneously as "asphalte,", its real name is "calcaire bituminous." The German name is "Asphaltkalkstein," and for brevity "asphaltstein."

Asphaltic Sandstone.—Asphaltic sandstone is composed of sandstone containing from less than one per cent to sometimes as high as seventy per cent of asphaltum. Maltha and petroleum are often present to the injury of the asphaltic sandstone for engineering uses.

This material is called by a few people in California by the name "asphalt rock," which is misleading. It has not yet proved itself very valuable. Experiments are being constantly made to establish its value and how to use it.

Asphalt Pavements.—Asphalt pavements are of three classes, laid upon concrete or other suitable foundation, and composed of either Trinidad asphalt, sand, powdered limestone, etc., or asphaltic limestone, or asphaltic sandstone, if laid upon roadways. But if laid upon walks, floors, roofs, etc., one or more of these materials are combined in various ways. I will define them separately.

Trinidad Asphalt Pavement, the standard of this class, is laid briefly as follows:—

The street is graded, rolled with a steam roller, supplied with a cement foundation, generally five or six inches thick. In case the street is covered with a Macadam roadbed or old stone block pavement, these are often used as foundations without the hydraulic cement concrete. If Macadam or old stone blocks are used for a foundation, a thin layer, about one and one-half inches thick, of asphaltic concrete, called "binder" is first laid upon these old surfaces. The asphalt paving mixture, called the "wearing surface," is laid upon the foundation prepared as described. The wearing surface is also laid on one and one-half inches of binder spread over five inches of hydraulic cement concrete.

This wearing surface mixture is spread with hot rakes and compressed with a steam roller to the thickness desired, one, one-half, two and in certain places two and one-half inches thick, as the street and system of pavement demands. The Trinidad asphalt paving mixture is compounded as follows: Asphaltic cement is first prepared by mixing refined asphalt and residuum oil, using from fifteen to eighteen pounds of residuum oil, according to the climate, etc. The asphaltic cement, serves to cement together the sand and powdered limestone entering into the wearing surface mixture, which is as follows:—

Asphaltic cement....12 to 15 per cent. by weight.
Sand................83 to 70 per cent.
Powered Limestone.. 5 to 15 per cent.

The mixing is done in special toothed mixing apparatus, the teeth interlacing and revolving on two shafts in opposite directions. The asphaltic cement and sand are first previously heated to above 250° F., which is the temperature at which the mixture should be laid upon the street.

The proportions vary according to the kind of sand used and the climate. The proportions of the oil with Asphaltic cement, as well as the proportions of the materials in the paving mixture, vary with the climate. More asphaltic cement is used in cold climates than in warm ones. This easy adaptability to climate is one of the reasons of the success of this pavement.

Asphaltic limestone pavement is a pavement laid upon a suitable concrete or other foundation, and composed of powdered asphaltic limestone spread upon the foundation previously prepared, and compressed in position by hot iron rammers, or otherwise. The process is briefly thus: The asphaltic limestone is mined in large lumps, crushed in a suitable crusher, powdered while cold in an appropriate pulverizer.

The powder is heated to a temperature a little above 275° F., spread upon the street and compressed in position. The natural asphaltic limestone, called sometimes rock-asphalt, must contain about 10% of asphaltum, more or less, according to the climate, for this class of paving.

Asphaltic sandstone pavement is a pavement of natural asphaltic sandstone, prepared briefly, as follows: The material is mined, crushed by a crusher, if not fine enough when arriving from the mine, and then disintegrated by the injection of steam into mass, or by passing the material through a revolving steam heated jacket, or cylinder. The material, when fine, is raked upon the prepared foundation and compressed into position generally by a steam roller. Asphaltic sandstone pavements, because still in an experimental stage, have not yet definite percentages of asphaltum decided upon as final and best.

Asphaltic mastic is the three kinds. First, that which is composed of a mixture of powered asphaltic limestone, Trinidad asphalt and possibly a little residum or other oil; second, that which is composed of sand, possibly a little lime stone dust and about 22% of asphaltic cement; third, that which is composed of crushed asphaltic sandstone, enriched by the addition of either asphaltic cement or a little maltha. All Asphalt mastics are mixed or prepared hot, and generally run into molds varying in shape at the different factories.

These are called asphaltic mastic blocks and are generally about six inches thick and either round, square or hexagonal. They are shipped to the place needed and there remelted and used in making walks, floors, roofs and in many other ways.

Asphaltic concretes are composed of either crushed stone cemented together with asphaltic limestone mastic; or of crushed stone cemented together with Trinidad asphalt mastic, that is, a mixture of crushed stone, sand, limestone dust and asphaltic cement, mixed hot and compressed in layers by hand. These concrete foundations have many uses.

The principal deposits of asphalts and its compounds are as follows:

Mixed with Earthy Matter.	Compounded with Limestone.	Compounded with Sandstone.
Bagdad	Austria	California
California	California	Colorado
Colorado	Dalmatia	Cuba
Columbia	France	France
Indian Territory	Germany	Germany
Mexico	Hungary	Indian Territory
Montana	Indian Territory	Kentucky
Palestine	New Mexico	New Mexico
Peru	Michigan	Russia
Texas	Italy	Texas
Trinidad	Russia	Utah
Utah	Sicily	
Venezuela	Spain	
	Switzerland	
	Texas	
	Utah	
	Washington	

Other bitumens as maltha, etc., and impurities, as sulphur, etc., are often present, sometimes to the commercial detriment of the material. One or more mines are worked in each of the above countries, except perhaps, Columbia, Indian Territory, Montana, New Mexico, Michigan, Washington and Dalmatia.

The following table gives the chemical analysis of representative asphalts, (chemically refined:)

Place.	Trinidad.	Mexico.	Peru.	Cuba.	Columbia.	Palestine.
Carbon	85.89	80.34	88.67	81.50	88.31	80.00
Oxygen	.56	10.09	1.65	8.90	1.68	.40
Hydrogen	11.06	9.57	9.68	9.60	9.64	9.00
Nitrogen	.....	....	.....	.....	.37	10.00
Sulphur	2.49	....	.....	.....	.....	.60
Ash	.....	....	.....	.....	.....	.....
Totals	100%	100%	100%	100%	100%	100%

The physical characteristics of the three principal asphaltic substances, asphaltum, asphaltic limestone and asphaltic sandstone are best given separately.

Trinidad asphalt supplies more than 90 per cent. of the asphaltum used in the world. It should therefore be described in detail, being, as it is, a standard asphalt and a type of all. Trinidad, an island at the mouth of the Orinoco, north of South America, is the property of Great Britian. Upon this Island is found the Trinidad Asphalt Lake, often referred to as the "Pitch Lake". It is a large, flat crater or deposit of asphaltum, mixed with much earthy and a little vegetable matter and water. This deposit is the property of the British Government. It is about a mile from the Southwest corner of the island, and covers almost 116 acres.

The surface is hard, very slightly undulating, intersected by shallow cracks, covered at spots with low temporary vegetation and pools of water. The asphalt is softer as we approach the center, where it is very soft.

Crude Trinidad asphalt is very brown and very brittle at ordinary temperatures. It is easily detached from the surface of the "Asphalt Lake", with a pick. Its irregular features show particles of clay and vegetable matter in it. When warmed in the hand it can be slightly bent. When broken it emits a bituminous odor. It is very porous. The chemical and physical tests indicated above and by the definition of asphaltum are easily made.

The complete analyses of the materials contained in crude Trinidad asphalt are as follows:

Authority,	Bowen.	Richardson.
Asphalt,	39.83	38.14
Earthy matter	33.99	26.38
Vegetable matter	9.31	7.63
Water	16.87	27.85
	100%	100%

These are mechanical mixtures. The variation in water is very great. lump of crude ashpalt in a room loses its water in a few weeks. Powder crude asphalt loses its water by evaporation in a few days. Samples asphalt from the center of the "Lake" vary slightly from those near th shore.

But for commercial uses the asphalt from within the Asphalt Lake i more constant, reliable and better than that which, for a short time, cam from places outside of the "Lake", and was known as "overflow or lan pitch."

The material contained many variable impurities, and lacked cementin qualities. It is no longer imported into the United States.

The analyses of representative asphaltic limestone from four mines i shown in the following:

Composition.	Val de Travers Switzerland.	Seyssel France.	Ragusa Sicily.	
Asphaltum	10.15	8.15	8.92	14.
Calcium Carbonate	88.40	91.30	88.21	67.
Clay and Iron Oxide	.25	.15	.91	
Sand			.60	17.5
Magnesium Carbonate	.30	.10	.96	
Insoluble	.45	.10		
Loss	.45	.20	.40	1.1
Total	100%	100%	100%	

Asphaltic sandstone is found in nature sometimes dense and sometime soft. The grain varies in some mines from very fine to others where it i very course.

Small particles of clay and other substances are often present. Littl shells, of several kinds are frequently found. Its color is almost invariabl black. A piece heated upon a stove quickly falls to pieces if above 6 pe cent. of asphaltum is present. Asphaltic sandstone very often contain petroleum and maltha in large quantities. In America, where crude asphaltum is cheap, it does not pay to extract the asphaltum from the asphaltic sandstone, as is done at a few places in Europe. Just after the success of asphalt pavements in Europe and the consequent reports sent to this country, a desire arose to have smooth and noiseless pavements in American cities.

At that time the public did not know what asphalt was, and because of the expense of transporting asphaltic limestone from Europe to America, many compounds artificial and imitations were proposed. Before long that which was previously known as gas tar or coal tar, began to be sold under the name of "Asphaltum." Tar was mixed with sand and gravel, first for walks and then for roadways. These mixtures were called "asphalt walks" and "asphalt pavements."

This went so far that at the time of the Tweed ring in New York City, many streets were spread over with coal tar imitations of asphalt pavements, which were soon called "tar poultice pavements" and rapidly disintegrated.

The use of coal tar preparations for pavements of streets has almost ceased. It is still too extensively used for roofs in this country, whereas, in Europe it is not used at all, except, perhaps, in cases of small temporary sheds where a temporary substance, like coal tar is admissible.

Genuine natural asphaltum retains its oil, hence its lasting qualities in pavements, concretes, etc. A practical method to tell asphalt from coal tar is as follows: Apply to any gas works or coal tar roofer and get a piece of coal tar.

Then apply to some dealer in natural asphaltum for a sample of refined Trinidad or other asphaltum. Having two samples, one of coal tar and one of asphalt, place them side by side. The coal tar emits a sharp, acrid odor, easily perceived. The asphalt must be rubbed hard to emit a strong odor. The asphalt odor will be quickly learned. It is entirely different from the coal tar odor. The asphalt odor is known as a bituminous odor. It is rather agreeable, perhaps a little sour smelling. Crude and refined asphalt when quite cold, emits a weak, clay like odor. Next taste one sample, then the other. Then attempt to chew a little of the refined asphalt. It will crumble. If it be an "asphalt paving cement" made from genuine asphalt and residuum oil of petroleum, a fairly agreeable chewing gum is the result.

Any attempt to chew coal tar, will result in disaster to the teeth, for it is very sticky and hard to remove from the teeth.

Expert analysis of mixtures to detect any coal tar present, if present in small quantities, is exceedingly difficult, but when large amounts of tar are present in the mixture, they are quickly discovered by the sharp odor and otherwise.

Much pavement was laid in one or two cities under the name "vulcanite," which was a mixture of about 25% of asphalt and 75% of coal tar, mixed with sand, powdered limestone, etc. This material is no longer laid where engineers have full control as to what shall enter into the pavements, unless they are influenced by other reasons than technical ones.

MINING, PREPARATION AND USES.

Trinidad asphalt as a standard, comes from the island of Trinidad, and from the Trinidad asphalt lake as previously described.

The carts and horses used to haul the crude asphalt from the deposit to the shore, drive upon the "Pitch Lake."

The material, which has already been described as dark brown, brittle and similar to dense, dry peat, is picked loose, thrown into the carts, hauled to the ocean beach, loaded upon scows and lighters and thence into the steamers and sailing vessels which bring it to America or take it to Europe.

The excavations made on the surface of the deposit soon disappear by being filled up with new material from the sides and bottom.

Upon arrival at a refinery in the United States or Europe, it is either unloaded and stored for future use, or conveyed directly into large refining tanks, called "asphalt stills" which are, in brief, cylindrical retorts set in brick work and heated by indirect fire. The crude asphalt is heated somewhat above boiling water and not above 400° F., for several days. By a few improvements this refining has been reduced to half the time it used to take. This process drives off the water, certain light volatile oils and gases.

The vegetable matter floats and is disposed of. A part of the earthy matter settles to the bottom. The product is refined Trinidad asphalt and is drawn off into light barrels with one end out. This refined asphalt is sent by rail or boat from the several refineries to the many cities using it.

Trinidad asphalt pavements absorb more than ninety per cent. of the asphalt imported from Trinidad. This pavement is laid, briefly, as follows:

The street is graded, rolled with a steam roller, supplied with a suitable foundation and then covered with a layer of a mixture of sand and powdered limestone, held together by asphaltic cement. A full description of Trinidad asphalt pavement has already been given previously in this paper.

The volume and importance of this class of pavement can be judged from the fact that it is found in ninety-two American cities, covering an area of almost 14,000,000 square yards, in January 1st, 1894, equivalent to more than 911 miles of roadway laid in sixteen years by thirty contractors or companies.

Asphaltic limestone is mined either by open quarrying or by tunneling. In warm weather open quarrying is difficult because of the tenacity of the material. Tunneling is carried on sometimes by drills and wedges alone, but generally with powder. Gas is not to be feared in most asphaltic limestone mines. Water, however, often appears and must be constantly pumped out. It is mined in large lumps, crushed generally in a rotary-toothed crusher into pieces three inches and less in size.

These pieces are passed through a pulverizer, generally the centrifugal type, sometimes in cylinders revolving and containing loose cast steel balls. The resulting powder is about as fine as mustard seed. It is stored up preferably under cover, for future use, unless needed at once.

One of the chief uses of this powder is for laying compressed asphaltic limestone pavement, known generally as "rock asphalt pavement," and like other pavements into which asphaltum enters, it is called simply, "asphalt pavement."

While technically incorrect, practically it is proper to call a pavement containing asphaltum, an asphalt pavement, because though asphalt forms but from ten per cent. to fifteen per cent. of the mixture, it is the characteristic which determines the name of the pavement. The process of laying asphaltic limestone pavements is briefly as follows: The powdered material is heated to about 275° F., either upon flat hearths or in revolving heaters. It is hauled to the streets in ordinary wagons, raked to such a depth over its concrete foundation that, when compressed by heated rammers or otherwise, it will have the required thickness, generally two or two and one-half inches.

This form of asphalt pavement for streets is not popular in the United States, and appears at but few places in cities near the Atlantic Coast.

Asphaltic limestone street pavements, like all limestone, become polished and prove somewhat slippery under heavy fog or a light rain. They are also slippery in clear, cold dry weather when clean. The class of asphalt pavement known as "Trinidad asphalt pavement," and others made on this system, are not slippery at any time, except when covered with a sheet of ice. This is due to the fact that Trinidad asphalt pavement is composed of about eighty per cent. of sand, making it gritty and not susceptible of becoming polished. Paris, London and Berlin keep large supplies of sand at several points near the streets paved with asphaltic limestone, for use during time of fog, slight rain, snow and ice. This is not necessary and is not done upon streets paved with Trinidad asphalt pavements, nor with pavements composed of asphaltic sandstone in California.

Asphaltic mastic, called often simply asphalt mastic, is made as follows:

The powdered asphaltic limestone is heated in revolving retorts, refined Trinidad asphalt is added until the mass contains about twenty per cent. of bituminous substances, including the original asphaltum, etc., in the asphaltic limestone.

The melted mass is constantly stirred until run into moulds. The mastic blocks or cakes thus formed are shipped all over the world.

Paris received its first successful asphalt street pavement in 1854; London, 1869; Buda-Pesth, 1891; Dresden. 1892; Hamburg, 1872; Berlin, 1873; Brussels, Geneva, Leipzig, Frankfort and others soon afterward.

The American cities were at first deceived, having expressed a desire for smooth, quiet pavements, and having heard of the asphalt pavements of Europe. The spurious article which encumbered our streets previous to 1873 was nothing more or less than coal tar, sand and gravel.

Because of the difficulty of transporting the asphaltic limestone from Europe to America, inventors were urged to seek some suitable substitute. Successful experiments made between 1871 and 1873 in Newark, N. J., and New York City, with an artificial mixture of Trinidad asphalt, residuum oil, limestone dust and sand, gave rise to the class of pavements now so successful, and called Trinidad asphalt pavements. In 1876, this pavement appeared in Washington as a pronounced success.

In 1878 it became the standard pavement of Washington. It has been rapidly adopted by more than seventy-one American cities. It is being laid at the rate of 140 miles annually.

Statistics January 1st, 1894, of Pavements of Europe and America:—
The asphalt pavements of all kinds in the United States and Canada on this date are as follows:—

Kind of Pavement.	Sqare Yards.	Miles.
Trinidad Asphalt*	13,900,000	911
Asphaltic limestone	151,000	10
Asphaltic sandstone and other asphaltic materials, experimental and otherwise	619,000	41
Totals	14,670,000	962

The total asphalt pavements of European cities at the same date were as follows:

City.	Square Yards.	Miles.
Berlin	1,280,796	83
Paris	401,617	26
London	370,000	24
Other Cities	271,000	18
Totals	2,323,413	151

In the preceding tables the roadway is reduced to a common width of twenty-six feet for comparisons. The roadways themselves vary above and below this width.

The heaviest travelled street in the world is Reu de Rivoli in Paris, which is paved with asphalt.

From the above you will see that according to Mr. Howard, the Trinidad asphalts supply more than 90 per cent. of the asphaltum used in the world, also that the lake asphalt is more constant, reliable and better than that which came from places outside of the lake and was known as the overflow or land asphalt.

I also find that the following gentlemen are emphatically in favor of Trinidad lake asphalt: A. W. Cook, Chief Engineer, Street Department, Chicago; City Engineer Ludden of Detroit, Mich., City Engineer Kinnear of Columbus, Ohio. In these cities overflow (California and Bermudez)

*Subject to slight corrections from final 1893 reports.

asphalts have been used but with poor success. In Indianapolis, the so-called vulcanite has been used with disastrous results. Mr. Clifford Richardson of the Engineer's Department District of Columbia, and the Commissioner of Public Works, City of New York, by the advice of Stevenson Towle, consulting Engineer, have reached similar conclusions.

Mr. Howard has probably visited every deposit in the world; Mr. Towle was sent to Trinidad and every opportunity has been given the engineers in Washington and New York to examine and test asphalts so as to get the best. Considering all, it is therefore no wonder that engineers in general, will consider nothing but the Trinidad Lake asphalt for paving purposes; but after studying Mr. S. P. Peckham's paper on petroleum in its relation to asphaltic pavement, read at the World's Congress of Chemists, Chicago, August 25th, 1893, I am of the opinion that the desired perfection in asphalt pavements has not yet been obtained, and think that the following from Mr. Peckham's paper will not fail to be of great interest:

Now within a few years and since all of this work was completed, the laying of asphalt pavements has become an industry involving the expenditure of vast sums of money and consequently presenting technical problems of vast importance.

The technology proceeds about as follows: A quantity of asphaltum is brought to New York from Trinidad. This asphaltum consists of water, twenty-seven per cent.; inorganic matter, consisting of fine aluminous sand, twenty-seven per cent., organic matter insoluble in carbon di-sulphide, eight per cent.; bitumen, thirty-eight per cent. This crude asphaltum is put into large open kettles and melted, the temperature being raised and maintained for some time in the neighborhood of 400° F. By this treatment, the water and lighter oils are expelled, and a portion of the mineral matter sinking out of the melted mass to the bottom of the kettle, a "refined pitch" is left, which is drawn off. This "refined pitch," consists in round numbers, of 56 per cent. of bitumen, 36 per cent of mineral matter and 8 per cent. of organic matter not bitumen. Of the 56 per cent. of bitumen, 36 per cent. is soluble in petroleum naphtha, leaving 20 per cent. that may be obtained from solution in carbon di-sulphide in brilliant black scales. Properly interpreted the refined pitch consists of:

Petrolene,	36 per cent.
Asphaltene,	20 per cent.
Organic matter not bitumen,	8 per cent.
Mineral matter,	36 per cent.
	64 per cent. 36 per cent.

Sixty-four per cent. of mineral without cohesion; that is, dry solids for the most part, are held together by 36 per cent. of a viscous adhesive tar. The 20 per cent. of the asphaltene dissolves in the petrolene, the remaining 44 per cent. being simply mineral and organic matter held together by the combined bitumens.

The next step in the process is the addition to the refined pitch of about 150 per cent. of its weight of some heavy petroleum residuum, which renders the bitumen softer, and at the same time, may or may not dissolve both petrolene and asphaltene. The mixture is made complete by blowing air through the melted mass for hours or even days. Sand is then added to the bituminous cement until the bitumen is only equal to ten per cent. of the mass. To incorporate the sand with the viscous mass, both are heated very hot and the sand and bitumen thoroughly mixed, when the mastic thus produced is put in its place and rolled.

Very great diversity is observed in the durability of the mastic thus prepared and laid, and to account for these diversities Mr. Clifford Richardson[*] would have us believe that it depends on the excavation on the Island of Trinidad from which the crude asphalt was taken. So too the Hon. Commissioner of Public Works of the City of New York, by the advice of Stephenson Towle[†], Consulting Engineer, has reached a similar conclusion. Speaking of the eighth avenue pavement, a part of which "showed indications of disintergration" before it was accepted Mr. Towle says, "the asphalt used in this work was submitted by the contractor and analyzed and approved by chemical experts;" and yet, further on in this same report, it is stated that

---

[*] Report of the operations of the Engineering Department of the District of Columbia for the fiscal year ending June 30th. 1892, pp. 96-123.

[†] Report of the Department of Public Works of the City of New York on Street Pavements with special reference to Asphalt Pavements. Thomas F. Gilroy, Commissioner, 1892. pp. 9 and 11 et seq.

the asphalt was condemned because it was taken from one excavation on the island rather than another.

It is a very significant fact that an asphalt pavement should show signs of disintergration within a few weeks of the time of its having been laid. Especially is this the case when it is considered that some of the most enduring constructions of antiquity, which have withstood the ravages of time for at least 3000 years, are constructions of asphalt. Modern constructions of asphalt in Europe are said to remain intact for more than twenty years. The asphalt pavement laid on Franklin avenue in the City of Buffalo, has stood, with almost no expense for repairs, for fifteen years, and I do not believe any one now knows from what exact spot the asphalt came. It is a very well known fact however, that the material used to soften the asphalt was one that would dissolve both constituents of the bitumen, and form a chemical union or solution rather than a mechanical mixture with them. No one knows how old the asphaltic sandstones are that contain ten per cent. of bitumen, but they do not disintegrate; they are solid, impervious and tough.

If sand could be cemented together with carbon, it would only be necessary to grind up anthracite slack and mix it with sand in order to produce a paving material. It is evident that the hydrogen combined with the carbon, gives its adhesive properties, and this within certain limits, for the moment a sufficient amount of hydrogen is removed to convert the petrolene into asphaltene, or whenever the softening material is a petroleum or residuum that will not dissolve asphaltene, the elements of disintegration are present.

The observation made by Boussingault, more than half a century ago, and quoted above, that prolonged heating at high temperature converts petrolene into asphaltene, which has no more adhesive properties than anthracite slack, appears to have been overlooked by Messrs. Richardson and Towle, for they do not appear to regard the manner of refining and blowing the asphaltic cement as of any importance.

Neither for the same reason do they appear to appreciate the bearing of my own researches or those of Messrs. Cabot and Jenny upon this question, although they have been on record now more than twenty years. While it has been well known for years that bitumen occurs in great variety, the selection of a proper material for softening the asphalt to the exclusion of others less desirable or wholly unfit, appears also to have escaped attention. A properly selected material for such a purpose would by entering into solution and chemical union with both the constituent parts of the bitumen of the asphalt, thereby increase its adhesive and binding properties upon the other constituents of the mastic. Added to all these omitted elements of the problem just enumerated, we have another of great importance, and that is the total proportion of bitumen to the total proportion of sand and other non-bituminous ingredients of the mastic.

Experience proves that less than ten or eleven per cent. to ninety or eighty-nine per cent. gives too little stability to the mass, while a larger percentage of bitumen makes the pavement too soft.

One more element of good asphaltic pavement is mechanical rather than chemical, and that is solidity. The pavement must be rolled as solid as asphaltic rock in order to keep out rain water and the action of the oxygen dissolved in it; for the effect of oxidation is to gradually convert the petrolene of the asphalt into asphaltene, leaving the small amount of softening material as the only binding constituent of the mixture. When this softening material is not a solvent for asphaltene, the pavement inevitably disintegrates.

I would therefore commend to the consideration of those who act as advisers to Commissioners of Public Works, not a less careful examination of the asphalt used from whatever source it may be obtained; but in addition an equally careful exercise of judgment regarding the softening material used, the chemical effects of refining and blowing, with careless management of either, and the composition and mechanical structure of the surface pavement.

It is far from complimentary to this age of scientific achievement that the ancient world should have furnished constructions of asphalt that have survived to the present centuries after their builders are forgotten, while with all our superfluities, of refining and blowing, with penetrating machines and chemical analyses, of which they knew nothing, we are only able to prepare constructions that endure but a few months or even years, at longest.

I am aware that this paper is inconclusive, but from the nature and present status of the problem it must be so. I think, however, it is, as it was intended to be, suggestive.

Mr. Peckhams suggestions are really being carried out, then the various asphalt paving companies have their laboratories and are continually experimenting and it would seem rational that if U. S. asphalts could be safely used, it would take the place of the Trinidad, particularly when a royalty of $2.00 per ton is charged at the lake.

The California Petroleum and Asphalt Company, of San Francisco, has informed me that it wishes to get an opportunity to bid on any asphalt paving that we might have here.

They claim that in the preparation of their socalled Alcatraz asphalt, no residuum oil is used. This company fluxes and prepares its products with Alcatraz liquid asphalt, a pure liquid bitumin produced at its purification works. This liquid asphalt is not produced by distilation, but by simple extraction by mechanical means from the clean sea sand which it saturates as it oozes from the immense beds of shale which are its source. It is semi-liquid and analizes over ninety-eight per cent. bitumen (so the company claims).

This liquid asphaltum takes the place of the residuum oil in the preparation of the asphaltic cements of the Trinidad class. The company above referred to, claims that this asphalt has been used with great success in South Omaha, Neb.

From personal observation, I fail however, to agree with the view of the company and think that the South Omaha asphalt pavement is very poor indeed, and I certainly will favor nothing but the lake asphalt until convinced that the manipulation of other asphalts will give as good results as those obtained from the lake deposits.

The following table will illustrate the relative cost of the various kinds of pavement for a term of say twenty years, total interest at five per cent. included, upon a street with a traffic like Nicellet Avenue, the tonnage of which is about 150.

Street Paving.

Class of Paving.	Cost per sq. yd.	Cost of Renewal.	Y'ars wore.	Actual cost per year.	Value of Pavement at the end of 20 yrs. per sq. yd.	Net cost per sq yd pr yr.
Cedar, 2 in. plank foundation without tar.......	$0.85	$0.79	6	$0 29	⅔ of 79=$ .53	$0.26
Cedar concrete foundation with tar..............	1.45	.74	8	.30	½ of .74= .37 concr'te .67 1.04	.25
Granite.........	1.88	1.88	20	.25	0.60	22
*Trinidad Lake Asphalt.	2.75			.42	2.00	.32
Small size Minnesota br'k.	1.80	1.13	10	.33	.80	.29
Large " " " .	1.90	1.23	10	.35	.80	.31
†Galesburg Brick.......	2.15	1.48	10	.40½	.80	.36
†Ohio Brick............	2.70	2.03	10	.52	.80	.48

The average cost for repairs for the last ten years for cedar, has been about 1-10 cents per yard per year.

*Kept in repair for ten years, after which add eight cents per yard per year.
†All rail rate for freight.

From the above table it will be seen that as far as net cost goes various kinds of pavements range as follows:

First, granite; second, cedar; third, brick; (with exception of the Ohio or Galesburg) fourth, asphalt.

As far as regards least resistance to traction the order would be, when the pavements are new and also old:
First, asphalt; second, brick; third, cedar; fourth, granite.

With reference to sanitation, asphalt would lead, with brick or granite next and cedar last. The same order will also prevail with reference to beauty. With reference to least noise; cedar is first, asphalt is second, brick is third, and granite is fourth.

When the pavements are all new, the most pleasant to drive over is evidently; first, cedar; second, asphalt; third, brick; fourth, granite; but after, say five years, asphalt would lead, with brick second, cedar third, and granite last.

The trouble with cedar is, that after four or five years of age, the blocks are rounded off and become rough and uneven, owing to the sap wearing off on the edges of the blocks.

Easiest on vehicles is asphalt first; cedar second, brick third; granite, fourth.

Easiest on horses—First, cedar; second, brick, third, asphalt; fourth, granite.

I believe statistics show that horses will slip more on granite than asphalt. Another matter of importance to be considered is the method required in renewing the pavements. Asphalt has a constant attention and is actually renewed during the course of years, without interfering with the traffic to any noticeable extent; whereas when the proper time comes for renewing the other pavements the entire streets will have to be closed for at least a block at a time. This is a great detriment to business, particularly on such streets as Nicollet avenue.

In Berlin, for instance, it is against the law to close a street for repairs on account of interference with business.

With reference to cleanliness, asphalt would lead, with brick, granite and cedar in succession. It will therefore be seen, that quite a number of points, besides the cost, are to be considered in ordering new pavements, and that as the property owners on the respective streets carry most of the burden, they should be given ample time and opportunity to consider the matter. But the city authorities should see that only the best of the various kinds of pavement is put down, and that all sewer and water mains with their respective connections and all other piping and conduits, are laid before any paving is done at all.

GRAVEL AND MCADAM ROADS.

For suburban and residence streets a good gravel road will answer all purposes until such time as traffic would increase enough to warrant paving to be put down.

To construct a good gravel road, it is absolutely necessary to provide for drainage of the foundation itself as well as of the finished roadway and upon the condition of the soil will depend the thickness required of the gravel covering, on good dry loam six inches of gravel will do, on sandy soil, ten to twelve inches might be required.

If possible, the soil proper should be first rolled and then covered in layers with gravel, which should be wetted and rolled. A gravel road will require constant attention. Some good work in this direction has been made in Minneapolis in the Eighth ward where plenty of gravel can be had.

McADAM ROADS——The usual method of constructing a good McAdam road, is in preparing the sub-foundation properly with, if nessessary, proper drainage pipes. A layer of from six (6) to twelve (12) inches good sized stones are placed on the sub-roadway by hand and leveled up with smaller stones. Upon this foundation, a covering of at least four (4) inches of broken or crushed stone is placed and rolled several times and leveled up with screenings from the stone crushers and sprinkled and rolled. The screenings are used for a binder between the crushed rock and gives the roadway a smooth appearance.

Great care should be taken in getting the proper binding material. Screenings from crushed limestone is good. In Chicago, gravel from Joliet is used. This gravel contains certain clay that has cementing qualities. I do not think that the native limestone is the proper material for McAdam. The trap rock near Duluth is used in the city extensively. McAdam roads must be constantly watched and repaired.

The Board of Public works of the city of Detroit, has most emphatically decided not to put down any more McAdam in that City and Superintendent Carter of the Boston Street Department says, with reference to the three hundred (300) miles of McAdam out of four hundred and fifty (450) miles of paved street in that city: "It is not to be wondered at that the expense of maintaining our streets, calls for a much larger outlay of money every year than in cities provided with more permanent forms of paving. Considering the cost and temporary results obtained by repairing our McAdam roads with a veneer of crushed stone, it would seem to be in the direct line of economy to adopt a more permanent form of roadway surface."

A good McAdam road will cost about from seventy-five cents (75) to one dollar($1.00) per square yard.

It may be of interest to know that the first Cedar pavement laid in Minneapolis was in the year 1882, on Washington Ave. S. between 2nd and 3rd Av's., and that it is still in place. The first granite paving was also laid in the same year. To-day we have about--

     71.6 miles of Cedar Paving
      9.5 " " Granite "
      2.77 " " Asphalt.

Before finishing I want to make a few remarks about street cleaning. Street cleaning can properly be divided into two categories, viz:--

Machine sweeping and hand sweeping.

Machine sweeping is generally done at night by revolving sweepers, drawn by horses, which carry the sweepings into the gutter where they are gathered up by men and put into carts and hauled off. In dry weather, a sprinkling cart should sprinkle the streets ahead of the sweeper just enough to lay the dust. Very often this is not done and blocks after blocks are simply deluged with dust raised by the sweepers.

This method (machine sweeping) is generally used on cedar, granite, brick and in some places on asphalt pavements. A good two horse sweeper will cost about four hundred dollars ($400.00). As of course, an asphalt paved street presents the smoothest surface, wind will raise the accumulated dust very easily, and for instance, on Park Avenue in this city, complaints are frequent with reference to the dust from the street. To properly clean an asphalt pavement, the second method viz:—

Hand sweeping should be applied continuously. The streets to be cleaned should be divided into districts, in charge of a proper lively man to attend to it at all times.

The necessary tools are:—

1st—One street scraper, which consists of a steel blade of No. 16 thickness, about 6 inches wide by 4 feet 6 inches long fastened in the middle to a 5 foot shaft $1\frac{3}{8}$ inches round; cost $3.00.

2nd—Two brooms with handles cost; 75 cents.

3rd—One scoop shovel   " 85 cents.

4th—One hand barrow   " $2.00.

During the day, the fresh droppings should be at once gathered in by the scrapers, or broom and shovel, and at least once in two days the entire street should be swept with the broom. In wet weather the entire work is done with scrapers. In dry weather sprinkling should be done, but only in such quantities as to lay the dust and this can best be done by having suitable connections with the City Water every 200 or 250 feet. The attendant can then with a hose give the street the proper wetting. Sprinkling carts generally flood the pavements making the crossings both wet and muddy.

There are sweeping machines made on the same principal as a carpet sweeper to prevent dust, but I am in doubt as to their success. Receptacles should be provided for either in alleys or at other places on the street-line where the sweepings can be deposited until hauled off by the teams.

In Minneapolis, the cost of cleaning our paved streets, outside of the asphalt pavement, varies from 2.2 to 3.2 cents per square yard, and in addition to this about 3 cents per yard for sprinkling.

The cleaning of our asphalt paved street, Park Avenue, is in the hands of a so-called improvement association. Each foot frontage on the street is assessed 10 cents, this money is turned over to the association, which is supposed to clean the same. The cost is about 2.7 cents per square yard, but would cost, say at least, 5 cents if properly done and would include sprinkling when necessary.

F. W. CAPPELEN, City Engineer,
Minneapolis.

[A lecture delivered before the Society of Engineers of the University of Minnesota, —EDITORS.]

PIONEER ELECTRICAL JOURNAL OF AMERICA.

THE ELECTRICAL WORLD
An Illustrated Weekly Review of Current Progress in Electricity and its Practical Applications

―――Most Popular of Technical Periodicals.

THE ELECTRICAL WORLD, weekly, is the largest, most handsomely illustrated and widest circulated journal of its kind in the world.

It is ably edited and is noted for popular treatment of subjects in simple and easy language, devoid of technicalities. No other technical journal has as many general readers.

THE ELECTRICAL WORLD devotes a large part of its space to alternating and multiphased currents; subjects that no student can afford to neglect, and which no other other electrical journal in the world treats so fully, while the weekly Digest of Current Technical Electrical Literature gives a complete resume of current foreign progress in electrical science and its applications.

Subscription, including U. S., Canadian or Mexican postage,
$3.00 a year. Of Newsdealers, 10 cents a week.
SAMPLE COPIES FREE. AGENTS WANTED.

Books on Electrical Subjects.

There is no work relating to Electricity, Street Railways or kindred subjects that is not either published or for sale at the office of THE ELECTRICAL WORLD, from which is also issued annually Johnston's **Electrical and Street Railway Directory**, price $5.00.

Books promptly mailed, postage prepaid, on receipt of price.
Catalogue and Information free.

THE W. J. JOHNSON COMPANY, Limited,
253 Broadway, New York.

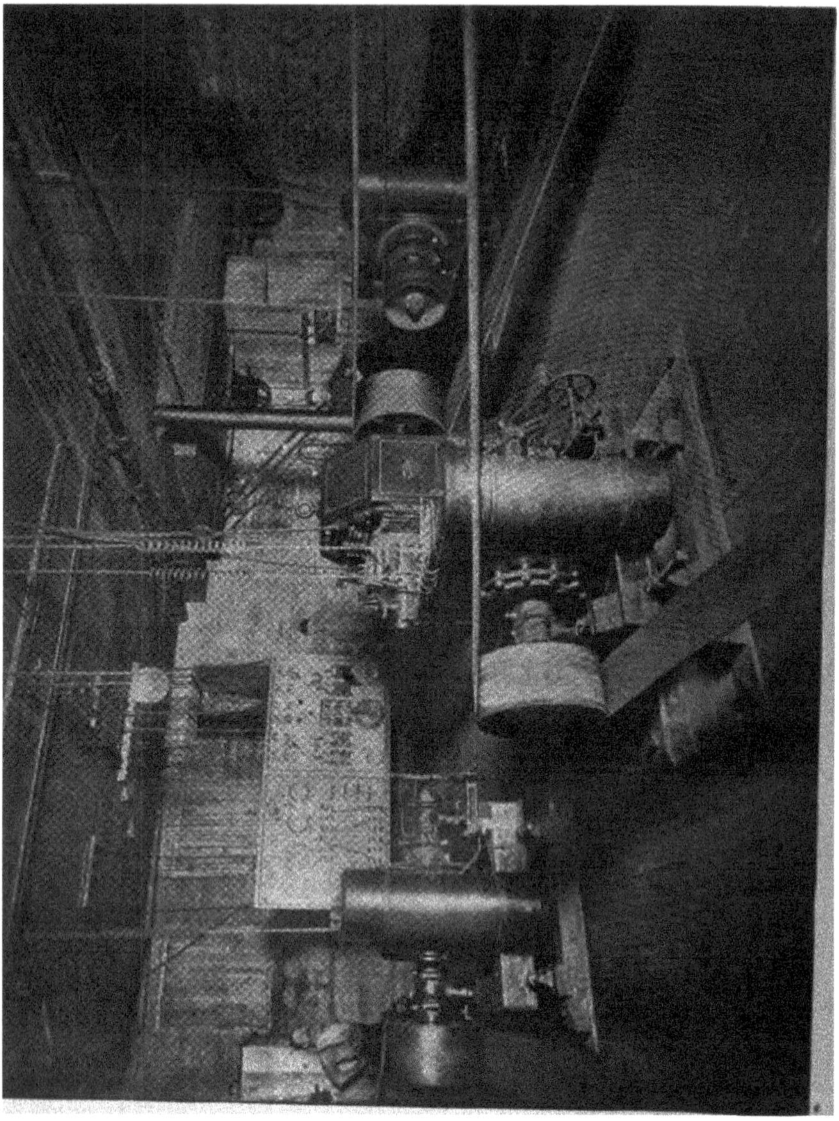

Electrical Engineering Company,

MINNEAPOLIS, MINN.

Headquarters for ....

Electric Light Supplies,

Electric House Goods,

Telephone Supplies and Parts,

COMPLETE TELEPHONE SETS,

FOR LONG OR SHORT DISTANCE SERVICE.

Electrical Construction Work

for central station or isolated lighting plants. (See isolated plant illustrated on opposite page.) We carry a full line of supplies and appliances for experimental work, and can make prompt shipments at very reasonable prices, with no extra charges for boxing and cartage.

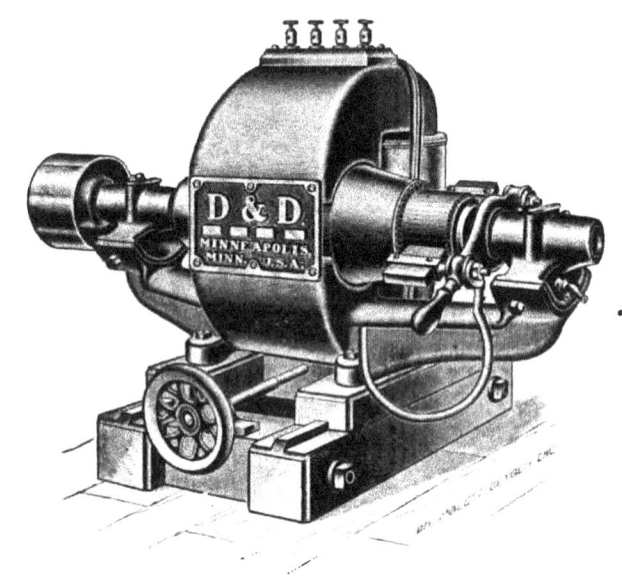

D. & D. Electric M'f'g. Co.

Manufacturers of

DYNAMOS
AND
MOTORS.

•••

Reasonable Prices.

Office and Factory 747-749 WASH. AVE. N.　～MINNEAPOLIS.

WHEN YOU ARE IN NEED OF →

Sewer and Culvert Pipe

Remember that the Best Quality of Ware
is Made in large Quantities, by

EVENS & HOWARD,

ST. LOUIS, MO.

WE MAKE PROMPT SHIPMENTS AND LOW PRICES AND OUR GOODS ARE POPULAR BECAUSE THEY GIVE SATISFACTION.

GEORGE D. SHEPARDSON,
Professor of Electrical Engineering.

THE YEAR BOOK

OF THE

SOCIETY OF ENGINEERS.

The YEAR BOOK is a scientific publication issued by the Society of Engineers in the College of Engineering, Metallurgy and the Mechanic Arts in the University of Minnesota. It is essentially technical in its scope and contains articles contributed by active and honorary members of the society and lectures before the society by eminent engineers.

This is the third publication of the YEAR BOOK, and as in the previous year, is in charge of an editorial committee representing the departments of Civil, Mechanical, Electrical and Mining Engineering.

The price of the YEAR BOOK is fifty cents per copy, or by mail (post paid) sixty cents. Copies of '93 and '94 Year Book may be had at same price. Address

The Year Book of the Society of Engineers,
University of Minnesota,
Minneapolis, Minn.

1895.
TRIBUNE JOB PRINTING CO.
MINNEAPOLIS, MINN.

The University of Minnesota.

COLLEGE
OF
ENGINEERING, METALLURGY
AND THE
MECHANIC ARTS.

CYRUS NORTHRUP, LL. D., President.

The College of Engineering, Metallurgy and the Mechanic Arts of the University of Minnesota offers a four years' course of preparation for the professions of

CIVIL ENGINEERING,   MINING ENGINEERING,
MECHANICAL ENGINEERING, METALLURGY,
ELECTRICAL ENGINEERING, CHEMISTRY.

and especial opportunities for the study of

INDUSTRIAL ART.

The work of the freshman and sophomore years lays the foundation in mathematics, physical sciences and shop practice for the more strictly technical professional work of the junior and senior years.

The large and increasing equipment of the college drawing rooms, laboratories and shops permits the instruction to be of a thoroughly modern and practical character. A considerable proportion of the student's time is employed in draughting, shop practice and testing.

The situation of the college in the manufacturing and commercial center of the great northwest is peculiarly favorable for technical instruction. Railway shops, manufacturing establishments, electric light and power stations, chemical and metallurgical works, ore docks and mines are always accessible. In them the student can see many of the largest engineering enterprises of the country conducted under business methods and in a scientific way.

Upon the completion of a full course of study the bachelor's degree is conferred.

Tuition is free. Shop and laboratory fees are low.

Requirements for admission: English grammar and composition, higher algebra, plain and solid geometry, history, physiology, natural philosophy, chemistry, botany, free-hand drawing, German or French. While four years' work in Latin may be offered, students are urged to present their preparation in German.

For further information address the dean of the college,

CHRISTOPHER W. HALL,
Minneapolis, Minn.

THE SOCIETY OF ENGINEERS

IN THE

UNIVERSITY OF MINNESOTA.

Officers, 1894-1895.

WILLIAM M. TILDERQUIST, '95, *President.*
ADAM C. BEYER, '96, *Vice President.*
FRANK ZIMMERMAN, '96, *Secretary.*
CHARLES D. HILFERTY, '96, *Treasurer.*
PLINEY E. HOLT, '96, *Business Manager.*

THE YEAR BOOK OF THE SOCIETY OF ENGINEERS.
1894-1895.

Editorial Committee.

FRED. M. ROUNDS, *Managing Editor.*
BURCHARD P. SHEPHERD, *Business Manager.*
WM. M. TILDERQUIST, *President of the Society.*
AUSTIN BURT, *Assistant Business Manager.*

Department Editors.

JAMES S. LANG,	*Mechanical Engineering.*
FRED. M. ROUNDS,	*Electrical Engineering.*
CHARLES D. WILKINSON,	*Mining Engineering.*
JOHN A. BOHLAND,	*Civil Engineering.*

THE NEW WAY WEST — GREAT NORTHERN RAILWAY

In Spring and early summer, the trans-continental traveler, if he be experienced, will avoid the heated and dust-laden desert wastes crossed by Southern routes, and travel by the

Great Northern Railway

over a dustless, rock-paved road-bed, amid the balsamic odors of pine and giant cedar, skirting tremendous peaks, and beside glacier-fed torrents. A trip through this phalanx of Nature's wonders is the event of a lifetime.

- GREAT FALLS, HELENA, BUTTE.
- KALISPELL, FLATHEAD VALLEY.
- FARGO, GRAND FORKS.
- THE KOOTENAI, SPOKANE.
- ALEXANDRIA, PARK REGION.
- LAKE CHELAN, OKANOGAN COUNTRY.
- DULUTH, WEST SUPERIOR.
- EVERETT, SEATTLE, PUGET SOUND.
- MINNEAPOLIS, (THE TWIN CITIES) ST. PAUL.
- ALASKA, CHINA, JAPAN.
- HOTEL LAFAYETTE, LAKE MINNETONKA.
- TACOMA, PORTLAND.
- SIOUX FALLS, YANKTON.
- SAN FRANCISCO.
- HONOLULU, SANDWICH ISLANDS.
- SIOUX CITY.

A trip across the continent without having seen the unrivaled attractions of this grandly scenic line would be incomplete. Pacific Coast tourists should not fail to take advantage of excursion rates, in effect from all Eastern points, which permit going and return by different routes, being sure their tickets read in one or both directions over the Great Northern Railway. For information, rates, etc., address

F. J. WHITNEY, G. P. & T. A., St. Paul, Minn.

YEAR BOOK

OF THE

SOCIETY OF ENGINEERS.

MAY, 1895.

PROFESSOR GEORGE D. SHEPARDSON.

George Defrees Shepardson was born at Cheviot, Ohio, in 1864. He is of the seventh generation from Daniel Shepherdson, who came to Salem, Massachusetts, in 1630. His mother, Eliza Smart Shepardson, is a descendant from John Smart, who came to Massachusetts in 1635 from Norfolk, England.

Both parents are now living at Granville, Ohio, where for twenty years they were at the head of " The Young Ladies' Institute," which by their endowment still flourishes as "The Shepardson College for Women." The value of these educational surroundings in the formation of early tastes and habits is proved not only by the position which the subject of our sketch now occupies in his chosen profession as well as among his fellow men, but by the eminence already gained by brothers in business and professions, and by the social distinction enjoyed by sisters.

While a mere lad our subject displayed a fondness for mechanics and machinery—prime essentials in his later chosen profession. While his schoolmates found recreation along the streams or with the ball, he would be found usually at a small machine shop near his home watching and frequently assisting the proprietor with various kinds of work, for which his quick perception and ready hand so well fitted him. Here it was that he first met and studied that subtle force which was destined to mould in no small degree his life work. Here his vacations were spent learning to handle batteries for electro-plating and while yet a school boy he assisted in making a small dynamo, whose successful working did much to further inspire him and direct his thoughts and energies to electrical studies.

While pursuing his regular college work at Denison university in the classical course he continued his interest in scientific work. In addition to his regular studies, he acted as engineer in charge of the Siemens series dynamo, supplying light for the physical laboratory. His first practical experience in running arc light dynamos was here obtained. The construction of an electric motor during his spare time, using pieces of lamp burners, clock frames, stove bolts, etc., shows the decided inclination of his mind to electrical studies.

Upon graduation from Denison in 1885, Mr. Shepardson had fully decided to prepare himself for some scientific specialty and to that end devoted the following year to physics, chemistry and mineralogy. Then followed a year's teaching in these branches at Granville, Ohio. Another year's work with the Edison Illuminating Company of Boston, as motor inspector, then as night dynamo operator furnished a valuable experience and revealed to the ambitious mind the vast field just opening in electricity. Appreciating this, and what was still more important, that to become a master in this work required a thorough knowledge in theoretical and experimental electricity, Mr. Shepardson spent the next two years at Cornell, graduating in 1889 with honor as Electrical Engineer. A summer spent as electrical expert with the Akron Electrical company, designing and constructing dynamos, permitted him to carry out some of his own ideas with flattering success. He also demonstrated the fallacy of their premises to another company proposing to revolutionize the ordinary methods by obtaining electrical energy directly from coal. Thesis work along the latter line had fitted our subject to render an opinion of high professional value, though very disappointing to the stockholders.

The unpleasantness and severe nervous strain attending such commercial ventures, as well as a call to an instructorship in physics at Cornell at this point, led Mr. Shepardson seriously to consider adopting teaching as his profession. His early experience as an instructor, as well as the unusually long and successful record of both father and mother as educators, led him to believe that, though commercial practice offered great pecuniary advantage, it was his duty to consecrate his life to training others.

Acting on this conviction, he remained at Cornell a year when he received the call from the University of Minnesota to the professorship in Electrical Engineering.

In the autumn of 1892 Professor Shepardson married Miss Harriet B. King, daughter of Mr. A. King, of King's Mills, Ohio.

Professor Shepardson's interest in all activity relating to his profession is shown by the numerous societies with which he has become identified. Of the eight, we mention the following: American Institute of Electrical Engineers, Society for the Promotion of Engineering Education, American Association for the Advancement of Science, honorary faculty of the National School of Electricity.

Perhaps in no department of his work has Professor Shepardson made more valuable contributions than to the current literature of his profession through his numerous writings. These have appeared not only in many of our leading American electrical and scientific publications, but also in several foreign reviews. Professor Shepardson combines to a remarkable degree the ability to reach higher planes of thought and research, and still be within reach of the student. He sees the student's difficulty and applies his own greater experience and knowledge, on the level of the student, to assist and encourage him.

WILLIAM R. HOAG.

THE MOST ECONOMICAL CUT-OFF FOR STEAM COMPRESSION.

DR. HENRY T. EDDY.

Some misapprehension seems to have prevailed as to the precise purport and relation of certain demonstrations that have been given as to the most economical point of cut-off. With a view to clear up this misapprehension the author has recently investigated the question anew, and proposes in the present brief review to state the conclusions to which this inquiry leads when based on the usual supposition that the expansion and compression curves of the indicator card are rectangular hyperbolas. The approximate results thus obtained will serve to make sufficiently clear the points to which it is desired to direct attention, and will leave to a future time certain more accurate, but more complicated methods of obtaining results, which will approximate still more closely to the actual indicator card.

1. Suppose it be required to find the best ratio of compression in case the data given are: the volume of the cylinder, the volume of the clearance, the pressure during admission, the back pressure of exhaust and the ratio of expansion.

This question has been solved in analytical form by Professor Cotterill;[1] and also by Professor J. Burkitt Webb,[2] who has shown that the results may be graphically represented upon the indicator card as follows: prolong the expansion curve until it intersects the exhaust pressure line, and measure the area thus added to the indicator card. Cut off an area equal to this from the top of the indicator card by a horizontal line drawn parallel to the line of zero pressures; then the point at which this line in-

---
[1] Steam Engine, 2d Ed., p. 257.
[2] Proc. A. A. A. S. for 1893, p. 119.

tersects the vertical clearance line will be a point of the compression curve of best compression cut-off to accompany the given admission line and expansion curve.

It can be shown at once from the analytical results that the ratio of compression thus obtained will always be greater than the given ratio of expansion in case the expansion is incomplete. The same result will be evident to any one who will make the graphical construction just mentioned; and it may be otherwise evident if it be duly considered that the best ratio of compression should lie between the ideal case of complete compression and a ratio equal to the given expansion.

2. But suppose, conversely, that we proceed to inquire what is the best ratio of compression in case the data given are: the volume of the steam cylinder, the volume of the clearance, the forward pressure of admission, the back pressure of exhaust and the ratio of compression. The analytical solution of this question leads to results analogous to those previously obtained, which also admit of graphical construction, as follows: prolong the compression curve until it cuts the admission pressure line, and measure the area thus added to the indicator card. Cut off an area equal to this from the bottom of the indicator card by a horizontal line parallel to the line of zero pressures, then the point at which this line intersects the vertical expressing the total volume of cylinder and clearance will be a point of the expansion curve of best admission cut-off to accompany the given back pressure and compression. It can be shown analytically or graphically or from general considerations that the ratio of expansion thus obtained will be greater than the given ratio of compression.

3. It is evident that neither of the foregoing results is such as to enable us to determine the best relative points of cut-off for expansion and compression. For, if we were to follow their indications and start with a given ratio of expansion, not complete, we should find the best ratio of compression to accompany it to be greater than it is; and if then the best ratio of expansion be found to accompany this compression, that must be still larger; and so on stroke by stroke. This requires that the cut-off for compression

and for expansion should become, step by step and stroke by stroke, more and more nearly complete. This conclusion is in entire accordance with our general knowledge of the fact that on the basis on which we are now proceeding, complete expansion and complete compression are theoretically best, and our results so far, show simply that starting with any ratio less than that, we should be led to approach complete expansion and compression step by step instead of changing suddenly. This step by step increase of ratio, however, in a cylinder of given volume involves the admission of a smaller and smaller amount of steam per stroke until complete expansion is attained. But since practical considerations forbid complete expansion, it is evident that such an investigation as has been made does not lead, as has heretofore been frequently assumed, to the solution of the question of the most advantageous relation between the ratios of compression and expansion. The same conclusion as to the applicability of results obtained in this way would evidently hold against a similar investigation based on some other and more accurate law of expansion and compression than that of the hyperbola. The difficulty lies, not in the law, but in the statement of the problem, which should evidently contain limitations nearly as follows:

4. To find the best ratios, both of expansion and compression, when the data are: the volume of the cylinder, the volume of the clearance and the volume of the steam admitted per stroke, together with the forward pressure during admission and the back pressure during exhaust. On investigating this question analytically the best adjustment is found to occur when the ratios of expansion and compression are equal. More work is therefore performed by a given volume of steam admitted, other things being equal, when this adjustment of ratios is made than when the adjustment is in accordance with either of the results previously given. The conclusion here reached will evidently require but very inconsiderable modifications to adapt it to the comparatively small deviations from hyperbolic expansion which occur in practice. The modification needed to give this still closer approximation cannot be given here, but must be reserved for a future occasion.

SOME ALTERNATE CURRENT NOTES.

H. L. TANNER, '95.

The following experiment is presented with the hope that at least a few of the readers may obtain a better idea of the value and use of some of the terms used in alternate current working. The following symbols and terms are used in the computations:

Ohmic resistance, R, is the resistance which would be met by a direct current in passing through a conductor.

Reactance or wattless resistance, $2\pi nL$, is equal to the component of impressed electro-motive force at right angles to the current, divided by the current. It tends to cause a difference of phase between the current and electro-motive force, and equals $2\pi nL$, where n is the frequency L the coefficient of self-induction, and $\pi = 3.1416$.

The apparent resistance, Ra, as here used, is the quantity by which the square of the virtual current must be multiplied to obtain the watts consumed in the coil if it contains an iron core. It is considerably larger than the true ohmic resistance of the coil, since power is consumed by hysteresis and eddy currents in the core. The apparent resistance Ra is equal to the ohmic resistance plus a quantity Rb, depending upon the losses in the iron core.

Since the reactance and apparent resistance act at right angles, they may be combined into a resultant called the *impedance*. This, since the components are at right angles, is equal to the square root of the sum of their squares, thus $imp. = \sqrt{Ra^2 + (2\pi nL)^2}$.

The angle of lag Θ is the angle which the impedance makes with the apparent resistance, or the angle by which the current lags behind the impressed electro-motive force.

The coefficient of self-induction L is the ratio of the product of the turns in the coil and the total induction through it to the exciting current, and is practically constant unless a high degree of magnetization is reached. It is defined by the equation $L=\frac{\Phi S}{I}$ where Φ is the total induction, S the number of turns, and I the current. To obtain the practical unit, the henry, the above value must be multiplied by 10^{-9}.

The frequency n of an alternating current is equal to the number of complete alternations per second.

The virtual current is equal to the square root of the mean of the squares of its instantaneous values, when it varies as a harmonic function of the time, or, in other words, the virtual current is equal to the maximum value divided by the $\sqrt{2}$. Since the heating and dynamometric effects of any current depend upon the mean square value, the virtual current is considered in alternate current working.

The virtual electro-motive force is found in the same manner as the virtual current, for a similar reason.

Illustration:—The secondary coil of a "Brush" one-light transformer and a non-inductive resistance consisting of four incandescent lamps in parallel were connected in series across the terminals of a 25-light Slatterly 52 volt transformer. See Fig. I. The

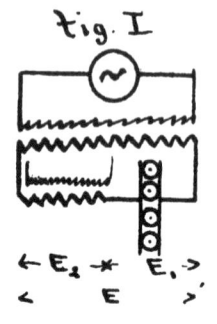

primary of the Slatterly was connected to a ten pole Fort Wayne alternator; the speed of the alternator was 1560 revolutions per minute, hence to find the frequency we have,

$$n = \frac{10}{2} \times \frac{1560}{60} = 130.$$

After connecting as above, volt-meter readings were taken as indicated in Fig. I, with a "Weston" direct and alternate current volt meter as follows:

(1) E_1 = voltage around lamps, = 24.4 volts.
(2) E_2 = voltage around transformer coil, = 36.2 volts.
(3) E = voltage around transformer coil and lamps = 56 volts.

The lamps and coil were next placed in series with a variable resistance across a constant potential direct current circuit and the resistance adjusted until the voltage around the lamps E_1 was the same as with the alternate current. The voltages E_2 and E were then found to be .51 and 24.8 volts respectively. The current through the circuit was 1.7 amperes, which is equal to the virtual current in the alternate current experiment. From the direct current volt and ampere readings we have, by Ohms law, the resistance of the lamps equal to 14.3 ohms and that of the transformer coil equal to .3 ohms.

Taking the three alternate current volt meter readings and laying them off to the same scale on E_1 as a base line,*(Bedell and Crehore, page 232) we have the triangle AOB. (See Fig. II.)

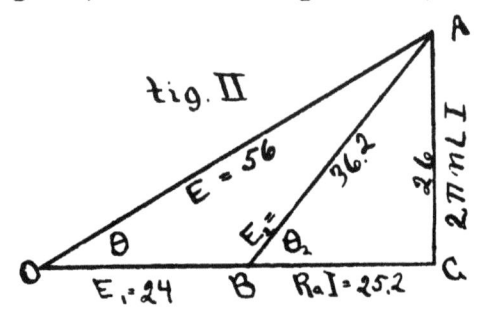

Completing the right angled triangle ACB we have on the same scale, AC = $2\pi n L I$, and BC = RaI, where L is the coefficient of self-induction in henrys, I the virtual current, Ra the apparent resistance, and n the frequency. By scale AC equals 26, and BC equals 25.2, from which L and Ra can be calculated, since all other terms are known.

$$2\pi nLI = 26.$$
$$\therefore L = \frac{26}{2 \times 3.1416 \times 130 \times 1.7} = .0118 \text{ henrys.}$$
$$R_a I = 25.2.$$
$$\therefore R_a = \frac{25.2}{1.7} = 14.7 \text{ ohms.}$$
$$\text{Wattless resistance} = 2\pi nL = \frac{26}{1.7} = 15.3 \text{ ohms.}$$

As a check upon the value of L just obtained, we can find the current, which corresponds to the maximum magnetization, by dividing the virtual current by $\sqrt{2}$. This gives 2.4 as the maximum current. Knowing this current and having the curve of magnetization obtained by the step-by-step method, the maximum magnetization was found to be about 7200 C. G. S. lines per square centimeter. From these values and the dimensions of the transform, the value of L can be calculated thus:

$\Phi = 7200 \times$ area of cross section of core in square centimeters $= 7200 \times 50 = 360,000.$

$$S = 72 \text{ turns.} \quad I = 2.4 \text{ amperes.}$$
$$\therefore L = \frac{\Phi S}{I} = \frac{7200 \times 50 \times 72}{2.4} = 10900000, \text{ C.G.S. units.}$$

or multiplying by 10^{-9} gives .0109 henrys, which compares favorably with that just obtained.

The impedance may be obtained by extracting the square root of the sum of the squares of the reactance and apparent resistance just found, or since by Fig. 2 AB is seen to be I times that value, we may put $I\sqrt{R_a^2 + (2\pi nL)^2}$ equal to the volt-meter reading $E_2 (= 36.2)$ and solve impedance thus

$$\sqrt{R_a^2 + (2\pi nL)^2} = \frac{36.2}{I} = \frac{36.2}{1.7} = 21.3 \text{ ohms.}$$

For the angle of lag we have by trigonometry,
$$\cos \Theta = \frac{OC}{OA} = \frac{24.4 + 25.2}{56} = .8857.$$
$$\therefore \Theta = 27° \, 40'$$
$$\text{also, } \cos \Theta_2 = \frac{BC}{BA} = \frac{25.2}{36.2} = .6951$$
$$\therefore \Theta_2 = 45° \, 58'.$$

The power expended in different parts of the circuit is found by multiplying the virtual current by the component of the virtual impressed electro-motive force which is used in overcoming the apparent resistance of that part. This component is equal to the cosine of the angle of lag multiplied by the impressed electromotive force. Hence we have watts consumed,

1st in coil= $W_2 = E_2 I \cos \theta = 36.2 \times 1.7 \times .695 = 42.9$ watts.

2nd, watts consumed in whole circuit are:

$W = EI \cos \theta = 56 \times 1.7 \times .8857 = 84.76$ watts.

From above results note that in computing the value of the current from the general equation $I = \dfrac{\text{E. M. F.}}{\text{impedance}} = \dfrac{E}{\sqrt{R^2 + (2\Pi nL)^2}}$ that unless the coefficient of self-induction is very large we must use the apparent resistance instead of the ohmic if there is a chance for loss by eddy currents or hysteresis.

The transformer used in the experiment was made up of L shaped punchings, so placed as to form a rectangle with closed iron circuit. These punchings were 3¼ in. × 7 in. with 1⅝ in. face, separated by thin paper and piled to a depth of 4¾ inches. Mean length of line of force 17 inches. Area of cross section of core 50 sq. cm. Number of turns on each leg 36, total turns 72. Weight of core 20 lbs. 10 oz.

The power consumed in the coil, which was found to be 42.96 watts, can be separated into that due to the resistance of the coil, and that due to hysteresis and eddy current losses. The hysteresis loss is the greatest of the three since the core is well laminated and the resistance of the coil is small. The loss due to the ohmic resistance, or $C^2 R$ loss is $1.7^2 \times .3 = .86$ watts, leaving a balance of 42.1 watts for hysteresis and eddy current losses, or 2.04 watts per lb. of iron when the frequency is 130 and the intensity of magnetization 7200 C. G. S. lines per square centimeter, which agrees closely with values given in Fleming's "Alternate Current Transformer," page 287, also in a table given in the "London Electrician" Nov. 23, 1895, and by Houston and Kennelly "Elec. World," Apr. 13, 1895.

CURRENT MOTORS.

JAMES S. LANG, '95.

The demand for cheap power is one of the factors which have located manufactories along water courses and crystallized civilization about well defined centers. To supply it, engineers and designers have devoted their energies unstintingly, until the modern steam engine now approaches that point where farther increase of efficiency becomes impossible from the inherent properties of the matter with which they are working. The water wheel employing head having, as now manufactured, an efficiency of nearly 90 per cent., also has little room left for material improvement. Thus it is seen that because the room for improvement is limited in the cases of the established prime movers, the demand for cheaper power compels engineers to search for some new principle which may be utilized with less either of first cost or maintenance.

Attention is naturally turned to flowing water as the cheapest source of power, for there is seen a constant waste òf energy for every pound of water and foot of flow. Here it is that the current motor stands pre-eminent, for while the water wheel employing head may only be utilized at points exceptionally favored by nature, the current motor is available at any point throughout the whole length of the stream. More important than its adaptibility is its economy,—economy which is mainly secured by low first cost; for whereas, in case of head water wheel, the cost of the wheel itself is a small fraction of the total cost of dam, tail-race, etc., in case of current motor, the wheel and connections comprise the total cost.

The main reason why current motors have not heretofore come into more favor is because of their faulty design in two essential points: First, the motor should by its large area of active surface make up for the low pressure available, and thus obtain sufficient hold upon the water. Second, it should be wholly below the surface, so that floating logs, ice, etc., will pass over without disturbing it.

The utilization of air currents as a source of power has increased rapidly within the last few years, and the modern windmill has proved one of the cheapest prime movers of its capacity.

We may compare air and water currents by finding the pressure produced by each at various velocities. The wind pressures in the following table are taken from the scale of the Smithsonian Institution; while the velocity of water producing an equal pressure per square foot is deduced from the formula:

$$P = ZFh\frac{v^2}{2g}$$

(Church's Mechanics.) Where P is the pressure produced, Z a co-efficient determined by experiment, h the heaviness of the fluid, v its velocity and g the acceleration of gravity.

Pressure. pounds per square foot.	Velocity of wind. Miles per hour.	Velocity of water. Feet per second.
2.65	16	1.58
3.	25	1.65
6.	35	2.33
10.	45	3.01
18.	60	4.04

Thus we see that, for example, the Mississippi between the Tenth avenue and Franklin avenue bridges, and for a long distance below, flowing at the rate of from 4 to 5 feet per second, produces a pressure upon stationary objects equivalent to that produced by a gale blowing at the rate of 60 to 80 miles per hour.

14   *James S. Lang.*

The best means of utilizing water currents, as of air currents, consists in a modification of the reversed screw propeller. A series of wheels having helical blades may be attached together and to a suitable transmitting jack by rods so linked as to form a flexible shaft by which the power of the whole may be transmitted, as in the accompaning figure.

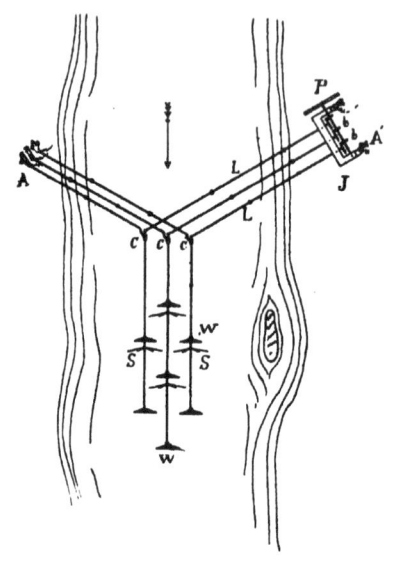

The blades are given a backward slant from center to tip, to prevent the lodgement of floating debris. The shafting, entirely flexible, and held up from the bottom by the spiders SS, is able to move sidewise if struck by any heavy floating object. In the figure, LL are the lines of shafting from the different gangs of wheels; CCC are Hooke's couplings by which the direction of the lines of shafting is changed from Cw to Cb. The bearings for the couplings are attached by cabling to the anchors A; J is the speed jack attached to anchors A'A', of which bb are bevel gears, by means of which the motion of the shafting LL is transmitted to the grooved pulley P. Thus we have the whole system suspended, as it were, against the action of the current, from the anchors AA'A'. Where the circumstances of the case require, heavy wire cable may be substituted for rodding in the shafting L.

Current Motors.

Let us now investigate the power of a reversed screw propeller of a diameter, say of six feet, pitch two feet, running in water flowing four feet per second, and so geared that it revolves sixty times per minute. We will suppose the blades to be of such width that their projection upon a plane perpendicular to the line of the shaft will embrace the whole area of the circle.

Then we have the radius of gyration $=\sqrt{\frac{(3)^2}{2}}=2.12$ feet.

Pressure upon projected blade area $=P=ZFh\frac{V^2}{2g}$

$$=\frac{1.13\times 3\times 2\times 62.5\times (2)^2}{64.4}=27.7 \text{ pounds.}$$

Let the velocity of a point of the wheel, whose distance from the center is equal to the radius of gyration, be $=v$; it will therefore be $3.1416\times 2.12\times 2=13.32$ feet per second.

Therefore the whole power developed by each wheel will be:

$$=\frac{Pv}{550}=\frac{13.32\times 27.7}{550}=.67 \text{ H.P.}$$

If 50 H.P. were required, we should use eighty such wheels, divided into several gangs, as indicated in the figure.

The cost of power by this means comes next for our consideration, which may be found as follows:

Cost of motor at $20 per wheel with its shafting=$1,600.00.

    Interest at   7 per cent.
    Depreciation, 7   "
    Repairs,   7   "
    Attendance, 3   "

    Total,   24   "   Cost per year=$384.00

Cost per H.P. hour $=\frac{384}{50\times 7200}=\$.00106$; on the supposition of power being used 7,200 hours per annum.

We may thus conclude that since the cost of power by this method is hardly one-tenth its cost by steam, that the current motor will find wide application wherever it is available.

MILITARY ENGINEERING.

WILLIAM WATTS FOLWELL.

The army regulations sum up the duties of the corps of engineers under these five heads:

(1.) Reconnoissance and surveys.

(2.) Planning and construction of all military works, whether permanent or temporary.

(3.) Planning and construction of all military roads and bridges.

(4.) Planning and execution of river and harbor improvements.

(5.) The collection, arrangement and preservation of all reports, memoirs, estimates, plans, drawings and models relating to these duties.

An examination of this compact list would soon reveal the vast extent and variety of the functions of the engineer corps, and the wide knowledge and skill necessary to their discharge. This corps is accordingly recognized as the most dignified branch o military service, and only a few of the graduates of the Military Academy highest in rank are admitted into it.

It is at once apparent that in their nature the duties of the military engineer are identical in the main with those of the civil engineer. Both have to make surveys, devise plans, and supervise construction. The elementary instruction given in the Military Academy is therefore similar in plan and scope with that offered in the courses in civil engineering in our best polytechnic schools. River and harbor improvements in the way of sea walls, breakwaters and lighthouses belong especially to the domain of the civil engineer, and it might be objected there is no propriety in assigning military officers the planning and construction of them. Two reasons may be suggested for such assignment—one, that our highly educated engineer officers may be employed in the long intervals of peace, which happily bless our country; the other, to save the government the loss and damage almost sure to follow the employment of civil employes, appointed rather for political reasons than for professional ability. The wisdom of this policy is shown by many splendid examples.

In European states in time of peace, the engineer corps find employment in the repair and extension of those systems of permanent fortification which virtually convert the whole state into a fortress. The whole frontier is beset with works so situated as to command every point at which an invading force might attempt to enter. These works are supported by inland fortified places, in which are maintained the magazines of arms, provisions and other war material. Another line of duty for peace is the perfection of maps of every probable theater of war, and the collection of information as to its people and resources. It has been reported that upon the outbreak of the Franco-German war, the German engineers had better maps of eastern France than the French themselves.

It may be, of course, understood that in all such operations carried on in times of peace, the best and latest results of science and instrument-making are employed.

Upon the outbreak of war, however, the operations and methods of the engineer corps at once take on new phases. The theodolite and base measuring bars, give place to the prismatic compass and the tape measure. Under some circumstances it will be comfortable and sufficiently accurate to determine distances by means of a lead pencil stadia, or the apparent sizes of men and animals.

Whenever a state of war threatens or breaks out, the chief of engineers will not fail to be found in the council which decides upon the objective point and the grand tactics of the campaign. This done, his subordinates will revise and if necessary prepare the maps of the region of operations, and examine the roads and waterways which intersect it. At every moment the engineer officers of the staff are the advisors of commanding officers as to the movement of troops and transportation of every kind.

The ease with which advancing columns may be delayed by the destruction of bridges, renders them of first importance in the eye of the engineer. He must be ready to replace a ruined structure or repair a damaged one with the utmost dispatch. A day's, an hour's delay may sacrifice an isolated division whose supports are cut off by the demolition of a single bridge over an unfordable stream. That engineer who knows how to seize on the first material at hand, and employ the men of the nearest regiment to extemporize a bridge, no matter how rude, provided it will carry men, is dear to the heart of a commander; all the better, if the engineer troops are present with their tool wagons and ready-made bridge material. The frequent demands for extemporized bridges long ago led to the invention of the ponton bridge, the peculiar character of which consists in the employment of boats for bridge supports. These boats built in the engineer depot are transported on wagons, which also carry the superstructure in the "knock-down." Aligned on one another between abutments the boats are connected by stringers, called "balks," on which the planks or "chess" are laid to form the floor. "Side rails" are placed atop of the chess directly over the outer balks. All connections are made by means of "lashings" of good hemp rope about a half-inch in diameter, forming joints at once secure and flexible. Anchors cast at suitable intervals above the bridge, sustain it against the current of a river. In tide-water "down stream anchors" are also required. Such a bridge will support any loads which are transported with an army in the field.

To insure the highest speed of construction certain companies or battalions of the engineer troops are usually detailed as pontoniers. By drill in the pontoniers' manual and practice in the field great dexterity is acquired. In ordinary circumstances, and on a bridge not exceeding five hundred feet in length, about one hundred men, divided into boat crews, balk carriers, chess carriers, etc., are needed for rapid construction. If no accident intervene a well drilled and well handled company should construct a bridge of four hundred feet in thirty minutes, and dismantle it in twenty.

Longer bridges, over swollen streams, with high and steep banks to retard the delivery of material require more men and more time. The writer constructed two bridges of 1,500 feet each over the Potomac at Berlin, Md., in July 1863, for the passage of Meade's army into Virginia after the battle of Gettysburg. It took several hours to complete them.

Arrived near the scene of expected action the engineer officer is occupied with selecting the positions for attack or defense, as the case may be, and with discovering and opening the routes by which the various corps and divisions may most conveniently reach their assigned places. Nor must he neglect the communication in rear by which supporting troops may be brought up, and supplies of all kinds reach the fighting line. Positions of safety for the hospitals and wagon trains need to be selected, and roads opened for access to them.

Since manœuvres on the field of battle have become infrequent, the engineer officer has less activity and responsibility in time of action than formerly, but whenever any tactical movement is attempted he is at once in requisition to advise as to the geographical situation and the route of march.

Like a game of chess, a battle may end in victory, defeat or a drawn game.

If his party has won the fight the engineer is called on to select the lines of pursuit and to take all necessary measures to facilitate the march.

If a retreat is necessary it is the melancholy duty of the engineer to select the successive positions to be held by the rear guard, while the army, with such impedimenta as may have been rescued, makes its way to a safe place of rendezvous.

In case of a drawn game the most usual procedure is to hold

on and fortify on the best lines afforded by the local topography.

The selection of such lines puts the talent of the engineer to the severest proof, although it sometimes happens, as it did to Grant in front of Petersburg, Va., that the positions and lines most acceptable to a commander are in the possession of the enemy. In such cases the engineer must make up if possible for disadvantage of situation by the strength and ingenious disposition of his defenses.

The plan usually adopted for holding an intermediate position is that of constructing great inclosed works at commanding points, and connecting them with lines of intrenchments. The name "rifle pits" was, in the war of the slave-holders' rebellion, improperly extended to such lines. It is a principle that all works and lines shall have such delineation and profile as to fully protect their garrisons, and command all the territory in front by cross fire.

Various exterior defenses such as abattis and slashings of forests must be looked after, and roads and covered ways in rear must not be forgotten. In case of an extensive line it may be necessary to build a railroad, as was done at Petersburg, to move supplies to the troops. If the rear is exposed to attack, or if it is desirable to liberate some portion of the troops for operation elsewhere, the line of intrenchments may be curved backward so as to envelop the whole force, and thus form what is called an intrenched camp. This was also done at Petersburg. In the location, planning, construction and maintenance of such works the engineer corps find abundant employment. It will be understood that the labor is performed by troops of the line, the engineer soldiers being employed in direction or in the construction of the magazines, embrasures and other portions requiring skill and science.

Here it may be remarked that in many instances in the late war the troops of the line were found so skillful and ingenious as to originate and construct works of great importance without troubling the engineers. As an example the mine under one of the enemy's forts in front of Petersburg may be cited. This notable operation was planned and executed by the officers and men of a Pennsylvania regiment which had been recruited in the mining districts. As a mine it was entirely successful, but the military effect, owing to neglect or misconduct was anything but a success.

It is not within the purpose of this contribution to enter into the details of fortification building. It was only intended to suggest the scope and importance of a branch of the engineering profession, lying apart from the ordinary practice and observation.

COMPARISON OF STRESSES PRODUCED IN A BRIDGE BY CONCENTRATED LOADS AND AN EQUIVALENT UNIFORM LOAD.

GEO. A. CASSEDAY, '95.

The Bridge is a Pratt truss, span 250 ft., 10 panels, height 40 ft. The load consists of two 104 ton engines followed by a uniform load of 3000 lbs. per ft. (Cooper's Class, Extra Heavy A.)

The positions for maximum stresses produced by concentrated loads are determined by the methods outlined, and by the table found in Johnson's "Theory and Practice of Modern Framed Structures."

Following is the development of formulæ for determining the positions of the uniform load when the maximum stresses are produced.

The two 104 ton engines together weigh 208 tons and extend over 103 feet of track. Each truss bears one-half of this load which amounts to $\frac{208 \times 2000}{2 \times 103} = 2020$ lbs. per ft. produced by the engines and 1500 lbs. per ft. by the train following.

CHORD STRESSES.

The maximum bending moment, at any section, occurs when the ratio of the load on one of the segments to the length of that segment equals the ratio of the total load to the length of the bridge.

In Fig. I suppose the train to move from right to left. Let "m" be the number of panels to the left of the section, "d" the length of one panel, "x" the number of feet from the head of the train to the section.

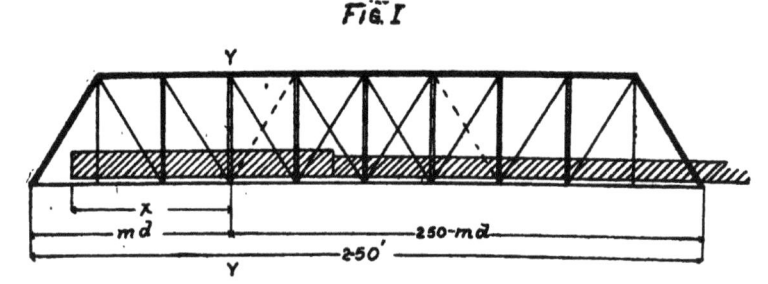

Fig. I

Then by the rule to produce the maximum bending moment at any section, as YY.

$$\frac{2020 \times x}{md} = \frac{103 \times 2020 + [250 - md - (103 - x)] 1500}{250}$$

$$x = \frac{(21425 - 1875 m)m}{1010 - 75m}$$

Let M = the maximum moment, and R_1 the left end reaction.

$$\text{Then } M = R_1 md - \frac{2020 x^2}{2}$$

To determine R_1:

The moment of the engines about their rear end equals

$$\frac{2020 \times \overline{103}^2}{2} = 10,715,000 \text{ ft. lbs.}$$

$$R_1 = \frac{10,715,000 + 208,000[250 - (103 + md - x)]}{250} + \frac{\frac{1500}{2}\left\{250 - (103 + md - x)\right\}^2}{250}$$

WEB STRESSES.

When the shear in any panel has its greatest value, the load upon that panel is to the total load on the truss as the length of the panel is to the length of the truss.

24 Geo. A. Casseday.

In Fig. II, supposing the train to move from right to left, as in the previous example, and using the same notation, the maximum shear in any panel is produced when

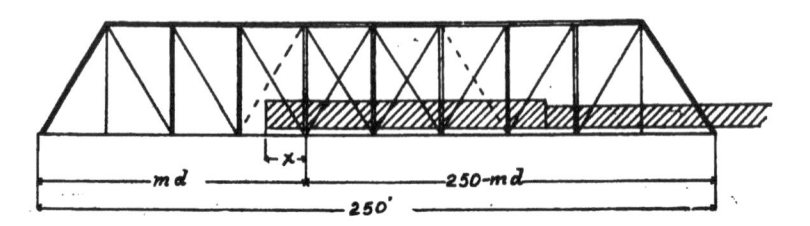

Fig. II

$$\frac{2020x}{25} = \frac{208,000 + (250 + x - md - 103)1500}{250}$$
$$x = 22.9144 - .0802\ md.$$

For the panels to the right in which the maximum shear is produced before the train load comes upon the bridge.

$$\frac{2020x}{25} = \frac{(250 - md + x)2020}{250}$$
$$x = \frac{250 - md}{9}$$

Let S = shear, and R_1 the left end reaction.

$$\text{Then } S = R_1 - \frac{2020x^2}{2 \times 25}$$

Following are tabulated the positions of the loads when the maximum moments and shears are produced; also the moment, shears and stresses. The tables refer to Fig. III.

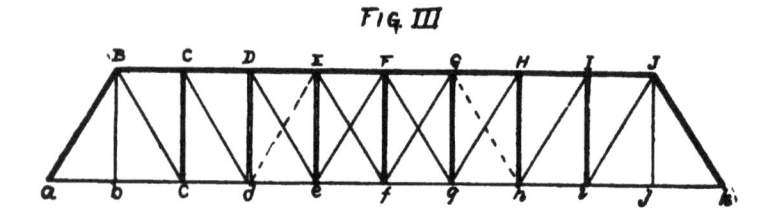

Fig. III

FOR CONCENTRATED LOADS.

\multicolumn{3}{c}{MAXIMUM MOMENTS.}	\multicolumn{3}{c}{MAXIMUM SHEARS.}
SECTION	WHEEL	MOMENT FOOT POUNDS	PANEL	WHEEL	SHEAR
Bb	4	5,153,152	ab	4	206,126 lbs.
Cc	7	8,875,706	bc	3	166,190 "
Dd	11	11,560,868	cd	3	130,630 "
Ee	13	13,093,080	de	3	98,812 "
Ff	16	13,323,260	ef	3	70,712 "
			fg	2	46,670 "
			gh	2	26,980 "
			hi	2	12,490 "

FOR UNIFORM LOADS.

\multicolumn{3}{c}{MAXIMUM MOMENTS.}	\multicolumn{3}{c}{MAXIMUM SHEARS.}
SECTION	x	MOMENTS FOOT POUNDS	PANEL	x	SHEAR
Bb	20.9	5,134,872	ab	20.91	204,370 lbs.
Cc	41.1	9,041,998	bc	18.9	165,150 "
Dd	60.4	11,967,348	cd	16.9	128,980 "
Ee	78.45	13,227,580	de	14.89	96,870 "
Ff	94.88	13,540,764	ef	12.88	68,800 "
			fg	10.88	44,810 "
			gh	8.33	25,250 "
			hi	5.55	

MEMBER	STRESSES IN 1000 LBS. Concentrated	Uniform	*PER CENT.	MEMBER	STRESSES IN 1000 LBS. Concentrated	Uniform	*PER CENT.
ab	+128.8	+128.4	−0.3	Bb	− 59.5	− 50.5	−15.1
bc	+128.8	+128.4	−0.3	Cc	−130.6	−129.0	− 1.2
cd	+221.9	+226.1	+1.9	Dd	− 98.8	− 96.9	− 1.9
de	+289.0	+299.2	+3.5	Ee	− 70.7	− 68.8	− 2.7
ef	+327.3	+330.7	+1.0	Ff	− 34.2	− 33.2	− 2.9
Bc	+196.0	+194.7	−0.7	AB	−243.0	−241.0	− 0.8
Cd	+154.0	+152.1	−1.2	BC	−221.9	−226.1	+ 1.9
De	+116.5	+114.1	−2.1	CD	−289.0	−299.2	+ 3.5
Ef	+ 83.4	+ 81.1	−2.8	DE	−327.3	−330.7	+ 1.0
eF	+ 40.3	+ 39.1	−3.0	EF	−333.1	−338.5	+ 1.6

*Per cent. increase or decrease of uniform over concentrated load stresses.

From the last table it is seen that the chord stresses produced by the uniform load are slightly in excess of those produced by the concentrated loads; while the web stresses produced by the uniform load are a small amount less than those produced by the concentrated loads; although in but one case, i.e. the hip vertical, is the difference as much as 4 per cent.

THE ETCHING OF IRON AND STEEL.

VICTOR HUGO, '96·

In consequence of the large extent to which iron and steel enter into modern engineering construction, a thorough knowledge of these materials is essential.

When a shaft is built up of scrap, necessitating many welds, and impure and foreign material is largely distributed throughout the mass, there is necessarily a great uncertainty as to its strength. A testing machine may show that its tensile strength is low, but we are left to guess as to whether its low strength is due to the quality of the scrap or to the piling and working.

A very simple method of ascertaining the condition of the piling and other characteristics of the iron, is to plane off one end of the shaft to be examined, and after carefully smoothing with a fine file to remove the tool marks, immerse the specimen in a bath of hydrochloric acid. The strength of the acid depends upon the material to be etched. For iron, the proportion giving the best satisfaction is one part acid to two parts water, and for steel, from one-half to full strength. The time of immersion will vary

28   *Victor Hugo.*

from six to twenty-four hours. After the material is etched sufficiently the acid is washed off and printer's ink is rubbed over the surface with a hand roller such as is commonly used by printers. An impression is then taken upon paper, the best results being obtained when the paper has a soft backing.

This method is extensively used in the laboratories of the Chicago, Milwaukee & Saint Paul railroad, and the writer is greatly indebted to Mr. George Gibbs, mechanical engineer of the road, for the accompanying etchings. The railroad company has very rigid specifications for all classes of material used, and a specimen from each invoice is tested both physically and chemically before the material is accepted. The etching is made part of the physical test and gives more information than could be obtained from a simple inspection of the fracture. The etchings are very clear and well defined, and from them an experienced person can read a great deal. The accompanying cuts are not as clear as the original etchings.

Fig. 1.

Fig. 1 is an etching of a scrap iron car axle; the figure shows it to be moderately homogeneous, the welds good, and it would pass as fair material.

The Etching of Iron and Steel. 29

Fig. 2.

Fig. 2 is a steel car axle, the steel being of the "Coffin's process," the figure showing it to be full of pin holes.

Fig. 3.

Fig. 3 is an etching of an axle from a foreign car, which broke and caused a wreck; the figure shows the breaking to be due to the lack of proper working. The dark portions are pieces of hard steel and the sharp, distinct line between the light and dark portions shows that the two are not welded. In figure 1 there are no sharp lines, the dark and light portions being blended together, showing thorough welding.

30   *Victor Hugo.*

Fig. 4.

Fig. 4 is a horizontal and end section of a hollow stay bolt; the fibre is very uniform and shows an excellent iron.

Fig. 5.

Fig. 5 is a bolt of "Ulster" iron; the figure showing it to be good iron, but lacking sufficient working.

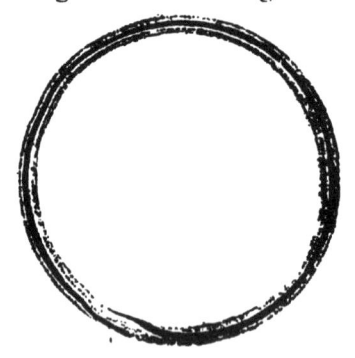

Fig. 6.

Fig. 6 is an etching of a composite boiler tube; showing a layer of steel. This figure gives some very valuable information, which would be almost impossible to obtain from any other source. Flue manufacturers sometimes work in steel instead of wrought iron. The reason is obvious, as the steel makes the tube show a higher test and renders it less liable to crack while being expanded into the boiler heads. On the other hand, steel is more susceptible to corrosion from the feed water, and the best engineering practice prohibits the use of steel tubes.

As an application of this etching process, let us suppose a railroad wreck to be caused by the breaking of an axle and the railroad company to be sued for damages. Naturally they will try to shift the blame on the company that manufactured the axle. The axle is tested in the testing machine and its strength is found to be low, but this proves nothing, for the axle company will say that the axle was strained in the wreck beyond its elastic limit, and of course could not now be expected to show its original strength. It is next analyzed for sulphur and phosphorus, but is found to be low in these elements, and it begins to look bad for the railroad company. An etching is then made, and supposing it to look like figure 3, it would show that the axle was not properly worked, and the axle company would be obliged to pay the railroad company's damage suits, because the material was not according to contract.

This method could also be used to show the effect on boiler plates when punched with the different kinds of rivet hole punches, and also to show the effect of drilling the hole or of reaming the punched one.

Though these etchings are generally of more interest than value, yet in some branches of manufacture they are most useful, and it is believed that as more and more "built up" material is used their value will be greatly increased.

GOOD ROADS.

C. PAUL JONES, '96.

The necessity of improved roads in both city and country is no longer doubted and the question is as to the best means of attaining the desired result. In 1884 when the good roads movement was started in this country, the great majority derided the enthusiasts who were to secure macadamized roads all over the country. It seems they wondered not so much at the object advocated as at the gigantic proportions of the undertaking. They said, "Look at the thousands of miles of country road in the United States." "Why, the government would become bankrupt before even a fair start had been made!"

Rome was not built in a day, nor were those superb roads of Germany and France, to which we point as examples, the work of a single year.

There is no one who does not see the need of better roads, especially when the roads over which we must travel are rough and muddy. Still there are many people who do not realize the immense amount of time and energy, both of which mean money, wasted on bad roads over which they must travel. Many of these people desire better roads but do not know how they may be had without the expenditure of a sum of money, in many cases, greater than the assessed valuation of the property benefited. Discouraged by this they continue to pay their taxes or work the necessary time on the road.

With the present condition of the public mind, one of the first steps must be a thorough education on matters relating to first cost, amount of aid to be expected from the state, and what a systematic method of road construction is. This step has been or is being taken. We have *Good Roads*, a magazine; a government bureau at Washington and in almost every state good roads conventions are held at short intervals.

The education is being carried on, and slowly people are realizing that their money may be spent to better advantage.

A new design or method of building a structure of any kind is necessarily an experiment and some one must make this experiment. The system of modern road building as applied to extensive areas went through the experimental stage in the state of New Jersey and although the method will of necessity be somewhat different in each state, yet New Jersey will serve as an example.

The agitation for good roads commenced in New Jersey about 1884; by 1887 enough interest had been aroused to call a convention at which the farmers were largely represented. This convention appointed a committee to examine the existing road laws and see what could be done in the way of legislation for better roads. After a good deal of labor the committee made its report and advised that the then existing system of road overseers be abolished and the care of the roads turned over to the township committee. A bill to this effect was introduced into the legislature in 1888 and, after failing three successive times, finally passed in 1891. Since its going into effect in 1892 it has given entire satisfaction and is conceded to be a much better plan.

The first practical operations for stone roads began in Essex county about 1876, these roads being built under a special law. Union county was the next to make an effort, and it secured the passage of a general law, known as the Union county law, allowing counties to bond themselves to build roads. Under this law the county borrowed money and covered the county with a complete system of telford or macadam roads and, with the interest of bonds added to the tax levy, the taxes were lower than before. A number of other counties took advantage of this law and improved their roads. The matter of good roads became of such interest and were found to be such valuable improvements in counties where they had been built, that the state government at last took a step toward helping in the good work.

The law known as "the state aid law of 1891" has been of great advantage to those who have partaken of its benefits. It offers an easy way of obtaining money for road purposes and the most equable way of distributing cost of improvements yet proposed.

The principal points of this law are as follows: On the presentation of a petition to the township committee by two-thirds of the property owners on any section of road not less than a mile in length, praying for the improvement of that road, the committee shall grant the petition unless the cost of road improvements has reached 5 per cent of the assessed valuation of the county in that year.

The law provides, that proper survey shall be made and bids be received for first-class telford or equally good stone roads; specifications and plans shall be approved by the president of the State Board of Agriculture. The cost of these improvements is to be paid, one-tenth by holders of the abutting property, two-thirds minus one-tenth by the county in which the improvement is made and one-third by the state. The sum of seventy-five thousand dollars is annually set aside for this purpose and if in any one year the state's portion is larger than this amount, it is divided *pro rata* among the counties making application for state aid.

One of the great objections to state aid is that the giving of state money for local improvements tends to make legislators strive to secure as large a slice of the state funds as possible for local purposes; moreover, it is unjust; the funds of the state are raised by general taxation and should be spent for general state improvement and not for the benefit of a small but favored section.

By the New Jersey law the amount paid by the state is comparatively small and serves mostly as an encouragement. The money is paid in such a way that there is no chance for ambitious legislators. The greater part of the cost comes on the county in general and a small share upon the property whose value is raised by the improvement. State aid in some such way as this is to be commended.

After securing money for roads, whether it be by state aid or otherwise, there must be a systematic method of expending it else it will be wasted, as is a large amount raised annually for road work. The system is still a matter of individual opinion, but personally I think the first step should be the election by the people, or the appointment by the governor, of a state road engineer and assistants, the chief to be paid by the state and the assistants by the district in which they are placed. The chief engineer's duty shall be to prepare plans and specifications for roads, bridges and culverts, to advise as to the best kinds of paving and road materials and the methods of laying them and to have general supervision over his assistants.

The first work of the assistants shall be to devise a complete system for the thorough drainage of the roads in their district, and beginning with the most important, drain all the roads and get rid of all water which would tend to stand at the side of the road or under the road bed. The great enemy of good roads is water; the road bed must be *dry*. After the roads have been thoroughly drained the surface can be put on, this being a mere matter of finish. The surface can be put on as rapidly as the district can afford it, and it has been proved that in New Jersey under the county bonding plan this can profitably be done at a cost of $3,460 a mile, and this under quite adverse circumstances.

With the experience of the states that have gone ahead in this matter and the increasing demand for better roads, it will not be long before Minnesota must take measures toward practical road improvement. In some cases the counties only await permission to issue bonds, in others they must have some encouragement. Road material of good quality is easy of access and near at hand in most parts of the state, and we have the engineering ability. There remains only the necessity of bringing the people to a full understanding of their need and how to satisfy it. When this is done Minnesota will have as good roads as those states that have been prime movers in this movement.

COMPETITORS OF ELECTRIC LIGHT.

GEO. D. SHEPARDSON.

Artificial light is a necessity in civilized life. Uncultured tribes retire with the fading daylight, but with civilization comes the desire to prolong the hours of business and social enjoyment. The production of a satisfactory light is a problem of universal importance, and the business of supplying light is one of enormous magnitude. England alone [1] requires eleven million tons of coal annually for the manufacture of illuminating gas and the annual exports of mineral oil from the United States amount to more than fifty million dollars. The rapid development of electrical science and industries is largely a result of the demand for better and cheaper light. Conversely, the recognized advantages of arc and incandescent lamps have greatly stimulated rival methods of illumination, and it is worth some study to learn of the history and present state of the art of lighting by these competitors.

Methods of artificial lighting are of great antiquity. A number of passages in the Bible indicate an early acquaintance with candles and lamps. Genesis [2] contains an account of Abram (cir. 1913, B. C.) seeing a smoking furnace and a burning lamp. Candles are mentioned several times in Job. The tabernacle [3] (cir. 1491, B. C.) contained an elaborate candlestick with seven lamps [4] burning pure olive oil. Herodotus [5] (cir. 450, B. C.) describes Egyptian lamps having wicks floating in vessels filled with olive oil and salt. The Greeks [6] and Romans used tallow

---
[1] Am. Gas Light Jour., LX: 618, Apr. 30, 1894.
[2] Gen. XV: 17.
[3] Exodus XXV: 31.
[4] Ex. XXVII: 20.
[5] Herod. II: 62.
[6] Smith Dict. G. & R. Antiq. 236.

and wax candles with rush wicks. Their early lamps were of unglazed pottery with one hole for the wick of flax-tow or rushes. Sometimes these lamps were of highly ornamented metal, having from one to twelve wicks. In the time of Homer [7] (cir. 1000, B. C.) lighted torches were used for nocturnal excursions. Lanterns and torches were carried [8] by the Jews and Romans who arrested the Savior. Elaborate lanterns with horn chimneys were owned by the Romans. Their torches [9] were ropes impregnated with inflammable materials. Ptolemy built a lighthouse [10] at Pharos, the site of Alexandria, Egypt, about 300 B. C. The tower was 450 feet high and was surmounted by a fire of burning wood that could be seen for forty-two miles. At an early date the streets of Antioch were lighted at night, and probably those of Rome also.

After the fall of the Roman Empire, little progress was made until near the close of the eighteenth century. Public lighting of streets seems to have been abandoned until about 1770, A. D., when lanterns containing oil lamps [11] were used in some of the streets of Paris, the first modern city to adopt this means of diminishing the frequency of robbery and murder.

Until about 1780, the candle and the ancient oil lamp with round wick were the only means [12] of artificial lighting. In 1783, Leger used a flat wick instead of a round one. The marked improvement in the combustion and brilliancy of its light led Argand [13] to carry the idea further. About 1784, he brought out his cylindrical burner in which the combustion of the oil was improved by the central draught. This seems to mark the distinction between ancient and modern lighting. In 1803, Carcel brought out a lamp with clockwork for feeding heavy oils by pressure, thus enabling the use of heavy vegetable oils. The "moderator" lamp of Franchot, in 1836, was much stronger than any of its predecessors.

Petroleum in different forms has been known and used since

---

7 Dict. G. & R. Antiq. 236.
8 John XVIII: 3.
9 Fosbroke Enc. Antiq. I: 323, 469.
10 Knight Dict., Mech., *Lighthouse.*
11 Sci. Am. Supp., No. 877, p. 14011.
12 Sci. Am. Supp. No. 877.
13 Chamb. Cyc., *Lamps.*

the earliest days of the human race. From time immemorial, it has been used in China [14] for lighting towns near oil springs and deep wells. The ruins of Nineveh and Babylon attest the accuracy of the account in Genesis XI: 3, that slime (asphalt) was used (cir. 2350, B. C.) as mortar in large edifices. The account in Job XXIX: 6, mentions "the rock pouring out rivers of oil." Near the Caspian Sea, oil springs have been flowing for certainly more than 2500 years.[15] At a temple of the Parsee fireworshippers near Baku on the Caspian Sea a bluish flame has burned continuously for centuries without human care, the gas coming from natural sources. Herodotus [16] (cir. 450, B. C.) mentions that liquid bitumen was found on Zacynthus, one of the Ionian Islands. This was one ingredient of the much feared Greek fire. Pliny (23-79, A. D.,) and Dioscorides [17] (cir. 150, A. D.?) state that at Agrigentum in Sicily, the oil floating upon the water of a spring was collected upon reeds and burned in lamps, and used for horse liniment. Petroleum has been collected on the surface of certain wells in the northern part of Italy since 1640. Natural gas was used [18] at Wigan, Great Britain, in 1667.

Petroleum has been found in America since prehistoric times. Indisputable evidence [19] shows that certain oil springs in Pennsylvania, Ohio and Canada were opened 500 to 1000 years ago. The first written account [20] of an oil spring in America was by a French missionary who visited in July, 1627, the one at what is now Cuba, Allegany county, New York. The Seneca Indians skimmed the oil from the surface of a large pool and used it as a linament. Later this was sold to the whites as "Seneka oil," "rock oil," or "coal oil." Considerable oil was obtained [21] from a well dug near Marietta, Ohio, in 1819, and was used locally for lighting. The disagreeable odor caused it to be considered a nuisance rather than a valuable commodity. Flowing springs and wells were found in West Virginia at an early date. The town of Fredonia, in western New York, was illuminated by

---
14 Knight Dict., Mech., p. 943.
15 Chamb. Cyc., *Baku*.
16 Amer. Cyc., *Baku*; Herod I: 119; II: 195 Amer. Cyc., *Petroleum*.
17 Diosc. I: 99. Knight Dict., Mech., p. 1673.
18 Knight Dict., Mech , I: 943.
19 Chamb. Cyc., *Petroleum*.
20 Ashburner, Trans. Am. Inst. Min. Eng., June, 1887.
21 S. P. Hildreth, Am. Jour. Sci., X: 5, 1826.

natural gas in 1821 and attracted great attention. About 1835 natural gas was found near Gambier, Ohio, where it has been used continuously for many years in the manufacture of lampblack. The petroleum business did not assume any considerable magnitude in America until 1860. In 1859, crude petroleum sold [22] at $20.00 a barrel. The following year the new wells in Northwestern Pennsylvania put 200,000 barrels upon the market, and the price fell to $10.00 per barrel, going down to 50 cents in April, 1861. In June, 1861, the first flowing well was struck and spouted 300 barrels per day. Soon a third well began spouting 3,000 barrels per day, and oil dropped to 10 cents per barrel, an empty barrel being worth fifteen times as much as the oil it would hold. In less than a year, 1,500,000 barrels of oil were produced along Oil Creek in Pennsylvania, more than half of it running to waste. Some wells yielded their owners $15,000 income per day. Prices fluctuated greatly and rapidly. Since 1869, the price has steadily dropped from $7.00 per barrel and, since 1878, has been below $2.00 per barrel, most of the time being below a dollar. Immense quantities of petroleum products are shipped from America, the export trade exceeding the home consumption. From 1864 to 1890, the petroleum exports[23] aggregated $1,043,474,435; 664,500,000 gallons of mineral oil were exported in 1890, with a valuation of $51,403,000.

Outside of the United States, the principal petroleum fields are in Burmah and near the Caspian Sea. The latter regions were taken from Persia by Russia in 1806,[24] but were not developed until 1872, after Russian engineers had been sent to America to investigate our methods. Upon their return, the fields about Baku were rapidly opened, and the number of wells increased from one in 1871 to 400 in 1883. The production in 1885 was 420,000,000 gallons of crude oil. When petroleum and its numerous derivatives became cheap enough for ordinary lighting, various forms of burner were invented. The flat and Argand burners, previously used for animal and vegetable oils, were easily adapted for kerosene. The modified Argand burner of the

---
22 Lebanon Gazette, Apr. 11, 1895.
23 Chamb. Cyclopedia, *Petroleum*.
24 Chamb. Cyc., *Baku*.

student lamp and the more recent and powerful "Electric" and "Rochester" burners are sufficiently familiar. Numerous forms of burner are extensively employed for lighting with gasolene and naphtha. All of the latter heat the liquid and convert it into vapor before burning. After the vapor has been generated, it may be burned with almost any form of tip suitable for ordinary gas. In the common form of plate burner, the vapor issues from a needle pointed orifice and strikes a curved plate, which spreads it out into a flat flame.

Although natural gas is generally found in connection with oil wells, it was not developed extensively until 1882, when large gas wells were struck in western Pennsylvania followed soon after by New York, Ohio, West Virginia and Indiana. The gas commonly had a pressure of 100 to 325[25] pounds per square inch, in one instance rising as high as 750 pounds. The natural gas caused the rapid growth of many towns and revolutionized several industries, being found much more economical and more easily managed than any other form of fuel. While the pressure is diminishing in many places, the supply promises to last for years. Artificial gas was first distilled[26] from coal in 1688 by Dr. Clayton of England. In 1750 Bishop Watson conveyed gas in pipes to some distance. In 1792 a Mr. Murdock lighted his house and office at Redruth, Cornwall, England, with coal gas. In 1798 he lighted some shops at Soho. In 1802 M. Lebon[27] lighted his own house and proposed to light the streets of Paris. In 1803 one of the theatres of London was lighted by gas. About 1807 Pall Mall in London was lighted. English Parliament chartered the first gas company in 1810. The first large installation of street gas lights was in 1813 on the Westminster bridge in London. In 1815 the streets of Paris and London were lighted by gas. Baltimore in 1816, Boston in 1820, and New York in 1825 were the first American cities to adopt gas.

The gas generally used in England and many of the older and smaller towns of America is made by the dry distillation of bituminous or "soft" coal in closed retorts. About 14 per cent. by

---

25 Sci. Am. Supp., No. 549, p. 8765.
26 Knight Dict., Mech., p. 944.
27 Knight Cyc. Arts and Sci., IV, 308; Fosbroke Enc. Antiq., I, 469.

weight[28] is realized as gas, and about 66 per cent. as coke, about 20 per cent. of the coke being afterward used for heating the retorts. The gas more commonly made in America, especially with recent works, is made by a different process and is known as fuel gas or water gas. The details vary in the apparatus of different makers, but the general principles are about the same. The manufacture of water gas is based upon the fact that many substances dissociate into their elements at very high temperatures and may form other compounds. Steam, with or without air,[29] is passed through a gasogene, a furnace containing a bed of highly incandescent coal, forming carbonic acid gas (CO_2) and hydrogen. The CO_2, in passing further through the incandescent mass, becomes reduced to carbon monoxide (CO). The gases leaving the furnace, or gasogene, consist of carbon monoxide and hydrogen with more or less of nitrogen and some carbon di-oxide. This part of the process simply changes the combustible from the solid to the gaseous state. The resultant gases would burn with a hot but non-luminous flame. They are next passed through a carbureter, in which they are intimately mixed with some hydrocarbon vapor, such as naphtha, benzene or naphthalene, which will make the flame luminous by the presence of carbon particles. The carbureted gas is then passed through red hot retorts, the heat fixing it so that the hydrocarbon vapor will not separate and condense. After passing through purifiers, which remove objectionable impurities, the gas goes into large receivers or holders from which the supply is drawn. Water gas costs[30] about 40 per cent. less than an equal volume of coal gas. Twelve pounds of anthracite coal will yield about 1,000 cubic feet.

Considerable interest is being taken at present in the recent discovery that acetylene can be produced cheaply by means of the electric furnace. This gas has high illuminating power and important developments may follow. The paper by Mr. Rounds in this volume contains an interesting resume.

The earliest type of gas burner was the "cockspur" with

---
28 Sci. Amer. Supp., No. 379.
29 Sci. Am. Supp., No. 898.
30 Sci. Am. Supp., Nos. 58 and 303.

three jets which gave long and unstable flames affected by the slightest atmospheric movement. The batswing burner[31] having a slit aperture and giving a flat fan like flame, was introduced about 1816. The fishtail burner, with two orifices at an angle of 60° so that the jets cross and flatten, was invented about 1822. These burners gave a much more steady flame, besides being more efficient. Later improvements have increased the efficiency of the fishtail burner from 1.66 to 3.09 candlepower per cubic foot of gas burned per hour, while the batswing has improved from 3.01 to 4.0 candles. The efficiency varies greatly with the adjustment and the pressure. For every form of burner there is one pressure at which it will yield greatest economy. In a given case, a fishtail burner gave 3 candlepower per cubic foot with gas at a pressure of 0 5 inches of water, but only 1.11 candlepower at 3.0 inches pressure. Between 80 and 90 per cent. of the burners used in England are of the ordinary type of brass-cased fishtails.

The Argand burner, brought out in 1784 for oil lamps, has been used for gas with many modifications. It gives more light than flat flame burners using the same amount of gas, but has the disadvantage of requiring a glass chimney. A number of regenerative or recuperative gas lamps have been invented with a view to increasing the efficiency of the flame, by heating the gas or the air or both before combustion. The principal burners of this type were by Siemens,[32] Wenham[32] and Schulke.[33] The relative efficiencies of these burners[34] is, that with equal rate of burning gas, a flat flame burner gives 14 candlepower, an Argand 17, and a Wenham 35 candlepower.

In order to appreciate the reasons for the next development, it is necessary to consider the real source of the light of the candle, oil-lamp and the common forms of gas burner. The flame is a column of heated gases which carry off by convection 80 per cent.[35] or more of the heat of combustion. The flame owes its luminosity to the comparatively few particles of unoxidized carbon near the burner that are heated to incandescence by the com-

---

[31] Knight Dict. Mech., *Gas Burner*, 946. Bellamy, Am. Gas Light Jour., LX: 619.
[32] Jour. Fkln. Inst., CXVIII: 298, Oct., 1884. Sci. Am. Supp., Nos. 219 and 301.
[33] Sci. Am. Supp., Nos. 454 and 482.
[34] W. J. Dibkin, Am. Gas Lt. Jour., LX: 624, Apr. 30, 1894.
[35] E. L. Nichols, El. Eng., X: 595; Elec. World, XVI: 387, 409.

bustion of the remaining portion. If combustion were perfect, the flame would be non-luminous, as with the Bunsen burner. The flame makes an inefficient furnace and the incandescent particles utilize only a very small fraction of the total amount of heat. Of the energy radiated by these particles, only a small part is of such wave-length as to affect the organ of vision and give light. If the net efficiency be taken as the ratio of the luminous energy radiated to the total energy consumed, the net efficiency of an oil or gas light varies from 0.1 to 0.3 per cent.

The recent progress in lighting (electricity not being considered in this paper), is in obtaining complete combustion of the gas, and heating to a high temperature some incandescing substance. The most familiar of these is the calcium[36] or Drummond light, in which a point in a lime cylinder is heated white hot by an oxyhydrogen flame, or by compressed oxygen and illuminating gas. After the invention of the Bunsen burner, many experimenters attempted to use it for lighting. In 1881, Clamond[37] heated a basket of magnesia to incandescence with a Bunsen burner. The basket was too short-lived. In 1882, Victor Popp heated a basket of platinum in a similar way. In 1883, Somzee[38] obtained a bright white light from a perforated capsule of lime or porous magnesia coated with zirconia. In 1883, Fahnehjelm[39] heated combs of magnesia rods held in an arch above a flat non-luminous flame of water gas. The combs lasted from 80 to 150 hours. This was used to a considerable extent in America. In 1890, Fahnehjelm made his pencils more durable by mixing magnesia with oxides of chromium, wolfram, manganese, cobalt, nickel or copper, singly or mixed. In 1891, Haitinger used alumina and oxide of chromium or of manganese, intimately mixed mechanically and subjected to a high temperature. This gave a bright, warm, reddish-yellow light. In 1893, Hirshfield[40] used a basket of sulphate of alumina and chromium oxide.

The Welsbach burner[41] seems to be the highest type of incandescent gas lamp. Carl Auer von Welsbach, a student of Profes-

---
36 Sci. Am. Supp., Nos. 281 and 328.
37 Sci. Am. Supp., No. 561.
38 Sci. Am. Supp., No. 561.
39 Sci. Am. Supp., No. 493. Trans. Am. Inst. Min. Eng., Feb., 1885.
40 Progressive Age, XII: 498.
41 Jour. Fkln. Inst., CXXV: 379.387, May, 1888.

sor Bunsen at the University of Heidelberg, became interested in illumination in 1880. His mantle burner is the result. He used a mantle knit from fine cotton thread into a cylinder. This was saturated in a solution of the salts of lanthanum, zirconium and yttrium. The cotton was then burned away, leaving a fragile skeleton or basket of the minerals. Welsbach's first English patent was applied for[42] in 1885. The next year he obtained another patent for using thorium oxide alone or with alumina. The thoria lengthened the life of the mantle and increased the amount of light by 20 to 30 per cent. Large numbers of these burners were placed upon the market in Europe and America. In this

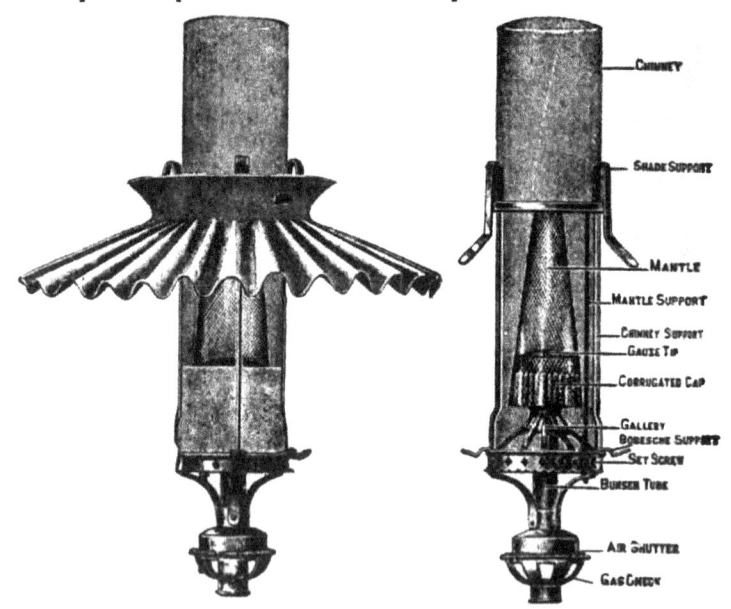

DETAILS OF THE WELSBACH BURNER.

country they soon lost favor and disappeared. About 1892, an improved Welsbach burner came out and is now meeting with great success, immense numbers being sold. The principal improvement is in the mantle. The cotton mantle is saturated[43] in a solution of 98 per cent. of thoria and 2 per cent. of cerium.

---

[42] Sci. Am. Supp., No. 606.
[43] Progressive Age, XII: 498.

After the cotton has been burned away, the basket is dipped in a weak solution of caoutchouc or collodion, in order to withstand transportation. This mantle is suspended over a Bunsen burner and becomes highly incandescent, giving a steady column of white light about 3 inches high and 1 inch in diameter. These give over 60 candlepower with a consumption of about 2.5 cubic feet of gas per hour. The two cuts showing exterior and interior views, give an excellent idea of the construction and general appearance of the most recent type of Welsbach or Auer burner. The chimney is roughened at the bottom to conceal the lower part of the burner. A glass or porcelain shade serves to throw the light downward. Sometimes a hemispherical opal globe, or sconce, is placed upon the lower supports for the purpose of diffusing the light so as to be more comfortable for reading or writing.

In comparing the desirability of various illuminants, we might well consider four principal items, namely, quality, efficiency, by-products and cost. Under quality should be considered steadiness, color and intensity. For reading or other special lighting, one of the most objectionable features of gas light is the continual flickering that seems unavoidable with flat flames. For this reason, as well as on account of greater cheapness, kerosene lamps will doubtless continue to be used for the bulk of domestic lighting. The color of the light is of importance for many purposes. The human eye is constructed for daylight, and the more nearly the character of artificial light approaches daylight, the better for vision and for the health of the eye. The white character of the electric arc, the calcium and the magnesium light, render them preferred where careful distinctions of color are desired. The ghastly greenish-yellow tinge of the Welsbach light is its greatest drawback. The constant aim is to rival the intensity of daylight. Each step of progress by one illuminant assists its rivals by fostering the demand for stronger light, so that the introduction of electric lights has generally been followed by an increased use of gas. The room for growth in this line is indicated by Bouguer's estimate[44] that the quantity of daylight ordinarily spread around a large city like Paris, for instance, represents more than ten

---

[44] Sci. Am. Supp., No. 877.

thousand times the entire amount of artificial light used there.

The efficiency of a light source is of interest from a scientific standpoint, and is of importance as it affects cost. The by-products of an illuminant are usually undesirable, being heat, smoke, smell, gases and danger from fire. The experienced agent for any particular illuminant will wisely select his favorable points and keep judicious silence on others. It might be noted here that the Welsbach burner does not give off poisonous monoxide[45] when properly adjusted, as has been reported. The cost of an illuminant should include original cost of fixtures and appliance, operation, maintenance and possible accidents.

---
45 Progressive Age, XII, 358, 409.

SYSTEMS OF MINING IN MINNESOTA IRON MINES.

CHARLES DEAN WILKINSON, '95.

The iron mines of Minnesota may be divided into two general classes according to the condition of the deposit:

I. Mines in which stripping will pay;

II. Mines in which stripping is not possible or will not pay.

The dividing line of these two classes is determined by a rule formulated by Mr. Denton, engineer of the Soudan mine at Tower:

"Stripping will pay when the volume of the ore is equal to, or greater than, the volume of material to be stripped." It must be noticed that volume and not depth is considered in this rule.

The first class comprises:

A. Mines in which the deposit is thin and covers a large area;

B. Mines in which the deposit is thicker and covers a more restricted area.

The second class comprises:

C. Mines in which there is a poor hanging wall, especially where the deposit is directly under glacial drift;

D. Mines which have a good hanging wall, or in which the deposit is at a great depth.

These divisions merge into one another, as for example the first general division into the second, or A into B, etc.

The systems applied to each of these conditions of deposit are as follows:

To A. Open-cut mining with steam shovel as in the Oliver mine.

To B. The "milling" system as used in the Auburn mine.

To C. The "caving" system as used in the Canton mine.

To D. The ordinary system of "stoping-out" as used in the Minnesota mine at Tower.

Mining by these different systems is carried on as follows:

First, open-cut mining with steam shovels.

The exploration preliminary to the location of mines on the Mesabi range is generally carried on by means of test-pits until the ore is reached. After striking the ore, investigation as to its general quality, the thickness of the deposit, etc., may be carried on with diamond drills.

When the location has been made the stripping is done either by steam shovels or by hand. The larger number of mines do the stripping by steam shovel, but where many bowlders are encountered, as in the Biwabik, it may be necessary to do the work by hand. While ore is being taken out benches must be left around the edges of the area already stripped, for the operation of the stripping shovels.

In removing the ore if the deposit is very soft the shovels may work directly into it, but if the ore is hard it must be loosened by blasts and loaded into the cars by the shovel. The working as carried forward gradually slopes downward, so that a great depth may be obtained without making the gradient of the tracks too heavy. The ore is loaded directly into cars by the shovel and the cars for one shovel may be handled by a single locomotive. The crew for a shovel consists of an engineer, crane-man and helper. The claim is made that a good shovel and locomotive can load and dispose of a 20-ton carload of ore every six minutes. This estimate is approximately true.

This system is peculiar to the Mesabi range and to some of the workings in the Alabama iron regions.

Second, the "milling" system.

Exploration and stripping for mines operated on the "milling" system is carried on in the same manner as for the steam shovel method.

When the deposit is stripped, a shaft is sunk at one side deep enough to run in a level to the lowest point of the ore. Another shaft is sunk to connect with the end of this level and at the bottom an ore-chute is placed. Blasts are now fired around the mouth of this second shaft in such positions that the loosened material will fall to the bottom into the ore-chute. In the Auburn mine the ore is trammed from this chute to the main shaft, the haulage being done by mules. The material is hoisted by skips and dumped either into small cars and conveyed to the stock piles, or loaded directly into the railroad cars.

As the work proceeds radially outward from the secondary shaft the hole presents the form of a volcanic crater. The men doing the drilling for the shots are suspended from the edge by ropes and the blasts are fired at change of shifts.

The only level driven is the one which connects the main shaft with the ore-chute, and the only timbers necessary are the pit frame, shaft-timbers of the main shaft and the level timbers.

With a correct gradient, gravity haulage may be substituted for mule haulage in the level.

The "milling" system is peculiar to the Mesabi range and was first introduced at the Berringer mine by Dr. J. A. Crowell and Mr. F. R. Whittelsey.

Third, the caving system.

Where the overlying material is too thick to strip and not firm enough to furnish a good roof for ordinary stoping out, it is necessary to use another system of mining, and this is provided in the "caving" system as used in the Canton mine at Biwabik and in part of the Chandler mine at Ely.

In operation the shaft is sunk and the top level driven first. At the extreme end of the level a room is cut out and the roof supported by light timbers or pillars of ore. When all the material from the room has been excavated, the pillars are "robbed" and the roof allowed to cave. The next room is then taken out, and so on till the level is completed. Then the second level is treated in the same manner. If the roof in any case refuses to cave, blasts are placed at the four corners and fired simultaneously. After caving, the material thus brought down may be excavated. Thus when the bottom of the deposit is reached there is nothing left but the glacial drift or other overburden. Trouble is often caused in mines operated on this system, by matting or wedging of timbers after caving down one or two levels, so that the roof will not fall.

Fourth, "stoping-out" system. Most of the work on the Vermillion range is carried on by means of the ordinary system of stoping-out.

The Minnesota or Soudan mine furnishes the best example of this system, as it has been the longest in operation and has the most finely equipped plant.

The exploration is done with diamond drills and after the deposit is located a shaft is sunk to the lowest part. Cross-cuts are run in for about 100 feet. From the ends of the cross-cuts, levels are run at right angles to them, as far as the ends of the deposit. Beginning at the end, the material is taken out by overhand stoping. The ore is milled to the level below, mills being provided at short intervals. The space opened is filled with broken country rock taken from the sidewalls or brought in from outside. A permanent roadway is left at the bottom of the level and is timbered with sets consisting of two posts and a cap, no sills being necessary. Round lagging is used. No other timbers are necessary in the levels as the hanging wall is very strong.

Ore-chutes are provided at the bottoms of the mills and open into the permanent roadway. The ore is trammed to the shaft by men and hoisted in cages. Both percussion and diamond drills are used in drilling for blasts. This use of the diamond drill for blast holes is peculiar to the Minnesota mine. Two roof holes running nearly parallel to the level are put in at opposite sides of the level and meet to form a V. The holes are about thirty feet long and are loaded with thirty pounds of dynamite which fills the hole to about two-thirds of its length. The two shots are fired simultaneously and bring down an immense amount of ore.

The following comparison of systems may be stated: That the open-cut steam shovel method is the cheapest for shallow deposits where the stripping is thin, cannot be questioned. However when the deposit becomes thicker, the milling system would seem the more efficient for two special reasons: First, if the ore body contains much water it will be necessary to drive shafts and tunnels for drainage, and such tunnels and shafts once driven the milling system will be cheaper. Second, as the working goes deeper and deeper the gradient of the railroad track becomes heavier and heavier until the engines will have great difficulty in hauling out trains.

The first of these reasons has much weight, as nearly all the Mesabi mines are very wet at the lower levels, and the second because a steam shovel and a locomotive are both machines of poor efficiency when compared with the results obtained from strong, well constructed and proportioned stationary engines, such as are used for winding in the milling mines.

The last two systems are the product of well defined conditions. To the present time no better methods have been found for such mining. They are certainly more costly than the open cut and milling systems but, with high grade ores and fair prices, can be operated with profit. The permanent plant necessary includes winding engines, compress or plant, machine shops, shaft-houses, pit-frames, pumping engines, and possibly crushers for hard ore mines, must be considered in comparing the cost of systems.

The open-cut mine with its machinery consisting of two or three steam shovels, a locomotive, small pumping engine, compressor plant if necessary, and a repair shop, certainly has the advantage as to first cost. The question remaining to be answered is, how will the expenses of operating compare. This can be answered when some of the open-cut mines have worked out a deposit.

TOPOGRAPHY IN THE UNITED STATES.

JOHN A. BOHLAND, '95.

Topography is the art of representing the features of the earth's surface by conventional signs and symbols, and of locating them in their true relative positions.

The question then arises, to what extent can the features of the earth's surface be represented on an intelligible and useful map. This naturally varies with the purpose for which the map is intended.

European experience tends to reduce the number of features which should be delineated upon a public map. In the United States it is decided to go even beyond the present practice of European nations, and limit the map to the representation of such natural features as are of sufficient magnitude to warrant representation with the scale. The artificial features are restricted by those of general or public importance.

Following this plan the map represents buildings, roads, streets, railroads, bridges, ferries, tunnels, fords, canals and boundaries of civil divisions. Property lines and private roads are not represented.

The topographic maps in European countries are as a rule made by army officers with the aid of some civil experts. These countries have very complete topographic maps of their territories; most of which have been renewed in the last few years by modern methods and accuracy, the old maps proving inefficient.

The scales adopted by European nations varies from $\frac{1}{12500}$ to $\frac{1}{100000}$ for general work, and $\frac{1}{2500}$ for special work.

In the United States the topographic work is carried on by the United States Coast and Geodetic Survey; the United States Lake Survey; the Army Engineer's Corps; the United States Geological Survey; in some states by a State Survey, and by private enterprises.

Topography in the United States. 53

From the above enumeration it would seem as though we had our topography duplicated time and again, but it is not so. We have, compared with other modern governments, no topography whatever. Although any of the above organizations are equipped with plans and methods to do the work, they have never been authorized to execute it except in scattered sections.

The United States Coast Survey was organized in 1807. In 1871 its work was extended to hydrography and topography of the great rivers of the country, except the Mississippi and Missouri which had been assigned to special commissions. One of its functions now is to assist the different states in carrying on their topographic surveys. In 1882 the Coast and Geodetic Survey entered upon a systematic triangulation with the view of making a complete topographic map, and executed a portion of such a map in Maryland and Virginia, being linked under a law of congress with eleven states in preliminary work.

Stations have been established throughout the country whose longitude, latitude and magnetic bearing have been accurately determined. Base lines have been measured with very great accuracy by the most modern methods and from them a network of triangulation has covered a large part of the country. One of these base-lines is located along Snelling avenue in Saint Paul. The triangulation lines are from one mile to two hundred miles long in mountainous districts, and from ten to forty miles long in ordinary rolling plains. The angles require from twelve to seventy-two repetitions for the desired accuracy.

The latitude and longitude of all stations are computed by geodetic formulas from the initial point whose latitude and azimuth are astronomically determined and longitude telegraphically determined. One of these initial points is situated on the campus of the university. Thus is completed the preliminary work.

The topography is then worked up on this skeleton. This is done by means of the plane table and is a graphic triangulation with minor distances taken by means of the stadia. It may be remarked that the plane table is not very generally used in Euro pean countries, the objection being the exposure of the drawing to the moist atmosphere.

The table is oriented in almost every case from previously plotted points; three such points being necessary to make an independent setting. A system of traversing is, however, often imperative in which only the last occupied station is necessary by making a foresight and backsight. The compass is also often resorted to.

The measurement of altitude is attained by means of the vertical arc of the alidade and the level upon the telescope. Readings are taken for index error and applied to the angle measured.

The difference in height is obtained by the solution of the right. angled triangle, thus: $h = d \tan x$, in which $d =$ the horizontal distance and $x =$ angle of elevation. This involves the use of tables of trignometric functions which is avoided by the use of the empirical formula deduced by assistant W. C. Hodgkins of the Coast and Geodetic Survey: $h = \pm 291 K a + .07 k^2$ which is further simplified: for the first term, [*(dist. in kilometer)* × *(angle in min.)* ÷ *3*]—[⅛ *of itself*], and for the second term *the square of the distance in kilometers* × *.07;* this gives *h* in *kilometers*. Any other unit may be used as well.

Another empirical formula much used is: $(h \text{ in ft.}) = ($*angle in min.* × *dist. in meters* ÷ *1000*$) - \frac{1}{20}$ *of the part in parenthesis.*

This height *h* is to be corrected for curvature of the earth and for refraction by the atmosphere.

The former correction is made sufficiently accurate by the following empirical rule: The curvature in feet is equal to ⅔ of the square of the distance in miles; always positive in sign.

Refraction is uncertain, varying during the day and with the elevation of the sight above the ground. It is greatest in the morning and evening and least at mid-day. It is usually assumed at ⅐ the curvature, and is always negative. Differences of height are not measured in this way at greater distances than 10 miles.

The Coast and Geodetic Survey has covered the greater part of the Atlantic, Gulf and Pacific coast including Alaska, with a strip of topography,—narrow where the coast is simple, and broad where it is complex. Besides some work in topography has been done in the interior. The scale commonly used is $\frac{1}{10000}$, $\frac{1}{20000}$ and

$\frac{1}{10500}$ with contour intervals ranging from 5 feet to 20 feet; for special work a scale of $\frac{1}{100}$, $\frac{1}{1200}$ and $\frac{1}{1800}$ is used.

The Engineers' Corps of the United States Army has also done work in topography, instances, the Lake Survey and the Mississippi and Missouri River Surveys. The scale is $\frac{1}{10500}$. The Lake Survey was organized in 1841.

The United States Geological Survey was organized in 1878, and although topography does not come exactly within the scope of a geological survey, the geologists found it impossible to carry on their work with full profit without some sort of a topographic map, and accordingly in 1882 topography was added to their survey at an average expense of $215,000.00 per year.

The topographical work of the Geological Survey is done on a triangulation system with the simplest possible plane table. The compass is used altogether in orienting.

The triangulation is comparatively cheap, from ten to twenty stations being occupied per month, and only from twelve to sixteen pointings upon each station are made.

Distances are measured by some form of odometer when the alidade is not provided with stadia wires.

Datum points for height are obtained mostly from railroad grades. Secondary heights are obtained by the vertical arc of the alidade; but elevations for interior work are generally measured by aneroids.

The scale used is $\frac{1}{62500}$ with contour intervals from 10 feet to 50 feet, $\frac{1}{125000}$ with contour intervals from 10 feet to 100 feet and $\frac{1}{250000}$, with contour intervals from 200 feet to 250 feet.

This is all the accuracy necessary for their particular work; but the public has use for something more exact, and in a few years we shall be following the footsteps of European countries and repeat the work of earlier surveys with modern exactness.

Several of the states have engaged in the preparation of a topographical map, either independently or in conjunction with the Coast and Geodetic Survey or the Geological Survey.

Kentucky was the first state to engage in topographic work. A survey was begun in 1854 on the plan of a traverse survey, making special note of topographical and geological features.

Up to the present time the following states have been aided by the Coast and Geodetic Survey in the manner provided by congress: Ohio, Indiana, Wisconsin, New Hampshire, Missouri, Pennsylvania, Kentucky, New Jersey, Connecticut, South Carolina, North Carolina, Tennessee, New York, Vermont, Minnesota and a few other southern states.

The state of New Jersey instituted a topographical survey in 1875. The plan of the work contemplated a map on the scale of 1 inch to the mile, with contours ranging from 5 feet to 20 feet. In 1884 the completion was undertaken by the United States Geological Survey and was finished in 1887. The legislation needed for the establishment and continuance of the survey by the state has been promptly and practically unanimously voted. Connecticut and Massachusetts have also completed their topographical surveys.

In Pennsylvania the work is carried on principally by the State Geological Survey.

New York for several years, terminating in 1885, supported a survey. The work done was of a high grade but it has been suspended.

Missouri is making a very elaborate topographical survey.

Minnesota has about 2,500 square miles of triangulation ready for topography. The past season has been spent in work upon a topographic sheet covering Minneapolis and Saint Paul, including about 225 square miles. A map including Duluth has been prepared by the Geological Survey.

The maps, as fast as produced, have found extended use in all sorts of industrial enterprises concerned with the surface of the ground; such as the preliminary location of railroads and highways, the study of water supply and drainage, irrigation and protection from floods, the relation of land to water in connection with navigation, as the project of connecting Lake Superior with the Mississippi river by a canal, cadastral detail for taxation, general engineering projects and scientific surveys.

THE MECHANICAL THEORY OF THE ORIGIN OF THE MOON.

DR. H. T. EDDY.

Abstract of a lecture before the Society of Engineers, October 29, 1894.

The problem of computing the elements of any eclipse of the sun or moon is the same whether that eclipse occurred in the remote past or will occur next year or in the next century. To the astronomer, mathematical analysis gives as clear and certain a light in one case as the other.

In like manner the established principles of mechanics have sufficed to unravel the tangled skein of interaction between the moon and earth, and have made it possible to trace backward through ages not measured by years, the ever progressive action of the tides, and lift the veil of a history that goes back to the time when earth and moon were young together. This history can now be read with certainty, since it demonstrates mathematically what that past must have been in order to have given rise to the present as it now exists, and it predicts with equal clearness what the future must be.

This has been done in great detail by Professor G. H. Darwin in a series of remarkable mathematical memoirs published in the Transactions of the London Philosophical Society. In them he has investigated the cumulative effect of the mutual tidal action of the earth and moon, with a mathematical genius that places his name beside those of Lagrange and Laplace, the immortal founders of celestial mechanics.

May I briefly sketch a few of the more salient points of this theory, which fortunately admit of elementary presentation. Many matters pertaining to this subject will be found with admirable clearness in a little book entitled *Time and Tide*, by Sir Robert Ball, Astronomer Royal of Ireland; but it would be an injustice to both these distinguished investigators to make them responsible for many of the suggestions and implications contained in this sketch.

In looking for those causes which have been the principal factors in evolving the present state of our earth and moon from some earlier condition, science has learned to have regard principally to those activities which are continuous in their manifestation, and operate age after age to work out their effects, rather than to look for efficient causes to any cataclysm or other sudden paroxysm of nature. In geology, the erosion of continents has produced the vast deposits of the sedimentary rocks. Before the earth had a solid crust its history was marked by successive stages of gradual cooling. Once entirely molten and fluid, it must have blazed with light as the sun now does. Later in the process of cooling it became viscous, then plastic and lava-like, and later still its atmosphere was dense with steam and vapors, as that of Jupiter is today. At last only occasional clouds float in a clear atmosphere. Long ages in the future when the earth has lost much more of its elemental heat the oceans and finally the atmosphere will sink into the solid crust of the earth. Then the earth will be bare and dead, and as devoid of life as the moon now is. These changes past and future are the necessary results of the perpetual flow of heat from the earth's interior and its dissipation into space.

Accompanying these changes the tides have been busy in producing others perhaps not less remarkable.

It is clear that when the moon raises tides in the ocean, its final effect must be to retard the earth's rotation; for when the moon's attraction produces high water in that part of the ocean under it, the rapid rotation of the earth carries the region of high

The Mechanical Theory of the Origin of the Moon. 59

water forward with it. The moon then acts to drag this area of high water backward, so that it shall continue to lie directly under itself.

Such action of the moon is like that of a huge brake, and tends to diminish the rotation of the earth on its axis. The moon has operated in this manner for untold ages, and although it is doubtful whether our great time-piece, the earth, has had its rotary velocity changed by an appreciable amount within historic times (just as it is doubtful whether the mean temperature of the earth has diminished perceptibly within the same time) yet it is not doubtful that there was a time when the earth rotated more rapidly than now, just as it is not doubtful that the earth once was hotter than at present.

So great must have been the length of time, however, which has elapsed during these changes that it is useless to attempt to assign measures of either of these periods in years.

In tracing back either of these series of changes toward a beginning we must at last arrive at a condition such that we can suppose no previous states or conditions like them. In considering an earlier and earlier state of more and more rapid rotation, the limiting possible rotation is reached when the earth was rotating with a rapidity such that the centrifugal force at the equator was equal to the force of gravity. The time of rotation must then have been somewhat less than two hours, instead of twenty-four, as at present.

Any state of the earth antecedent to this, it is not our intention to consider at present, but rather to learn how it has fared with the moon meanwhile.

Mechanics tell us that action and reaction are ever equal and opposite. Hence if the moon has exerted a pull on the tides such as to retard the earth's rotation, then *per contra*, the earth has exerted a pull on the moon to make it go faster in its orbit. This may at first seem incredible, but such is nevertheless the fact, for the area of high water occupied by the tides is carried forward by the rotation of the earth and does not lie immediately under the moon as it would do in case the earth had no rotary velocity. The tidal mass is then so situated as to accelerate the moon in its orbit. Such acceleration can occur only by the moon's moving in a larger and larger orbit; for by Kepler's laws the velocity, period and size of its orbit have fixed relations to each other. The consequence of this tidal pull then has been that in all past ages since the moon first began to circle about the earth it has gradually moved further and further away from it, and has revolved in a wider and wider orbit about it.

You already anticipate my argument and see that, in the more and more remote past, the moon must have been nearer and nearer to the earth, until, in some immeasureably remote past, that now dead and cold orb was circling as a blazing hot mass in actual contact with an equally hot earth, from which it was just beginning to separate by reason of mutual tidal actions such as we have already seen to be still in operation.

Let us return and think for a moment of the previously considered state of the earth, to see what, on mechanical principles, may have occurred at the beginning to effect the separation of the moon from the bosom of the earth.

Suppose the materials now constituting the earth and moon to have been revolving as a single molten mass in a period of some two hours or less, and to have been acted on by the attraction of the sun alone. The sun's attraction as a tide-producer is about half as efficient as that of the moon now is. Its magnitude is sufficient, however, to cause a gradual retardation of the earth's rotation. Yet that fact is not of itself so important as one or two others which in conjunction with it must have exerted a controlling influence in this beginning of the natural history of the earth-moon system.

Every elastic body, be it molecule, tuning fork, pendulum or world, has one or more natural periods of bodily vibration. The solid earth necessarily has such a period, and the plastic or fluid mass of the original earth had such a period also, the length of which, not to assign it very exactly, may be taken as four hours, perhaps. Its actual length is unimportant. The tides caused by the sun in such a mass would occur as bodily deformations, propagated throughout its entire figure in the period just mentioned. They would not be mere superficial disturbances, such as arise in shallow oceans lying on a solid globe, as at present. A tide in such a case would be an elongation of the entire mass of the earth by the sun's attraction. This elongation would have an opportunity to subside as the mass rotates, but in so doing it must subside nearly as far below the mean level as it had previously risen above the mean level.

These alterations of level, or oscillations in the fluid mass, must recur several times before the effects of the tide originating in one spot could finally die away and be dissipated by friction. However, the final effect must be gradually to diminish the rotary velocity of the earth mass and lengthen its period of rotation, until, in course of time it became equal (or very nearly equal) to half the natural period of oscillation of the mass. Then would come a critical era in the earth's mechanical history; for then the effects of successive tides must become cumulative. For, by reason of its rotary velocity, each successive oscillation would be superposed on the remnents of preceding oscillations in such a manner that the actual total oscillation must continuously increase in magnitude. It is evident that a molten mass cannot oscillate more and more violently without end. A final disruption is inevitable. Such disruption might or might not result in complete separation of the mass into parts such that they would remain apart. That would depend on the manner in which the total mass and the total energy was divided between the two.

What is evident is, that this critical era would continue through such a prolonged period of time that numberless convulsions of like character must occur during the long time in which the periods of rotation and oscillation are nearly commensurable. Our moon is doubtless the outcome of one successful separation among thousands of abortive attempts, each of which might have closely approached the conditions necessary to accomplish a like result. It is interesting to notice that a similar critical era would again recur in the history of the earth, when its period of rotation should have been reduced nearly or exactly to equality with that of its period of oscillation, and still another when its rotation should become nearly twice that of its period of oscillation, and others following these.

Whether the earth-mass would be in danger of disruption during any or all these eras would depend partly upon its relative fluidity, plasticity or solidity, and partly upon circumstances which should control the relation between the subdivision of its mass and that of its energy between the two parts.

In the case of the large superior planets, we notice that each of them have several moons that have evidently been given off at different critical eras. The size of these planets is so great that they have apparently retained so much of their original fluidity and heat as to render this possible during several successive eras. It is a question whether these planets have any of them become solid even yet. It must for the present remain an open question as to which of our eras gave birth to our moon, and whether there may not have been later moons also.

Perhaps the case of Mars, which is smaller than the earth, may enable us to form a better judgment as to the probabilities of the case than we could otherwise have attained. Mars has two moons: a distant one retarding its rotation in the same manner as our moon retards the earth; and another so near that its period of revolution about Mars is a little more than seven and a half hours, which is more than three times as fast as Mars rotates. The action of this moon is therefore exactly the opposite of our moon. It must accelerate the rotation of Mars on its axis, and be itself retarded in its orbit. That means, as has been shown, that it will circle about Mars in a smaller and smaller orbit continually until it will at last touch the surface of Mars and lose its separate existence.

It thus appears that our present moon may possibly be the sole surviver of several which once circled about the earth, but which have been caught in mid career by the retardation of the earth and brought again to her surface.

As before stated, Professor Darwin is not responsible for such deductions from his theory as this. His analysis deals with a multitude of other matters too difficult to be treated without mathematical formulæ. He deals with the future of the earth-moon system as well as with its past, and sketches a final condition when the earth shall immovably face the moon as the moon

The Mechanical Theory of the Origin of the Moon. 63

now faces the earth, and both shall then revolve as if rigidly fastened face to face at some three times their present distance in a period of about fifty-seven days. His analysis shows why the moon has lost its own axial rotary velocity before the earth has done so. It treats of the eccentricity and inclination of the moon's orbit, as well as the tilt of the earth and moon in their respective orbits. It treats further of more general questions of planetary motion, and especially has he made it clear that the planets themselves have had no genesis, such as he has been shown to be mechanically possible for the moon. They cannot have arisen in any such way. How they did arise still remains to be solved. That this question will be solved we may well believe from Darwin's success with the earth-moon problem.

RAILROAD LOCATION.

W. W. RICH.

In a former talk I brought to your attention the subject of railway reconnoissance. Following this, come the preliminary and location surveys. As in the consideration of the reconnoissance I endeavored as far as possible to take matters not elaborated in text-books, so of surveys I shall hope to touch upon some matters which are sometimes overlooked. Even in the elementary work of establishing a line for construction there is need for painstaking attention to little things.

More than twenty-five years ago, it was my good fortune to serve for several years as Civil Assistant in the United States Engineering Department, under General G. K. Warren, that brilliant but unfortunate officer who had been relieved of the command of the Fifth Corps at the closing battle of our civil war and, at the time to which I refer, was in charge of river and harbor work in the Northwest. He possessed a remarkable knowledge of details of all work in field and office, keeping closely in touch with operations extending over a wide territory. In a higher degree than any other man whom I have known. he had the power of concentrating and holding indefinitely all his energies and learning upon a single point. This is a very great and rare advantage, a talent of more value than genius, and I may say, in passing, that I have more faith in the genius of hard work as a factor in success, than the sort which is semi-inspirational. In the rare cases where the two are combined, a century now and then is accented with a giant.

To General Warren's kind forebearance and practical suggestions I owe much; even when the wind was decidedly in the east,

I usually learned a lesson. At 10 o'clock one night, he, his principal assistant and myself, were at a long draughting table busy over sketches and maps, in the preparation of an estimate for a canal in the valley of one of our large rivers. At one important place, an original note-book was in demand. A moment later I was glad that my own was not applicable to that particular case. The book was found, the field notes of an engineer of artistic tastes, whose topographical sketching was a thing of beauty, but in this particular instance not a joy, even for a brief moment. Where there should have been full information as to the slope of the ground, its height above the river and a note as to whether it was a part of the gravel terrace or the rocky debris from the bluff, there was a waterline sketch of the river, a few hatchings adjacent to the margin, a picture in high art of a triangulation station and in faint penciling, at a distance of several hundred feet from the water, the words "high ground." General Warren's wrath was great. "Whose book is this?" he demanded. "Mr. Blank's" I answered. "Where is he?" "With Mr. Cheeseboro, in Chicago." Said he, "I will write to Cheeseboro to discharge him at once. Look at it! High ground! Why doesn't he say how high it is? Is it thirty feet or three hundred feet? Why doesn't he give his best judgment of the fact, as he saw it?" And then came, with a military accompaniment of theological hot shot, the very pertinent inquiry: "*What is the use of a man's going out to survey unless he finds out something?*" and he might have added, "and records it." An important lesson in a nutshell which I never forgot. My friend of the artistic note-book was hung, drawn and quartered once more, but Mr. Cheeseboro was never asked to reduce his force.

One reason of deficient information in the note-books is that young men frequently, and older ones occasionally, do not understand fully what they are looking for or do not appreciate just what bearing the multitudinous things in sight have upon the main purpose for which they are sent out. Another thing: allowing an engineer not to be lazy and concerned chiefly in making a successful union with his salary at the close of the month, he may consider some item under his eye too trivial to note or thinks he will remember it and enter it when he has more leisure; or perhaps, on the other hand, fritters away the hours in recording matters close at hand, to the neglect of more distant and important things.

It is thought desirable by most young engineers to get their first practical knowledge in railway work upon a survey. Speaking from my own experience as well as observation, I think this a mistake and that it will be of value to him if he can secure a situation for his first season as a rodman or other assistant in the construction of a road. He will thus become familiar with the relation of grade lines to surface lines; the "pay" quantities of earthwork per mile; the relation of indicated light center line profile quantities on steep side hills to the greatly larger actual quantities of finished work; the requirements for structures and their foundations and how the construction over swamps, or their avoidance, may affect unfavorably or otherwise, the cost of the construction and the item of alignment. These, I am sure, will be of assistance to every man if they precede his work upon the reconnoissance and instrument surveys.

Right here, perhaps we may properly have a little more talk about note-books. Great carelessness exists among many engineers in the matter of field records of surveys. They fully understand just what they are doing, know the location of all the places and things mentioned, and times of doing the work, but forget that what is as plain as A, B, C to them will not be so to others, and that their notes are likely to pass through various hands, possibly years afterwards, and so there is a paucity of information as to daily doing, a vagueness in the titles, or perhaps an entire omission of them; and dates, too, are frequently omitted.

I once had placed in my hands a set of transit and level books (there were no topography records) of one hundred and fifty miles of a railway survey. It was supposed that these books might be of use in a new survey to be made over a portion of the same route, as they were the records of a line run six years before by an engineer who had had considerable service with one of our largest railway companies. I had known him very well—a man of education, having good judgment in location and construction, but "easy going," as we say, and deficient in the organization and control of men; careless in record-keeping, liable to error by ignoring the usual and necessary checking of computations, slouchy and slow in manner and speech.

In none of the books was there a date, name of an engineer, or of any place, excepting that a search showed names of the three or four small towns on the route, but not of the place whence the line started. Other than the foregoing, there was nothing to indicate in what part of the globe the work had been done. There was not a suggestion whether the record had been made in "good old colony times" or in the nineteenth century. Pages together of the transit-book were as clear of disfiguring information, excepting station numbers and angles, as when they left the bindery. I recall that at long intervals there was an oasis in the agnostic desert, of a feebly drawn parallelogram, and by it was scrawled "house" or "barn" and occasionally the word "creek" near a line which, without direction arrows, straggled across the center line. These were absolutely all the items in the transit notes. The survey was not on the prairie and presumably there was some topography worthy of mention. The level books were equally reticent. The notes were crowded into four columns and the monotony of numerals was only broken by the words "peg," "bench" and "creek." It would have been comforting to have found a suggestion as to whether the creeks had width, depth or direction, where and what were the benches, or even whether pegs were of pine or oak. This is an exact case, though an extreme one, but very many surveys are made where records are puzzling and imperfect.

The gentleman who had charge of this survey subsequently published a book on surveying. I have said that he was habitually careless. As an instance—I once had occasion to check a simple right-angled triangulation which he had made, as the notes, when plotted, indicated the probability of an error. The hubs of his base line being still in place, I measured it and the angles. The right angle was correct, the other one wrong and the base line had an excess of fifteen feet.

I need not remind you that all observations and computations should be carefully verified. When possible, this should be done by some method other than the first; it is not always safe to depend upon a repetition of one process.

Among my assistants of a few years ago, was a young man of fine social and moral qualities and excellent scholarship, a graduate of a good school of civil engineering. He was energetic, and possessed a very active mind. Before you could finish instructions upon any particular point, he would say, "Yes, I understand, I see just what you want," but he rarely did. He was almost certain to slip something necessary, and which his first conception had not grasped. His mind seemed to have a hair trigger attachment, and was pretty certain to go off at half-cock. His field work was often wrong, his mapping incorrect, his estimates unreliable—all for want of a thorough understanding of the thing to be done and a failure to check his work. I parted with him with sincere regret, for I had a real fondness for him, and tried to make him over before I let him go. He failed wherever he went and I believe finally left the profession.

Every note-book of any survey should show on its title page the name of the corporation, firm or individual ordering the work, and a statement of the termini and other known places in the route; a full description of the initial point and its ties to land lines, block corners, or some other known landmark; also names of engineer in charge, transitman, topographer, leveler and rodman; and the date of the beginning of work. These should be in every book, each class of which should be numbered consecutively 1, 2, 3, etc. The level books should show, in addition to the foregoing, the datum of elevations—sea level, if possible—or if necessary an assumed one with a description of some permanent bench-mark from which the line of levels originates. Such a set of books can always be identified as belonging to one survey. Nothing should be put on the outside of their covers; these can best be labelled in the office, to conform to the general system of numbering and indexing there in use. Each morning at the beginning of work, should first be written in strong lines, in each book, the day of the week, month and year. These should never be omitted. Within the past few years such dates in the location

Railroad Location. 69

and construction note-books of the road I have the honor to serve have enabled us to win several right of way lawsuits.

Pages of each book should be numbered and one or two pages next following the title be left for indexing. Here, when the book is full, should be set down, in order with page references, its contents, describing the various lines of their letters, offsets and other important information so that a glance will tell what the book contains.

The chief of the party should keep a book containing notes of progress, such memoranda as pertain to trade and business likely to develop at places where he plans to have stations and briefs of agreements for gifts of land or right of way which he may make, for often more favorable contracts for these can be made by the engineer "striking while the iron is hot" than can be done later by the right of way agent. I may add that the chief of location should always be provided with blank contracts for such use. In the back part of his book will be a list of his men, rates of pay, and times of employment; a statement of his instruments and camp outfit and an account of money received and expended. These will be needed at the close of the month for the preparation of his pay-rolls and vouchers. This book, like the others, will be paged and indexed. Every man keeping notes should be instructed to make his figures strong and loud; faint pencil lines are suggestive of timidity and uncertainty and, besides, they soon grow dim and make hard reading.

It is not a good plan to copy notes for there is great chance for errors and omissions. I would rather have an original book, blotched by dead flies and mosquitoes or a dash of rain, than the cleanest copy, which is never free from the suspicion of error, and never indicates anything of the personality of the man in the field.

Methods of keeping record in a note-book vary. I will not undertake to specify any in detail, but say that transit notes of lines run by angles or curve deflections should be definite and plain, checked by the needle and a column of corrected bearings. In level-books I see an advantage in keeping notes in not less than six columns. They should certainly be kept so that the plus and minus sights which affect the integrity of the line of elevations can be added to prove the work, and footings left at the bottom of each page. Squares in the topography book represent usually one hundred feet but this scale is sometimes, for convenience in certain places, halved or doubled. When this is done, it should be noted in writing; it is a good plan to set down distances frequently in figures, as a check on the sketching.

The report and estimate of the reconnoissance having shown the resources which the explored territory contains, the characteristics and the approximate cost of a line in the proposed route, the preliminary and location surveys are next in order. These are for the purpose of fixing the alignment and grades of the final line and their relation to the various trade centers and industries existing or to be developed. It is obvious that these will necessitate the making of extended and copious notes and sketches, so that the work of a compenent topographer is of much more importance in trial surveys than for the located line. In the one case it is indispensable, in the other desirable, but the necessary topography for the location can be picked up after the line is fixed.

If the country has wagon roads or is an open prairie, the camp outfit and supplies may be carried upon one or two wagons, as may be required by the extent of the work or the convenience of depots of supplies. If the line must be cut through a wilderness, packers (about six in number) must be used, if it be in summer, or if in winter an ox team and sled, or perhaps a team of horses and a sleigh. Two good axmen can brush out daily the two or three miles of road needed, where the frozen lakes and marshes can be made available. If the region be mountainous, pack men or pack mules, perhaps both, must be used.

The organization of a party, exclusive of what force is required for the transportation of supplies and baggage and the cutting out of roads, will be about as follows:

1. An engineer in charge, preferably the one who made the reconnoissance, but this is not important, as he has maps upon which the route is indicated within comparatively narrow limits.

2. A transitman who, to be thoroughly efficient, must have good executive qualities. He has immediate charge of most of the field force.

3. One or two topographers (each with one or two assistants, if on mountain work.)
4. A leveler.
5. A rodman—occasionally two.
6. A rear flagman.
7. Two chainmen, the head chainman to receive the larger salary. As the instrument men must see that instruments are kept in adjustment, so must the head chainman watch his chain, adjusting its length as often as once a week. It should have an excess length of one-tenth of one foot.
8. One head axman, and from two to four additional axmen. The head axman should receive extra compensation. He must have good judgment, energy and "push," as upon his efficiency, in a timber country, the rate of progress depends in a large degree.
9. A first-class cook. No other sort should be employed if peace and good-will are to prevail. He ought to have the skill of a Delmonico and the virtues of a Melancthon.
10. The cook's assistant, called in lumber camps the "cookee," whose duties are obvious. A small party can do without him.
11. A campman, who cuts firewood, and assists teamsters or packers in making camp.

I have included a rear flagman, but if the line be run with a compass, he will not be wanted; and in all transit lines, even those of final location, he is considered no longer a necessity except in a rocky region. On the prairie the transitman, before leaving any point, sets up behind it a lath in the top of which he fastens a bit of white paper, and a number of these can frequently be seen in the field of the telescope, furnishing a most satisfactory check upon the transitman's work on long tangents. In the timber, straight blazed poles six or seven feet in height will be found equally convenient and fully as reliable as the youthful back flagman, who sometimes wearies in well doing and lets his rod swing out of plumb, or drops it occasionally to fight mosquitoes or kill snakes.

The exact size of the party will depend on the work to be done. On the prairie it may sometimes be as few as eight or ten; in a heavily timbered, roadless region seventeen to twenty-one, and in mountain districts as many as twenty-five to thirty are sometimes required.

Unless the line lies in a well-settled country, where it is easy to hire help when needed, there should always be an extra man on the pay-roll to replace a sick or disabled one. It never pays to be short-handed, especially where there is much timber and brush to be cut. Besides, men will work more faithfully and cheerfully if it be evident that one can be discharged without crippling the force.

If the work be at a distance from civilization, the engineer in charge ought to know something of the more common medicinal remedies, and also of simple surgery, for he must sometimes be both surgeon and doctor for months together. I have had occasion to practice in both these lines more than once, to the manifest sorrow of my patients, who were compelled to accept my surgery as barely better than none, or take my decoctions and boluses, modo prescripto.

A word regarding the domestic affairs of the camp. Abundant supplies of the best quality obtainable must be provided and well cooked. The monthly expenses will be but slightly greater than for an indifferent mess. If you give men plenty of excellent food, they will repay it with usury in cheerful and efficient service, and stand pushing without protest. Poor or short rations will cause the best of workmen to grumble and shirk. Feed and care for them well, then promptly discharge the first man who complains.

Begin business early each day and quit early, so that map and profile making may be done in season and the cook finish his work before bed time. The reverse of this practice takes the vim out of any party. I used to have breakfast in summer at five o'clock, luncheon at eleven, and quit work at five P. M. In winter, all daylight must be used in the field. Unless in great and manifest need, do not work on Sundays—it does not pay.

No gambling must be permitted in camp; if allowed, it will surely breed trouble. Card playing or reading should stop before nine o'clock, at which hour all lights must be extinguished, except in the office tent, where work may compel a later hour. No man can be fresh for the morrow's work who sits up late at night to play cinch or seven-up, or to read the thrilling story of "Reinard the Red Revenger, or the Pirate King of the Floridas."

Railroad Location. 73

In such a party, there will, of course, be great differences in the ability and character of both officers and men. The two principal classes represented will, in some respects, be as wide apart as the shores of the ocean. Those of the one class do their work more or less faithfully, but without enthusiasm or much interest. The other class—the one from which promotions are always made—is not crowded. The few who compose it feel almost the same interest in business as does the chief of the party. They are loyal to him and to the corporation that he represents. They never "wax fat and kick," but by example and precept, help him to maintain authority. They do not hesitate to take up some task not specifically theirs, if thereby business may be forwarded. They are not sticklers for rank or place, but are open-eyed and ready-handed to see and do anything and everything which may help to advance the work in hand to a successful conclusion. They do not forget that the time of two minutes and forty seconds is now historical and that in these days of pneumatic sulkies and kite-shaped tracks when the record is close to the two-minute mark, that the horse whose nose is thrust under the wire a twentieth of a second before his competitor, wins both applause and the purse. The University of Minnesota has had representatives of both classes in the field within the past five years.

Upon preliminary lines, especially in the timber, more rapid progress can be made by using a compass instead of a transit, unless there be great local attraction, and even then the difficulty can be obviated by taking back sights from every instrument station and noting the angle with the forward line. The compass is sufficiently accurate and, in the woods, the line requires much less brushing out than a transit line, but on all compass line the leveler must be furnished with a lively axman. The prismatic compass I consider convenient above all others. I have run many miles of preliminary lines with one having a two-and-a-quarter inch needle. It can be set on the top of a long walking stick or held to the eye by the hand, and liability to error in recording a course is reduced to a minimum, as but one bearing is in view at the moment of observation.

The chief of party orders the direction of the line, sometimes from the rear, oftener from the front—in thick timber by calling from some point beyond view, and in open country by taking with him the head axman and the transit pole or flag, which in timber is carried by the head chainman, but now is set by the head axman on some prominent point as a foresight for the chainmen and compassman. If the line be run with the transit, a plug and tack is set at the foresight. It is not to be expected that this first line will at all points lie upon the most desirable ground, but no time should be spent in "backing up" in order to secure a good profile, one approximating what a location would show. This is frequently done and is a useless expense. I once engaged an engineer highly recommended for such work, to run a preliminary line in a timbered country. His map came in, and at one place in a valley of a stream less than sixteen hundred feet wide, he had three lines for a considerable distance, where one with some points of topography properly secured, would have shown just what could be done. It was an index of indolence or bad judgment, and one of the reasons preventing his re-engagement.

The compassman or transitman enters in his book, from bottom to top, the station of the survey; in another column the angles to the right or left; in another, the magnetic bearing of the line; and in a fourth, the courses corrected by azimuths. These latter are a check on the work and may be useful in plotting the survey, which should never be done by deflection angles. He also sketches such topography as is in view from his instrument stations and from intermediate ones, which he notes in passing from point to point. This leaves the topographers free to observe and record the topography at some distance from the line. This is best done, I think, by pacing and the use of the hand level or barometer, tape measurements being required only occasionally where the way is narrowed by obstacles which shut the probable final line within close limits. These are termed controlling points and, if no other passages for a line can be found, they become fixed points in the definite location, and the topography must be accurately taken so that the situation where a top or bottom contour or a few hatchings, if supplemented by figures showing the estimated elevation above or depression below the general level, will serve every purpose and save time.

From stations on the line, the relative elevation of points on an offset to it may be indicated by fractions in which the denominator represents, in feet, the distance from the line and the numerator, preceded by a plus or minus sign, indicates the number of feet above or below the origin of the offset. These notes are of great value.

The topographer must understand just what he should show; that he must take in as wide a range as possible on the first preliminary, indicating the greater elevations and depressions, not forgetting anything really important, but spending no time on fancy sketches or noting minor irregularities of surface or things near the line, which the instrumentman is to gather. Of course, if the country is so rough that enough cross-sections must be taken on either side to determine continuous contours, his task, though slow and laborious, does not greatly tax his judgment, for he can omit nothing in the comparatively limited range of his work.

The topographer should make all his notes in heavy lines with a No. 3 pencil (not faintly, as if he were not sure of being right,) and must beware of distorting his scale or finishing his sketches by anticipating and recording what is only partly in sight but which, as he progresses, may take on quite another form. I recall my first efforts in this line on an actual survey, and the trouble I had from the last named cause. One hundred feet on the scale on the book was so small that my notes overflowed in a dreadful distortion of facts.

The first preliminary line is plotted on a scale of two or three hundred feet to the inch and the topography is then penciled upon it. From this can be laid down a second preliminary line to be run in the field by angles, or sometimes a preliminary location, which can be run without much delay on account of curves, as intersections need not be generally made, and some parts of such a line are liable to need no revising. On either side of this second line, for short distances, more exact topography is gathered to determine adjustments, if any be required, for the final line in the interest of alignment or economy of construction. The approximate line being decided upon between any two points which are pivotal, it will frequently save time and expense in moving the camping outfit twice over the same ground instead of once, if the final location be made before proceeding.

In an open country the location may sometimes be made without first plotting it on the preliminary map, but on most lines there are some places, especially in rock and in running to a grade along hillsides, where great accuracy is required, and there mapped lines are indispensible. At such places, portions of the preliminary map are platted on a scale on one hundred feet to one inch, the topography carefully drawn, the grade contour defined, and the center line adjusted with such exactness as is possible from the map and computation. Frequently, especially along rocky cliffs, minor changes in alignment are made in the interest of safety or economy after construction is well under way. "Good enough" is an expression sometimes applied to some part of his line by an engineer who, perhaps, is impatient to get ahead more rapidly, but it should have no place in his vocabulary so long as intelligent service can reduce the earthwork or rock quantities a sufficient number of cubic yards per day to more than pay all expense of the survey, without detriment to grades or alignment.

The final line is tied by steel tape measurements to property lines and corners and careful measurements and sketches will show its relation to highways, buildings, fields, fences, streams, etc., which are crossed, or are near enough to it, to in any way affect the cost of right of way or the safety or convenience of the public.

The engineer in charge can usually select the ground which will be near the final line, but he may have to try two or more lines between points of control before he is certain of the best one, but in each there will almost always be some well-marked topographical reason for his running them.

There is, however, one variety of timbered country to which I alluded in my former talk—one of irregularly broken surface, without streams worthy the name, or definite lines suggestive of

Railroad Location. 77

a route,—one of hills, pot holes swamps and ponds, without method in their badness, which requires special treatment. In such a country, the reconnoissance will probably have shown at intervals of several miles, certain points which must govern; it may be a lake, a high hill, or end of an extensive morass. Between these it is difficult to find the best line, if it be sought by the usual method, for a preliminary line may swing for one obstacle and then another, perhaps leading the engineer at some distance from a direct course or necessitating other lines on the "cut and try" principle. Instead of this sort of a survey, it is better to run a direct line between the known fixed points, not turning aside for any elevation or depression, however great. This will constitute a base line for topography which must extend over a belt of country wide enough so that somewhere within it will lie the desired line. Usually this belt will be from one-half to one mile in width or from twelve to twenty-five hundred feet on either side of the survey. The direction having been determined from land lines and the magnetic declination, the compassman pushes forward his line as rapidly as possible, sketching such topography as is in view from the instrument stations and what he can see as he walks forward. From the summit of the more formidable hills he measures and stakes with designating letters side lines along their crests until lower ground is reached, and the leveler takes elevations at each station thereon as on the main line. The chief of the party and the topographer pace offsets from the lesser elevations and depressions; each, recording the direction of his paced line, usually at right angle to the survey, proceeds from one thousand to two thousand feet or more, as may be necessary, sketching the important things on either side of him as if he were on a staked line. He observes the undulations which he passes over, recording as before described, in fractional form, distances from and elevations above or below his starting point;—or as it will not be convenient, probably, to keep in mind the initial point, he may make his record without regard to it.

Thus $\frac{-16}{300}$ then $\frac{0}{200}$ then $\frac{+12}{500}$ then $\frac{-20}{100}$, etc.

From the end of his offset he paces parallel with the survey from eight hundred to one thousand feet, then back to the instrument a line perpendicularly thereto, keeping notes as on the outgoing line.

This kind of work gives healthful exercise, and if the topographers be "diligent in business" they will walk daily a goodly number of miles further than others of the party. In the manner I have described, from one and one-half to two and one-half miles of daily progress can be made with the survey.

At night the instrumentman makes a profile and map, generally on a scale of four hundred feet to one inch, plotting on the latter all the side lines run by the instrument and all topography taken. The topographers then take the map, draw their offsets thereon and sketch the topography taken along them; then, using three offsets as datum lines, plot upon a convenient vertical scale, usually twenty feet to the inch, a profile of the elevations and distances which have been recorded in fractional form. Thus are secured approximately correct cross-sections at the more important points.

The engineer in charge, with the main line profile before him, can lay down on the map a preliminary location, to be afterwards amended and corrected, and may construct a profile which will indicate about what gradients can be had and, approximately, the earthwork. To cover the territory with contour lines would require much more labor in the field and at the table and would not so well, certainly not better, suggest to the eye the location, as do the sectional profiles described. I have known lines to be established in this manner, in a forbidding country, for distances from twelve to twenty miles which, when ready for construction, did not vary at any point more than half a mile from a direct course, had no objectionable curvature and only a small percentage of added length.

The ruling gradient will have received full consideration in the reconnoissance, but its importance must still be foremost, for in proportion as it is favorable or otherwise, will depend, in large measure, the value of the property. A large expenditure will be warranted at a few points if thereby it may be materially reduced below what nature suggests. What would be considered large outlay at these places, if distributed over the entire road, will not raise the average cost per mile to an extent sufficient very largely to increase the fixed charges.

For instance: On a division of one hundred and fifty miles, there may be two or three summits which by an extra outlay of one hundred and fifty thousand dollars could be cut down so as to reduce the ruling grade from forty to thirty feet per mile; but this amount distributed over the entire division would increase the funded debt about eleven hundred and eighty dollars per mile, allowing for the usual discount on bonds, which for a new enterprise is usually not less than fifteen per cent. At five per cent interest, this would increase the annual fixed charge per mile by fifty-nine dollars; or, if we include, as we should, the extra amount required annually for sinking fund on fifty-year bonds, a total of seventy dollars annually. Now the cost of conducting transportation, by reason of reduced cost per loaded car mile for fuel, oil, waste and wages of engine and train men, will be reduced by seventeen per cent, and for all expenses, not less than six per cent. This, upon a tonage of three hundred thousand tons per annum each way, allowing five hundred tons average trainloads and assuming train expenses at one dollar per mile, would equal seventy-two dollars per mile, annually, or a slight gain over the investment. This saving would increase with the growing traffic. I have assumed a tonnage of three hundred thousand tons each way, because it was the tonnage assumed by me for some calculations made in 1883, before beginning the location of the "Soo" line, as an amount likely to be transported within five years from date of its completion, between Minneapolis and Lake Michigan, a distance of three hundred and forty-two miles. I may add that the actual tonnage hauled between these points six years after the completion of the road was more than seventy per cent greater than that upon which my estimates were based.

Maps of the location will be usually made on a scale of eight hundred feet to an inch for an uninhabited region; for a settled country on a scale of four hundred feet, and in towns and cities on a scale of two hundred feet to the inch. Station grounds will be mapped on a scale of one hundred feet, while large yards, shop grounds, etc., will require a scale of fifty feet to the inch to clearly show tracks and buildings. The right of way maps will be made in camp, on brown paper, and two tracings made, each covering about six or eight miles, or across a township of surveyed lands. These tracings will be sent, as soon as completed, to the chief engineer so that he may at once,—for delays here are apt to be expensive,—take steps to acquire the right of way. The charter, except for land-grant roads, usually allows a company but one hundred feet and such extra widths as are needed for borrow pits, spoil banks, station grounds and kindred necessities.

The engineer in charge of the survey must be sure that he has indicated on the right-of-way maps, all extra widths required for the foregoing purposes; also for bottom widths of large embankments, and not less than two hundred feet for fire protection at all bridges of much importance.

The early practice, so far as I have knowledge, was to have the profiles and transit books sent to headquarters where the draughtsman made the maps. If an error was found, great delay and expense resulted. When maps are made daily in the field, errors or omissions can be at once discovered and remedied at trifling cost.

Frequently location profiles show little but station numbers, grades and surface lines, but they ought also to indicate the location and elevation of all bench marks, the nature of soil and rock, the angles of slopes on hillsides, varieties of timber, if there be such, the beginning and end of all marshes and the soundings therein, with a note explaining the nature and consistency of the material; the size, depth and direction of watercourses, and their flood marks; the probable nature of foundations of bridges, approximate length of the latter, and the location and size of culverts.

These latter will represent the *opinions* of the locating engineer, but will, doubtless, be modified by the constructing engineer who must give special attention to drainage areas.

The profile should be made on paper wide enough so that below it may be drawn, in a heavy line, the alignment with all curves shown so as to readily catch the eye, the beginning and the end of each and the degree of curve being set down in plain figures. Along this center line should be sketched the topography for short distances on either side, including the crossings of all railways, highways and property lines. In short, put sufficient notes on the profile so that it will not be necessary to consult any note-book whatever, or right of way map. Such a profile, when all the grades have finally been adjusted, will be inked in at the headquarters office, and a half dozen blueprint copies made for use of division and assistant engineers and for contractors. Before this is done, the profile will be divided for convenience into mile sections, and the latter numbered, and sometimes approximate quantities of grading, bridging, etc., will be given on each section. This will be alluded to hereafter.

A few words about grade lines. When I see a man sit down before a virgin profile, with a long thread in his hand, I am apprehensive that he is about to do some mischief and, remembering my own deplorable sins, am full of pity for the corporation whose bank account is in jeopardy. An innocent bit of black thread not more than eighteen inches long, has sometimes caused trouble and financial waste.

Long, swinging grades look well but are an expensive luxury. Undulating grades, if modified by suitable vertical curves, can be operated without any difficulty at little cost, and may save considerable earthwork; but for the slight savings, it is not well to break a grade line into many sags.

There is occasion on every mile of every profile, for intelligent and painstaking study in the fixing of grade lines. Unless in very uneven ground, where economy will usually be found in making fills from cuts which are within reasonable haul, there should be an excess of embankment. The ideal roadway would have no cuts, for then there would be little chance for trouble on account of drifting snow or defective drainage. Fighting snow is costly, and the heaving of the soil in cuts makes rough track which expensive shimming can only mitigate in part. All clay cuts ought to be underdrained with tiling. Those which are apparently dry in summer will discharge water constantly through tiles laid three feet below the surface ditches.

I will not consider here the relative value of curvature, and rise and fall, which are treated of at length by Mr. Wellington, and in part by other writers, and with which you are familiar; only remarking that it appears to me that some engineers give too much value to rise and fall where the same is on easy gradients considerably below the maxium, and curvature has often proved too great a bugbear, frightening some into useless and costly construction to avoid it. If thorough compensation of grade be made for the limiting effect of curvature and tangents of not less than one hundred and fifty feet are used between the ends of easement curves, we can usually afford to accept the extra track expenses rather than to largely increase the construction account to secure more favorable alignment.

As to compensation for curvature. In recent years I have had to modify my practice of ten years ago. We have demonstrated by the performance of our freight trains, that five-hundredths feet per degree, used in 1886-1887 is not quite sufficient where maximum tangent gradient is eight-tenths of one per cent, or forty-two feet per mile, but that six-hundredths of a foot per degree is required, and for a tangent gradient of six-tenths per cent, or thirty-two feet per mile, about seven-hundredths of a foot is necessary.

Easement curves are important and I use them for anything above a curve of forty minutes per station of one hundred feet; these and vertical curves at changes of gradients have much to do with the easy riding of passenger trains while vertical curves greatly lessen the liability of accidents to freight trains, occasioned by the breaking of couplings, and should always be used where the grade angle tangents for any change exceeds two or three tenths of one foot per station.

There are some other practical considerations worthy of attention.

First. It often happens that to shorten the line fifty or one hundred feet and reduce by a few degrees a necessary curve

Railroad Location. 83

excavated to form embankments on either side. Now this often results *first*, in keeping the grade line higher than necessary and in increasing the cost; *second*, in uncovering some material difficult and therefore expensive to dig, or possibly starting a slide, with its long legacy of trouble and expense, and *third*, in an increased yearly cost from drifting snows and the heaving of track, by reason of defective drainage. Errors in alignment, of this kind, are plentiful, and are continually adding something to the annual cost of maintenance of railways. Usually, but not always, the line may be thrown out far enough to lower the grade line to a maximum of cost for construction on either side of the hill, the grade line barely touching the latter, or better still, being in embankment of a foot or two.

SECOND. Upon steep side-hill work, the locating engineer frequently adjusts alignment and grades for a light profile, making roadbed partly in cutting and partly in embankment. Unless the material is dry and no springs are likely to be developed (and they appear sometimes where least expected,) or in solid rock, this is liable to result in disaster, and though the first cost will be increased by setting the roadway at most points into the hill, it should generally be done. I had to learn through unhappy experiences that this kind of mistaken economy was a poor investment.

THIRD. At localities where there are liable to be stations, care should be taken to put them, if possible, on level grades with a considerable distance on either side approximating a level. If necessary to locate them on long grades, see that the grade between switches and for a long train length on either side of the station grounds, does not exceed one-half of the maximum, so that trains which stop to enter the lower end of sidings will have no trouble in getting in and no train will have difficulty in leaving a station. It will not always be possible to do this, but it should be kept in mind as of very great importance; and, at least, avoid locating stations where the gradients will limit the size of trains. Never establish stations on curves unless in extreme necessity. Such position is inconvenient, and invites disaster.

FOURTH. At crossings of other railways, every chance for an

under or over crossing should be examined and one selected, if it exists. The cost of interlocking signals and appliances represents a capital account of about twenty-eight thousand dollars. This will more than construct and maintain a first-class under-crossing, where the conditions are at all favorable, and this, without considering the question of accidents which sometimes occur at well-signaled grade crossings. However, new railway companies are singularly averse to spending the extra amounts immediately needed for other than the latter.

FIFTH. It is not always best to cross a river at its narrowest point if piers are to obstruct the current, unless its beds and sides be solid rock or material that will not scour and the under side of trusses be well above possible high water.

SIXTH. Where there is likely to be developed some manufacturing industry, as a saw-mill, flour-mill, or other factory, the line must be located, if possible, far enough back from the bank of the stream or lake to afford ample room for storage yards, buildings and side tracks sufficient so that all business of that particular plant can be kept on one side of the main line.

SEVENTH. The length of operating divisions is an important matter. Unless fixed by the grouping of the different ruling grades or the distance between terminals, such length should not be less than one hundred and fifty miles. Formerly divisions were placed about one hundred miles apart, but the tendency is now to longer distances, thus reducing the number of division organizations and employes as well as the saving of several miles of tracks in yards. In thickly settled localities, the way freight trains might be unable to cover an entire division of the length proposed in one day, but the work could be apportioned between two trains, each handling through business for a part of the run, and so require no over-night station for any train crew between division points.

EIGHTH. Another very common error is the failure to secure sufficient ground for station and yard purposes. We all have sins of this kind to bring to confessional. Having made serious mistakes myself, I particularly note the necessity of making ample provision at the outset, when land is usually cheap and can often be had without cost. For ordinary stations I would suggest a plat not less than three hundred feet by two thousand or twenty-five hundred feet; while at division points and at important terminals, from eighty to two hundred acres are none too many. In Northeast Minneapolis, the "Soo" railway has one hundred and eighty acres; at Gladstone, Mich., one hundred seventy-two acres, twelve acres of which were purchased within three years after the first selection of ground. As time advances there are increasing demands for more room to accommodate the various lines of business which center at a railway station, so that there is little danger of getting too much ground for railway use but a very great liability of taking too little.

Before closing, I should again call your attention to the profile of the located line. After its grades have been carefully revised with respect to their relation to the items of drainage, drifting snows, compensation for curvature, the elmination of grade angles, the momentum of trains, the location of sidetracks and stations and due economy of construction, there must be made from it an estimate of the quantities of material necessary to be moved or furnished in the clearing and grubbing of the right of way and the preparation of sub-structure of the permanent way.

The purpose of the estimate made at the close of the reconnoissance was to inform the projectors of the probable total cost of making and equipping a railway; the chief use of the estimate we are now considering, is to determine the approximate quantities of the various materials below the track and their probable cost. Such a statement will be a guide to contractors who are to submit tenders and a basis of comparison by the chief engineer, of the bids received. Incidentally, it will be a rough check on the reports of resident engineers, who will be directly in charge of laying out the work and making measurements and computations for final estimates.

This closes the subject of preliminary and location surveys, with their resulting maps and profiles. The chief engineer, having prepared specifications and forms of contracts, is now ready to solicit bids for the construction work and to organize his forces for its supervision.

THE ORE TESTING PLANT.

H. C. CUTLER, B. E. M.

The Ore Testing Plant, recently erected on the university campus, is a valuable addition to the equipment of the School of Mining and Metallurgy. The development of this school made it necessary to have some place where ores could be tested, their value determined and the best method of treatment ascertained, in order to give students practical experience in the different processes used in the treatment of ores and in handling various kinds of machines. This plant fully answers all requirements and has been pronounced by experts the best equipped and arranged plant in the country if not in the world.

The building is of brick and stone and cost about $5,000 to erect. It is situated upon the bank of the Mississippi river, the slope of which affords the necessary drop for the series of levels, thus allowing the force of gravity to move the ore through the different machines. The machinery costs about $6,000, is full size and representative of that found in any of the western mills. The following is a list of the principal machines: A ten by four Blake crusher; a Bridgeman sampler, size B; a preliminary screen; a series of trommels; a twenty foot link belt elevator; a pair of twelve by twelve inch geared rolls; a four-compartment spitzkasten; a three-compartment Hartz jig; a Collum jig; a three and one-half foot Huntington mill, a three and a five stamp battery with feeders; a four foot Frue vanner; a twelve foot buddle; a three foot amalgamating pan; a five foot settler; a Bruckner

:l;
ng
y,

ng
od
en
st
a
he
ly

di-
nd
a
nt
for
ed
nt

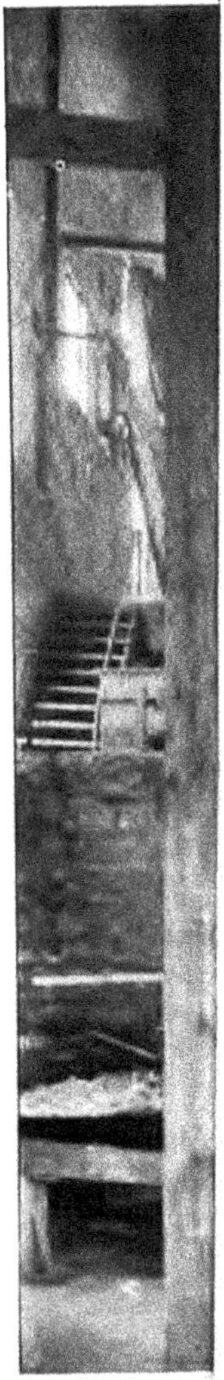

the
ket
sh
ed,
to
led
the
the
ori-
ses,

fter
olls
ing
re-
od-
evel

below where they are treated in the spitkasten, the two compartment Collum jig or the Hartz jig.

Third. If the ore is to be treated by the third method it is shoveled through an opening in the sampling floor into a car on the next lower level. From this car it is dumped into the feeder of either the three stamp, five stamp or Huntington mill. Upon the addition of water a pulp is formed. This pulp, after passing over an amalgamated copper plate, flows through launders to the Frue vanner on the next lower level. This machine yields two products, concentrates and tailings. The latter are carried either to the buddle where they are reconcentrated or to the amalgamating pan and settler or run to waste as the case may require. The concentrates from the vanner are taken to a steam dryer and after drying are roasted either in a reverberatory or Bruckner furnace for subsequent treatment by chlorination or lixiviation. The chlorination work is effected in a forty-eight inch Thies barrel chlorinator where the gold is subjected to chlorine gas under pressure. The chloride of gold is leached out into tanks where the metallic gold is precipitated and afterwards taken out and refined.

Lots varying from five hundred pounds to a car load may be handled at this plant and all known processes applied. The work is practical and gives training towards accuracy in methods and close economy in working, results which the modern professions of mining and metallurgy demand.

ECONOMY OF CONDUCTORS USED IN THE TRANSMISSION OF CURRENTS FOR ELECTRIC RAILWAYS.

EDWARD P. BURCH.

(Abstract from a Lecture before the Society of Engineers, University of Minnesota, Feb. 11, 1895.)

In all the principal cities of the United States there are large electric railway systems. These systems consist of power houses in which electric generators, driven by steam engines or turbines, furnish current at from 500 to 600 volts potential to a number of overhead or underground insulated copper feeders which in turn supply current to insulated trolley line sections. From the trolley the current passes through the motors to the rails. The greater part of the current follows the track to a point near the power station, from whence conductors, commonly old rails, carry it to the negative dynamo bus bar. The proper size of these feeders and of the conductors for the return circuit is of importance both from an engineering and a financial standpoint.

On account of the heavy currents used in electric railway work, the conductors are large. The heating effect of the current and the power wasted in the conductor varies *inversely* with the area. The first cost of the conductor varies *directly* as its area. If we use small wires they will heat and too much energy will be lost in the transmission. On the other hand, if we install conductors which are too large, the interest on the money invested will be too great. Evidently the expenditure for maintaining power to overcome the resistance of the line, and the expenditure for first cost, follow opposite laws.* Between extremes, there must be an economical area,—an area such that the sum of the cost of power wasted and the interest on the investment will be a minimum.

Let us assume,—

I, the strength of the current in amperes.
r, the resistance of the conductors in ohms, and
m, the resistance of one mil-foot of the conductor.
a, the cost at the power house of one E. H. P. for one year.
l, the length of the conductor in feet.
d, the area of the conductor in circular mils.
n, the weight of one mil-foot of the conductor in pounds.
k, the cost of the conductor per pound.
b, the annual rate of interest and depreciation on the conductor.
v, the E. M. F. of the generator.
w, the watts transmitted to the motors.

Then as stated above the economical area is such that

$$\frac{I^2 r \, a}{746} + p \, l \, d \, k \, n \text{ is to be a minimum, } u. \text{ Equation (1)}.$$

Noting that $r = \frac{lm}{d}$ Equation (1) becomes,—

$$\frac{I^2 l \, m \, a}{746 \, d} + p \, l \, d \, k \, n = u$$

Assuming that the current is fixed by the conditions of the problem, e. g. incandescent or arc lighting,† we differentiate with respect to the variable, d. Equating the coefficient to zero we have

$$\frac{I^2 l \, m \, a}{746 \, d} = p \, l \, d \, k \, n \qquad (2)$$

Or the most economical area will be that for which the annual cost of the energy wasted just equals the annual interest on the capital invested. This is known as Kelvin's Law.

From equation (2) we have

$$d = I \sqrt{\frac{l \, m \, a}{746 \, p \, l \, k \, n}} \qquad (3)$$

We note that the economical area, d, is independent of the length of the circuit and of the voltage employed.

*Kapp, 4th Edition p. 372.
†See article by E. P. Roberts, Elec. World, July 18, 1891.

In order to illustrate the application of this formula, let us, as before, assume that the potential of the receiving end of the circuit is constant, i. e., is fixed by the conditions of the problem. Let the rate of interest and depreciation be 6%. Let $m=9.5$ ohms, $a=$1c. per E. H. P. hour$=$$87.60 per E. H. P. year, and let $n=.00000303$ pounds for a copper conductor, and $h=$13c. per pound.

Then $d = I\sqrt{\dfrac{9.5 \times 87.60}{746 \times .06 \times .000003 \times .13}} = I \times 6900.$

That is, in order to work at the greatest economy we must provide 6900 c. m. per ampere flowing. You will note that 6900 c. m. per ampere is larger than a No. 12 wire, and that it would not melt under 260 amperes, yet this is the economical area for one ampere or per ampere. Electric railway feeders do not heat enough to become even warm.

Equation (2), $\dfrac{I^2 r a}{746} = p\, l\, d\, k\, n$, which expresses Kelvin's Law, is not applicable to cases of power transmission. The fact that this has not been generally known has led to some absurd engineering, especially so, in low potential transmissions at very great distances.[*]

In the transmission of power to distant motors, where the power house potential is a fixed quantity the current transmitted, I, and the resistance of the line are *dependent* variables. In other words, if the motors under our electric trains or elsewhere are to do the work *at a certain rate*, the current delivered will depend upon the motor potential; and the motor potential depends upon the resistance of the line. If the resistance of the line is high, the current used by the motors will actually be greater than if the resistance of the line was less.

It may be suggested that this is peculiar, since it is well known that, in the series wound railway motors, the running current is entirely independent of the motor potential. But the fact is, if the motors are to be worked at a certain rate and the motor potential is lower in one case than in another, the current must be held on the motors longer, i. e. the motors must be worked

---

[*] Elec. Power Transmissions. Bell, Elec. World, Feb. 16, 1895, p 203.

harder, must stop and start quicker, run faster and "coast" less, or, if this is impossible, more trains must be put on in order to give the same passenger service. The result in any case of lower motor potential due to increased line resistance is that the total current going from the power house is increased.

To return to our original equation, $\dfrac{I^2 r a}{746} + pldkn = u.$ (1)

I and v are dependent variables.

$$w = vI - I^2 r, \text{ whence } r = \dfrac{vI - w}{I^2} \quad (4)$$

Eliminating r from equation (1).

$$\dfrac{pl^2 mnkI^2}{vI - w} + (vI - w)\dfrac{a}{746} = u. \quad (5)$$

Differentiating and solving for the economical current,

$$I = \dfrac{w}{v}\left(1 + \dfrac{\sqrt{\varDelta}}{\sqrt{\dfrac{v^2}{I^2} + \varDelta}}\right) \quad (6), \text{ where } \varDelta = \dfrac{746\, pmnk}{a}$$

The general problem has been studied by Professor Wm. A. Anthony; see Electrical Engineer, October 31, 1894.

This equation expressing the value of the economical current, shows that *it is dependent upon the power house potential and the length of the conductor*, which is not the case where I is fixed by the conditions of the problem, where the potential at the receiving end is constant.

If we wish to transmit w watts and v and l are known, we first use equation (6) and find the economical current. From the equation $w = vI - I^2 r$, r is at once found, and, l and m being known, the economical area of the copper conductor d is found from the equation $d = \dfrac{n_1 l}{r}.$

In railway work, unfortunately, w is never known and a mean or average value for either w or l can not be even roughly estimated. In order to make use of this valuable "correction" to Kelvin's law, I have prepared practical tables showing the difference between the economical area as determined by the real solution, equation (6), and the economical area as shown approximately by Kelvin's law, equation (3), for different values of l, v being constant.

Assume that we are to transmit 30,000 watts to motors, the power house potential v being 600 volts. We will use the same constants as before,—
$p=.06$. $m=9.5$. $n=.000003$. $a=87.60$.
Then Δ becomes .000001893.

$$I=\frac{w}{v}\left(1+\sqrt{\frac{\Delta}{I^2}+\Delta}\right) \text{ eq. (6), and } d=\frac{mII^2}{vI-w}$$

l	I	d	Loss on Line Per Cent	$d=6904.5I$	Add to Correct eq. (3).
1	50.00011	345,200		345,200	
1,000	50.11463	346,915	.227	345,990	.26%
10,000	51.14625	362,725	2.24	353,370	2.64%
100,000	51.17550	530,225	10.83	422,355	25.53%
	61.17550				
From equation (6)				From eq. (3)	

50,000 feet is a common maximum distance of transmission in electric railway work, and the table shows that the common maximum correction will not be over 13%.

Starting again with Kelvin's law, the approximate relation of the economical area, d, to the current I, $d=I\sqrt{\frac{m\,a}{746\,pkn}}$. m is the resistance of one mil-foot of the conductor. For pure copper at 0° Cent., this is given as 9.612 ohms. The corrections to be placed on this constant are those due to conductivity and to flaws and joints of the copper, all of which may be taken at 97%, and a correction due to temperature, which is placed at $(1+.004\ t°\text{ Cent.})$* In order to get at a mean value of t, we must take $t_{\text{mean}}=\frac{t_1 I_1 + t_2 I_2 + t_3 I_3 + \text{etc.}}{I_1 + I_2 + I_3 + \text{etc.}}$, as has often been shown. For overhead feeders running from the Minneapolis and Saint Paul power houses, I have estimated that $t_{\text{mean}}=8°$ Cent.$=46°$ Fahr. Then $m=9,612\times(1+.004\times8)\div.97=10.226$. For steel rails this constant is taken most easily from data of actual resistance tests, since it is found to vary greatly with the bonding of the rail joints. $m=\frac{r\,d}{l}$.

a is the cost of an E. H. P. year. If current is not used during the 8,760 hours of the year, we multiply the actual number of

---
*Kennelly, Elec. World,—— Feb. 16, 1895, page 190.

hours per year that current is used by the cost of an E. H. P. hour. a includes fuel, labor, repairs, interest on the plant, taxes, etc.

p is the annual rate of interest on the investment in the conductors, plus the rate of depreciation. By depreciation we mean repairs, deterioration, and loss (or increase) in value due to change in market prices. Depreciation is often a larger item than interest. I have in mind an example in the lead covered cable used in a railway feeder system where 300,000 pounds depreciated at the rate of 20% per annum. Again, four years ago copper feeders were 18 cents per pound; the present price is near 12 cents—depreciation over 8% per annum.

k, the cost of the conductor per pound, includes its supports, insulators, insulation, labor of erection, freight, etc., and all items which vary with the weight of the conducting material.

n of course varies with the material and the insulation. Bare copper may be taken as .000003 pounds. In electric railway problems the study may first be confined to copper, and corrections may be made for the rail return circuit. The rails of course are intended for the roadway. However, by bonding the joints thoroughly we get a most excellent conductor. Steel has but 18% of the conductivity of copper, yet the total area of the steel in a 60 pound double track road is equivalent to 5,350,000 c. m. of copper. The old form of rail bond was a No. 4 galvanized iron wire, about three feet long, around each joint. The form used by the T. C. R. T. Co. during the past two years is a copper plate bond about four inches long, 450,000 c. m. area, cross-section $2\frac{1}{4}'' \times \frac{5}{32}''$, a plate which is riveted to the rails near the ends. It is placed under the fish plates. A double track 60 pound T rail bonded with these plates has a resistance of about .006 ohms per 1000 feet, where the old iron wire bonds and track had eight times as much, (the results of many tests on the tracks in Minneapolis and Saint Paul).

Knowing the economical "copper" area of the conductors, when the length of the circuit is given, we are at once able to figure the economical resistance of the total circuit. If the resistance of the track return is given, or fixed, and thus cannot be decreased, we are able by simple subtraction to state the econom-

, we are able by simple subtraction to state the economiical resistance and area of the copper feeders forming half of the circuit. Often the economical resistance of the track is so much lower than it need be that the size of the copper feeders may be made very small.

Next the current I. It fluctuates enormously. Trains of different sizes are starting and stopping, running along with or without current, with different loads, up and down grades, with different forms of controllers, various types of motors, different motormen, etc., etc.

Figure 1 shows this fluctuation for a part of a trip on an Interurban car. I have previously shown curves for other lines and also for the current in feeders.* Figure 2 is a common load diagram of feeders which run to the heart of the city.

The current which we wish to get is not the average but a mean current, $I_{mean} = \sqrt{\dfrac{I_1^2 t_1 + I_2^2 t_2 + I_3^2 t_3 + \text{etc.}}{t_1 + t_2 + t_3 + \text{etc.}}}$ as has often been shown.

One would soon become weary if an attempt was made to square the ordinates of figure 2 for the different fluctuations during a minute, an hour, a day, a year, several years, etc. Unfortunately none of us could stand the drudgery. We will show later how this current is obtained.

Suppose for a moment that after all our calculations, we determine the economical amount and size of our copper feeders and of our return circuit,—if the conductors are not distributed to the best advantage there is evidently a waste of money and power.

Perhaps 95% of the street railway systems of the United States use the same general wiring scheme. Feeders run somewhat radially from the power house to distant sections, as for example, in Minneapolis to the University line, First Avenue line, Hennepin Avenue, the Loop, etc. The city is divided into lines or sections and the feeders, trolleys and mains of one section are thoroughly insulated from each and every other section, so that trouble on one cannot "tie up" others.

Nearly two years ago a new feeder plan was put in service in Minneapolis and Saint Paul; a method of connecting street rail-

*See Trans. Am. Inst. Elect. Eng., June, 1892.

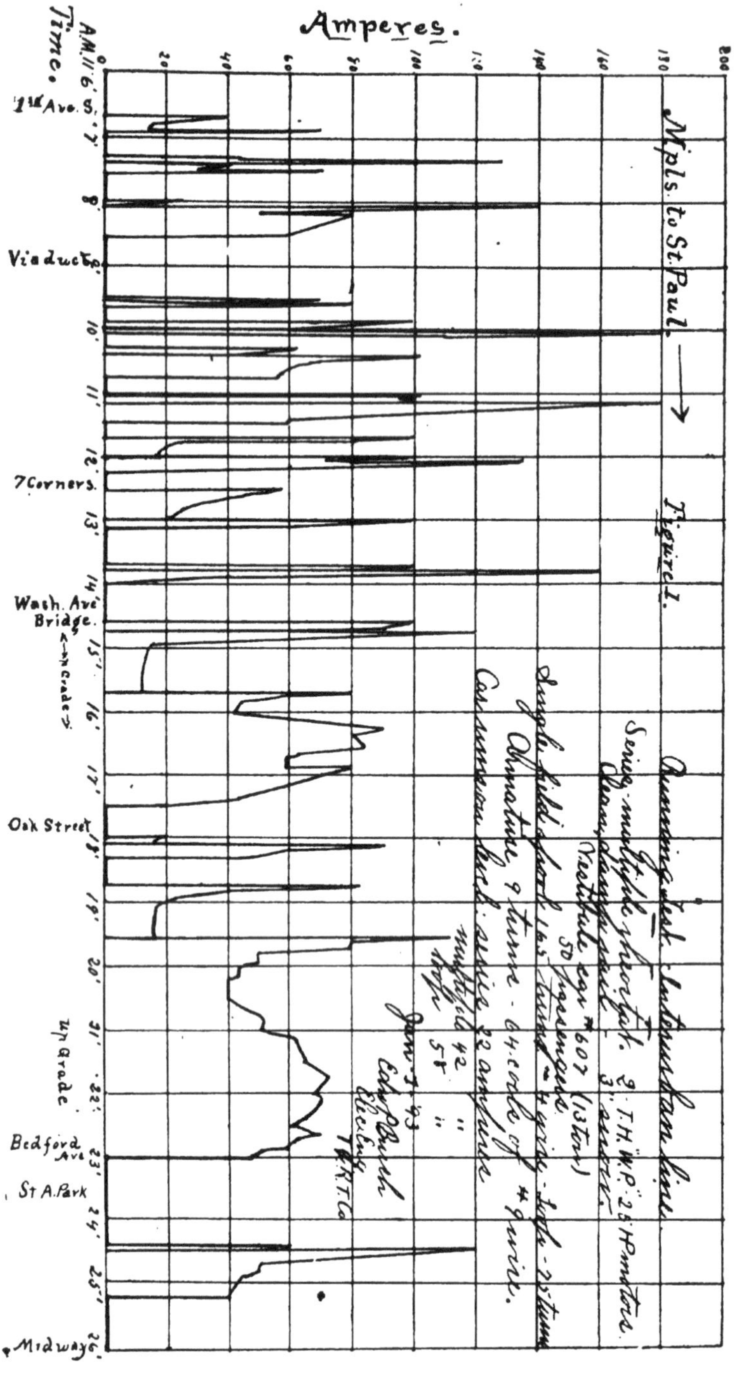

Amperes.
Time.
A.M. 11 6'
1st Ave. S.
Viaduct
7 Corners.
Wash. Ave Bridge.
Oak Street
Up Grade
Bedford Ave
St A. Park
Midway
Mpls. to St. Paul
Figure I.
Jan 7 93

way feeders, which may be of interest as an account of it has not been published. We believe it is not in service in any other street railway system. It has been called a Jumper System. It consists of connecting the sections to each other through No. 12 copper wire fuses, in such a way that the load on the feeders and sections is equalized and the feeders made to help each other and pull together.

When the trolleys of different sections are "jumpered," the fuse is simply connected round the circuit breaker. Feeders running parallel to each other but feeding different sections, are connected, *i. e.*, jumpered, by means of fuses which are placed where the feeders diverge or as far from the power house as possible. The 65 feeder-sections and the three power houses of Minneapolis and Saint Paul are thus connected and pull together. In case of the wire being down or other trouble, the jumpers, or equalizers, simply blow and the section is cleared.

The advantages of this system are:

1. It avoids the great fluctuation of load on any one feeder. This is clearly shown by comparing figures 2 and 3 where a Hennepin avenue feeder, No. 14, running from power house No. 1, is shown in figure 2 working on the unjumpered plan, and in figure 3 working jumpered to other feeders and sections. As a result of greater steadiness of the current in each feeder, the line potential changes slowly. It is noted at once that the lamps in the cars give a very steady light. Each feeder being in parallel with others, the current is much more economically distributed.

2. The total current at the power house varies less, since the power houses, now in parallel circuit with each other, work together and divide the fluctuation of load. We are thus able to work with fewer reserve units, and work machinery more economically, and more nearly up to capacity, saving in friction, fuel, labor, repairs and general efficiency.

3. Trouble at the switch board is greatly decreased. There are no heavy pulls and consequent blowing of the automatic circuit breakers, due to fluctuations on any one feeder. This is a matter of great importance.

4. The power houses being jumpered through feeders as well as connected by the rails, points between have much less drop jumpered than unjumpered, because each power house supplies part of the current and thus the area of the conductor is practically increased.

In general, feeder loads are equalized—made more steady, and the drop and loss on the line is greatly decreased. Copper is so worked, that a feeder which is too large helps one which is too small and the current is distributed more advantageously. Again, the feeder current can be measured with a Weston ammeter without trouble. Frequent tests during the year give us the percentage of the total output which goes through each feeder. The station watt meters give the total output. Readings are taken every hour from which the switch board operators plot curves showing the daily, weekly and monthly load diagrams.

The form of these curves for one of the Minneapolis and the Saint Paul station is given in figure 4. The power house load is not steady as shown, but is continually fluctuating, the fluctuating for full loads being 30 per cent. on either side of the average. During the light loads after midnight a large part of the load is the running of motors which drive arc dynamos, and all power houses in the two cities, except one, are shut down. This gives a very small variation of load during the night. Other load diagrams show the variation for the days of the week, and, during the different months, the effect of the weather, darkness, the opera, travel to summer resorts, etc.

In many cases we are called upon to make estimates for new feeder lines and extensions, and to look ahead and figure on expected growth of business. Experience and a study of this complicated problem always point to a very conservative estimate, for we know that the interest on the investment grows at all hours, 365 days in the year, whether expected increased traffic materializes or not. Business may change rapidly in a city or in one section of it, so that lines may require changing or a new power house may be necessary. Important changes may effect the system e. g. the series parallel controller has saved 20 per

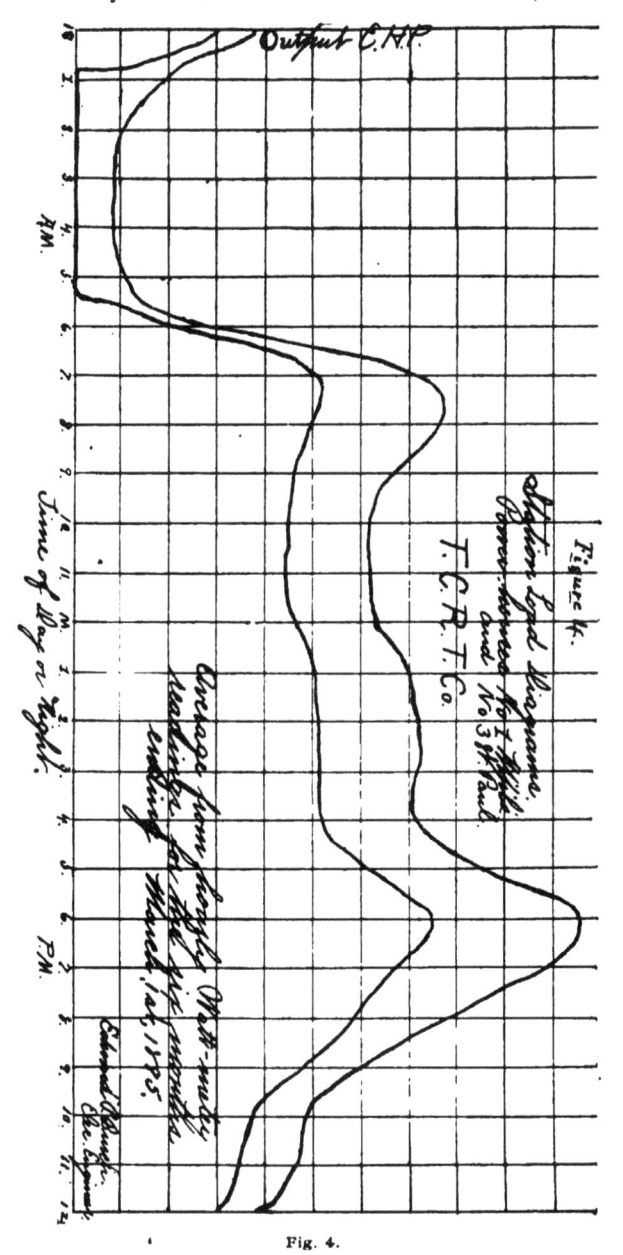

Fig. 4.

cent. of our fuel bills; the now common use of 600 volts, in place of 500 volt power house potential; the introduction of the jumper system of distribution and the use of heavy short rail bonds, all decrease the line loss, the output and cost of power. Then there is the possibility of cheap water power, improvement in stokers, and fuel oil burning devices, which may decrease the cost of power faster than business may increase. Competition and adverse municipal legislation may change the conditions of economy. In general a corporation should be conservative, and invest in conductors only when they are necessary for actual economy of operation. Wild guesses and the use of tables showing the "proper size for a *given loss*" may be avoided by studying the general subject of the economical area and the *economical loss*, and by collecting data on current strength and the exact value of the constants used in Kelvin's formulæ.

RATING A PRICE CURRENT METER.

LESLIE H. CHAPMAN, '95.

The Price Current Meter is an instrument generally used in obtaining the velocity of currents in rivers and harbors. It consists of three distinct parts:
1st. The wheel on which the current acts,
2nd. The vane,
3rd. The counting apparatus.

The wheel has five conical buckets arranged so as to revolve in a horizontal plane. To reduce the friction to a minimum the axle of the wheel revolves in agate bearings contained in air chambers which protect them from the water or any injurious material it might contain. The form of the wheel and buckets is such as to combine considerable strength with a minimum liability to obstruction by any debris the water may contain.

The wheel is made to face the current by means of two vanes, intersecting at right angles and arranged on a central rod which forms the frame of the meter. The meter frame is pivoted to an inner ring which slips over the rod by which the meter is held and is free to move both vertically and horizontally. The rod, a hollow brass tube, is made in sections and is graduated so as to read to feet and tenths.

The axle of the wheel is arranged so as to "make and break" an electric circuit, thus registering the number of revolutions of the wheel on an electric register. The meter is rated by moving it through still water at known rates and observing the number of revolutions of the wheel.

In the case under consideration the apparatus used consisted of a current meter with battery and electric register, a stop-watch, a boat and a wire rope which was stretched over the surface of the water and marked off at intervals of 100 feet.

The meter was placed in the water about three feet from the boat and, as the latter was not turned around, the meter was half the time ahead of the boat and half the time behind it. The boat was moved over the course by pulling on the wire rope. The following notes were taken, meter at the depth of 1.75 feet:

No. of Trial.	1st Time.	2nd Time.	3rd Time.	4th Time.	REVOLUTIONS. 100 ft. S Base	100 ft. S.Mid.	100 ft. N.Mid.	100 ft. N.Base	
1	0	2 m. 12 s.	4 m. 4.4 s.	6 m. 1.6 s.	8094	8125	8155	8183	Up.
2	0	2 m. 36 s.	0	2 m. 03 s.	8359	8333.5	....	....	Down.
3	0	1 m. 18 s.	2 m. 33 s.	3 m. 49.6 s.	8377	8405	8434	8462	Up.
4	0	1 m. 25.6 s.	....	4 m. 3.6 s.	8568.5	8539	8511	8482	Down.
5	0	1 m. 57.4 s.	3 m. 36.4 s.	5 m. 12.8 s.	8589.5	8612	8639.5	8666	Up.
6	0	1 m. 44.6 s.	3 m. 4.8 s.	4 m. 36.4 s.	8788	8760	8730	8702.5	D wn.

South to North "up," North to South "down."

The meter was then placed at a depth of 2.25 feet below the surface of the water and 18 sets of observations, similar in form to those just given, were taken. The notes were then reduced to feet and revolutions per second. Taking the velocities in feet per second as the x ordinates and the revolutions per second as the y ordinates, and plotting them to scale, as shown in the plate, it is found that they form a straight line (nearly), $y = ax + b$, showing that they are linear functions of each other. This gives for the values of the constants, $a = 0.3401$, $b = -0.06$. Plotting the observations taken at depth 1.75 feet separately, it is found that $a = 0.3398$ and $b = -0.06$. For the observations taken at depth 2.25 feet, $a = 0.3454$ and $b = -0.19$.

To check the values of a and b as found by the graphical method, they were computed by the analytical method which is based on the assumption that the most probable values of a and b are found by taking a properly weighted arithmetical mean.

CURRENT METER RATING
TAKEN AT
LAKE CALHOUN MAY '93.

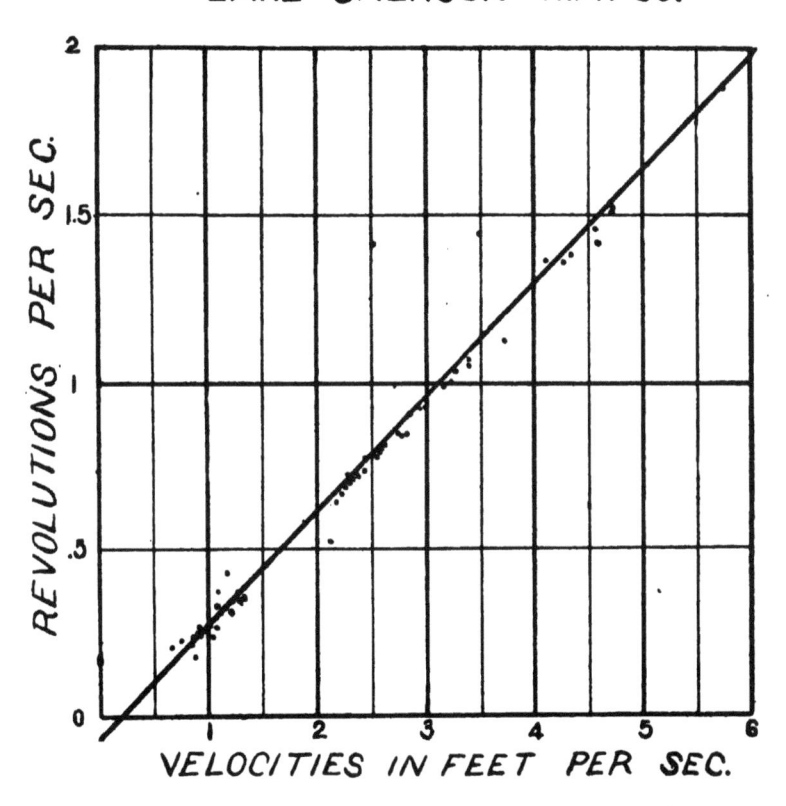

The equation for the determination of a then takes the following form:

$$[(x-x_0)^2 a + (x-x_0)(y-y_0)] = 0;$$

in which x_0 = mean of observed values of x; y_0 = mean of observed values of y. From the equation of a straight line $y_0 = ax_0 + b$, b is then found.

Taking all the observations, this method gives for the constants a and b; $a=0.34307$, $b=-0.12160$. These values give for wheel velocities slightly above zero revolutions per second, a velocity of current differing by nearly 100 per cent from those given in Mr. Price's table, the difference decreasing until at 2.01 revolutions per second they give the same velocity of current.

In order to see if the depth at which the meter is placed has any effect on the rate of the meter, the observations taken at a depth of 1.75 feet were taken separately giving $a=0.3098$ and $b=-0.0304$. Then computing those observations taken at a depth of 2.25 feet, having velocities corresponding to those taken at a depth of 1.75 feet, we find $a=0.398$, $b=-0.098$. Taking these constants and computing the velocities for a given number of revolutions per second, it is found that for zero revolutions per second the velocity at a depth of 1.75 feet gives a current velocity of 0.0979 feet per second, while at a depth of 2.25 feet the current velocity is 0.246 foot per second.

On increasing the number of revolutions per second, the equation for a depth of 1.75 feet gives a greater increase of current velocity than the equation for a depth of 2.25 feet. This seems to indicate that the water is pushed or pulled along, to a certain extent, by the boat as it moves along over the course.

To ascertain if the position of the meter ahead or behind the boat has any effect on the rate of the meter, the observations were divided into two groups, those taken when the meter was in front of the boat forming one group, and the observations taken when the meter was behind forming the other. The equations derived from these observations show that for a given number of revolutions per second the velocity obtained when the meter is behind is less than when it is in front. This is probably the effect of the water as it moves in to replace that carried along by the friction of the boat.

Comparing the table given by the first equation, i. e. with the meter in front, with a table given by Mr. Price, which is the mean of a large number of ratings, it is found that for zero revolutions per second his rating gives a velocity of about 10 per cent. less. The difference between the ratings then diminishes until it reaches 0.23 revolutions per second where they are the same. After this the difference very gradually increases until at one revolution per minute his table gives a velocity 0.4 of 1 per cent. greater than was obtained from the table given by the first equation.

CONSTRUCTION OF A SEWER SYSTEM.

GEO. L. WILSON.

[Extract from lectures given before the Society of Engineers.]

THE DESIGN OF A SEWER SYSTEM.

Assuming that the problem before us is the design of a system of underground channels to collect and to convey from a given thickly populated area all the liquid wastes from houses, stores, manufactories, laundries, etc., together with the storm-water, in one set of pipes or conduits, we have to design a system of sewers on the "Combined system" as it is called, in distinction from the "Separate system" which does not provide for the storm-water but only for the liquid waste from buildings.

There are four things in this problem to be considered with particular attention:

I. The area, physical outlines and controlling features of the district to be drained; its geological character and the depth to which the drainage should extend.

II. The rainfall in the district with reference to its *maximum* rate and the duration of the heaviest storms, also the amount of such rainfall which experience has shown must be provided for.

III. The character and extent of the water supply.

IV. Final disposal of the sewage. (Adams, Sewers and Drains, p. 21.)

MAP.

The greatest assistance in the design of a sewer system is a good contour map of the area to be sewered and if circumstances will allow, one should be prepared; yet, it will generally be found that there is no such map ready before one needs it, and the authorities will not appreciate its value sufficiently to pay for making it.

It is a matter of necessity, however, that the grade of the streets should be fixed before the design of the sewerage system and, if this has been done, a good working substitute for the contour map is to take a good-sized map of the area under consideration and go over it, placing the elevations of the established grades at all street intersections, and marking the length of each block on the map. If the grades of the streets have not been established or determined, this will be the first step to take, and we shall very likely find this question will have to be settled before any other matter is taken up. Passing over this important subject by supposing that so much of the work has been done as is necessary to prepare the topographical map spoken of, work may be begun.

The location of the outlet is first to be settled and if we assume that the town lies on a stream of sufficient size to dispose of the sewage, i. e. whose flow at a minimum is 150 to 200 cubic feet per minute per 1,000 people, a proper outlet will be the nearest point where the sewage can be discharged into running water. It must be remembered that sewers cannot be emptied into stagnant water without producing a nuisance. Probably it will be found best to divide the whole area to be sewered into a number of small districts; each with its separate outlet and small main sewers, for these two reasons:

1st. By a division into small districts the expense is greatly reduced since it is the large main sewers that make the largest part of the cost of a system, reducing the districts cuts down the main sewers.

2nd. It will be much easier to construct the sewers for small districts, thus providing sewer facilities at the minimum expense as fast as they are needed.

In dividing up a given area into small districts, each with separate sewers, keep in sight that sanitary science is moving towards the purification of all sewage before it is cast into running streams, and it is more than probable that the near future will see all towns required to purify, to some extent, their sewage either by chemical precipitation or filtration, especially in the more thickly settled portions of the country. With this in mind, so arrange the outlets of the different districts that, when it becomes necessary, an intercepting sewer can be built which will collect all the sewage and convey it to some easily accessible point for treatment; the old outlets then will be useful as storm-water overflows. With the location of the outlet fixed, the route of the principal or main sewer will generally be determined by the line of lowest land in the streets. If the streets or alleys do not offer a reasonable route, a line can be located across private property. This will require condemnation of an easement for the sewer and unless there will be a large saving in first cost, the great advantage of having sewers on public grounds for ease of access in maintenance and repairs will decide the location in the public ways.

Reasons for not following the lowest line, such as important buildings, or a business street requiring sewerage at once, may necessitate the main sewer being built past such points first, providing it can be done without too much increase of expense. Other reasons may occur, such as non-assessable property on lowest line, influence of property owners to change the location, etc.

The location of main having been settled, the question of grade is next in order. This is to be determined first by considering that the main must be low enough to drain the lowest ground. It must be below the lateral sewers draining into it, at least enough to give them a proper fall, and must have sufficient grade to give, if possible, a self cleansing flow, at the same time being no deeper than is necessary on account of economy of construction.

In case these conditions indicate a steep grade, design a section of solid construction with granite invert, if necessary, and bear in mind that the steeper the grade the smaller does the sewer and also the cost become.

The minimum rate of grade is a difficult question. The following table will show what some authorities consider as approved grades:

MINIMUM GRADES—REQUIRED VELOCITY.

Size.	Latham.	Adams.
6 to 9 inches.	3 feet per second.	4.5 feet per second.
12 to 24 inches.	2½ feet per second.	3.5 feet per second.
Above 24 inches.	2 feet per second.	3.5 feet per second.

See table in Trautwine, page 91, edition of 1888.

The grades in most cases will depend on the character of the ground. For sewers in comparatively level country with no sand hills and steep grades, from which sand and gravel will wash in to clog the sewers, quite light grades may be used. From 0.1 to 0.2 feet per 100 feet should be considered as a minimum for sewers above 4 feet, carrying street refuse and washings; while from 0.5 upwards is an approved rate and should be obtained when possible, particularly when sand and gravel may be expected to wash into the sewers. Guard against flat grades at the foot of steep ones, as these are the places where deposits are sure to occur. Raise the lower grades as much as possible to make the sewers self-cleansing. Flat grades will occur, but get as much fall as practical for sewers on flat ground at the foot of steep grades.

Having the route and the grade of the main sewer fixed, lay out the lines for lateral sewers so that they will accommodate the property to the best advantage; that is, locate them along the streets upon which the lots face. Or it may be advisable to place the sewers in the alleys of the town, particularly so if the grades have been arranged so that the surface water can be collected in catch-basins placed at the alley corners instead of street corners. If the town-site is laid out regularly with alleys in every block, the placing of the sewers in the alleys not only prevents tearing up the main streets, but reduces materially the distance from building to the sewer and consequently the cost of making connections.

DEPTH.

The lines of the lateral sewers having been fixed, each of these may be taken up separately, in which case to fix the depth from the surface is the first consideration. Before the days of deep basements and sub-basements, a sewer 10 feet deep was supposed to be of ample depth. The tendency is to build deeper into the ground as well as higher, and for principal streets of large cities depths of 14 feet to be drained are not uncommon. It is hardly good practice, however, to build an entire system to provide for these deep cellars but the sewers on one or two main streets may be put at an extra depth if the improvements already made demand it. To settle this point have depth of all existing cellars marked on the profile. A depth of 13 feet is as deep as the lateral or pipe sewers should be laid in general practice for cities, while for all small towns and especially for residence districts a depth of 8½ to 9½ feet is ample. There are places where even a less depth is sufficient, but this like all other details of the work depends more on good judgment applied to each particular place than on any rule.

The determination of the grades for the laterals will be governed greatly by the street grades, but in the fixing of the grades for the laterals, keep in mind that all sewage should be kept moving with a continuous flow at as nearly a uniform velocity as practical from the time it is emptied into the sewer until the outlet is reached. If this is done a deposit of sediment is formed and the sewage reaches the outfall in a fresh condition and comparatively little gas is evolved.

When the lines of laterals are fixed, go over each and calculate the area that will be drained at each block, from the upper end of the main sewer, also the total area drained by each at the main and mark these areas on the map. Now we have indicated how the lines of mains and laterals are to be determined and the grade of each. As the area to be drained at successive points are known, condition (I) of the problem may be considered as stated, and the next subject to be considered is that of rainfall and probable maximum flow from the areas to be drained. As this is a matter that has been much discussed and as great varieties of opinion have existed and do exist, we will take this point up in some detail.

FIRST. Looking for the earlier investigations of the subject, we find that in 1857 the noted English engineers, Messrs. Bidder, Hawksley and Bazalgette, engaged in the municipal works of

London, had gaugings made in the sewers of the most populous districts of the city, and from these they "felt warranted in including as a rule of averages, that 0 25 inch of rain-fall would not contribute more than 0.125 inch to the sewer and that a fall of 0.40 inch would not yield more than 0.25 inch to the sewers." These gaugings are the first of which we have record as being taken for the purpose of finding out the percentage of flow during a storm.

In 1865, Col. Wm. Haywood, Engineer of the Sewer Commissioners of the city of London, in a paper before the Institute of Civil Engineers, mentioned the results of gaugings made in 1857 and 1858 during severe rains when with 2.75 inches of rain in 36 hours, the London bridge sewer discharged 53 per cent of the fall in the same time. In '58 the same sewer discharged 74 per cent. of a fall of 0.24 inch in 1.5 hours. Again two months after 78 per cent of a fall of 0.48 inch in 1.67 hours. This report gave no information as to the character of the rainfall, the manner in which it was observed, or other details.

About the same time John Roe, surveyor of Holborn & Finsbury sewer, made gaugings from which he concluded that of a rainfall of 1.0 inch per hour, 54 per cent. of the precipitation will reach the sewer. Yet the conclusions of the London engineers have been made the basis for much work in the United States and until recently it was regarded, on this doubtful authority, that about 50 per cent. of the rainfall would run off from urban surfaces during the progress of the storm. This too in a country like the United States where the annual rainfall is much heavier than in England.

For most of the earlier work in the United States a maximum rate of 1 inch per hour was used. J. W. Adams (in S. & D. Pop. Districts, 1888) says, "the frequency of rainfalls of 1 inch per hour when compared with such falls as 2 or 3 inches, renders its consideration a more practical question, and the possibility of this ratio of fall occurring for shorter intervals of time is so apparent from observation on the seaboard and in the interior that we may regard it as a very proper maximum. It has been adopted as such in England and so far as we can learn also in this country."

Cities whose sewerage systems have been designed on this assumption have found its error, and special stormwater or enlarged mains have had to be provided to relieve districts often overflowed.

In Providence, R. I., records of every considerable rainfall for 40 years, from '34 to '74 were kept and published in the report of J. H. Shedd, city engineer, for the year 1874.

These were believed to be entirely reliable and from these records of 324 storms, only three lasting less than one hour, it was found that a rainfall of 2.0 inches or over per hour may be expected to occur only once in 13 years; one of 1.5 inches once in 7 years, and one of 1.0 inch per hour once in about 4 years. Such records do not show what we wish to know; the time and total depth of fall will give the average rate. The important matter is the rate of rainfall during the maximum intensity and the duration of that period, as these are the things we need to know to determine the maximum flood capacity of sewers. The valuable records for this purpose are those made by the automatic self-registering gauges to be found in the weather bureau offices in the larger cities. These give a continuous record for the entire period covered by the storm. Many more of these gauges are needed and it is to be urged upon the attention of the public that the records are of great practical use.

There are four well-known formulas in use for determining the quantity of storm-water that a given area will yield. The best known is that of the celebrated Swiss engineer Burkli-Ziegler, City Engineer of Zurich, proposed in 1880. The English engineer Hawskley worked out the one called by his name, and there are two American formulas, that of R. E. McMath published in the Trans. Am. Soc. C. E. for 1887, modeled after that of Burkli-Ziegler but with different coefficients so that quite different results are obtained, and the Adams formula, by J. W. Adams of Brooklyn.

In a paper in the Trans. Am. Soc. C. E. for 1889, E. Kuichling of Rochester, N. Y., gives a reduction of these formulas to a common notation, which we will adopt for the purpose of comparison.

Construction of a Sewer System. 113

The notation is: 2=maximum discharge in cubic feet per second.

r=maximum rate of rainfall in inches per hour, which is also practically the same as if given in cubic feet per acre per second.

A=area drained in acres.

S=sine of general slope of surface or average fall divided by the average length.

(1) Hawksley: $2 = 3.946 \, Ar \sqrt[4]{\frac{S}{Ar}}$

(2) Burkli-Ziegler: $2 = (1.757 \text{ to } 4.218 \text{ average } 3.515) \, Ar \sqrt[4]{\frac{S}{A}}$

(3) Adams: $2 = 1.035 \, Ar \sqrt[12]{\frac{S}{A^2 r^2}}$

(4) McMath: $2 = (1.234 \text{ to } 2.986 \text{ average } 2.488) \, Ar \sqrt[5]{\frac{S}{A}}$

The first and third of these formulas have reference to the ordinary conditions of city territory and are designed to apply best for a rainfall of (r=) 1.0 inch per hour.

The second and fourth formulas are more flexible, and adapted to a wider range of conditions. The smaller coefficients apply to suburban districts and the larger to densely populated areas solidly built up, with all streets and alleys paved.

In the fall of 1892, A. J. Grover made a series of gaugings in Omaha, Neb., of the discharge from an area of 2100 acres with surface grades varying from 1% to 18%. The area was about 11,000 feet by 7,000 feet, and contained a population of about 40,000, or 19 per acre, and it is of interest that while the results of the gaugings vary widely, from 5% upward, the maximum discharge was 60% of a rainfall of 0.32 inch that fell during 2.55 hours.

For the account of these gaugings and the ingenious device used in them, see Trans. Am. Soc. C. E. for Jan., 1893, Vol. XXVIII, pp. 1-12.

DETERMINATION OF SIZES OF SEWERS.

From the application of the Burkli-Ziegler formula, as mentioned (p. 10) we find the maximum amount of storm-water to be provided for at each junction point. Knowing "Q" and the rate of grade the different sizes will be determined by a simple application of Kutter's formula, which is derived from the formula of Chezy (1775) $V = c\sqrt{RS}$ in which

$$C = \left\{ \frac{41.6 + \frac{1.811}{n} + \frac{.00281}{S}}{1 + \left((41.6 + \frac{.00281}{S}) \times \frac{n}{\sqrt{r}}\right)} \right\}$$

"N" = Coefficient depending on the lining of cannel for sewers 0.13 or 0.15.

V = mean velocity in feet for second; c = coefficient of velocity.

$S = \frac{h}{l} = \frac{\text{fall of surface}}{\text{distance}} = $ line of slope.

r = hydraulic mean radius (depth) = $\frac{\text{area of cross section}}{\text{wetted perimeter}}$

For the constant n in this formula, a value of 0.13 is recommended. This is the standard formula and is recognized as such by engineers generally. While the formula looks at first glance slightly complicated, it can be reduced to a comparatively simple form. See for tables, Flynn's Tables, Van Nostrand's Sci. Series, Nos. 67 and 84. Also Gauguillet & Kutter on the flow of water, as translated by Hering & Trautwine—1889—published by Wiley & Son.

Having determined sizes, locate manholes from 150 to 250 feet apart. In fixing grades and alignment between manholes, keep in mind what is called "Rawlinson's Principle," from Mr. R. Rawlinson who first called attention to the correct method of laying all pipe sewers in *perfectly straight lines* from manhole to manhole, or other points at which the pipes can be observed. If built in this way the whole system is under control and can be reached without digging up the streets.

Finally, make plans showing both the map and profiles of the proposed sewers and write a report which shall state clearly the work proposed to be done and the reasons therefor.

Specifications should state what is to be done and how the work is to be executed. They should be so clearly written and definitely expressed that no loop holes for law suits or the neglect of dishonest contractors will be left, keeping in mind that as engineers, your duty will be to see that both parties to the contract get full and even justice.

SEWER ASSESSMENTS.

In any problem it must be understood at the beginning what is the especial thing to be done. This time it is to apportion the cost of a public improvement upon the property benefited, so that each individual piece shall bear its just and equitable proportion of the whole cost.

Having, generally, to deal with a large number of persons, each having his particular views as to his own interests, many difficulties arise.

METHODS.

The first question is as to the plan that shall be adopted as the basis of assessment. The various customs in use may be classed as follows:

(1). The city pays the entire cost from the proceeds of a loan or an issue of bonds.

(2). The city pays the cost of the largest (main) sewers from bonds or a general tax, and the private property abutting pays for the smaller (lateral) sewers.

(3). The entire system is paid for by the abutting property.

Of these plans the first is objectionable from the fact that it tends to overburden the city with too large a debt. For if the town, as a whole, pays the bills from borrowed money, each street clamors to have its sewer built as soon as possible, and the work goes on as long as bonds continue to furnish the money.

The second plan is that most generally adopted in large cities. The main sewers drain large areas and are considered as of general benefit. Their construction is expensive and usually can be paid for only by the city at large. This is where the municipality as a whole must step in to assume the cost of large public works. The lateral sewers are of only local benefit and as such should be paid for by the abutting property.

The third method is popular with small cities, or cities where the area can be divided into districts, a complete system worked out for each and an estimate made of the cost for the whole district. The assessment can be figured for the entire work and be equally distributed over the whole area included in each district.

In cases where none of the sewers are large, often the cost of each separate contract is assessed on the property fronting on the improvement. This plan gives the most unequal and unsatisfactory results. The method is followed where it is undesirable to pay any part of the cost from the sale of bonds or general taxation and is based upon the idea that the property drained should bear the whole expense as the work progresses. The result of this method is that lots of the same size within a short distance of each other may differ 200 to 400 per cent. in assessment.

DIVISION OF COST.

The method in which the city pays a portion of the expense and the benefited property the balance is, as already mentioned, the usual practice in large cities covering as they do extensive areas and requiring some very large and expensive main sewers. In the smaller towns and cities the whole cost is borne by the abutting property.

To decide among the different plans for dividing the cost among separate estates and equitably to assess the expense upon the persons benefited is often a perplexing question, and the decision is almost sure to be unsatisfactory to a large number of property owners. The business of the assessor and tax collector in all history has been an extremely unpopular one.

The different plans for dividing the expense are as follows:

(1). By frontage.

(2). By area.

(3). By a combination of frontage and area.

(4). By an annual tax for construction and maintenance.

(5). By a fee to be paid when property is connected with the sewer.

The first or frontage plan is that in most general use in towns laid out on a rectangular plan. It is the easiest to apply as the property frontages are generally known and matters of record. If the separate estates or lots are of uniform depth the total cost divided by the total assessable frontage gives at once a uniform rate. In case of varying depths and irregularly shaped lots or unplatted property, it is found that the second method, by fixing the rate by the area of the property drained, gives most equitable results. This is especially true in the old cities of the Eastern States and in particular in the New England towns. The area to be assessed is fixed at a certain depth from the street which will be about the usual depth of a single lot, 125 to 150 feet.

The third method by combining area and frontage is approved in some cities. It is now used in Boston and Providence. In all cases provision should be made for property having two street frontages, as corner lots. Usually one side only is assessed or else the 50 or 60 feet of the total frontage is exempted. The area is assessed but once.

The fourth method, that of an annual tax for the use of the sewer or a *rental* price, is as yet a novelty. It is being tried in a few places, among which are Portland, Me., Brockton and Marlborough, Mass. Its advocates claim that the payment for the sewer in this way is proportional to the amount of water used, and therefore to the benefits derived. This requires that the water connections should be metered and on the meter quantities are based the prices to be paid.

The fifth plan, that of a fee to be paid when connection is made with the sewer is used in but few places, and there are so many evident objections to it that it is mentioned only as a curiosity.

The method which gives the best satisfaction, and which can safely be recommended as just and as giving the fairest results to all parties, is to make the rate uniform over as large an area as possible, this to be fixed by city ordinance. To decide on the proportion which shall be paid by the city and by the property owners requires especial knowledge of the local conditions in a large town and of the sentiment of the people.

In small places, or where the "separate" system is to be used, the whole cost can easily be borne by the property. Especially so if it is arranged, as it should be, that the individual assessments may be paid in four or five annual installments, which is the present custom in most states. The usual maximum rate is from $1.00 to $1.50 per front foot; over this the cost is so arranged as to be paid by the city.

From the experience of the older cities and the changes they have made from time to time, also from the advice of engineers of wide experience, it may be considered as settled that the best practice is to establish a uniform rate per lineal foot of frontage, based upon the estimated cost of the sewers for the whole town if possible; otherwise, for separate districts. This fixed charge being known in advance people wishing sewers constructed in front of their property will know positively the price to be paid and be able to decide upon petitioning for the improvement.

While the position of one who fixes taxes or assessments to be paid is, as already mentioned, never a popular one, still by following out a method giving just results without depending upon individual judgment, one will have at least the satisfaction of a duty well done.

ATOM-ARRANGEMENT IN SPACE.

DR. GEORGE B. FRANKFORTER.

The fact that tartaric acid exists in several distinctly modified forms was first carefully noted by Pasteur. Similar and more marked examples of isomerism than tartaric acid were known, but no attempts were made to explain them until the last quarter of this century. Pasteur, in his classical work on tartaric acid, showed that the different forms of the acid were due to the different arrangements of the atoms in the molecule. He showed by treating racemic acid, commonly known as inactive tartaric acid, with soda and ammonia that a sodium—ammonium salt was formed with two sharply defined crystals, the one bearing the same relation to the other in form that the right hand glove bears to the left. Upon examination of these two forms of crystals Pasteur found that one form represented an acid which turned the plane of polarization to the right, and that the other form represented an acid which turned the plane of polarization to the left. In his "*Lecon sur la dissymetrie moleculare,*" Pasteur explained the phenomenon by an assumption that certain atoms of carbon in their relation and combination with other parts of the molecule were the true causes of the optical properties. The theory was accepted and within the last decade brought into great prominence by Le Bel and van't Hoff in their development of the stereo-chemical theory. Simultaneously and quite independent of each other, the two above named men published the result of an optical study of a number of carbon compounds, concluding that these newly discovered properties were due to an assymetric carbon atom, that is an atom, the four affinities of which were held by different atoms or groups of atoms. For an illustration methane, the simplest of the hydrocarbons, will serve well. Represented graphically methane will appear with the carbon atom as a center around which are clustered the hydrogen atoms—

$$\begin{array}{c} H \\ | \\ H-C-H \\ | \\ H \end{array}$$

When three hydrogen atoms are replaced by three different groups of atoms, the carbon becomes assymetric. A simple and historical example is the common lactic acid or alpha oxypropionic acid. Here the central carbon atom holds the groups CH_3, CO_2H and OH while the fourth affinity holds hydrogen thus giving the molecule assymetry—

$$\begin{array}{c} H \\ | \\ CO_2H-C-OH \\ | \\ CH_3 \end{array}$$

Even before the "Pasteur Experiment," lactic acid had been carefully investigated by J. Wislicenus, one of the pioneers of structural and stereo-chemical theories. Wislicenus showed that common lactic acid existed in two forms, one an exact image of the other. Giving the common form three dimensions as set forth in the tetragonal theory, the molecule would appear as a solid figure in space—

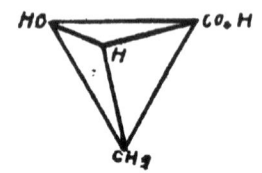

The second form would appear as an exact image of the above figure. A third form of the acid, of scarcely less importance in the development of the tetragonal theory, does not contain the assymetric atom, but represents the hydroxyl group as filling

Atom-Arrangement in Space. 121

one of the spaces in the methyl group and making the central atom hold two hydrogen atoms. Assymetry is therefore destroyed as shown by an examination of the acid. Indeed a new reaction alone in the formation of the "lactone," is sufficient reason for representing these acids by the above glyptic forms.

The wonderful properties of tartaric acid had been noted ever since its discovery, but the hope of looking into the physics of the atom was not entertained until the discoveries of Pasteur. Since then, by the persistent work of Le Bel and van't Hoff, its structure has been accepted as final. Tartaric acid contains two assymetric carbon atoms indicated by the number of isomeric forms. These two carbon atoms represent a double nucleus in which the mechanical center of the molecule is at the point of juncture of the axes of the two tetrahedrons. From this imaginary point the isomeric forms of the acid may readily be conceived:

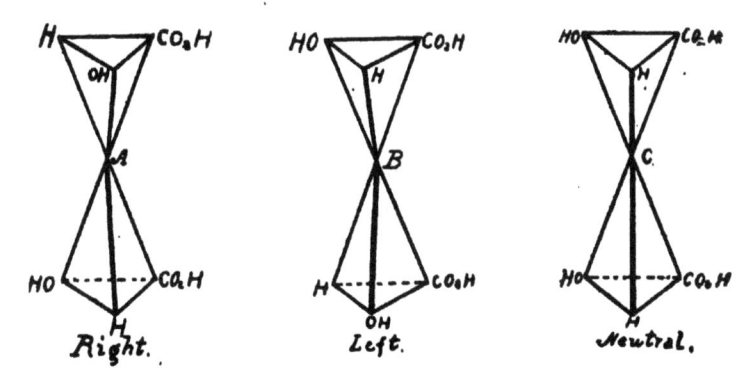

Right. *Left.* *Neutral.*

A glance at these geometrical figures indicates the existence of the acid in several forms, in which the different parts of the molecule bear different relations to each other. If the eye be placed at the imaginary center A, and the H made to turn towards O H, O H towards CO_2H, then H, O H and CO_2H, will appear to turn in the same direction as the hands of a clock. This would represent a dextro-rotary acid. If under similar conditions the central point be at B, the direction, passing from simpler to more complex radicals, will be opposite to the hands of a clock, and the acid will turn the plane of polarization to the left. If the point be at

C, the direction in one tetrahedron is to the right, while in the other it is to the left. Neutralization is the result and the third form of the acid is optically inactive.

Another acid not less important in support of the tetragonal theory is fumaric and its isomer, maleïc acid. Fumaric and maleïc acids may be regarded as derivations of either tartaric or succinic acids. They differ, however, from the two latter acids inasmuch as the two central carbon atoms are joined by the edges of the tetrahedron instead of the angles or summits in tartaric acid. The following glyptic formulæ will show at a glance the two possible isomeric forms.

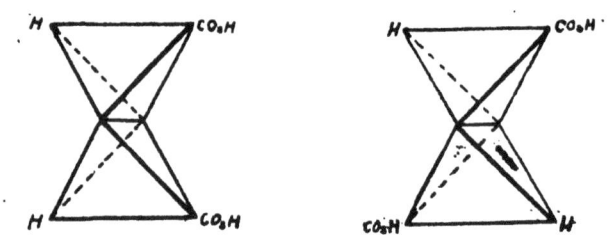

There is a third form of linking the central carbon atoms not represented in the tartaric or fumaric acid union. Instead of axial and lateral juncture, a facial juncture occurs and is represented by the acetylene di-carboxylic acid.

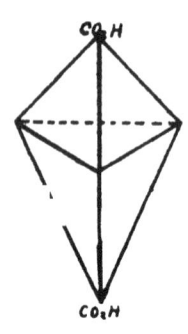

Here assymetry is impossible. The tetragonal theory admits but one form. It is thus evident that the theory of Le Bel and van't Hoff harmonizes perfectly so far as mechanical arrangement is concerned. From old axiomatic facts to new ones fresh from the hands of the investigator, the theory has a perfect explanation. The chemist seems satisfied. He desires, wisely, to accept all theories which harmonize with facts.

As to assymetry and its relation to the optical properties as indicated under tartaric acid, Kekulé first observed that certain organic compounds produced peculiar optical effects upon light. Le Bel and van't Hoff took up the idea, naturally associating it with the assymetric theory. So remarkable were the results of their work that even with only a comparatively few definitely known assymetric compounds upon which to work, assymetry was immediately pronounced the cause of optical properties.

Assymetry has not stopped with optical phenomena. It unquestionably explains physiological properties of substances. Although at present only comparatively few physiologically active substances are known structurally, still, I believe that enough is known to warrant the above assertion.

The alkaloids, of which little is yet known, will no doubt conform strictly to the above statement. A few have already been examined. Nicotine, the physiological action of which is well known, contains an assymetric carbon atom and exists, like tartaric acid, in three distinct forms.

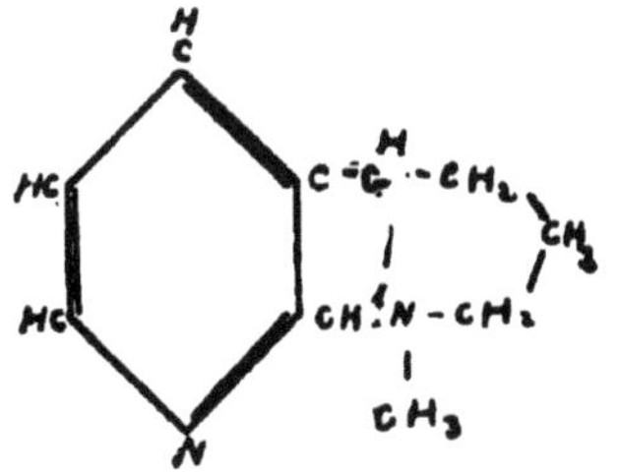

The assymetric carbon atom is here marked by a heavy-faced letter. A glance at the figure will show that nicotine may exist in at least five distinctly modified forms. Common nicotine, as will be seen by the above graphic formula, is lævo-rotary, while the salts of nicotine are dextro-rotary.

The structure of cocaine, another common active alkaloid, is known, and may be represented by the following graphic form:

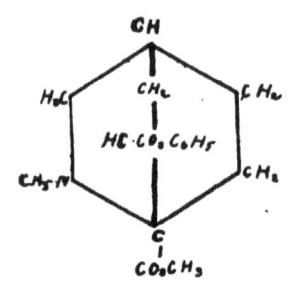

$$\begin{array}{c} CH \\ H_2C \diagup CH_2 \diagdown CH_2 \\ | \quad HC\cdot CO_2C_6H_5 \quad | \\ CH_3N \diagdown \quad \diagup CH_2 \\ C \\ | \\ CO_2CH_3 \end{array}$$

Cocaine contains three assymetric carbon atoms. It therefore exists in several distinctly modified forms. The dextro, lævo and inactive forms occur in nature while others have been made synthetically. Other alkaloids, narceine and berberine, contain no assymetric carbon atom, and have recently been shown to be optically and physiologically inactive. So marvelously have facts harmonized with theory that perhaps many have gone too far, losing sight of the fact that the tetragonal theory is still a theory. Van't Hoff in his over-enthusiasm was perhaps justly criticised by the great Kolbe who expressed his keen disgust in the modified words of Faust:

> "Nachher von allen anderen Sachen
> Müsst' Ihr Euch an die 'Structurchemie' machen,
> Da seht, dass Ihr *tiefsinnig* fasst,
> *Was in des Menschen Him nicht passt;*
> Für was drein geht und nicht drein geht,
> Ein prächtig wort zu Diensten steht."

The above criticism is certainly just, so far as accepting bodily the structural and tetragonal theories. Nevertheless these theories have served well as temporary structures in which to store modern chemical facts. They have served a noble purpose, even should they be rejected to-morrow. Without them, organic chemistry would have been a conglomerate mass of facts. As it is, organic chemistry is without doubt the most systematic branch of modern science. These theories have done much to formulate the science, though they are still only temporary theories. Indeed, such weighty contradictions arise when they are put to strictly physical test, that like Kolbe, one is surprised that they should have satisfied the minds of greatest scientists of the age. One of the chief objections to-day is molecular motion. It is an axiomatic fact that there is a constant molecular motion. It is quite impossible to conceive of the tetrahedron giving perfect freedom of motion. The theory itself assumes the atoms to be stationary. This is perhaps the most objectionable point in the whole theory, and a new theory which will admit perfect freedom to atoms in the molecule must finally be presented. In my judgment, the theory which is to supplant the tetragonal theory is yet to be formulated from Helmholtz's theory of "Vortex motion in liquids" and Sir William Thomson's vortex theory and vortex atom. With such a theory harmonizing with the accepted theories of matter, the science of chemistry must soon become one of the exact sciences.

ACETYLENE AND ITS PRODUCTION.

FRED. M. ROUNDS, '95.

Upon the initial appearance of the Welsbach burner, by virtue of its many advantages, opinions were expressed as to its replacing to a large extent the electric incandescent light.

While this has been true to some extent, yet the public have not lost sight of the sanitary advantages of the electric incandescent lamp, nor been blinded by the brilliant glow of the Welsbach burner.

And now electricity has indiscreetly become an active agent in the production of a gas of high illuminating power, and which may prove to be its serious rival.

Thomas L. Willson with a view of reducing lime to calcium, heated a powdered mixture of chalk and charcoal in an electric furnace. The mixture fused into a mass resembling metal which, upon testing, showed no traces of calcium. By chance a piece of the mass fell into a bucket of water and a very heavy gas was given off. This upon testing proved to be acetylene (C_2H_2), a previously known hydrocarbon.

The metallic mass formed by the fusing of the chalk and charcoal was calcium carbide (CaC_2). It had united with the water according to the reaction expressed by the formula,

$$CaC_2 + 2H_2O = C_2H_2 + Ca(HO)_2,$$

the hydrogen having united with the carbon to form acetylene gas, and the oxygen with the calcium to form slaked lime.

Travers made calcium carbide from a mixture of chloride of lime, carbon and sodium. It is a dark gray substance with specific gravity of 2.262 at 18°C. When pure a pound will yield more than 5 cubic feet of acetylene. The carbide does not deli-

Acetylene and its Productions. 127

quesce rapidly when in lumps. For commercial purposes it is cast into rods 12 inches long and 1.25 inches in diameter. The rods weigh one pound each, and will give off 5 cubic feet of gas 98 per cent. pure.

Carbides of barium, potassium and other elements will produce acetylene gas, but the chief advantage in producing the gas from calcic carbide is that the raw materials may be used, and the discovery of this electrical method offers the cheapest process at present available.

The properties of acetylene have been known some time. It was first produced by Wöhler, who fused an alloy of zinc and calcium with carbon, and then produced the gas from the fused mass. It is one of the richest illuminating gases, being more luminous than olefiant gas and fifteen times as luminous as ordinary gas. It has a penetrating odor strongly resembling that of garlic.

Acetylene may be stored in liquid form at a temperature of 1°C., under a pressure of 725 pounds per square inch. When liquified and sprayed, it forms into a whitish vapor which burns with a clear flame having a temperature of 1,000°C., that of ordinary gas being 1,360°C. It is soluble, volume for volume, in water, and is convertible into benzine, aniline, ethylene, alcohol and other useful products. If stored in liquid form, the gas is especially adapted for isolated lighting, such as country hotels, buoys, and railway cars. When burned in a suitable jet at a rate of 0.5 cubic feet per hour, a single rod weighing a pound will give a light of over 20 candle power for about 10 hours. Mixed with an equal volume of air before being burned, the diluting action of the nitrogen reduces the illuminating value from 240 to 130 candle power.

The furnace in which the calcium carbide is produced does not differ materially from the ordinary electric furnace. The upper electrode is movable and may be lowered into a carbon crucible which serves as the other electrode.

To produce 2,000 pounds of the carbide, 1,200 pounds of fine coal dust and 2,000 pounds of burnt lime are needed; the current being from 4,000 to 5,000 amperes, with a total of 180 electrical H. P. for 12 hours.

The cost where there is cheap water power, coal, lime, etc., is about $15 per ton. In fact, the success of this process depends chiefly upon cheap water power, for only 5 per cent. of the heat produced by coal is recoverable in an electric furnace.

One ton of calcium carbide will produce about 10,500 cubic feet of acetylene, which would be equal in illumination to about 100,000 cubic feet of ordinary city gas of from 22 to 25 c. p. per 5 foot burner.

It is stated that 1 H. P. of electrical energy, if consumed in incandescent lighting, would give light equal to 1 as compared with 1.57 if consumed in the production of the acetylene light. Central station men claim that this would not be a sufficient difference to make it possible for the acetylene light to compete with the electric light, as the cost of storing, packing and selling would diminish this difference materially.

Furthermore, the price quoted for calcium carbide, $12 per ton, rests upon a cost of only $5 per yearly electrical H. P., while the price of power generated at Niagara is placed at $18 for the same amount, and this price is at the generator terminals.

The brilliancy of the acetylene flame is more than counterbalanced by the lack of sanitary advantages which are found in the incandescent electric lamp. It consumes the oxygen in the air and gives off smoke and carbonic acid.

The Welsbach burner has not materially affected the electric light, and since acetylene, for the same illumination, would have to be sold at about half the price of ordinary gas, it is scarcely probable that the electric light will find an active rival in it.

If, however, there should be a demand for this light, there would also be an additional demand for power, and should it prove to be a serious rival of the electric light, it undoubtedly would be profitable for central stations to undertake its manufacture and thus obtain the much needed and long desired day-load.

SPECIFICATIONS FOR ENGINEERING STRUCTURES.

CHAS. F. LOWETH.

A specification is an expression in writing and of record, setting forth the requirements to be met and fulfilled in furnishing material for, or constructing a piece of work.

The subject matter of engineering specifications may be of material only, as cement, bricks, steel rails and lumber; for a machine, as a steam engine, either simply delivered f. o. b. cars, or set up in place complete and ready for use; or for a completed structure more or less complicated, as a bridge or building, or even a railroad fully built and equipped ready for use.

The requirements in a specification will generally cover those matters pertaining to the amount and extent of the work, the quality and character of the work to be done, and materials employed or furnished and, to a greater or less degree, the manner in which the work is to be done or the material made; in fact, any condition or requirement which the purchaser may deem essential and desires to have observed in the execution of the work may be incorporated in the specifications.

A specification will generally become eventually a part of an agreement between two parties; one the purchaser, the other the seller or contractor. It will serve previously, however, as a basis for a bid or proposition from a bidder or contractor, to the prospective purchaser, as upon its conditions and requirements the purchase price must necessarily be based.

It is seen that the specifications will frequently comprise the larger part of a contract, and generally very much the larger part, providing they make any pretention whatever to being complete.

The specifications may be made a part of the contract either by being attached thereto, properly designated and referred to as a part of the agreement; or, the two may be so incorporated one with the other as to have no dividing line. The former is perhaps the more general, though there is an advantage in the latter in that all of the terms and conditions are set forth in writing so as to have proper consideration before the bid is made.

The question arises as to what should be incorporated in the specifications. In our own practice it has seemed best to put into them all of the requirements, technical and otherwise, which would in any way affect the cost and desirability of the work, leaving for what we may term the contract part of the agreement, the simple statement of the agreement mutually entered into, the names of the parties, the prices to be paid and promise to pay.

All technical conditions will of course be set forth in the specifications in any event, and all others which have any bearing upon the cost and desirability of the work should likewise be stated, so that, when a bid is accepted, the making of a contract will not bring up for consideration requirements not previously stated or understood. Hence, conditions as to bond, time of completion, damages for delay, manner, time of and the kind of payment and all such kindred matters should, and generally can be determined beforehand and be so stated that they may be taken into consideration at the first, by the bidder, at his own estimate of their worth. The introduction of new conditions with the acceptance of a bid and when a contract is about to be entered into, is quite likely to create suspicion and a lack of harmony at the inception of the work.

No essential conditions should ever be omitted and considered as understood, implied, or as matters of custom. If the specifications are minute and explicit in all matters, they will frequently seem unnecessarily bulky and the document bear more resemblance to a treatise upon the subject than anything else. On the other hand, vague and incomplete specifications are quite likely to bring, sooner or later, much trouble and annoyance to all parties, and especially to the engineer. As in most other things, the happy mean will only be determined as a result of experience and good judgment. Perhaps on the whole it will be better to be too explicit rather than otherwise, providing, of course, that the writer knows what he is talking about; if he does not, the more indefinite and general his specification the better he will get along. It will usually be unwise to require that the work be according to any customary, or equal to any ideal but not definitely stated, standard of quality or excellence, as Custom will likely prove a fickle goddess, eventually only found in a court room siding with the greater number of expert (?) witnesses. On the other hand, any requirements which are contrary to ordinary and customary practice should be explicitly stated.

An engineer writing specifications for work, of which a large amount of similar character has been done by him in that vicinity, and which serves, to some extent, as a standard, and where the parties likely to bid on the work are few and familiar with him and his methods, and the character of the work he requires, may make his specifications less definite and explicit than an engineer who will have to deal with new or strange contractors. Important work is sometimes contracted for upon specifications that are very general in their nature, but this practice can only be a safe one under the circumstances just mentioned, and is always to be avoided if possible.

Avoid uncertain or ambiguous expressions; also any which will carry an impression that there is favoritism for a certain make or brand of goods; if there is good reason for such preferences, let them be clearly stated so that all bidders may figure accordingly. Such expressions as that the work is to be done "as directed by the engineer," or "satisfactory to the engineer," which bristle all through some specifications, appear to indicate that the engineer does not know in advance what he wants, and are otherwise objectionable in that the bidder cannot determine definitely what will be required. The entire work must of course be completed satisfactorily to the engineer, and it will be so required, but it is unwise to repeat this requirement for special items of work or material lest it lose its force as applicable to the whole.

It is best to leave as little to be inferred and understood or to be referred to verbally, as possible; as only by this means can bids be made on the same basis, and all bidders have an equal opportunity, and the chance of forgetting to make some explanation verbally, which was not thought worth while to put in writing, will be eliminated, and with it the possibility of much unpleasantness.

All of the physical conditions which will influence in the design and cost of the work should be determined beforehand, and be fully stated in the specifications, so that the bidder may estimate the cost intelligently. The expense of the preliminary investigations necessary to determine these conditions will be fully warranted in the saving in cost, and freedom from annoyance, incident to letting work in which the uncertain elements are reduced to a minimum. When the physical conditions cannot be fully determined in advance, the work will frequently be best let on the basis of itemized prices, and the risks due to unfavorable conditions or material assumed by the purchaser.

Specifications should be so worded that the engineer will hold the reins, so to speak, and be able, under all conditions, natural or otherwise, and with competent or incompetent, honest or dishonest contractors, to secure the class of work intended in the contract, and for which the purchaser is obligated to pay. If the work is subject to open competition, bids will be received from and the work awarded to contractors of more or less experience and greater or less honesty. For this reason the specifications should be thoroughly complete and will necessarily include some requirements unnecessary in most cases, and tending, perhaps, to cast unpleasant reflections upon an honest contractor, but very essential if the work is done by one of the other class. There are some contractors who know no law except that clearly stated in the specifications.

Avoid so far as practicable, all requirements or conditions likely to prove dead letters. To ask for something not essential and which is quite sure not to be insisted upon in the execution of the work, will leave the bidder and contractor in doubt as to how many other requirements can be considered by him as of like nature. And yet the specifications must make reasonable provisions for possible yet improbable contingencies, which would increase the difficulty and cost of the work if met with, but he should not expect to enforce them if not necessary. Where to draw the line in this and similar respects, will call for the exercise of all the engineer's skill and experience, in order that his work may command the respect of both his clients and the bidders, and secure for the former, good and durable work, and at reasonable cost, and for the latter, fair and honest treatment.

Generally, the specifications are supplemented by plans, maps or profiles, etc., which are an essential part of the agreement. The relation between plans and specifications will vary; sometimes the specifications can so minutely describe the work and material that no plans will be required, or the plans may be so complete in detail as to require little additional description. Generally the plans and specifications will supplement each other, and the former will most conveniently give the required information as to form, size and dimensions, leaving for the latter the requirements as to quality of work and material, and description of such details as are too small, or for other reasons not indicated upon the plans. Care will be necessary to see that plans and specifications, or either within themselves, do not conflict.

Generally, the engineer will be concerned only with results to be attained; the operation or method of obtaining such will rest with the contractor. As a rule, the specifications place upon the contractor all risks incident to the construction of the work and from the action of the elements, and it is but justice that he should have the right to execute it in such a manner as will insure him the greatest economy and safety in construction.

There are, however, occasions where the engineer must look as well to the manner in which the results are to be obtained and insist upon such methods and conduct of the work as will insure proper and desired results. Such plans as are likely to lead to disaster and imperfect work should not be permitted, for it is far less easy to replace imperfect work than to prevent it in the beginning. If there was time for reconstructing or rebuilding imperfect work or if such reconstruction would not injure subsequent work, then the contractor might be allowed to go on and practically demonstrate the inefficiency of any particular plan, but such permissive conditions rarely exist.

Originality, to any large degree, can seldom be obtained in writing specifications; many requirements for ordinary work and materials cannot be better stated in any other phraseology than that which has been used in perhaps hundreds of previous instances. It will be better in most cases to preserve old and well understood terms and phrases than to introduce new ones, the exact meaning of which may be subject to doubt or disagreement. The design of the work will usually offer ample opportunity for the exercise of originality.

I once heard it stated that the design of a truss span was practically finished as soon as the stress diagram was completed, and that any one could, with the aid of the diagram, make the plans for a perfect structure; but such is not the case, since the design of the details may be so lacking and imperfect that the structure would fall as soon as the false works upon which it was erected were removed and this, notwithstanding that the general dimensions and other features of the design shown on the stress diagram provide for ample safety. It is so with specifications; they must not only be well written, but properly, wisely and fairly interpreted and lived up to, or, however perfect they may be the work will be lacking. Herein may come the most immediate practical benefit of these remarks, as it will fall to the lot of the young engineer to have more to do at the first with comparing specifications for the work with which he is connected, with the actual results that are obtained than with the preparation of specifications themselves. This is true, not only of the young engineer, but of all, since the specifications which can be written in a few hours or a few days will be the guide and the law to be lived up to during the succeeding month, or perhaps many months, during which the work is in process of construction.

As young men, your first work will likely be as inspectors or assistants on construction work, and this will present many opportunities of learning not to be obtained elsewhere. Make the best of them in observing methods, costs and results. Be helpful about the work and avoid the appearance of being a spy. Be quick to observe and learn, be friendly and fair and you will get along; if supercilious, officious, fault-finding, or careless, the workmen will find many ways of making your work disagreeable, you will keep yourself and superiors in hot water and miss golden opportunities of learning things that cannot be obtained from books or in any other way.

SOCIETY OF ENGINEERS

IN THE

UNIVERSITY OF MINNESOTA.

LIST OF MEMBERS

MAY, 1895.

HONORARY MEMBERS.

C. W. Hall, M. A.
Wm. R. Hoag, C. E.
J. E. Wadsworth, C. E.
Geo. D. Shepardson, M. E.
Wm. R. Appleby, B. A.

H. E. Smith, M. E.
W. H. Kirchner, B. S.
J. H. Gill, B. M. E.
B. E. Trask, C. E.
G. H. Morse, B. E. E.

ACTIVE MEMBERS.

Abbott, A. L.
Adams, G. F.
Beyer, A. C.
Becker, Geo.
Blake R. P.
Bohland, J. A.
Bishman, A. E.
Buck, D.
Burch, A. M.
Burgner, L. P.
Cassedy, G. A.
Carswell, R. S.
Chapman, L. H.
Chesnut, G. L.
Coleman, L. M.
Cross, C. H.
Cutler, H.
Dahl, H. M. F.

Dustin, F. G.
Erickson, H. A.
Garvey, J. J.
Hastings, C.
Hibbard, T.
Hilferty, C. D.
Holt, P. E.
Hughes, T. M.
Hugo, V.
Johnson, N.
Jones, C. P.
Jurgensen, D. F.
Klein, Wm.
Lang, J. S.
Long, F. W.
Magnusson, C.
Markhus, O. G. F.
McCrea, M. W.

McIntosh, J. B.
Neil, V. A.
Rounds, F. M.
Savage, E. S.
Shepherd, B. P.
Silliman, H. D.
Stewart, N. P.
Tanner, H. L.
Tanner, W. N.
Tilderquist, W.
von Schlegell, F.
Weaver, A. C.
Wheeler, H. M.
Wilkinson, C. D.
Yale, W., Jr.
Zimmerman, F.
Zintheo, C. J.

CORRESPONDING MEMBERS.

1875

NAME AND RESIDENCE.	DEGREE.	OCCUPATION.
LEONARD, HENRY C. (B. S., '78) Minneapolis, Minn.	B. C. E.	Physician.
RANK, SAMUEL A. Central City, Colo.	B. C. E.	Civil and Mining Engineer.
STEWART, J. CLARK Minneapolis, Minn.	B. C. E.	Physician: Prof. of Pathology at University of Minnesota.

1876

GILLETTE, LEWIS S. Minneapolis, Minn.	B. C. E.	President of the Gillette-Herzog Manufacturing Co.
HENDRICKSON, EUGENE A. St. Paul, Minn.	B. C. E.	Lawyer.
THAYER, CHARLES E. Minneapolis, Minn.	B. C. E.	Grain Dealer.

1877

PARDEE, WALTER S. Minneapolis, Minn.	B. ARCH.	Architect.

1878

BUSHNELL, CHARLES S. Minneapolis, Minn.	B. M. E.	Secretary and Treasurer of N. W. Stove Works.

1879

DAWLEY, WILLIAM S. Danville, Ill.	B. C. E.	Assistant Chief Engineer C. & E. I. Railway Co.
*FURBER, PIERCE P.	B. C. E.	Died, April 6, 1899.

1883

BARR, JOHN H. Ithaca, N. Y.	B. M. E.	Ass't Prof. Mechanical Engineering, Cornell University.
PETERS, WILLIAM. G. Tacoma, Wash.	B. C. E.	Vice-President Columbia National Bank.
SMITH, LOUIS O. Le Sueur, Minn.	B. C. E.	Civil Engineer.

1884

LOY, GEORGE J. Spokane Falls, Wash.	B. C. E.	Supt. of Construction of New Water Works.
MATTHEWS, IRVING W. Waterville, Wash.	B. C. E.	Real Estate, Insurance and Abstracts of Titles and Civil Engineer.

1885

BUSHNELL, ELBERT E. New York City.	B. M. E.	Proprietor of the Hooper Typewriter Prisms.
*FITZGERALD, PATRICK T.	B. C. E.	Died, April 2, 1887.
REED, ALBERT I. Hastings, Minn.	B. C. E.	Civil Engineer.

1886

WOODMANSEE, CHARLES C. Midway Park, St. Paul.	B. ARCH.	Book-keeper.

1887

NAME AND RESIDENCE.	DEGREE.	OCCUPATION.
ANDREWS, GEORGE C. Minneapolis, Minn.	B. M. E.	Heating Manufacturer and Contractor.
CRANE, FREMONT (B. S. '86) Prescott, Arizona.	B. C. E.	Civil Engineer.

1888

ANDERSON, CHRISTIAN Portland, Ore.	B. C. E.	Civil Engineer.
LOE, ERIC H. Minneapolis, Minn.	B. M. E.	Mechanical Engineer, with Nordyke & Marman.
MORRIS, JOHN West Pullman, Ill.	B. M. E.	Assistant Superintendent and Mechanical Engineer of Plano Mfg. Co.
HOAG, WILLIAM R. (B. C. E. '84) Minneapolis, Minn.	C. E.	Professor of Civil Engineering, University of Minnesota.

1889

COE, Clarence S. Wenatschee, Wash.	B. C. E.	Civil Engineer, in Engineering Dept. of C., M. & St. P. Ry. Co.

1890

BURT, JOHN L. Minneapolis, Minn.	B. C. E.	Commission Merchant.
DANN, WILBER W. Minneapolis, Minn.	B. C. E.	Civil Engineer, Supt. of Water Works Construction at St. James, Minn.
GILMAN, FRED H. Minneapolis, Minn.	B. C. E.	Editor Mississippi Lumberman.
GREENWOOD, WILLISTON Minneapolis, Minn.	B. C. E.	Civil Engineer.
HAYDEN, JOHN F. Minneapolis, Minn.	B. C. E.	Manager Mississippi Lumberman.
HIGGINS, JOHN T. St. Paul, Minn.	B. C. E.	Physician.
HOYT, WILLIAM H. Duluth, Minn.	B. C. E.	Assistant Engineer, Duluth and Iron Range R. R.
NILSON, THORWALD E. Minneapolis, Minn.	B. M. E.	Manager of the Norgren Distilling Co.
SMITH, WILLIAM C. St. Cloud, Minn.	B. C. E.	Civil Engineer.
WOODWARD, HERBERT M. Boston, Mass.	B. M. E.	Professor of Manual Training.

1891

ASLAKSON, BAXTER M. Dayton, Ohio.	B. M. E.	With the Stillwell-Bierce & Smith-Vaile Co.
CARROLL, JAMES E. Minneapolis, Minn.	B. C. E.	Civil Engineer.
CHOWEN, WALTER A. Browns Valley, Minn.	B. C. E.	Civil Engineer.
DOUGLAS, FRED L. New York City.	B. C. E.	Civil Engineer, with L. L. Buck.
GERRY, MARTIN H. (B. M. E. '90) Chicago, Ill.	B. E. E.	Electrician, with the Metropolitan West Side Elevated Railroad Co.
HUHN, GEORGE P. Minneapolis, Minn.	B. E. E.	Flour City National Bank.

1892

NAME AND RESIDENCE.	DEGREE.	OCCUPATION.
BURCH, EDWARD P. Minneapolis, Minn.	B. E. E.	Electrician, Twin City Rapid Transit Company.
BURTIS, WILLIAM H. Minneapolis, Minn.	B. E. E.	Electrical Engineer and Contractor.
FELTON, RALPH P. Minneapolis, Minn.	B. M. E.	Book-keeper.
GOODKIND, LEO. St. Paul, Minn.	B. ARCH.	Architect, with Reed & Stem.
GRAY, WILLIAM I. Minneapolis, Minn.	B. E. E.	Electrical Engineer and Contractor.
HANKENSON, JOHN J. Minneapolis, Minn.	B. C. E.	Bridge and Sanitary Engineer.
HIGGINS, ELVIN L. Hutchinson, Minn.	B. C. E.	Civil Engineer.
HOWARD, MONROE S. Minneapolis, Minn.	B. E. E.	Electrical Engineer and Contractor.
MANN, FRED M. Boston, Mass.	B. C. E.	Post-Graduate Student in Architecture, Massachusetts Institute of Technology.
PLOWMAN, GEORGE T. Minneapolis, Minn.	B. ARCH.	Draughtsman.

1893

ANDERSON, OLE J. Nicollet, Minn.	B. C. E.	Nicollet County Surveyor.
AVERY, HENRY B. Minneapolis, Minn.	B. M. E.	Draughtsman for N. W. Wheel and Foundry Co.
BATCHELDER, FRANK L. St. Paul, Minn.	B. C. E.	With C. F. Loweth.
CHASE, ARTHUR W. Hastings, Minn.	B. E. E.	Electrician.
COUPER, GEO. B. Northfield, Minn.	B. M. E.	Superintendent Northfield Electric Light Co.
DEWEY, WILLIAM H. New York City.	B. E. E.	Assistant Engineer for the American Boiler Co.
ERF, JOHN W. Minneapolis, Minn.	B. C. E.	With Gillette-Herzog Mfg. Co.
GUTHRIE, JOHN D. Minneapolis, Minn.	B. E. E.	Medical Student, University of Minnesota.
HOYT, HIRAM P. Minneapolis, Minn.	B. C. E.	With Gillette-Herzog Mfg. Co.
MORSE, GEORGE H. Minneapolis, Minn.	B. E. E.	Instructor in National School of Electricity.
REIDHEAD, FRANK E. Minneapolis, Minn.	B. E. E.	Electrician, Minneapolis General Electric Co.
SPRINGER, FRANK W. Minneapolis, Minn.	B. E. E.	Electrician.
WASHBURN, DELOS C. Minneapolis, Minn.	B. ARCH.	With J. T. Fanning.

1894

NAME AND RESIDENCE.	DEGREE.	OCCUPATION.
BRAY, GEO. E. Minneapolis, Minn.	B. M. E.	With D. & D. Electric Mfg. Co.
CHALMERS, CHARLES H. Minneapolis, Minn.	B. E. E.	Electrician and Assistant Manager D. & D. Electric Mfg. Co.
CUNNINGHAM, ANDREW O. New Orleans, La.	B. E. E.	With Gillette-Herzog Mfg. Co.
CUTLER, HARRY C. Minneapolis, Minn.	B. E. M.	Scholar in Mining at University of Minnesota.
CHRISTIANSON, PETER Minneapolis, Minn.	B. E. M.	Instructor in Mining at University of Minnesota.
GILMAN, JAS. B. Minneapolis, Minn.	B. C. E.	With Gillette-Herzog Mfg. Co.
JOHNSON, NOAH Minneapolis, Minn.	B. C. E.	Scholar University of Minnesota.
TRASK, BIRNEY E. (B. C. E.'90) Highland Park, Ill.	C. E.	Commandant of Cadets, Northwestern Military Academy.
WEEKS, WILLIAM C. Minneapolis, Minn.	B. C. E.	With the "Soo" line.

"The best instruments, even though their first cost is greater, will render better service and last enough longer to make them decidedly the cheapest."

Alteneder
DRAWING INSTRUMENTS

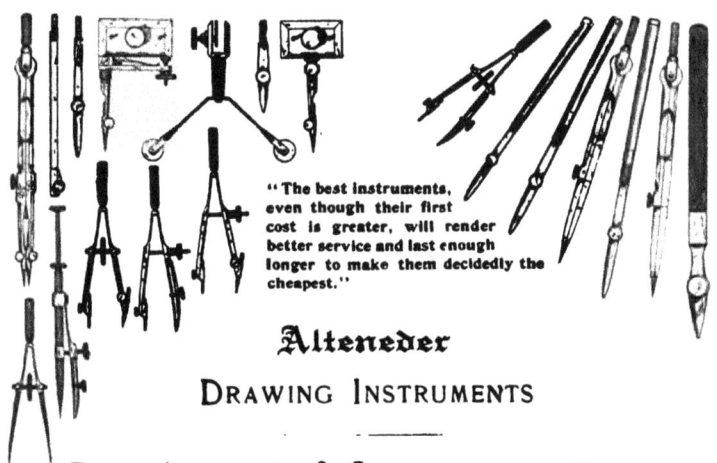

THEO. ALTENEDER & SONS MANUFACTURERS PHILADELPHIA

Send five cents in postage stamps for new Catalogue

| The Chloride Electrical Storage Syndicate, ——Limited,—— Manchester, England. | Societe Anonymê pour le Travail Electric de Metaux, Paris, France. |

The Only Storage Battery in Use in Central Stations of American Manufacture.

THE ELECTRIC
Storage Battery Company
Drexel Building, Philadelphia, Pa.

TRADE MARK
CHLORIDE ACCUMULATOR
REGISTERED SEPT. 11TH, 1894.

| Electrical Storage Cells —of— Any Desired Capacity. | Catalogues Giving Capacities, Dimensions, Weights Prices, etc., on Application. |

J. T. FANNING

Consulting Civil Engineer

Kasota Block
MINNEAPOLIS

W. W. RICH

Chief Engineer
M., St. P. & S. Ste. M. Ry.

Guaranty Loan Building
MINNEAPOLIS

GEORGE L. WILSON
M. Am. Soc C. E.

Ass't City Engineer and Supt. of Sewers

City Hall
SAINT PAUL

Wm. A. PIKE

Consulting Engineer

Guaranty Loan Building
MINNEAPOLIS

F. F. SHARPLESS

Analytical Chemist and Assayer

811 Wright Block
MINNEAPOLIS

MORGAN BROOKS

President Electrical Engineering Co.

249 Second Avenue South
MINNEAPOLIS

EDW. P. BURCH

Electrician
Twin City Rapid Transit Co.

MINNEAPOLIS

H. E. BURT

Superintendent Wis. & Minn. Ry.

923 S. E. Eighth Street
MINNEAPOLIS

"OTHER THINGS BEING EQUAL"

We do not expect trade to come our way when any one else offers a better price, but, "Other things being equal," we do feel entitled to student patronage because we in turn patronize student enterpises. We can supply you with anything in the line of Books or Mathematical Instruments.

OPEN DURING SUMMER VACATION.

The University Book Store.

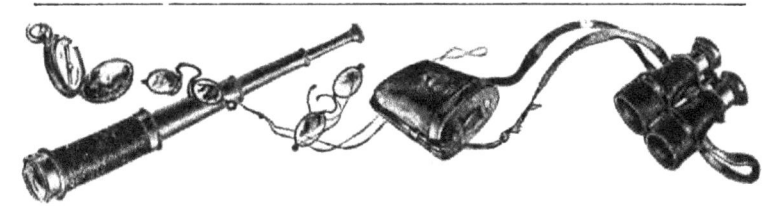

BOERINGER & SON, Engineers' Architects' and Surveyors' Instruments. Technical Drawing Material.

Opticians, 54 East 3d St., ST. PAUL, MINN.

Scientifically Designed and Constructed

Minneapolis Radiator & Iron Co. Radiators
829 Ninth St. S. E.,
MINNEAPOLIS, - MINN.

WILLANS & ROBINSON

PATENT HIGH SPEED

Central Valve Engines

SIMPLE
COMPOUND
TRIPLE EXPANSION

Extreme Steam Economy
    Silence and Durability
        Small Space Required
            Minimum of Attendance

105,000 HORSE POWER IN USE OR ON ORDER.

■ ■ ■ ■ ■ ■

M. C. BULLOCK MFG. CO.

Sole American Licensees and Builders,

1170 W. LAKE ST.,

CHICAGO.

THE UNIVERSITY OF MINNESOTA
COLLEGE OF
ENGINEERING, METALLURGY AND THE MECHANIC ARTS

CYRUS NORTHROP, LL. D., President
CHRISTOPHER W. HALL, M. A., Dean

The College of Engineering, Metallurgy and the Mechanic Arts offers four years' courses in

> Civil Engineering,
> Mechanical Engineering,
> Electrical Engineering,
> Mining Engineering,
> Metallurgy,
> Chemistry,

with Special Courses in the above-named departments and in

Industrial Art

The large and increasing equipment of the college drawing rooms, shops and laboratories gives **a modern and practical character** to the work. Continually increasing **prominence is given to the theoretical** phases of every line of professional investigation.

The more recent lines of work opened to students are the following:

Structural Engineering, with especial emphasis on bridge building and the designing of steel and stone structures generally;

In **Mining Engineering** a four weeks' field course in mine surveying and mine engineering.

The **Ore Milling and Testing Laboratories** are the most complete in the United States.

Just opened is the attractive field of **Locomotive Engineering** comprising complete courses in the construction and operation of locomotive engines; also courses in **car design.**

The situation of the College in the manufacturing and commercial center of the upper Mississippi region is particularly favorable. Railway shops, manufactories, electric light and power stations, chemical and metallurgical works, ore docks and mines are open not only for inspection but for study and research work. Thus the student can become acquainted with large engineering enterprises conducted under business methods and in a scientific way.

Requirements for admission: English grammar and composition, with essay; algebra, plane and solid geometry; history or civil government; natural sciences; free-hand drawing; German or French. While four years' work in Latin may be offered students are urged to present their preparation in German. Persons of mature years may be admitted to special lines of study provided they give evidence of ability to pursue the same with profit.

Tuition is free. Shop and laboratory fees are very low.

Upon the completion of a full course of study the bachelor's degree is conferred.

For catalogue and further information address the Dean of the College at Minneapolis, Minnesota.

> For
>
> ## Engineers
>
> We will have an improved stock another year and at prices that cannot be duplicated.
>
> ## The University Book Store.

Improved Thompson Steam Engine Indicator, adapted for all speeds.

Pressure Gauges for all purposes.

Metropolitan and Columbia Recording Gauges.

Prof. R. C. Carpenter's Throttling and Separating Steam Calorimeters.

Prof. R. C. Carpenter's Coal Calorimeter.

Injectors and Ejectors.

Exhaust Steam Injector.

Thermometers for all industrial purposes.

Chime and Siren Whistles.

Water Gauges.

Steam Traps, and boiler and engine appliances in general.

For Catalogue and prices address

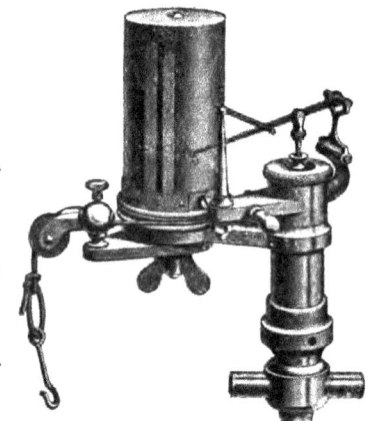

Schaffer & Budenberg,

Works and General Offices: BROOKLYN, N. Y.

Sales Offices: 22 W. Lake St., Chicago; 66 John St., New York.

PROFESSOR H. E. SMITH.

The YEAR BOOK

OF THE
SOCIETY OF ENGINEERS.

MAY, 1896.

PROFESSOR SMITH.

Harry Ezra Smith was born January 16, 1865, in Pike, Wyoming County, New York. He was the only son in a family of five children, the ninth generation from Rev. Henry Smith, who was the first settled minister of Wethersfield, Connecticut, having come from England about 1636. His mother, Amanda Adams Smith, was a descendant of John Adams, who landed at Plymouth, Massachusetts, from "The good ship Fortune" in November, 1621. Although the father, who was a woolen manufacturer, died when the subject of this sketch was only fourteen years old, the mother managed to educate her children. After completing the course at Pike Seminary in his native town, the young man was appointed to a scholarship at Cornell University in 1882. Here he maintained an excellent record, taking the First Prize in Sibley College during his Junior year for "greatest merit in college work." He graduated from the Mechanical Engineering course in 1887, receiving honors for general excellence, and honorable mention for his graduating thesis on a "Trial of a Babcock and Wilcox Boiler."

While pursuing the technical course he was also wise in gaining practical experience by actual work in shops. An interval in his college course was spent in the employ of the Straight Line Engine Company, of Syracuse, N. Y. After graduation he followed this up by several months' experience with the Brown & Sharpe Manufacturing Company whose reputation is world wide for accurate and precise workmanship in standard gauges, tools, and fine machinery. After a thorough study of their shop methods of doing work and keeping records, he resigned to take a position as assistant foreman with the Woodbury Engine Company, of Rochester, N. Y. This position in turn was given up in order to become an assistant in the erecting department of Wm. Sellers & Company, of Philadelphia.

In the spring of 1888 Mr. Smith had the good fortune to secure as a life companion Miss Minnie R. Ballard, of Centerville, Allegany County, N. Y.

In the fall of 1888 he was tendered an instructorship in Sibley College, Cornell University, which position was held for one year, when he was called to the University of Minnesota as Instructor in Woodworking and Foundry Practice. From this position he has risen step by step until, upon the resignation of Professor W. A. Pike in 1892, he was given full charge of the Department of Mechanical Engineering with the rank of Assistant Professor. Since the advent of his present associate, Assistant Professor Hibbard, Mr. Smith has been able to devote his attention more largely to his chosen specialty of Experimental Engineering.

Professor Smith's reputation for accurate work in experimental engineering has created frequent demand for his services in conducting efficiency and capacity tests of steam engines, boilers, and other mechanical products. These calls afford unexcelled opportunities for the advanced Mechanical and Electrical Engineering students to take part in the observations and calculations of commercial tests and to become familiar with some of the problems presented to engineers in active professional practice. Such extra-mural laboratory work is highly prized by the students as supplementing the working equipment of the University and enlarging the practical development of the Engineering courses. During the past year, the students have assisted Professor Smith in efficiency and capacity tests of three one hundred and twenty-five horse-power compound high-speed Ames engines, directly connected with electrical generators, a twelve hundred horse-power vertical triple expansion compound Schichau engine, and a sixty horse-power Ball engine, besides a number of minor tests.

Professor Smith keeps in close touch with educational and professional colleagues. He is a junior member of the American Society of Mechanical Engineers, and a member of the Society for the Promotion of Engineering Education. He was a charter member of the first chapter of the Society of Sigma Xi, founded at Cornell University in 1886, and is also a charter member of the chapter just established at the University of Minnesota.

GEO. D. SHEPARDSON.

MODERN FOUNDRY PRACTICE IN CONNECTION WITH MANUFACTURE.

JOHN MORRIS, '88·

Though the working of metals was practiced by the ancients early in the history of civilization, the practice of founding was not very extensively carried on until within the last three centuries. We find a mention of working iron recorded in the Scriptures at an early period (Gen. 4:22). Whatever may have been unrecorded in the history of iron manufacture in the early ages, enough has been written to establish the importance of the iron industry throughout the civilized world since its early discovery.

The method of working iron, as well as the difference in product, varied in the early history from what we find in mediæval ages; and all this differed greatly fro n what we find to be the history of this important industry during the last few centuries. It is not our object, in this short article, to even attempt to give the early history of this widely diffused metal, only as much as is necessary to get at the growth of this industry during past ages, and thus assist in showing the rapid development during our own present century. It seems from good authority that the reduction of the iron ores has been practiced since quite a remote period in various countries, and that, too, with considerable success in producing iron and steel, at different periods in their history. A few uncertain mentions are made of cast-iron having been produced several centuries preceding the Christian era; also, a few scattering references are made to casts of various kinds produced during the early centuries of our present dispensation.

This is certain, though records are practically silent, that very little was done except the production of iron and steel from open hearths, until the eighteenth century, when the practice of founding became generally known. Much credit is due to the early promoters of the iron industry, but doubtless a greater advancement has been made in the manufacture, both in kind and quality, of this most useful metal, during the last quarter of a century, than in all preceding time. While there has been a very great development in this line, possibly the greater advancement which has made this first possible, is due to the general improvement in the method of making castings and the introduction of moulding machines. Up to the last twenty-five years, nearly all the moulding was done by hand after the original method; and, indeed, much of the work is being done after this method in our modern foundries, and in many the work is still carried on exclusively after this method. The fact that it was the only method originally pursued, and that it is still extensively practiced in nearly all foundries, speaks well of this method in more than one way. While it has the advantage of being inexpensive when only a small number of castings are required, it always has the disadvantage of producing imperfect work. When a pattern of a desired casting is put in a flask, and the sand is properly "rammed" around it, the most important part of the task remains yet to be accomplished. The removing of a simple pattern from the sand mould is not attended with any great degree of difficulty, but with a complex piece the reverse is the case. The first operation is to "sponge" the pattern. This is done by following the outline of the pattern with a very small stream of water from a sponge. After this is done the pattern is rapped by either tapping the pattern itself, or placing an iron piece in a hole or socket in the pattern, which serves also as a lifter, and rapping this sufficiently to loosen the pattern in the sand. It will readily be seen that the rapping of the pattern produces most of the imperfection. For example, take an ordinary bevel pinion, which is moulded on end. This is rapped sideways, right and left, and possibly a little in a direction at right angles. It will be observed that the teeth, on the right and left sides, will be long by reason of the rapping sideways, while the teeth on the sides at right angles will appear thicker and not as long, a condition which renders a pinion entirely unfit for use. This is not always the case, as many patterns are but little affected by such a variation as this; and this variation depends largely on the care and skill of the moulder; but in general it will apply to all small patterns. The "carding" of patterns on small work eliminates this difficulty to a certain extent, while it also has another merit, of producing from six to a dozen pieces as cheaply as if only one pattern was used. The patterns are gated together, two, four, six, or a dozen, depending on the size of the pattern, with runners in between. A "sand match" is often used with a card, and this greatly facilitates the work. The card is taken and a "sand match" is made to fit the pattern up to the parting line. This enables the moulder to work rapidly and produce comparatively perfect castings. With large castings, the difficulty attendant upon small patterns is to a great extent removed and is supplemented by still other difficulties. The continual practice and experience in handling large castings has been a means of bringing about methods that now enable workmen to produce a great many difficult castings with comparative certainty.

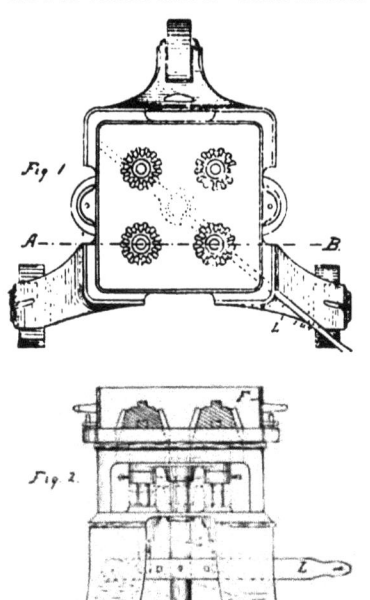

The advantages of the moulding machines over the hand method are twofold; principally in producing better moulds and, consequently, better castings; and, subordinately, in reducing the cost of the work. The construction of moulding machines varies with the class of work and the kind of castings. Different manufacturers also make such machines as are best adapted to their individual need, but the general principle of the moulding machines is much the same. Briefly described, it is as follows: A moulding machine consists of an iron stand of suitable dimensions to accommodate the pattern or patterns, and of such height as is found convenient for an ordinary workman to manipulate and handle the flask.

Figure 1 shows the plan of an ordinary moulding machine, and Figure 2 is an elevation in section on line A—B. The patterns are raised into position, as shown in Figure 2, and held in place by means of the lever L, and the vertical standard S. The moulding sand is put into the flask F, and properly "rammed" in the usual way, and now the patterns are mechanically removed, by means of lever L, and are dropped down to the position showed in dotted lines; then the flask is taken up off of the dowel pins, inverted and examined, to see that all parts are perfect, and then is placed in position on the moulding floor. Figure 3 is a plan, and Figure 4 the elevation of the stand for the counterpart; and in this is only a "match plate". As is seen from construction, it is not necessary to have the patterns movable in Figures 3 and 4, though often the patterns are such that both parts must be made movable. The moulding sand is put into the flask and "rammed" as before, and now the flask F' is taken up, examined, and then put into position on flask F. Generally the moulder runs ten or twenty-five flasks of the "drag", putting same in position on the floor, and now changes over and uses the machine for the "cope" shown in Figures 3 and 4, and making the mould for the counterpart and putting these into position on the drag flasks, and so on for the day's work. It is now obvious that the removing of the patterns from the sand, mechanically, does not require any rapping, and, consequently, leaves the mould as perfect as the pattern. Often patterns have only one or two projecting parts, while the rest of the surface is comparatively uniform, or at least has plenty of "draw", and the flask can be readily removed from the surface. In such a case the projecting parts only are made movable and the pattern is made on the top plate of the stand as a "match plate", and the projecting parts are made to fit closely when in place; and when the sand is properly "rammed" they are mechanically removed, and dropped down entirely out of the mould by means of a lever, practically as shown in Figure 2.

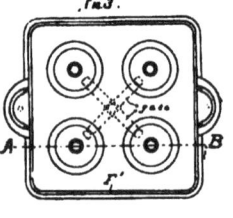

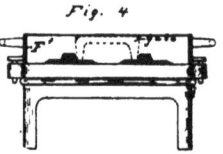

The carding of patterns on moulding machines may be and is practiced whenever the size of a pattern will admit. This, of course, refers only to small patterns though castings of considerable size are successfully made on machines with the use of mechanical or pneumatic hoists. When large patterns and heavy flasks are used, the air hoists produce a very satisfactory result, and are being introduced wherever the amount and quality of work requires such an appliance.

Having now produced the moulds, either by hand or with a moulding machine, we are ready for the second important operation that the moulder has to perform, and that is pouring off the metal.

However, before proceeding with this part of the work, a brief outline of the work in the foundry will be here given, with a few formulæ of mixtures that are successfully used in practice.

The charging of the cupola requires close attention, and this varies with the class of work; and the success of this work depends largely on the good judgment of the foundryman in charge. The following will illustrate a method of charging in a 32-inch cupola:

On the bed 500 pounds of coke, and on this 2,000 pounds of iron; next, 100 pounds of coke, and on this 1,500 pounds of iron; continuing these amounts for the remainder of the charges. A cupola, charged like or similar to the above, will melt four to five tons per hour, and, if everything goes well, eight or ten tons may be melted in one heat, and even more has been accomplished under favorable circumstances.

In a cupola melting sixty-five tons daily, the following charges were used: Nine hundred or 1,000 pounds of coke on the bed, 3,000 pounds of iron; then 200 pounds of coke and 2,500 pounds of iron, and so on for ten charges, and the remainder of the charges, 200 pounds each of coke, with 2,000 pounds of iron.

A quantity of crushed limestone is thrown in with the charges during the process of melting for "flux".

For light miscellaneous castings, such as are used for agricultural machinery, the following formula will produce good results: Forty per cent. No. 1 soft pig; 30 per cent. No. 1 Bessemer; 30 per cent. good machine scrap, or, in lieu, No. 2 Bessemer.

There should be at least three kinds of iron, of different grades, in order to secure good results. For a heavier class of castings, the following is a good mixture: Forty per cent. No. 2 soft pig; 20 per cent. No. 1 Bessemer; 20 per cent. No. 2 Bessemer; 20 per cent. good scrap.

The quality of the soft irons has more to do with getting good results and suitable castings than that of the hard irons. A good casting should possess these conditions: Sound casting, free from blowholes and flaws; strength to resist strain; and soft enough to work well with a tool. Following will be given the mixtures A, B, C, recently used; and the result in each with three test bars taken from the beginning, middle, and end of heat. It will be noticed that the three mixtures are alike, only differing in proportion. It will also be noticed that the A mixture produced test bars that were low in transverse strength, and the same showed weak castings in the construction, while B and C showed well and produced good results.

A.

33⅓ per cent. Bessemer   } 2200.
26⅔   "   scrap   } 2380.
23⅓   "   No. 2 Foundry   } 2220.
16⅔   "   soft

B.

20 per cent. Bessemer   } 2100.
33   "   scrap   } 2580.
27   "   No. 2 Foundry   } 2820.
20   "   soft

C.

26 per cent. Bessemer   } 2618.
28   "   scrap   } 2690.
26   "   No. 2 Foundry   } 2700.
20   "   soft

The following mixture was made for especially high-grade casting and proved to be very good, giving a strong and sound casting, soft enough to work well. This would be very suitable for large engine cylinders and such classes of work:

25 per cent. Imported Scotch Summerlee   }
50   "   Hanging Rock, charcoal (Ohio)   } Average strength,
15   "   Lake Superior   }   2800 lbs.
10   "   good scrap   }

Other mixtures now being used for light castings produce very satisfactory results in every way, of which is the following:

40 per cent. No. 2 soft   }
15   "   No. 1 Bay View   }
10   "   No. 2 Bay View   } Average strength, 2475 lbs.
33   "   scrap   }
 2   "   Ferro Silicon   }

The chemical analysis of No. 1 Bay View is as follows:

Silicon,	2 to 2.50
Phosphorus,	.20
Manganese,	.32
Sulphur,	.023
Carbon,	3 to 3.25
Iron,	93.80

Having now mixed and melted the iron, it is ready for the last operation of pouring the metal into the moulds previously prepared. This is attended with some difficulty, and must be done with care to avoid washing away the wall of the mould, and the blowing and swelling of the casting. With large casting the transporting of metal and pouring off is done mechanically, by means of foundry cranes, and often large ladles weighing many tons are used and handled with success and safety.

The castings are allowed to remain in the sand until sufficiently cooled, and are then taken out, ready to be sent to the tumbling mills for cleaning, and the sand is sprinkled and shoveled over, ready for the next day's work.

FIRE-PROOF FLOORS IN MODERN BUILDINGS.

F. L. DOUGLAS, B. C. E., '91.

The remarkable and continually increasing use of the "steel skeleton" type of building construction since 1887 or 1888, at which time the advantages of iron and steel in building construction became generally recognized, has necessarily created a demand for a suitable material for the floors, ceilings and partitions of buildings, to serve the twofold purpose of protecting the expensive iron work from fire, and of safely sustaining the loads likely to be imposed.

In order to protect the iron work against fire this material must itself be thoroughly fire resisting. It must also cover the iron work to such a depth that in case of fire the iron will not be heated to a dangerous temperature. The common end in view in the discussion of fire-proof floors is to ascertain what system most nearly fulfills the conditions of efficiency, minimum cost of manufacture and erection, and maximum speed in erection.

Fire-proof floors may be divided into two general classes:

First, floors composed of separate pieces, generally brick or terra cotta, joined together along nearly vertical joints by cement or mortar.

Second, floors of a monolithic construction, made generally of concrete, in which iron may or may not be imbedded, to act in conjunction with the concrete in performing its work.

In the first general class the use of brick arches may be dealt with in a few words, since it is becoming more and more obsolete. The objections to brick arches are their great weight, limited strength, and difficulty of protecting the bottom flange of the supporting beam against fire.

The weight of a brick arch, together with the concrete filling on top, is about 70 lbs., or varying from 20 lbs. to 35 lbs. greater than the hollow tile, and 40 lbs. greater than some of the lightest types of floor construction. This excess of weight requires heavier beams, girders and columns at a correspondingly greater cost. Tests have shown that when brick arches are loaded eccentrically, as all floors are liable to be, they are surprisingly weak. The bottom flange of the beam or girder, supporting brick arches, is left exposed, since the lower part of the web and the top of the bottom flange form the skewback for the arches. Brick arches are poorly adapted to take ceilings. Brick arches when made of a good quality of red pressed brick have a very neat appearance from below and when this consideration is a prominent one they may be advantageously used.

HOLLOW TILE ARCHES.

Hollow tile seems to have been first used for floors in Europe. Mr. F. Von Empberger speaks of the Liverpool flags used in 1853 consisting of a single tube running from beam to beam, the spans being very short. Although taken up in the United States at a later date, we find this system of fireproof flooring developed to a much higher state of perfection and far more extensively used than elsewhere. In Europe the single tube is still used and a liberal thickness of concrete is placed on top. This furnishes most of the strength to the combination, reducing the function of the tile practically to that of ceiling tile.

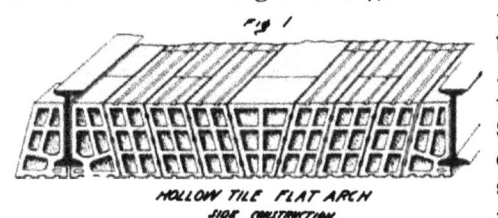

Not until the era of "skeleton construction" began, however, was hollow tile or any other form of fire-proofing extensively used. The number of different forms of hollow tile which have been used or proposed in the meantime is very large, but they may all be comprised at present under the following heads, viz.: Side construction (Fig. 1.) End construction (Fig. 2). The distinction between the two being that in the former the voids run parallel with the beams, and in the latter the voids are perpendicular to them. In both of these types all the ribs are straight. Many designs, have appeared on the market in which the ribs were more or less curved in the endeavor to conform to the curve of equilibrium of a loaded arch. These forms may or may not have a flat ceiling. In the former case the inclined ribs are carried down below the arch ribs to the plane of the intended ceiling, supporting the horizontal part of the block and thus forming a level ceiling. In the latter case the arrangement of the blocks is very similar to the ordinary brick or stone arch having a curved soffit. The floors having curved ribs meet more fully the engineering aim to distribute the material in the curve of equilibrium so as to produce the maximum efficiency. The great objection to curved ribs is that the blocks made for one span of arch are not adapted to other spans, and since, in buildings of any considerable size, a great number of different spans exist, an enormous number of different shapes are required, increasing the cost of manufacture and the difficulty and time required in getting the right block to the right place. A compromise between the various elements of the problem is effected by the use of the two types illustrated and referred to above. In these, comparatively few shapes are required; a small variation in span being provided for by using a key of different thickness and greater variations by using a greater or less number of Voussoir blocks. These types furnish a flat surface both above and below, a very desirable feature.

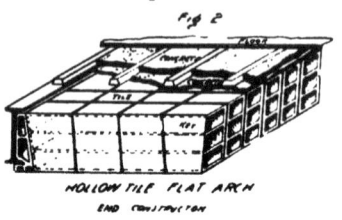

In all hollow tile arches a layer of concrete is placed on top in which are imbedded nailing strips for the floor boards.

A large number of tests have been made on hollow tile arches, giving a very wide range of results.

A series of six tests of the strength of hollow tile arches of the side construction type, made at Trenton, N. J., in 1894, gave a breaking load of from 298 lbs. to 839 lbs. per square foot uniformly distributed. Tests were made on hollow tile segmental arches, span 15 ft. 4 in. One test made with load extending up to the middle of the arch, from one beam, gave a breaking load of 1,000 lbs. per sq. ft. over the loaded area, or 500 lbs. per sq. ft. over the entire area of the arch. The concrete filling in the haunches was green, otherwise the test would probably have been more satisfactory.

Another test of a similar arch, uniformly loaded, carried 1,200 lbs. per sq. ft. on the loaded area with only a slight permanent set, though it is certain that a portion of this load was carried by adjacent unloaded portions of the arch; hence the above figures should be considerably reduced for comparison.

A series of twenty-seven tests recently made by Geo. Hill, Mem. Am. Soc. of C. E., on hollow tile arches, led him to conclude, so far as this limited number of tests would indicate, that "side construction" arches were very weak in the skewback and that "end construction" arches, when properly imbedded in good concrete against the beams, were very much stronger. Mr. Hill claims that a properly designed skewback has carried 5,000 pounds per lineal foot without failure.

As a protection for the iron work against fire, hollow tile has proved itself to be very efficient. A fire occurred April 2nd, 1893, in the Temple Court building, New York City. The wood floors, window and door framing and contents of offices furnished the combustible material and the fire raged fiercely for hours unchecked. When it was finally put out, an examination showed the hollow tile itself not only to be intact, but the iron was in no way injured.

This type of floor can be rapidly erected and for this purpose 2 inch planks are supported on timbers hung from the beams. The top of the planking comes about 1 inch below the bottom of the I-beams and the arch blocks are laid upon the plank, beginning of course with the skewback and working towards the center from both directions. No great degree of skill is required nor is a high grade of mortar necessary. The work of erection, to secure good results, should be carefully superintended. This, like any other work, will suffer for want of proper execution.

Because of the short time required for the mortar to harden, the floor can be used soon after erection. The weight of hollow tile arches, together with the concrete filling, varies according to the depth of arch and concrete, from 35 lbs. to 50 or 55 lbs. per square ft.

Hollow tile floors are poorly adapted to cutting away for the passage of pipes, etc., and where the floor beams and girders come together on a skew, the blocks require trimming to a level, affording opportunity for poor workmanship and uncertain results.

The majority of tests upon hollow tile arches show them to fail suddenly, without preliminary deflection. Cracks occasionally appear and give warning of weakness, before failure occurs.

Where the hollow tile flooring is used, the floor beams are "caped" into the girders supporting them so as to make all beams flush on the bottom. This involves an additional cost to the beams of about $4.00 per ton over the cost of beams simply framed together.

It is likely that more attention will be shown in the future, than in the past, to the proper design and manufacture of hollow tile arches, resulting in greater strength and more definite knowledge of their properties.

The recent appearance in the field of competition of various types of floor which show much greater strength, more uniformity of results, and having in some cases less weight, will be a strong factor in bringing this about, and force the manufacturers of hollow tile, who have hitherto had little competition, to improve the strength and uniformity of their products.

CONCRETE IRON FLOORS IN GENERAL.

A book was published in London in 1877 by Thaddeus Hyatt, describing a building in which the combination of iron and concrete was used; also giving results of fifty tests. Mr. Hyatt refers to a still earlier publication, describing the use of the combination in France. Tests were made in America, as early as 1855, by R. G. Hatfield, of combination beams. In 1875 a dwelling was constructed in Port Chester, N. Y., in which Beton was used in combination with iron.

Whether or not the combination of iron and concrete for floors was first used in the United States is of less importance than the fact that in Europe it has attained an extensive commercial development not approached in this country. Its use in Europe developed in somewhat the same way, and under similar circumstances as hollow tile in this country.

The great advantage in the use of iron in combination with concrete is in securing a higher strength on the tensile side of combined concrete-iron beams than can be obtained with concrete alone.

Evidence is quite conclusive that concrete is remarkably well adapted to preserve iron, when the latter is imbedded completely in the former. A piece of iron was found to be in a perfect state of preservation, after being imbedded in concrete over 400 years. The modulus of elasticity of concrete, according to Prof. Boeck, is one-fortieth that of steel, enabling the steel to take a forty times greater load under the same strain or deformation. The cohesion between iron and concrete exceeds the strength of concrete itself. The thermic expansion of both is the same; therefore no secondary stresses occur when the temperature changes.

The time required for the setting of some types of concrete-iron arches before they are capable of sustaining a heavy load, is an undesirable quality, since a much larger quantity of centering is required, and often times arches are called upon to carry their heaviest loads very soon after completion, for the storage of materials.

Any floor system consisting of concrete and iron in combination, will show the effect of excessive loading or faulty construction generally by the appearance of cracks, or undue deflection, in time to remove the loading and prevent collapse of the floor, while with hollow tile construction failure is generally the first evidence of weakness. Concrete is not generally regarded as possessing as good fire-resisting qualities as hollow tile, yet many maintain that it will not disintegrate under a severe fire and water test. At a recent meeting of the American Society of Civil Engineers, the discussion which followed the reading of a paper on fire-proof floors showed a general belief that concrete was not as good as hollow tile in resisting fire and water.

METROPOLITAN SYSTEM

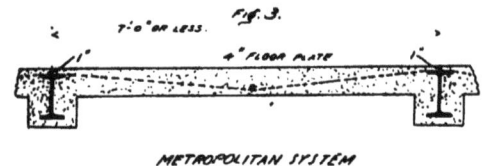

The metropolitan system of fire-proof floors as such, has been on the market about two years. It consists, briefly, of iron cables running continuously over the tops of the floor beams and imbedded in a "floor plate" usually four inches thick, made principally of plaster of paris and sawdust, the top of the plate being one inch above the top of the floor beams. When a flat ceiling is not required (Fig. 3), the wire cables consisting of two No. 12 gauge twisted galvanized iron wires are run continuously over the tops of the floor beams and secured by hooks to the last beams. The cables sag between the beams and a uniform deflection is given by laying a ⅝ in. iron rod on the wires, midway between the beams. A level "center" is next placed so that the top is 3 in. below the tops of the beams and the composition, as above described, is applied after being mixed together in the proportion of five to one by weight, with enough water to become plastic. The center extends down around the bottom part of the beam so that after the composition is in place the beams and wires are entirely imbedded.

In cases where a flat ceiling is required (Fig. 4), 2x⅛ in. iron flats placed on edge and running at right angles with the beams are suspended by hooks from the latter so as to come immediately under the bottom flanges, wire netting is secured to these bars and the composition is applied as above described, but is only about 1½ in. thick. The main floor plate which is about 4 in. thick and the covering for the beams are then applied. All iron is covered by at least one inch of the composition. The hollow space thus formed between the "ceiling plate" and the "floor plate" is convenient for laying pipes and wires, as well as for the non-conduction of heat and sound.

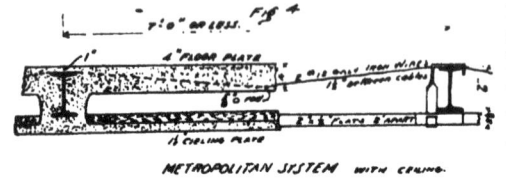

It is claimed by the manufacturers that the composition will harden so as to be at once capable of sustaining, with safety, the load for which it was calculated. Plastering is applied directly to the composition and boards for the floor are nailed to strips imbedded in it.

Quite extensive tests have been made on this system, which show very satisfactory results. About six tests carried to partial or total failure, made by the manufacturers in the presence of a number of architects, showed the floor to have an ultimate capacity of from 1,300 lbs. to 1,900 lbs. per sq. ft.

This floor has also withstood severe drop, fire and water tests satisfactorily.

The weight of this system of flooring is, the writer believes, the lightest on the market, being, when thoroughly dried, only 28 lbs per sq. ft. of floor, exclusive of beams, plaster, ceiling and boarding. The cost varies from 16c to 24c per sq. ft. without a level ceiling, and 18c to 26c with a level ceiling. This system possesses better fire-proof than water-proof qualities. Heavy rains have occurred during erection which penetrated the top floors, dripping down upon the floors below with a portion of the plaster of paris in solution. The composition is very disagreeable to handle, it is said. The principal objection to the system, however, is the discoloration of the ceiling caused by the gradual working to the surface of the water used in the mixture after it has taken up the coloring matter contained in the wood shavings and iron rust. The retention for a long time of moisture by the sawdust is also an undesirable feature.

MELAN SYSTEM.

This system was first used in Europe in 1893, principally for floors and vaults. Although the use of the Melan system in this country has thus far been confined to the construction of highway bridges, a brief description will not be out of place.

Fig. 5

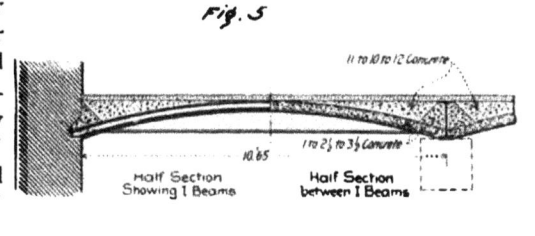

Half Section Showing I Beams — *Half Section between I Beams*

MELAN SYSTEM

This system consists, briefly, of curved I-beams imbedded in concrete. (See Fig. 5.)

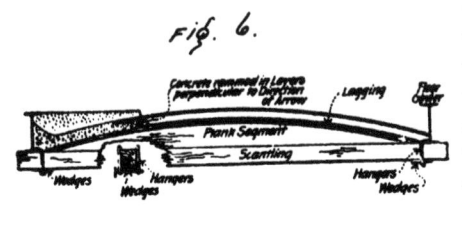

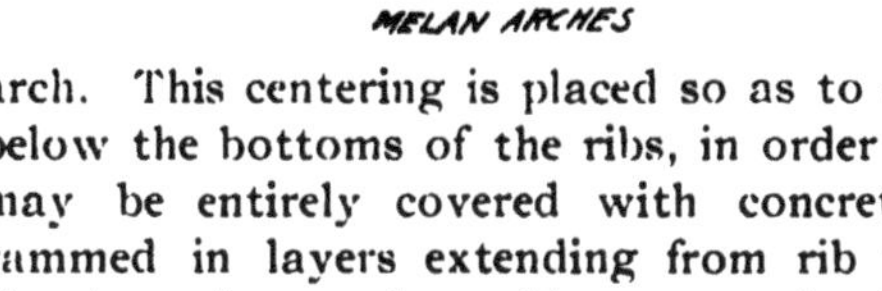

These curved beams abut against and rest upon the webs and lower flanges respectively of the girders, and are arched so that the top of the ribs are flush with the tops of the girders; the ribs are made to fit tight between the girders by the use of wedges driven between the webs of the ribs and girders. No riveted connections are required. After the ribs are in place a centering is provided with tight lagging (see Fig. 6), which conforms to the intrados of the arch. This centering is placed so as to come about one inch below the bottoms of the ribs, in order that the iron work may be entirely covered with concrete. The concrete is rammed in layers extending from rib to rib, beginning at the haunches and working toward the crown from both directions. This concrete, which must be of good quality, extends only to the top of the iron ribs. The space above is filled in with poorer concrete and cinders or other refractory material to form a level top and imbed the nailing strips which re ceive the floor. The size and arrangement of I-beams and depth of concrete for various distances between girders is given in the accompanying table.

	I-Beams				
Space, feet with rise of 1-12 to 1-15.	Depth, ins.	Weight.	Dis. apart, ins.	Weight, lbs. per sq. ft. of floor.	Depth of concr. in ins.
10 to 12	3	6	40	1.8	4
12 to 16	4	6	40	1.8	4½
16 to 20	4	7½	40	2.25	4½
20 to 24	5	10	50	2.4	5½

The speed attained in laying according to an example given in the *Engineering News*, was two cubic feet per man per hour, or four to six square feet per man per hour. It is stated that this is a low speed record.

Tests show this system to possess very great strength. The relative strength of several systems of arches when of same thickness is stated by Mr. Von Empberger, who represents the Melan system in this country, to be as follows: Brick arch, 1; Concrete, 5; Manier, 16; Melan, 36.

The claim is made by some that the Melan system depends for efficiency upon too many elements. The time required for setting of the concrete before centers can be removed is at least one week.

Care is required to secure good mortar, thoroughly mixed and properly rammed into place, necessitating careful superintendence where common labor is employed. The ribs must be free from rust when erected, to secure a proper bond between the concrete and the iron. The system is not so well adapted to form a flat ceiling as some others. The concrete in this system is called upon to transmit stresses in two directions—a condition which should be avoided where possible in sound engineering—the concrete acts as a beam between the ribs and also as an arch between the girders. The Melan system seems better adapted to use in warehouses, etc., where great strength is required, than in office buildings and others subjected to a moderate loading.

ROEBLING SYSTEM.

This system is identical in the main with the Manier system of Europe.

The system as adapted to floors consists of a "wire cloth arch, stiffened by steel rods which are sprung between the floor

Fire-Proof Floors. 19

beams and abut against the seat formed by the web and lower flange of the I-beams." On this wire arch Portland cement concrete is deposited and allowed to harden. The concrete is filled in so as to form a level surface on top, to receive the nailing strips and floor boards. A level ceiling is provided, as shown in the illustration, by running a system of iron rods from beam to beam and attached to them by means of patent clamps.

The ceiling, which is suspended below the bottom of the floor beams by means of the patent clamps, allows free circulation of air under the beams after the plastering is finished, a feature upon which considerable stress is laid by the manufacturers. The advantage claimed is that, in case of fire in a particular spot, the air, being free to circulate, is not apt to become so heated as to injure the iron, while, if confined to a small space, it is liable to become highly heated and communicate the same temperature to the iron beams. The continuous air space is a feature possessed by no other system.

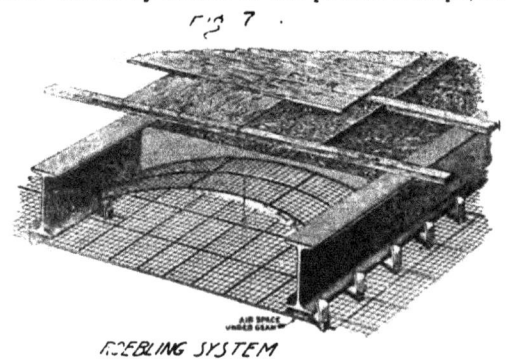

The cinder concrete which is generally used, is combined with sand and cement in variable proportions. At a building visited by the writer the proportion was one part Portland cement, two parts sand, and five parts cinders.

When extra heavy loads are to be provided for, as in warehouses, broken stone is used in place of cinders, increasing the strength as well as the weight.

The results of tests given in the manufacturer's pamphlet upon floors in buildings show the arches to sustain from 1,000 to 1,200 lbs. and in one case 2,490 lbs. per square foot, without sign of weakness in the former and without failure in the latter.

As to the ability of concrete to withstand fire and water, there seems to be a difference of opinion.

The writer tested a piece of concrete used in this system, taken from the floor of a building about two minutes before the test, and after heating it to almost a white heat immersed it in cold water. The cinders were unaffected, but a considerable portion of the sand and cement separated from the mass, leaving the latter considerably weakened. Others, who have made the fire and water test and who would have preferred poor results, say that it stood the test well.

The floors are used two days after being made and are guaranteed to withstand a test after being ten days old. Rapid progress can be made in laying this floor. The wire cloth is cut to the proper length at the mills, is quickly put in place, serving the purpose of a center, and the concrete can at once be put on.

The weight of the floor, as given in the pamphlet, varies from 47 to 59 lbs. per sq. ft., exclusive of iron beams. The cost varies from 18 to 23 cts. per sq. ft.

In the Roebling system the distribution of the material to withstand stress is such as to quite fully satisfy the engineering aim. The concrete being laid in the form of an arch, is subjected to little or no tension. If tension is developed in any part of the arch it will be at the intrados, where the wire cloth is placed to resist it. The wire cloth, having considerable strength itself, will protect any weak spot due to possible defective laying of concrete. Care must be exercised in laying the wire cloth, that the crown does not come too high, in which case the concrete at the crown, where it should be about 2½ in., will be too thin and thus impair the strength of the arch.

This system is well suited to cutting away for pipes, etc., and the hollow space is convenient for laying them.

COLUMBIAN SYSTEM.

This system was invented in this country and has been in use about three years, during which time 27 different buildings or groups of buildings have been supplied with it.

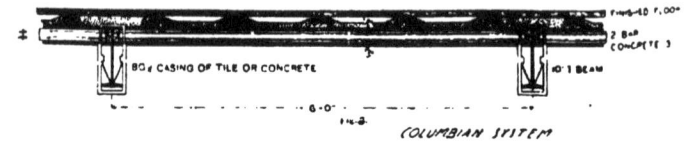

The Columbian system (see Fig. 8) consists of bars of steel, shaped so as to give a large area, for the concrete to adhere to. These bars rest in "U" shaped stirrups which lie on the top of the I-beams. The bars and stirrups are imbedded in the middle of a concrete slab, in much the same way that the wires are imbedded in the composition in the Manhattan system.

Accounts of tests given in the catalogue, show very satisfactory results. One floor sustained 1,154 lbs. per sq. ft. without deflection. A remarkably severe fire test was made upon this system in Pittsburg in 1895, under conditions specified by the Board of Fire Underwriters of Allegheny County, Pa. A section of floor in an enclosed space was subjected to intense heat for an hour and then drenched with water. During the test the arch carried a load of 750 lbs. per sq. ft. It is stated that the arch was uninjured.

Like concrete in any other form, it requires some time to set before it is safe to use—probably two or three days, depending on circumstances. The weight of this system, as given in the catalogue, is from 30 to 50 lbs. per sq. ft. The cost varies from 15 to 25 cts. per sq. ft.

The Columbian system is open to one objection from which many others are free, viz.: The steel used is of a special shape made by one concern, and a delay for any cause in obtaining the steel would necessarily delay the progress of the work; while a floor made of material which can be procured in the open market is less liable to interruption.

The advantage of combining iron and concrete, as has been stated, is to place the iron so as to strengthen the concrete against tension, and to do this, the iron must be near the soffit of the arch or bottom of the beam, where the tension exists. This is accomplished in a greater or less degree in the Roebling, Metropolitan, and Melan systems, but is evidently not aimed at in the Columbian, since the bar of steel is in the middle of the concrete slab.

Two other systems may be briefly mentioned in conclusion, namely, the Ransome system, and the St. Louis Iron Wire and Expanded Metal Company's system. Neither have thus far been used in the East. The former has been used in San Francisco, and consists of twisted square iron bars, imbedded in concrete near the bottom. The rods cross at right angles from side to side of the room at varying intervals, affording a good opportunity for making a very ornamental beamed ceiling.

The latter is much like the Melan system, with the addition of expanded metal, which is imbedded in the concrete over the entire surface of the floor, insuring the integrity of the latter. Channels are used for ribs in place of the I-beams of the Melan system. These are placed with the trough up, which is filled with concrete up to the level of the girders. The expanded metal is then laid on and imbedded in concrete.

No system possesses all of the desirable features of a fireproof floor, nor all of the objectionable ones, but each combines them in varying proportions. The choice of a system depends upon so many conditions, some of which vary from day to day and others with the locality, that each case must be investigated for itself in determining the system best adapted to it.

IRRIGATION ENGINEERING IN THE UNITED STATES.

C. H. KENDALL, C. E.

The practice of irrigation has come to be one of the most important questions of the day. Its development on a scientific and practical basis in this country has been so rapid that Irrigation Engineering is now a recognized profession, and, at the present time, no branch of Civil Engineering is receiving more marked attention throughout the Western states.

Previous to 1882, no irrigation work was designed or constructed on sound engineering principles, but the development of this art has been so progressive in the past few years that now our works, while not of such magnitude as the English works in India, surpass those of Egypt, France, Spain, Italy, Mexico, and South America, which countries have irrigated for centuries, and do compare very favorably with those of India.

To illustrate the phenomenal growth of this new profession in the United States, mention may be made of the following: the flourishing American Society of Irrigation Engineers, organized in 1891; the transactions of the four National Irrigation Congresses; the numerous Irrigation Commissions and Conventions held annually; that California, Colorado, and Wyoming have their State Irrigation Engineers; that many of the Agricultural Experiment Stations of the State Universities have irrigation engineers enrolled on their staffs; and that now several periodicals are published devoted to the interests of irrigation.

This recent rapid growth is due to the recognition by the Federal Government of the importance of the subject to the future growth and prosperity of its people. Hundreds of thousands of dollars have been appropriated for furthering investigations on a scientific basis. In connection with the United States Geological Survey, hydrographical and hydrological surveys and investigations have been made that are of incalculable value. It is the official reports of these surveys, together with the investigations of the Department of Agriculture and of individual states, that afford us the best literature upon the subject.

The magnitude and intricacy of the problems met with require the highest degree of engineering ability, and the Irrigation Engineer, in order to cope successfully with these problems, must be especially well qualified in the principles of Hydraulics, Meteorology, Hydrology, Geology, Topographical and Hydrographical surveying, and Water Rights Legislation. The field is a large one, and though much has been done, the amount is but a small proportion of what will be done in the future. This is at once apparent when we remember that the population of the United States doubles every thirty years; that the center of population is steadily moving westward at the rate of fifty miles every ten years; and that now there is less than four million acres of land irrigated, while there still remains about three hundred million acres of irrigable land awaiting development.

We will now discuss, in a general way, as far as the present limits of this paper will allow, the present practice and methods employed in the West, taking up the following subjects:—

I. Division of territory.
II. Quantity of water needed.
III. Source of supply.
IV. Classes of works.
V. Distribution of water.
VI. Application of water.
VII. Economical and financial aspects.

Division of Territory.

As a matter of convenience, the United States has been divided into sections according to the amount of annual precipitation. These divisions are termed the "arid," "humid," and "semi-humid" regions. Where the rainfall is more than twenty inches, it is classed as a humid region; and here irrigation is not absolutely necessary, but very often so increases the yield and insures against crop failure that it becomes a very profitable consideration. In this region we may class the states and portions of states east of the 97th meridian, and also the western parts of northern California, Oregon, and Washington.

The region where the rainfall is from 12 to 20 inches is termed the semi-humid, and to this class belong the states of North and South Dakota, Nebraska, Kansas, most of Texas, and Oklahoma Territory. Here some years rainfall is abundant, the crops luxuriant, and prosperity evident; then come dry years and failure of crops, which lead to discouragement, if not actual starvation. It is in this region of variable and uncertain rainfall that irrigation should be more studied and practiced; then certainty of crops would result, with a corresponding increase in yield, and not, as now, be a mere matter of speculation depending upon nature watering the land.

The states and territories west of the above region, having a rate of rainfall below twelve inches, belong to the arid region and no attempt is here made to cultivate the land without irrigation. We find here our most advanced irrigation works in operation; giving abundant crops every year and, where climatic conditions favor, often two or more crops are produced a year. The area of this section is about nine hundred million acres, much of which will never produce crops, even with water, because the climate and soil are unfavorable.

QUANTITY OF WATER NEEDED.

The term "duty of water" is used to express the amount of land a given quantity of water will irrigate. This duty is by no means constant, but varies throughout the arid region according to the character of the water supply, the methods of employing it, the character of the soil and crops, and the skill and experience of the irrigator.

The average duty of water is one hundred acres to the second-foot. Besides this generally accepted term for the duty, one meets with the following expressions in different localities: "Acre-inches," "acre-feet," "California miner's inches," and "Colorado miner's inches." One cubic foot of water per second is equal to: 86,400 cubic feet per day, 646,317 gallons per day, 2,700 tons per day, 24 acre-inches per day, 50 California miner's inches, 38.4 Colorado miner's inches. It will flood one hundred acres in one hundred days, twenty-four inches deep, or one hundred and fifty acres in one hundred days, eighteen inches deep, and so on.

The amount of rainfall necessary for raising a successful crop is about sixteen inches. In Utah, the average duty is one hundred acres per second-foot, and the following table, published by Professor Fortier, shows the various depths of water applied to the land in producing the crops mentioned,—calculated on the basis of one hundred days:

Crop.	Depth of Water.	Duty in acres per sec.-ft.
Strawberries	27.50 inches	93
Cauliflower	8.25 "	291
Tomatoes	24.75 "	97
Mixed crop	23.00 '	103
Barley	7.25 '	330
Corn	3.75 '	660
Potatoes	16.63 '	143
Onions	35.50 '	67
Peach Orchard	12 00 "	213
MEAN	16.40 "	256

SOURCES OF SUPPLY.

The climate, geology, and topography are the chief factors in determining the sources of supply, which we may class under the following heads: Rainfall; Running Streams and Springs; Storage Reservoirs; Ground Water, or Sub-surface Supplies; Artesian Wells; Ordinary Wells.

So far, all attempts to produce rainfall artificially have been unsuccessful. The first five sources furnish water for "gravity irrigation," which supplies ninety-nine one-hundredths of the irrigated land. The supply from wells is termed "lift irrigation," and though this method is almost inappreciable in extent, it is becoming more and more popular, due to modern improvements and special designs of pumping plants for the purpose.

CLASSES OF WORKS.

Herbert M. Wilson, C. E., in his work on "American Irrigation Engineering," includes five great classes of works in gravity irrigation, viz.: Perennial Works; Periodical Works; Storage Works; Irrigation from Sub-surface Sources; Irrigation from Artesian Wells. By perennial works are meant those canals which receive their supply from streams of sufficient discharge to afford irrigation at all times to the lands commanded by them. Among works of this class, the following may be mentioned as of particular interest and magnificence:—

(a) The Turlock Canal, diverted from the Tuolumne River in California. It has a total length of 180 miles and commands 176,110 acres. It has a capacity of 1,500 second-feet and the estimated cost was about $1,110,000.

(b) The Idaho Mining and Irrigation Company's canal, diverted from the Boise River, is 70 miles long and irrigates 350,000 acres.

(c) The Pescos Canal System, diverted from the Pescos River in New Mexico, commands, with several hundred miles of laterals, about 400,000 acres.

(d) The Bear River canal, from the Bear River in Utah, has 150 miles of main line and commands 236,000 acres. Its estimated total cost was $3,000,000.

Periodical works are canals taking their supply from streams which furnish water for a portion of the irrigation season only. Such works may be found supplying a limited territory throughout the West where the conditions are such that they supplement the natural supply in the soil which alone is almost sufficient for the cultivation of crops.

Storage works are constructed in intermittent streams, impounding the flood waters to supplement the flow, so as to insure a constant supply during the irrigation season regardless of rainfall.

There are various classes of these works according to character and location of the storage basins and sites of the dams. Among the most important may be mentioned:—

The Carayamaca earthern dam, in California, impounding 11,500 acre-feet of water and covering an area of 1000 acres. It is 635 feet long and 40 feet high. The water is conducted from it through a wooden flume 36 miles long, which passes over some 315 trestles.

The Bear Valley Reservoir, of California, has surface area of 2,252 acres and capacity of 40,550 acre-feet (10,000,000,000 cu. ft.). The dam is of ashlar masonry, 300 feet in length on crest, 64 feet high, and arched up stream with a radius of 335 feet. The new dam, located just below, is 120 feet high and also of curved form.

The Sweetwater Reservoir Dam, of San Diego, is of rubble masonry, 94 feet high, 380 feet long, and impounds 18,000 acre-feet (770,000,000 cu. ft.).

The Buchanan Reservoir dam is of uncoursed rubble masonry, 780 feet long, 100 feet high, curved with radius at centre of 1,146 feet. It impounds 42,400 acre-feet over an area of 1000 acres.

Irrigation from ground water sources is by tunnels under stream beds, or into the hillside to tap some water bearing stratum, or by open cuts in the sloping ground, or by wells to collect the ground water. These supplies are situated at various depths and large volumes of water are obtained from these sources in various portions of the West. In California, submerged dams have been built across dry stream beds to cut off the under-flow and bring it to the surface. A great deal of water has been "mined" in both Colorado and California by constructing sub-surface canals or tunnels into the underground storage. Also a few depressed canals have been excavated along slopes.

Only a small amount of water used for irrigation is obtained from artesian wells. Still there are about 9,000 wells widely distributed, but mostly in California, Colorado, Utah, the Dakotas, and Texas. Artesian water is not as suitable for irrigation as surface and sub-surface water and is mostly employed for stock, small gardens, fruit trees, and grass. The cost is more than twice that of the ordinary method of obtaining water.

Water supplied by "lift" irrigation is one of the most promising methods of the future. Though relatively small in amount when compared with that from streams, it has great importance from the fact that dependence must be placed upon it in many localities where running water cannot be had.

Wind-mills have been extensively used and in connection with a small reservoir furnish a constant supply to a small acreage. Where good supplies are to be found, it is the cheapest way. Steam pumps are not yet fully appreciated, but pumping plants are coming into use and a large number are now in operation in all portions of the West; in California, Colorado, Wyoming, and Arizona in particular. Each pump supplies water for from fifty to one hundred acres. The use of the gasoline engine is finding favor, as it does not require constant attention, but will work automatically for several hours, or all day after once started.

DISTRIBUTION OF WATER.

Methods of carrying water to the points of application are:

1. By canals or ditches, with or without masonry lining. The unlined is the cheapest, but much washing of banks occurs and the loss by percolation and evaporation is very large. When lined, they give excellent satisfaction, but are more costly than timber flumes.

2. Wooden flumes have been quite extensively employed in some sections and have been found to be the most economical when well made.

3. Concrete or stoneware pipe laid on regular grade gives

the least loss from evaporation, but the cost for maintenance is sometimes high, as the pipe is liable to be disarranged or choked by roots.

4. Wrought iron or steel riveted pipes, coated with asphaltum, to prevent rusting, are coming into use and have the advantage of wood or stoneware as regards evaporation. These can be used under pressure and consequently can be run in a direct line to the land to be irrigated. They cost more than wood and deteriorate rapidly in certain soils, but sometimes the length of line saved more than compensates for extra cost.

APPLICATION OF WATER.

The water is usually delivered, and where possible, at the highest point of lot, so it will gravitate to any desired point.

One method of irrigating, where water is plentiful, is to run the water in unlined open ditches and flood the entire surface. It is very wasteful and requires level land. When water is not so abundant, ditches are lined and the water is turned into basins about the trees or furrows about the crops and allowed to soak down into the soil.

On uneven ground it is necessary to use small iron pipes or timber flumes. The flow of the water is regulated by mechanically devised troughs and weirs.

Sub-surface irrigation is practiced to some extent and consists of a series of concrete pipes laid in the ground deep enough to escape disturbance by cultivation. This method avoids surface evaporation, but is costly and the pipes get choked by roots.

ECONOMICAL AND FINANCIAL ASPECTS.

The average annual cost of applying water, per acre, is from $0.75 to $2. However, in Southern California, $10 is not uncommonly paid where land is valued at $1,000 per acre for horticultural purposes. The average first cost of water supply is at the rate of $8.15 and its average value is $26 per acre. The following table taken from the United States Census Report for 1890, as prepared by F. H. Newell, of the Geological Survey, is very valuable as showing the extent and cost of irrigation and furnishing accurate statistics on the subject.

30   C. H. Kendall.

EXTENT AND COST OF IRRIGATION.

States and Territories employing irrigation.	Crop irrigated.	Per cent. of area irrigated to total land area.	Total number of farms with irrigated crops.	Per cent of area of irrigated crops to whole area owned by irrigators.	Average size of irrigated crops per farm.	Average first cost of water per acre.	Average value of water per acre as estimated by irrigators.	Average annual cost of water per acre.	Average cost of preparing land for cultivation per acre.	Average value of land irrigated per acre.	Average value of products from irrigated land per acre.
	Acres.				Acres.						
Total United States	3,564,416	0·50	52,584	20·72	67	$8.15	$26.00	$0.99	$12.12	$83.28	$14.89
Arizona	65,821	0·09	1,075	43·21	61	7.07	12.58	1.55	8.60	48.68	13.92
California	218,249	1·01	13,732	17·86	73	15.84	52.28	1.60	22.27	150.00	19.00
Colorado	350,583	1·34	9,859	31·09	92	7.15	28.46	.79	9.72	67.02	13.12
Idaho	224,403	0·40	4,333	26·08	53	4.74	13.18	.90	9.31	46.50	1.93
Montana	91,746	0·38	3,706	23·05	95	4.63	15.04	.05	8. 9	49.40	12.96
Nevada	177,944	0·32	1,467	14·13	192	7.58	24.60	.84	10.57	41.00	12.92
New Mexico	263,473	0·11	3,085	17·98	30	5.58	18.30	1.54	11.71	50.98	12.80
Oregon		0·29	3,150	15·89	56	4.64	15.48	.94	12.59	57.00	13.90
Utah		0·50	9,724	22·02	27	10.55	'6.84	.91	14.85	84.25	18.03
Washington	49,399	0·12	1,050	17·21	47	4.03	13.15	.75	10.27	50.00	17.09
Wyoming	229,676	0·37	1,917	15·24	119	3.62	8.69	.44	8.23	31.40	8.25
Total for eastern sub-humid region	67,295	0·02	1,557	6·43	43	4.07	14.81	1.21	4.62		
North Dakota	446	0·001	7	34·76	63	3.20	17.90	.25	6.42		
South Dakota	15,717	0·03	189	29·95	83	4.42	7.59	.66	7.81		
Nebraska	11,744	0·02	214	14·44	55	4.40	21.12	.44	4.86		
Kansas	20,818	0·04	639	13·92	40	6.14	23.57	1.40	3.05		
Texas	18,571	0·01	628	2·47	30			1.10	11.60		

SOME EXPERIMENTS WITH BROM-CYAN.

H. C. CUTLER, B. E. M., '94.

Some few months ago the metallurgical world was stirred by the announcement of a new solvent for gold and, consequently, of a new process for its extraction from ores. The new solvent was a mixture of potassium cyanide and brom-cyan. Brom-cyan is a compound of bromine and cyanogen, having the formula of BrCN. A one-half per cent. solution of potassium cyanide and a one-quarter per cent. solution of brom-cyan will dissolve gold leaf nearly as quickly as aqua regia.

The writer has made a large number of experiments in the laboratory and on a working scale with the solution. On some classes of ores very favorable results were obtained, while on others the solution worked no better than the potassium cyanide alone.

The chemist who claimed the discovery of the new solvent used in his experiments crystals of pure brom-cyan, which he stated could be manufactured cheaply. At the time of making these experiments, the crystals of pure brom-cyan were not obtainable, hence a solution made by mixing bromine and a solution of potassium cyanide was used. In making this solution a number of interesting facts were noticed.

Two methods were used. In the first, a saturated solution of potassium cyanide and pure bromine were mixed. In the second, a weak solution of potassium cyanide (not over 5 per cent.) was added to water saturated with bromine.

In the first method, there was a violent action when the two chemicals were mixed and a large number of reactions took place. Potassium bromide (K Br) and caustic potash (K O H) were formed. The cyanogen radical (C N) was split up. The carbon and nitrogen thus formed, uniting with the oxygen and hydrogen of the water, formed a number of organic compounds. Some of these organic compounds were thrown down in the state of a heavy black precipitate. The remainder were soluble and imparted a deep brown color to the resulting solution. The precipitate varied in amount each time the solution was made. Just what conditions are necessary for the least amount was not ascertained.

In the second method, the results were different. The violent action seen in the first method was entirely absent. Upon adding the solution of bromide in water to the weak potassium cyanide solution the resulting liquid remained perfectly clear until the bromide was in excess, when it assumed the light brown color of bromine.

The formula for this reaction was determined by the writer in the following manner: The strength of a saturated solution of bromine in water was found, by titrating with silver nitrate, to be 1.245 per cent. A burrette was then quickly filled with this solution. Ten cubic centimeters of a .3 per cent solution of potassium cyanide was placed in a small flask. The flask was then closed with a rubber cork in which there was a hole just large enough for the end of the burrette. The cyanide solution was then titrated with the bromine water. It took 2.95+ cubic centimeters of the bromine solution to neutralize the 10 cubic centimeters of the potassium cyanide solution. From the molecular weights it will be found that there was just enough bromine in the 2.95+ cubic centimeters of solution to unite with the cyanogen of the potassium cyanide to form brom-cyan (BrCN). Caustic potash (KOH) and no potassium bromide (KBr) was found in the resulting solution. The reaction, therefore, may be as follows:

$$2KCN + 2Br + O + H_2O = 2BrCN + 2KOH.$$

Greater economy would result in making brom-cyan by the second method.

These methods were used only for the purpose of experiment. If the process was to be used on a practical scale the crystals of brom-cyan should be obtained. The solution could then be made and handled much more conveniently.

The following table contains results of some experiments with potassium cyanide alone, and a mixture of potassium cyanide, and brom-cyan:

Some Experiments with Brom-Cyan.

TABLE OF EXPERIMENTS.

Number.	CHARACTERISTICS OF THE ORE	Kind and per cent. solution used.	Time of Treatment.	Origin'l Assay. Ounces Ag.	Ounces Au.	Per Cent. Extrat'd Ag.	Au.	Str'gth of KCN after treatment.	REMARKS
1.	Concentrates. Heavy iron sulphide partially oxidized from being exposed.	a. 1. .5 KCN 2. " 3. " b. 1. .5 KCN+.25 BrCN 2. " 3. "	20 hrs. 40 " 60 " 20 " 40 " 60 "	38.5 38.5 38.5 38.5 38.5 38.5	1.9 1.9 1.9 1.9 1.9 1.9	1.2 4.5 6.6 13.4 19.6 26.3	14.2 18.7 21.3 47.6 74.8 92.5	.32 .28 .23 .31 .28 .27	100 c.c. of solution used and about 50 grams of ore alkali wash. 60 c.c. of KCN solution and 50 c.c. of BrCN solution and about 50 grams of ore. Alkali wash.
2.	From pyrites panned out of ore after being treated by cyanide process for 48 hours.	a. .3 KCN b. .45 KCN+.2 BrCN	72 " 72 "	38.5 2.3	1.9 2.28	78.3 81.8	86.85 87.72	.125 .139	250 c.c. KCN used in a and 100c.c. of each in b. About 100 grams of ore. Alkali wash.
3.	Heavy iron pyrites direct from mine.	.45 KCN+.2 BrCN	48 "	2.0	0.64	15.9	62.5	.0325	100 c.c. of solution and 50 grams of ore.
4.	Same ore as No. 3 partially wasted	1. .45 KCN 2. " 3. " 4. " 5. .45 KCN+.2 BrCN 6. " 7. " 8. " 9. .45 KCN+1 gram sodium dioxide. 10. " 11. " 12. "	20 " 40 " 60 " 20 " 40 " 60 " 20 " 60 " 40 " 20 "	2.2 2.2 2.2 2.2 2.2 2.2 2.2 2.2 2.2 2.2 2.2 2.2	0.68 0.68 0.68 0.68 0.68 0.68 0.68 0.68 0.68 0.68 0.68 0.68	9.1 2.3 0.21 9.09 18.1 4.3 0.2 10. 13.6 4.2 1.3 1.4	11.7 9.67 6.00 12.1 17.6 9.86 8.75 18.94 9.67 5.88 2.13 17.7	.025 .065 .08 .39 .02 .038 .072 .343 .15 .175 .192 .35	100 c.c. of solution and 50 grams of ore. [treatment. Water wash previous to 50 c.c. of each solution and 50 grams of ore. Water wash. 100c.c.solution and about 50 grams of ore. Water wash.
5.	Silious ore containing 13.2 per cent of iron pyrites	a. 1. .5 KCN 2. .3 KCN+.25 BrCN b. 1. "	48 " 48 " 48 "	14.24 14.24 14.24	4.72 4.72 4.72	27.2 56.7 71.4	48.3 73.4 87.3	.325 .27 .243	Alkali wash previous to treatment.

It will be noticed in the foregoing table that the loss of brom-cyan (BrCN) in treatment is not given. This is due to the fact that no satisfactory method of determining this loss could be devised. The potassium bromide (KBr) formed in dissolving the gold according to equation,

$$BrCN + 3KCN + 2Au = 2(KAuCN_2) + KBr,$$

interfered in the titration of the BrCN.

The experiments show that in some cases brom-cyan solution gives better results than simple potassium cyanide.

As the apparatus which was available for making brom-cyan on a large scale was very crude, the results obtained were not as satisfactory as the laboratory tests. There is no doubt that, if the brom-cyan solution could have been applied in the proper way, laboratory results could have been duplicated on a working scale.

NOTES ON MACHINE DESIGNING.

BY JOHN H. BARR, '83.

Assistant Professor of Machine Design, Sibley College, Cornell University.

The steam engine, like all other engineering constructions, is strictly subject to the laws of mechanics; but the operation of the engine presents such a complex problem in dynamics that it is practically impossible to base its design on purely rational methods. The varying steam pressure upon the piston is but one element in the complicated system of forces acting, as acceleration (linear and angular), friction, variation of the external load, gravity, etc., etc., all exert an influence upon the stresses produced in the members. These alone make the exact computation of dimensions for strength and rigidity exceedingly difficult; and, moreover, other considerations, both "theoretical" and "practical", are of equal importance, and must receive proper attention from the practical designer. He has to deal with the pressures upon the bearing surfaces and their velocities of rubbing, which affect the mechanical efficiency, durability, and freedom from heating. In many cases, the stopping of the engine during working hours is a more serious source of loss than is extravagance in the use of fuel. Thus, while thermal efficiency is of the highest importance, generally, it is only one factor in the final efficiency.

Apart from the engineering elements in the problem, there is always the commercial element. This requires the general use of regular standard forms and sizes when feasible, and often modifies the computations based upon pure mechanics. Economy of construction dictates the adoption of forms and dimensions which can be readily, accurately, and certainly produced in the shops. The limitations of the mechanic arts, as practiced in the foundry and machine shop, must ever be in the mind of the designer. As between the ideal form which can be produced only at great expense, and the "good enough" form which is a great deal cheaper, the latter is frequently the only one practicable.

Then again, a builder often has calls for engines which do not differ greatly in capacity, and he meets this demand by building two engines having the same stroke but somewhat different diameters of piston; for example, one is 11 x 12 inches and the other is 12 x 12 inches. The almost universal practice in such cases is to use the same frame, crank-shaft, connecting rod, crosshead, etc., for both sizes. This results in relatively stronger members for the smaller engine; but the saving in construction outweighs the small gain in material which would result if all the members of the smaller engine were reduced in proportion to the loads upon them.

It is owing to such considerations as these, which can only be appreciated by one who has observed thoughtfully for years, that the designer must acquire much experience before he can become thoroughly successful. The traditional worthlessness of the young technical graduate is due to his lack of familiarity with the so-called practical considerations. If, when he steps from the college into the shop or office, he is not at once a successful designer, it is not because his ideas are bad, but because he has not enough of them. The school shops and laboratories can be made to do much toward remedying these defects, but the time available for such drill as they afford can never produce the mature judgment required of the well rounded engineer. The education begun in the technical school must be carried on through years of practice. This is no argument against the courses of the best schools; for they give a training which one can scarcely acquire in practice; while the experience and judgment essential to the practical engineer are just the things that are most surely attained in his professional life.

This dual training, that of the college and that of the shop and office, are not in themselves sufficient to produce the highest order of engineer. The engineering genius, like the musical genius, has an inborn aptitude for the work of his profession. But mechanical intuition can be cultivated by drill and observation. Among the most useful methods to this end may be mentioned the practice of sketching and noting dimensions of existing constructions, especially of the product of highly successful designers. When a peculiar form is noticed in a machine member, study it particularly to see if this form was adopted for good reasons, and if some better construction—all essentials considered—could not be substituted. For a most helpful discourse on this subject, the reader is referred to a paper by Mr. John T. Hawkins on "The Education of Intuition in Machine Designing," *Transactions of the American Society of Mechanical Engineers*, Vol. VIII., page 458.

Few engineers can rely safely upon their own experience alone; the successes and failures of others furnish food for reflection. Machine design is a composite of art and science; the science suggests the lines of design, and should be applied when feasible as a check; but the element of art often dictates the final forms and dimensions.

Returning to our illustration of the steam engine, we find that certain dimensions can be subjected to analysis and more or less exact computation; but the true value of most of the calculations as to strength is of the nature of insurance. Anyone who has ever compared the proportions of a modern high-speed engine with dimensions calculated by the ordinary formulas of mechanics has been impressed with the apparently high "factor of safety" which is frequently observed; and yet these engines are not free from break-downs. The explanation is that many of the general dimensions actually used have been found necessary in service. They have been adopted to provide for the various elements which can hardly be treated analytically. There are, to be sure, instances in which the dimensions are much beyond the requirements, for it is the practice of some builders to make certain parts much heavier than those of equally successful competitors. This practice may be due to ignorance of the requirements, or is a bid for popular favor through comparison; the presumption usually being in favor of the heavier machine when judged by partially informed buyers. In general, it may be said that mass in the frame and stationary parts of a machine are desirable where there is liability of severe shock; while the moving members should be as light as is consistent with strength and rigidity. Of course this does not apply to such moving members as fly wheels, etc., and there are many exceptions to this rule.

The exercise of individual judgment leads to a wide diversity in the proportions adopted by different designers of similar machines; which is in marked contrast with the general agreement as to certain other proportions.

The writer has been engaged during the past year or two in comparing the proportions of high-speed engines. A partial report of this examination was presented to the American Society of Mechanical Engineers at the recent meeting in New York. The method used was to write to various builders, enclosing a blank form to be filled out with the required data. The information collected was classified, and values substituted in standard formulas, (of a rational form when possible), and the constants were then derived. For example, in studying crank-shafts, those engines having center, or inside, cranks were treated by themselves, using the formula $d = C\sqrt[3]{H.P. \div N}$; in which $d =$ diam. of shaft, H. P. = rated horse-power, N = revs. per minute. From the data for each engine the value $\sqrt[3]{H.P. \div N}$ was calculated, and plotted as an abscissa; the value given for d being used as the ordinate; these co-ordinates gave a point. About fifty points were obtained in this way, from as many different engines, and lines were drawn to represent the average and extremes of practice. The equations of these lines give values of the constant C for the average and the extremes of practice. The values of the constants as thus obtained are C = 7.56 for the mean; and C = 5.98, and C = 8.76, for the minimum and maximum, respectively.

In Unwin's Machine Design (Part I, page 225), a similar formula is given, with the value of C assigned as 4.55. This is for marine engine practice, and it serves to show that we must design the smaller high-speed engines on the basis of high-speed engine experience.

Many other examples could be cited to show that any particular constant will not give satisfactory results under widely varying conditions; but the limits of the present paper do not permit further extension, even if such comparisons were within its scope. The above example of the crank-shaft is introduced simply as an illustration. It may be said in passing, however, that the above mean value of the constant for crank-shafts represents very satisfactorily the practice of many leading builders; while in the examination of other proportions, as crank-pin diameters, no such general agreement was discovered.

An investigation is now under way upon the proportions of slow-speed stationary engines, and further work of a similar character is contemplated on other classes of machinery.

It is gratifying to report that a large number of the most progressive builders have co-operated cordially in this examination of the current practice in engine construction. The spirit shown is in marked contrast to that prevalent a few years ago, and it is encouraging to the young engineer as he enters the profession, to feel that his older associates are not all bent upon keeping from him the trade secrets. Of course each manufacturer has much information, acquired, perhaps, by expensive experiment, which he must guard out of self-protection; but the observing young engineer of today, who shows a proper regard for the rights of others, has great opportunities.

DIRECT AND ALTERNATING ELECTRO-MOTIVE FORCES IN SERIES.

BY HORACE T. EDDY, B. E. E.

About a year ago, the attention of the writer was called to a case in which an alternating e. m. f. was superposed on a direct incandescent light circuit. The brightness of the light was not increased very greatly by the addition of an alternating e. m. f. of about half the voltage of the direct current.

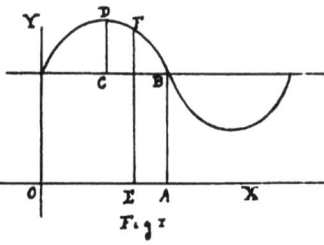

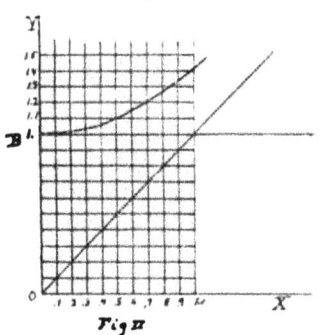

On measuring with an alternating voltmeter the e. m. f. of such pulsating currents formed by superposing an alternating on a direct e. m. f., it was found that an alternating e. m. f. of considerable voltage did not greatly increase the voltage of the direct current.

In order to compute the effective e. m. f. of such a pulsating current, let A B in Fig. 1 represent the magnitude of the direct e. m. f., CD the maximum value of the alternating e. m. f., and EF the e. m. f. of the pulsating current at any instant, supposing the alternating e. m. f. to be simply periodic.

The reading on an alternating voltmeter is the square root of the mean square of the instantaneous values of the e. m. f. When applied to a simple alternating current whose maximum value, CD, is equal to a, the reading of the voltmeter is a divided by $\sqrt{2}$. This quantity is positive during one half of the alternation and negative during the other half: consequently, when it is superposed on a direct e. m. f. it increases it during one half of an alternation and decreases it during the other half. This is shown in the figure by the fact that the ordinate, EF, is greater than AB during one half of the time and less than AB during the other half.

Now the effective e. m. f., ^{-}y of a pulsating current is, as before stated, the square root of the mean square of the instantaneous values, y of the e. m. f. To compute the numerical value of ^{-}y we have the expression—

$$^{-}y^2 = \frac{\int(b+a\sin.x)dx}{\int dx} \quad \text{To be taken between the limits 0 and } 2\pi.$$

$$^{-}y^2 = \frac{1}{2\pi}[b^2 \int dx + 2ab \int \sin. xdx + a^2 \int \sin.^2 xdx]$$

$$^{-}y^2 = b^2 + \frac{a^2}{2}.$$

Let us illustrate this result by the case in which $a=100$ and $b=100$ volts. The reading of the alternating voltmeter for $a=100$ will be $100 \div \sqrt{2} = 70.7$, and the reading for the pulsating current will be $\sqrt{100^2 + 70.7^2} = 122.5$ volts.

The relation between the pulsating, alternating, and direct voltages can be shown best graphically. Call the voltage of the alternating e. m. f. ^{-}x.

Then $^{-}x^2 = \frac{a^2}{2}$; hence $^{-}y^2 - ^{-}x^2 = b^2$,

which is the equation of a rectangular hyperbola as shown in Fig. 2, in which $OB = b$.

It is possible to use a figure like this, drawn on cross-section paper, instead of a table, to find the voltage of a pulsating current arising from the superposition of alternating and direct voltages of any magnitude, for we can write the equation,

$$\frac{^{-}y^2}{b^2} - \frac{^{-}x^2}{b^2} = 1$$

Then if we take $\frac{^{-}x}{b}$, the ratio of the alternating to the direct e. m. f., as the abcissa of any point of the curve, the corresponding ordinate is $\frac{^{-}y}{b}$, the ratio of the pulsating to the direct voltage, provided OB is taken as unity. But the equation can also be written in the form $\frac{^{-}y^2}{^{-}x^2} - \frac{b^2}{^{-}x^2} = 1$. If now OB be taken as unity, $\frac{b}{^{-}x}$, the ratio of the direct to the alternating e. m. f.'s, will be the abscissa and $\frac{-y}{-x}$, the ratio of the pulsating to the alternating voltages, will be the corresponding ordinate to the curve at any point. We can thus read the voltage of the pulsating current, whether the direct voltage is greater or less than the alternating.

The simplest method of obtaining the pulsating e. m. f. is to lay off the direct and alternating voltages at right angles on any scale. The hypothenuse of the triangle with these voltages as sides will be the effective e. m. f.

THE PRINCIPLES OF ARTIFICIAL LIGHTING

BY PROFESSOR GEO. D. SHEPARDSON.

Artificial lighting may be considered under two classes, according as the object sought is *illumination appearance* or *illumination*. In one case it is desired to see the lights themselves. In the other case the objects illuminated by the lights are observed. For light-houses, beacons, and some sorts of advertising, it is desired to have the lights themselves conspicuous. For many other purposes, such as lighting reading desks, pictures, and work-benches, illumination is desired rather than illumination appearance. For some purposes, both illumination and illumination appearance are desired, as for show-window advertising, lighting ornamental stairways and halls, or principal streets.

Illumination appearance is obtained by making the sources of light conspicuous. Familiar examples in Minneapolis are the circles of lights on the towers above the Glass Block and Olson's, the incandescent crosses on the tower of the Wesley Methodist Church, the rows of incandescent lamps along the cornice of the Plymouth and the head-light above the Metropolitan Theatre. A variation is in the use of rows of incandescent lamps forming letters for advertising purposes, such as the signs on the Bijou and Metropolitan theatres, the mammoth Ceresota sign above one of the flouring mills, and similar temporary signs used during the holidays and other festival seasons.

A class of lighting that might be considered as being illumination or illumination appearance or both, is the lighting of translucent signs such as are numerous on any of the business streets, the real source of light being concealed in a translucent enclosure.

Illumination combined with illumination appearance is desired in many cases. Here, again, the advertiser has developed excellent examples, such as the rows of arc lamps along the sidewalks around the Glass Block, the Syndicate Block, and the Plymouth. The arcs call attention from a distance and also light the store-fronts, being behind passers-by on the sidewalk and high enough not to interfere seriously with the view of those in carriages. Fancy designs in store windows also serve the double purpose, when the lights themselves are not too brilliant, as when colored or frosted lamps are used. An excellent combination of illumination and appearance is found in the street cars. The lights near the ceiling give perfect illumination to those within the car, and also, by shining through the translucent signs, indicate, to those outside, the line to which the car belongs.

For lighting interiors of buildings, especially corridors and halls with high ceilings, excellent effects may be produced by combining illumination with illumination appearance. The rotunda of the New York Life building, and the main entrance to the Lumber Exchange, are examples where the arrangement of lights attracts attention, and yet the illumination is good. The chapel at the University and the lower rooms in the Phœnix Block also illustrate satisfactory illumination for desk work, combined with striking illumination appearance for general effect.

Illumination appearance is often unintentionally and undesirably obtained, when only illumination is desired. In other cases, a certain amount of illumination appearance is desired, but too much is secured.

Illumination without illumination appearance is found in diffused daylight. The prime source of light, the sun, is not seen, but the light is evenly distributed in all directions. The ideal of artificial illumination is to approach sunlight.

The distinction between illumination and illumination appearance, is not as well or as commonly understood as it should be, and frequent blunders are the result.* For instance, the ambitious country town insists on having its streets lighted by brilliant arc lamps, in order, thereby, to obtain a certain metropolitan aspect. Arc lamps are excellent for street lighting if placed not further than one or two blocks apart and if twenty-five to thirty-five feet above the level of the roadway. But, with the not infrequent practice of hanging them eighteen to twenty feet above the roadway and long distances apart, the lamps serve as beacons, to indicate directions and to blind persons in the streets, more than as sources of illumination.

---

*See paper by A. Scheible on Illumination vs. Glare, *N. Y. Elec. Eng.*, Vol. XX., page 565, Dec. 11, 1895.

An equal amount of money spent in erecting and operating a much larger number of smaller lights, placed at shorter intervals, would better serve the purpose of lighting the roadway with some uniformity and of enabling persons to see the way clearly and safely. For lighting streets and passage-ways, one needs illumination rather than illumination appearance. Similar considerations generally hold true for architectural lighting. One should recognize distinctly the effect sought and then consider carefully the best means of obtaining it.

Having recognized the two more or less distinct classes of lighting, the principles to be followed for obtaining satisfactory results are not difficult to discern.

For illumination appearance, the sources of light should be conspicuous. If dazzling effects are desired, nothing can be more satisfactory than arc lights, either with clear glass globes or with no globes. Calcium or magnesium lights are more troublesome and expensive in maintenance, but may be used in some places. Blinding effects may be enhanced by the use of parallel beams of light from electric search lights of fabulous candle-power. When a general blaze of light is desired, large and numerous gas flames and fireworks have a field almost their own, although the equally liberal use of arc lamps with ground or opal globes, or of incandescent electric lamps, gives magnificent effects. Witness the profusion of light on parts of Nicollet and Wabasha Avenues, or the principal shopping streets of other large cities, also the display illumination appearance at exhibitions and fairs in cities.

A quite different class of illumination appearance, such as is required for illuminated signs, where the lights trace letters or other outlines, involves the use of comparatively large numbers of smaller and less intense lights, so arranged as to give the impression of being continuous lines of light. For such lighting, gas is suitable only in places free from excessive wind and from the presence of inflammable substances. Incandescent lamps with either clear, frosted or colored globes, are peculiarly adapted for such lighting. A convenient method is to mount the lamp sockets in any desired position upon a screen of wire netting.

The principles involved in securing satisfactory illumination with absence of illumination appearance are quite simple, although not always recognized, and although sometimes difficult to apply in practice. The ideal is to keep the source of light unseen, while the objects of view are sufficiently and evenly lighted. The eye sees various objects by means of the light coming from those objects. If the light from the object is weak, the eye does not receive sufficient light to form distinct images. If the object is too light, the eye partially closes to protect itself. Two similar objects may be illuminated with equal amounts of light, but if one is in the neighborhood of other objects more strongly illuminated, or if a strong source of light comes within the angle of vision, the object with more strongly lighted surroundings will be less distinctly seen than the other one equally bright but with a darker background. If strong lights are in the field of view, the iris of the eye closes so as to limit the total amount of light received. This automatic protective device regulates the opening of the curtain so as to cut the maximum light down to that required by the eye; consequently the light from all objects less bright is reduced in the same proportion.

What constitutes sufficient light depends both upon the purpose of the lighting and the disposition of the sources of light. If the lights are arranged to the best advantage, being suitably distributed and not being seen themselves, satisfaction is obtained if the illumination is equal to that given by one candle at distances indicated in the following table:*

Street pavements or sidewalks............	10 feet.
Walls of buildings............	8 "
Public halls, churches, theatres, etc., (general light),	3 to 5 "
Workshops (general light)............	5 "
Work benches............	0.3 "
Tables, reading, eating, etc.,............	0.3 to 0.5 "
Corridors, halls, etc.,............	2.5 "
Living-rooms............	2 to 4 "

The number and size of lights required to give sufficient illumination will vary considerably, being affected by the arrangement of the lights and by reflection from walls, ceiling and objects in room. If the sources of light are in view while one is reading or working, stronger illumination is required than if the lights are out of sight.

For obtaining even illumination, one of the first requisites is that the sources of light shall be out of sight. Unless this is secured, the iris closes so as to accommodate the eye to the strongest light in view, and, consequently, less total light comes

*See paper by Richards in *London Electrical Review*, Vol. XXIX., p. 269, Sept. 4, 1891; also book by Webber on Science and Practice of Lighting, p. 25.

to the eye from the objects of vision. The field of vision is a cone which has the eye for its apex and the outer edges of which make angles with the center line, varying between 50° and 95°. As the eye changes position frequently, the sources of light should therefore be removed outside of a cone considerably larger than the angle of vision; otherwise the eye would be continually accommodating itself to widely different intensities of illumination and would soon tire.

It is too common to have public halls and churches lighted in such a way that one can hardly see the speaker or performer without squinting between or under lights that blind rather than illuminate. A common source of difficulty is that the lights are too low. They should be high enough to be out of the angle of vision of the majority of spectators. Doubtless the drowsiness that regularly creeps over some evening audiences is due quite as much to the brilliancy of the lights as to the dullness of the speaker.

Electric lights are peculiarly adapted to being placed high out of the angle of vision. Incandescent lamps may be studded around the ceiling or along the edges. Recently it has become more or less common to place incandescent lamps behind a translucent cove, or above projecting cornices, so that the lamps themselves are entirely invisible. Another plan, used to some extent with both arc and incandescent lamps, is to have reflecting screens beneath the lamps so as to throw all of the light upward against the whitened ceiling, which becomes a secondary source of light. When arc lamps are used in this way, the lower carbon is made positive, so that the crater throws light directly upon the ceiling. Lamps thus arranged give an evenly diffused daylight effect that is very pleasing, if made strong enough.*

A second requisite for even illumination is to avoid the regular reflection of light to the eye. All bodies reflect light, but with varying intensity. Polished surfaces usually reflect regularly, while rough surfaces reflect irregularly or, in other words, diffuse the light. If one is looking at a polished surface, it is desirable that the source of light be so placed that it cannot be regularly reflected to the eye from the observed surface. Otherwise the reflected source of light comes within the angle

---

*See paper by B. A. Dobson on Artificial Lighting of Workshops, *London Electrician*, Oct. 27 and Nov. 3, 1893; *N. Y. Elec. Eng.*, Vol. XVI., pp. 513 and 548; also, Dobson in *Cassier's Magazine*, Vol. V., page 417.

of vision. Hence the familiar rule to have the direct light come from over one's shoulder.

This suggests a third desirable condition, namely, that the light come from several sources or from one source of large area. By spreading the source of light over a large area, the regularly reflected light from each element becomes less intense and the glare is correspondingly reduced. For this reason arc lamps for interior lighting often give much better satisfaction when surrounded by opal globes. Although the opal or ground glass globes cut off about half of the total light, yet the apparent source of light becomes many times larger and regular reflection is greatly reduced. Also, if the lamp comes within the angle of vision, the intensity of the source of light is greatly reduced, so that even with less total illumination, objects are more clearly and easily seen. For the same reason, Welsbach incandescent gas lamps are much more comfortable for reading if they are surrounded by diffusing globes. Incandescent electric lamps also give a softer light, when close to one's work, if they are in porcelain or opal globes.

A fourth condition of satisfactory illumination is not to have too great variations in the illumination of different parts of the same area. The human eye was made for long-distance vision*, and, when used for close work, such as reading, it is necessary to rest it occasionally by brief glances toward more distant objects. If the illumination at a distance is far less than that close by, the eye must constantly change the adaptation of the iris and so quickly tire. For this reason, reading rooms and work-shops should have a good general light in addition to the special lights for individual desks or tools.

When illumination appearance is desired as well as illumination, two plans may be followed. Have the lights bright and arranged in striking positions, but so as to be out of the angle of vision, for instance, ceiling lights or arc lamps on sidewalk. Or, if the lamps are necessarily within the angle of vision, have numbers of dim or diffused sources of light, so that no intense light may come from any particular spots.

The foregoing are a few of the principles to be observed in securing satisfactory artificial lighting. Electric lights have

---

*See paper on Teleopsis, *Denisons' Quarterly*, Vol. II., page 41, 1894.

great advantages over other illuminants*, and render possible certain styles of lighting that are greatly in advance of earlier methods. Further suggestions for applying the principles above noted may be found in the excellent treatises of Palaz† and Webber‡.

*See paper by Geo. D. Shepardson on Some Advantages of Electric Light, read before Minnesota chapter of American Institute of Architects; *Improvement Bulletin*, Vol. 6, No. 20, April 17, 1896.

†Palaz, Industrial Photometry; especially Chap. 6 on Distribution and Measurement of Illumination.

‡Webber, Science and Practice of Lighting.

ELEMENTS OF METHODS OF METAL MINING, BASED UPON LAKE SUPERIOR PRACTICE.

BY PROFESSOR F. W. DENTON.

Mining operations may be divided in a broad and general way into two classes. The first, technically called "prospecting," has as its object the discovery of a marketable deposit of mineral, and includes the determination of the size, shape, quality, and other characteristics of the deposit. The second class of operations is devoted to the removal of the deposit from its position in the earth to the surface, where it passes from the hands of the miner to those of the ore-dresser, metallurgist or salesman. It is only with a part of the second class of operations that this paper has to do.

In order to lift the mineral to the surface, it is first necessary to establish ways of communication between the surface and the deposit. These are technically termed "shafts" and "adits,"* and as they may be arranged in several ways, are expensive to establish, and must be maintained until the deposit is exhausted, their location will depend upon a variety of considerations. This first step in the removal of the mineral to the surface is of such importance as to usually require a special study. It is called in the text-books the "Winning of the Deposit" or the "Preparatory Work." To obtain a large daily output, it is necessary to make connections with the deposits at numerous points. This is done by making side or branch connections with the deposit from one or more main lines of communication with the surface. Additional points from which the deposit may be attacked are established by extending these branch connections into and through the deposit itself.

As a result of the preparatory work the deposit becomes divided into a series of stories or blocks. If the deposit be narrow, as it usually is in native metal mines, there will be one main drift† for each story; but if wide, as in the case of some

---
*Shafts are vertical or steeply inclined openings. Adits are horizontal openings connecting directly with the surface.

†Main drifts are the horizontal openings made at the bottom of each story and used as main roads to connect with the shafts and adits.

iron ore deposits, or if in the form of large beds as on the Mesabi iron range, there will be a number of main drifts parallel or at various angles with one another at the bottom of each block or story. When connection with the surface has been established and the deposit has been penetrated by one or more series of these main drifts, the actual work of removal begins and is called "the exploitation" of the deposit, and the system of exploitation followed is commonly known as the "Method" or "System of Mining." It is the object of this article to describe this part of the work of the removal of the deposit, and the considerations which should influence the establishment of a method of mining.

Every method of mining must provide for the following operations:

1st. Breaking the ore, which includes the drilling and blasting.

2nd. Filling or maintaining the cavities formed by the removal of the ore.

3d. Transporting the ore to a shaft or some other connection with the surface.

The order in which these operations have been named indicates their relative importance and also their relative cost.

BREAKING ORE.

This is usually accomplished by drilling holes in the solid ore, into which dynamite or other explosives is placed and fired. The efficiency of the operation is measured by the number of tons of ore satisfactorily broken per dollar expended for the labor, drilling, and explosives necessary. This efficiency will be a minimum when there is but one "free face", and therefore only one direction in which the force of the explosive can act, and when that direction is upwards. The conditions of minimum efficiency for any given material occur in sinking vertical shafts. The conditions of maximum efficiency are several "free faces", large blasts, and an opportunity for the force of gravity to have its fullest effect. Intermediate conditions will give intermediate efficiencies.

The blocks or stories formed by the preparatory work are always mined in descending order; that is, the top block will be completely removed before the second, and the second before the third, etc. The work of removal may be going on simultaneously in several stories, but the top story will always be

Methods of Metal Mining. 51

the most nearly mined out. This is a natural order, since work will begin first at the top of a deposit considered as a whole, and therefore work should end there first. This order is also the best one for the common systems of mining, and is absolutely necessary in those systems which include the complete removal of any individual block of ore, since such removal will destroy the drifts at the bottom of the block next above. The

PLATE I.—View, from the surface, of the Auburn iron mine on the Mesabi range, showing the miners at work "underhand stoping" or blasting the ore into openings (*raises*) which connect with drifts in the ore body, which, in turn, connect with a shaft not appearing. The ore lies near the surface and is first stripped of its covering.

individual blocks, however, may be mined from the bottom up or from the top down.

The first method of breaking ore to be described is termed "stoping"* and the place where stoping is carried on is called a

* By stoping is meant removing ore in horizontal slices which are usually about eight feet thick; either the top or bottom of the slice is a "free face."

"stope." If stoping is begun at the top of a block it is called "underhand stoping" and if it is begun at the bottom, "overhand stoping." In both cases it is necessary to make a vertical cut or opening from which to begin stoping. In overhand stoping, the opening is begun at the bottom of the block and is called a "raise", and may be extended to the level above, in which event it would be called a "winze."* Overhand stopes are seldom carried at once to the top of the block, but a layer of ore from five to fifteen feet deep is left to serve as a floor for the level† above, or to assist in keeping the walls of the vein apart, or for both purposes. This layer is known as the "floor pillar." Small openings are blasted through these floor pillars at intervals to secure ventilation in the stope. These openings may also be used as a means of getting to and from the stope, and as passages for air pipes, etc. The ore left in the floor pillars may be subsequently wholly or partly stoped, or may be abandoned entirely.

In underhand stopes, the vertical opening from which stoping begins may be driven from the top down, in which case it is said to have been "sunk", or it may be raised as in overhand stoping. Raising would be practiced, if possible, as it is cheaper than sinking, since gravity favors in one case and opposes in the other. In any event the raise or sink will usually become a winze to permit sending the broken ore to the lower level and thus save the extra handling necessary to get the broken ore into cars on the upper level.

Of the two methods of stoping, overhand stoping will give the greater efficiency, since gravity has full play. Another important difference is, that in overhand work the broken ore may be left in the stope to keep the walls from caving and also to serve as a support for the miners and drilling machines. In underhand stoping the ore must be moved as fast as it is broken, as otherwise the solid ore would soon be covered by the broken ore. This results in the formation of a large cavity above the miners, which is always objectionable. Such a place cannot be examined readily for loose pieces of hanging wall rock which are always the most common source of accidents. If the upper level is to be used a timber floor must be put in. Underhand stoping, therefore, is used only to a very small extent. It is employed for removing the floor and other pillars left by the

---
*A winze is a vertical or inclined opening connecting two main drifts.
†"Level" is synonymous with main drift.

first overhand stope and in open-pit mining. Nearly all stoping is carried on overhand.

Plate I is a surface view of the Auburn mine of the Minnesota Iron Company on the Mesabi range. The deposit is first stripped of its covering and then by underhand stoping blasted and allowed to run into raises placed sixty to eighty feet apart. The raises are about sixty feet deep and connect with drifts

PLATE II.—View in a drift at the top of a raise in the East End mine of the Pittsburg & Lake Angeline Iron Mining Company at Ishpeming, Michigan. A layer of ore has been left above the drift, and the miners are preparing to blast it down. At the right is shown solid ore, and at the back the caved ground which follows the miners down as they remove the ore.

which in turn connect with an inclined double skip shaft. This method of mining is known locally as the "milling" system.

The second method of breaking ore is known as "drifting," and consists in "drifting" or driving comparatively small horizontal openings into the ore. The openings formed are termed "drifts." The first drifts made in a block of ore have but one free face and since the opening is horizontal, gravity is almost neutral, neither assisting nor retarding the work. Therefore, for any given ore, the efficiency of drifting will be less than that of stoping, and greater than that of sinking. If a second drift be made adjacent to the first, there will be two free faces in this second drift, and other conditions remaining the same, the efficiency of the second drift will be considerably greater than that of the first. Drifting with one side of the drift free is termed "slicing." If directly under a series of drifts, no ore being left between, other drifts be run, these lower drifts will have the top free and some of them one side as well. In the last case the efficiency will be further increased to a small extent. Where the drifting method is used in breaking a block of ore all of the preceding conditions are met with and the efficiency of the whole work will depend upon the relative amounts of favorable and unfavorable conditions.

A third and last method of breaking ore is termed "caving."

The principle in caving is to undercut or undermine a body of ore until, no longer able to support itself, it falls or caves, and in falling becomes broken up, thus accomplishing the results ordinarily obtained by drilling and blasting. The undercutting is usually done by driving a series of drifts in the bottom of the block to be caved, the drifts being adjacent to one another, or with small pillars of ore between. When undercutting has progressed far enough to cause the overlying ore to show signs of settling, the miners will be taken away and the ore left to fall as it will, or more commonly the timber and pillars will be blasted down to hasten the caving. The caving method therefore, involves the use of stoping or drifting for the undercutting. In practice both may be used, although caving is usually combined with drifting. Its efficiency will be a maximum for a particular ore when the drifting or undercutting necessary is a minimum.

A modification of this method is much used in mining the so-called "soft" hematite ores. Instead of undercutting a large block of ore until it falls by its own weight, a small block is only partially undercut, and the overlying ore is blasted down by holes drilled into it from the drifts below. Gravity is thus given an opportunity to assist in breaking the upper portion of the block. Plate II is a view of such work and shows the miners in the act of drilling an "upper" into a layer of ore that has been left above the drift in which they are standing. This modification permits of more regular and systematic work, since the ore is brought down in small quantities at a time and the fall of the ore is more under the control of the miners.

Methods of Metal Mining.   55

It is less efficient than the first method, however, and is only applicable to shallow blocks. Probably not more than ten feet of ore over the drifts can be advantageously mined in this way.

Caving hard ground, in order to break it up and thus avoid the expense of drilling and blasting or filling the cavities formed by the removal of ore, has reached a high state of development in the Lake Superior region. This method has saved many tons of good ore which otherwise would have been lost, owing to the greater cost of removing it by the older methods. It is applicable even to the hardest ores under favorable conditions. Unless well applied however it may cause the loss of considerable ore. Such loss is usually caused by the upper portion of the caved ore becoming mixed with the overlying earth, sand, or barren rock, which, of course, settles or falls with the ore. When the block of ore to be caved is thick, the overlying rock weak, and there is nothing between the rock and the ore to separate them, the loss from this cause may be very large. If, however, by previous work, a mattress of crushed and broken timber has been formed between the top of the block and the overlying rock, the loss of ore from mixing may be no greater than in other methods of breaking.

Of the three methods of breaking ore, stoping, drifting and caving, overhand stoping will give the greatest efficiency if we consider only the breaking of the ore. This is especially true in the case of very hard ores which are difficult to break, and with such ores the method of breaking is the most important factor in the system of mining. The greater efficiency of stoping becomes less marked as the ore becomes softer, and finally stoping ceases to be applicable to very soft ores, which either have a tendency to run, or under which it would be dangerous for men to work. In soft ores, therefore, drifting is used, and becomes the only practicable way of removing the ore.

For ores intermediate between very hard and soft, the breaking may be done by any of the three methods, and it is in dealing with ores of this class that the greatest skill and experience are called for. In such ores the superior efficiency of stoping is not so marked that the breaking of the ore is always the most important factor in the system of mining, and therefore efficiency in breaking may be sacrificed to increase the efficiency somewhere else. The efficiency of caving is intermediate between that of stoping and drifting, and caving is used chiefly to reduce the cost of breaking ore when the conditions are unfavorable for stoping, and the ore hard and therefore expensive to drift in.

FILLING AND MAINTAINING CAVITIES.

The removal of ore produces cavities, or chambers, in the deposit which must be taken care of in some way, for, if allowed to increase indefinitely, Nature will surely fill them eventually; perhaps by sudden and extensive caving, which would be both dangerous and costly.

Cavities are filled naturally in one of two ways: Either a

PLATE III.—View of square set timbering cut by machinery and formerly used in the Lake mine of the Cleveland Iron Mining Company at Ishpeming, Michigan.

large mass of the overlying ground settles, filling the cavity completely, or comparatively small pieces of the roof and walls fall at intervals and, by accumulating, eventually fill the cavity. As soft and hard ores may occur, surrounded by either hard or soft ground, the treatment of cavities formed by the removal of ore may become a difficult and important matter.

In dealing with cavities in any kind of material, the object should be to make them as self-maintaining as possible. This is accomplished by leaving sufficiently solid side walls and arching the back or roof. There is a large class of soft grounds in

Methods of Metal Mining. 57

which, if a chamber with a flat roof be excavated, the roof will fall piece by piece until it has reached an arched form, and then remain intact for an indefinite period. In a similar manner it is possible that a wide drift with a flat roof may be run in a soft ore-body with safety, and yet after the drift is timbered, a considerable weight may be brought upon the timber by the settling of the material below a natural arch in the roof of the drift.

Ground is strongest usually when freshly exposed, which

PLATE IV.—View of square set timbering employed in the Calumet and Hecla mines, Michigan. The original negative was made by Mr. J. M. Vickers, of Ishpeming, who has kindly consented to its reproduction.

explains why flat roofs will stand for a while and then fall. If soft ground therefore, is to be left unsupported for a considerable time after it has first been exposed, the roof should be arched. Even when timbering follows closely after mining, it may often be advantageous to arch the roof to avoid pressure upon the timber. It is by the formation of these natural arches, combined with the cohesive strength of the ground, that all large cavities are maintained.

A simple calculation will show the impossibility of supporting ground by timber alone at depths of several hundred feet below the surface. The maintenance of cavities is no more difficult at depths of several thousand feet than at depths of three or four hundred feet. In fact, the greater difficulty is likely to be experienced near the surface, owing to the presence of water and lack of homogeneity of the ground. The limits of cavities in any given material will be fixed, therefore, by the self-sustaining qualities of the ore and the surrounding ground, and when these limits are reached the cavity must be caved by the blasting of the roof, filled by introducing broken rock and sand, or simply abandoned. When the conditions render caving practicable, this method of dealing with cavities will usually be followed as being the cheapest. Caving large chambers, also called "rooms," is, however, more or less uncertain in its results. When practiced, the caving is accomplished by blasting out the timbering of the room a short distance above the floor, if the timber shows considerable weight, and if it does not, by blasting both timber and roof. If the cavity is filled with broken rock or sand, the filling is obtained as cheaply as possible, and is introduced with a minimum amount of handling.

The filling material may be obtained from the surface, or it may come from some point in the mine where the rock is soft and easy to blast. If the hanging wall be easy to break, raises may be carried up into it dipping 45° with the horizontal and located at small intervals along the strike, and from these raises rock may be blasted and allowed to run into the adjoining cavity. At a mine in Michigan, sand filling is obtained on the surface close to the shaft and carried by an endless rope system of haulage to the top of the shaft where it is dumped into a large iron pipe which delivers it at the different levels in the mine, whence it is trammed to a raise connecting with the cavity to be filled. If the conditions allow the filling material to be obtained directly at the top of the raises leading to the places to be filled, the filling method of dealing with cavities may be the cheapest. If, however, this favorable condition should not exist, or if it be necessary to pack the filling close to the top of the cavity, or for any other reason to handle it extensively, this method may become too expensive to be used, and if caving be

Methods of Metal Mining. 59

impracticable the cavity must simply be abandoned as soon as the natural limits are reached. If abandoned, the walls, which are usually composed of the ore or valuable mineral, must be abandoned also, and unless removed by subsequent work of a different nature the loss from this source may reach as high as sixty per cent. of the total mineral. Thus far it has been assumed that only large masses were to be dealt with. It may also be necessary to support small pieces that have become detached from the roof and walls by blasting, movement of the ground, or weathering. This kind of support is given chiefly by various styles of timbering, the functions of which may be classed under the following heads:

First, to protect men and to maintain openings by keeping back loose pieces of ground.

Second, to act as a staging or support for the miners and their tools.

Third, to give warning of approaching danger from caving.

The style of the timbering will depend upon the special conditions of each mine. The common forms of timbering are known as props, stulls, cribbing, square sets, and drift sets. Props are usually temporary timbers used for holding up isolated pieces of loose ground while the main timbering is being put in, or the ground blasted; they are generally placed vertically and are of light timber. Stulls are large, permanent, carefully fitted timbers that are used to support isolated pieces of ground that have been exposed by the removal of the ore; they always extend from foot to hanging wall, and are placed at angles slightly above the perpendicular to the plane of the dip. If the ground is very heavy these stulls may be placed in groups, called "batteries". In setting up both props and stulls the pressure upon the timber should be uniformly distributed, in order to get the maximum resistance from the timber and avoid splitting. This is accomplished by the use of wedges, which are tightly driven between the ends of the timber and the rock. If well set up, stulls will not fail by splitting or falling. If a battery of stulls is not sufficient to keep the ground up, cribbing may be used. Cribbing consists of timber from 6 to 12 inches in diameter and 6 to 20 feet long, piled log-cabin fashion. The pile is made at right angles to the plane of the dip, and if the dip is steep the timbers are slightly notched or spiked, to keep them in place until the pressure comes upon them. This support is made many times stronger by filling the interior of

60   F. W. Denton.

the pile with broken rock, which converts it into an artificial pillar. If, when filled, cribbing will not hold the ground, the case comes under the conditions of the first part of the discussion of cavities and must be handled accordingly.

When the vein becomes too wide to permit the use of timbers extending from wall to wall, or when the whole roof needs support, a different style of timbering is used and is known as

PLATE V.—The view is taken at the end of the ninth level of the Minnesota hard hematite mine on the Vermilion range. The first "drift stope", 18 feet high and the full width of the deposit, has been made, and drift timbering is being erected to maintain an opening through the filling which will precede the taking off of another stope or slice.

"square sets". Square sets are of two general classes; one having the main timbers always horizontal and vertical, the other having them parallel and at right angles to the plane of the dip. Plate III shows the first class. Horizontal timbers are either called "caps", if they are all alike, or "caps" and "studdles" if of two kinds, and the vertical timbers are called "legs." Plate IV shows the second class. In this the nomenclature is not so simple since there is a greater variety of pieces which require many names to distinguish them, and such names will generally have a local significance only. In the first class of square sets the practice on Lake Superior is to leave the timbers round. The joints are generally cut by hand although one very complete mill has been erected for cutting timbers like that shown in Plate III. The style of joint varies at each mine, the chief difference being in the matter of tenons. Some have tenons on both ends of the legs, some on the bottom or top only, and some have no tenons at all. In the one example of the second class of square sets that exists on Lake Superior shown in Plate IV, the timber is sawed to 12 x 12 inches. There is no hard and fast rule for deciding which class of square sets should be used. Two considerations would probably influence a decision; first, the probable direction of the maximum pressure upon the timber, and second the ease of fitting the timber to the cavities formed. It is impossible to predict exactly in what direction the maximum pressure will come, but if we have a hard deposit 20 to 40 feet thick and dipping at an angle of 45° or less, with a weak hanging wall, we can safely say that there would be considerable side pressure upon a vertical system of timbering.

To fit the first class of square set timber between parallel walls dipping at 45° or less and only twenty to forty feet apart, a great deal of special cutting and fitting will be necessary to make a good contact between the walls and the timber. On the other hand, if the second class of timber is used for such conditions, the fitting will consist of short props and blocking inserted between the limits of the regular system of timber and the walls. When the walls are not well defined or are not parallel, and the cavity is wide, and both top and side pressure are to be expected, the first form of square sets will be employed with the addition of diagonal bracing placed to oppose the side pressure.

Drift sets, as the name implies, are used to support ground in drifts and usually consist of three pieces, two legs and a cap. Formerly, a fourth piece, known as a "sill" or "mud sill," was used, upon which the legs of a set rested, to prevent them from sinking into the bottom of the drift. At present, however, sills have practically gone out of use, except in cases of very soft ground, it having been found that legs sawed square at the bottom stand well enough. The sills were placed either across the drift, the two legs of any one set resting on the same sill, or parallel to the sides of the drift. The method of placing the sills depended upon the method of supporting the tram rails and of timbering rooms opened out from the sides of the drift. The joints of drift sets are designed to suit the pressure expected, which may come from the top or side, or from both directions.

For main drifts or levels, which must be maintained for a long time, the largest timber available is often used, the diameter varying from ten to thirty inches. Plate V shows main drift timbering. For smaller and secondary drifts, often called sub-drifts, and for drifts used to extract ore only, the timber seldom exceeds nine or ten inches in diameter. The pressure upon the caps of the main drift sets may often be so great that considerable timber has to be replaced before the drifts can be abandoned. It would seem that for such cases the use of iron or steel I beams, or of common rails, would be advantageous. These iron beams will support a great deal more pressure than the timber usually available, without consuming so much head room. Such iron and steel beams have been used in foreign countries with apparent success. In all kinds of timbering small sticks of timber called "lagging" may be placed between the heavy timbers and the ground to distribute the pressure, or to support small, loose pieces, that otherwise might fall. Lagging consists either of natural round poles, four to ten inches in diameter at the butt, or of such poles split into halves, and extends from one timber to the next.

The first object of timbering, to protect the men and maintain the cavities, is accomplished by some of the ways which have just been briefly described. It may be made to serve as a staging for the miners and their tools by the addition of a few planks laid across the timbers or spiked to them, and in this way its second object is accomplished. The third object, to give warning of approaching heavy caving, is attained by erecting the timber in such a manner that it will show any movement in the surrounding ground. To do this there must be as many points of contact between the timbering and the surrounding ground as possible. This contact can be best obtained by the liberal use of blocking between the lagging, or the main timbers, and the ground. If large cavities occur back of the timber they should be promptly filled with cordwood or similar material. Ground rarely falls or caves without giving some warning by cracking, moving, or spalling, but such warning can easily pass unnoticed unless the timbering is affected. Well blocked timber will not only show by incipient crushing or cracking that the

Methods of Metal Mining. 63

pressure upon it is increasing, but will also resist such increased pressure longer than loosely erected timber.

Dimensions of timber naturally vary with the character of the timbering, weight to be borne and available supply. Props are usually small, 8 to 12 inches in diameter, and 3 to 10 feet long, and of any kind of timber. Stulls may be from 8 to 40 inches in diameter and 4 to 20 feet long. When the largest diam-

PLATE VI.—The view is taken in the Minnesota mine, on the Vermilion range, and shows the method of loading cars in the main drift at the bottom of the filling. These loading chutes are established at intervals of 25 to 40 feet.

eters are needed, the weight of the stull becomes an important factor and limits the kinds of timber to be used to those of low specific gravity, usually to white pine. In square set timbering the diameter of the timber will be from 10 to 30 inches, with an average of about 14 to 16 inches. As previously stated, the inclined square sets described are made of sawed lumber 12x12 inches. The most important dimensions in square set timber of any kind are the lengths of caps and legs. These regulate the size of a set of timber and the size determines the number of tons of ore procured per set of timber, and therefore the cost per ton for timber. Sets are generally cubical in form, and the length of a side, center to center of timbers, varies from 5 to 8 feet. Five-foot sets will displace 125 cubic feet each, or 7 tons, allowing 18 cubic feet per ton. An eight-foot set, making the same assumptions, would displace about 28 tons with the same number of joints to be cut and only a small amount of additional timber. The cost per ton for timber of this description therefore will decrease quite rapidly as the size increases. Large sets require ground that will stand well while the timber is being erected, and are more difficult to handle than the small sets. Economy, however, has encouraged their use and the modern tendency, although against the square set system of timbering as a whole, favors the large sets where the system is still employed. If properly laid out and executed, square sets are reliable and economical especially for the so-called "soft" hematites which are expensive to break in any way except by overhand stoping. Cases of sudden and complete collapses of this kind of timbering have given it a reputation for treachery which is not altogether warranted. Theoretical considerations tell us that such timber must be carefully erected, and that it is strongest when in perfect alignment. A pressure that is great enough to move the timber a small amount out of line will usually be sufficient to effect its complete collapse eventually.

The expense of cutting tenons upon the ends of the legs has caused tenons to be abandoned at some mines. Whether tenons are really necessary or not is difficult to decide, but in the writer's opinion the strength and reliability of square sets are largely dependent upon the stiffness and accuracy of the joints, and stiffness is certainly increased by tenons on the legs. Whatever the form of the joint, however, it should be accurately cut and the timber erected with care and well blocked against solid ground on all sides.

The handling of timber about a mine is a detail of considerable importance. The fundamental rule is, never to hoist timber unless absolutely necessary, but if possible connect the place to be timbered with the level above and send the timber down from this level to its final position. This is not only the cheapest method of handling the timber when it is practicable, but it is the most convenient as the ore mined is sent to the level below and the two do not interfere.

There is a special method of maintaining cavities which is used as a substitute for timbering and which should be mentioned in this connection. If the vein is hard and of low grade, and the hanging wall weak, a layer of from 2 to 5 feet of the vein may be left attached to the hanging wall to support it, and thus save the cost of timbering. This method is practiced in some of the low grade copper bearing conglomerates of Michigan, which are invariably associated with weak hanging walls. The adoption of the system, of course, will depend simply upon the relation of the cost of timbering to the value of the mineral left.

TRANSPORTATION OF ORE TO THE SHAFT.

The general principles in handling ore are to hoist it but once and then only in a regularly equipped shaft, and to avoid shoveling as much as possible. To comply with these rules ore broken between any two levels must be sent down to the lower level and thence to the shaft with as little handling as possible, gravity being utilized to the fullest extent.

The method of transportation, it is evident, will depend upon the dip of the deposit and the method of breaking the ore. When overhand stoping is used and the dip is steep enough, the ore is allowed to fall to the lower level where it is loaded into cars and trammed to the shaft. The loading will be cheapest if the broken ore is stopped just above the level of the top of the cars and then by means of chutes run into the cars by gravity. The conditions which usually prevent such an arrangement are too flat a dip and pieces of ore too large to be dropped into the cars without injuring the latter. A flat dip is sometimes impossible to overcome, but modern practice seems to favor overcoming the second difficulty by building cars of special construction to withstand the great shock of the falling pieces.

As several stopes may be opened from one level it is usually necessary to maintain a clear road under the several stopes. This is done by the use of timbers. If the vein be narrow enough and the walls good, a row of stull timbers will be used backed by lagging; if wide, drift sets. Plate V shows the timbering used to maintain an opening under the filling in the Minnesota iron mine. Blasting may be safely carried on above such timber if the timber be kept covered with a few feet (5 to 15) of broken ore or rock. This broken ore acts as a cushion and absorbs the shocks of the falling pieces, keeping the timber intact and making it possible to break and tram ore from a given place at the same time. If the dip be too flat or other conditions unfavorable for the establishment of loading chutes at regular intervals, a platform may be erected along the side of the drift at the level of the top of the cars, and the broken ore allowed to run out upon this platform, from which it may be shoveled or rolled into the cars.

When the ore is broken from the top down, as in the drifting method, it is necessary to transport the ore from the end or "breast" of the drift, where it has been broken, to the top of some raise or chute leading to the level below. For such transportation wheelbarrows are commonly used and occasionally small and light tram-cars holding from one-half to one ton. It is perhaps unnecessary to state that this kind of transportation should be a minimum and, if possible, the raises or chutes should be near enough together to permit shoveling the broken ore directly into them.

The cars used for transporting ore, as already indicated, are designed to suit the method of loading and the character of the ore. The material used in car construction is of high grade, and more attention is now being paid to car construction than heretofore. Steel plate, varying in thickness from $3/16$ to $1/2$ of an inch, is being used, the bottom being made of plate $1/8$ of an inch thicker than that of the sides. If the cars are to be loaded with hard ore from chutes, the truck is usually made up of two forged axles and two heavy longitudinal timbers, which are intended to reinforce the bottom plate and absorb the shock. Also, the box of the car may in such cases be made shallow, in order to lessen the free fall of the ore. If the ore be allowed to roll out on the floor of the level, it must be lifted into the ore cars, and in such cases the cars will be as low as possible, and either open at both ends or provided with doors, one of which at least will be hinged at its lower side and thus open downwards, permitting its use as an inclined plane to assist in loading large pieces. In soft ore mines, and in hard ore mines where the loading is not done from chutes, cars made entirely of metal are used. Both inside and outside bearings are used, and generally self-oiling wheels.

To further diminish the car resistance, wheels of large diameter are employed, 14 and 16-inch wheels being used. More attention is being paid in the Lake Superior district to tram-cars and tramming than ever before, and a tramming cost higher than five to seven cents per ton is considered warranted only by very exceptional circumstances.

Mechanical transportation is represented by electric motors, compressed air motors, and endless rope haulage. Mules are also employed in the iron mines of Minnesota. The independence of the compressed air motor gives it a decided advantage in metal mines where a block of ground may be mined out in a short time, or where it may be difficult and expensive to install or keep running a rope or electric system.

At the shaft the ore may remain in the cars and be hoisted directly to the surface by cages, or the ore may be dumped into pockets at the shaft and subsequently loaded into skips by gravity. The shaft-pocket-and-skip method is a favorite one among the iron mines and seems to be growing in popularity. The method makes tramming and hoisting independent of each other, at least for a short time, and permits each to be arranged to the best advantage.

It will be seen from this brief and general description of the three main operations included under "Methods of Mining," that when a special deposit is under consideration, the adoption of a method of mining it may be either a very simple matter or a complicated one. Conditions favorable to cheap breaking may be very expensive to maintain or may be directly opposed to cheap work elsewhere, and so a method has generally to be adopted which is a compromise between conflicting conditions. The efficiency of any method as a whole must finally be determined by the cost of a ton of ore delivered at the shaft.

NOTES ON THE DESIGN AND MANUFACTURE OF DYNAMO ELECTRIC MACHINERY.

BY C. H. CHALMERS, '94.

It is the purpose of this article to treat briefly of a few points in the design and manufacture of dynamo electric machinery that have come to the notice of the writer. Particular attention will be paid to considerations which the technical press and text books regard as of minor importance.

No logical order or arrangement will be attempted, the different items being discussed without reference to each other.

ARMATURE DISCS.

These discs are usually made from soft charcoal iron or a mild steel and generally run about fifty to the inch. Little or nothing seems to be gained by making them thinner, while on the other hand, if this thickness is exceeded, the losses from foucault currents cause excessive heating. Much has been said about insulating the discs from each other by means of paper, varnish, etc. It is now the practice of the leading American manufacturers to build up their armature cores without any other insulation than the oxide which is on the discs. The writer found his company using paper insulation on all their cores when he took charge of the design and testing of the apparatus. In order to be absolutely certain he had two ten H. P. armatures built, one with paper and the other without. These, before being wound, he ran for several hours in a strong field and noted carefully the rise in temperature of each. He found no difference whatever. They have used no paper in their cores since, and have noted no change in the action of the armatures.

Armature discs are sometimes punched and sometimes cut out with a set of tinners' circular shears. Unless the punch is kept in first class order, the edges of the disc are apt to be frayed and rough, while the shears make a clean cut every time. These shears are much less expensive than the punches, and, when once adjusted for a particular disc, can be operated by cheap labor.

Dynamo Design and Manufacture. 69

For small armatures, the discs may be cut out in squares and turned off in the lathe. A side cutting tool can be used so that two or three cuts will finish the armature. The finishing cut should be a thin one, to avoid burring the discs together. Toothed armatures are made in a variety of ways. The larger factories use punches, while the smaller ones make the slots with a shaper, planer, or milling machine.

HEADS OF ARMATURES.

Closely allied to the subject of armature discs is that of the heavy discs or heads which are usually put on each end of the armature to give it strength. For this question there are three solutions; I. A heavy disc of mild steel, wrought iron, or cast iron; II. A similar one of brass, gun metal, or some other non-magnetic metal; III. The use of no head at all. The use of the iron disc lowers the magnetic density in the armature and in this way tends to decrease the losses due to hysteresis and foucault currents; on the other hand, being thick, it is in itself subject to heavy parasitic currents and consequent loss. A little reflection will reveal the fact that these losses will be principally due to eddies.

The permeability of the iron disc increases the voltage available for causing eddies, while its ohmic resistance tends to reduce the currents which naturally flow. Taking the armature head by itself, it is evident that the lower the permeability and the higher the ohmic resistance, the less will be the heating of the head.

The loss due to eddies is:

(1) $$W = \frac{E^2}{R}$$

Where W equals Watts, E equals volts, and R equals ohms. But E in a given case varies directly with the permeability, so that

(2) $$W \propto \frac{U^2}{R}$$

Where U equals permeability. Equation (2) shows the great advantage of the brass or gun-metal head.

The permeability is reduced from the square of a very considerable quantity to unity, while the ohmic resistance is increased an amount which is negligible when compared with the gain due to the decreased permeability.

These considerations have led to the adoption of brass heads on many machines, and to a subsequent decrease in the

70   C. H. Chalmers.

heating of the armature. It has long been recognized that an armature without heavy discs at the ends, would be the best and only true solution of this question, but it is only recently that manufacturers have built their armatures without heads or heavy discs at the ends. Of course only ring type armatures are built in this latter way.

RELATIVE LENGTH OF FIELD AND ARMATURE.

As machines are built nowadays, every little detail that will tend to make the machine run cooler is worthy the attention of the designer. In well designed machines the heating, rather than sparking, is the limiting factor in the load, which the machine will safely carry. The armature must be laminated in planes that are parallel to the induction, for by so doing the interstices are placed directly in the path of the foucault currents. An examination of Figs. 1 and 2 will show the effect of the relative length of pole piece and armature core, so far as concerns the magnetic induction. In Fig. 1, the lines of force are shown as entering the end of the armature in a direction normal to the lamination. This will evidently be a source of loss from foucault currents, which may be materially decreased by the construction in Fig. 2. Fig. 2 has the further advantage of increasing the cross section of the air-gap with a consequent decrease of the energy needed for exciting the field. On the other hand, Fig. 1 has an advantage in that the conductors on the surface of the armatures are shorter and the C^2R loss less. There seems to be no definite rule among authorities regarding the ratio of length of armature

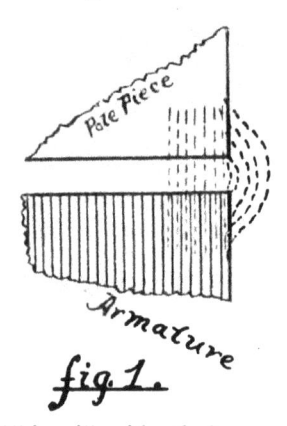

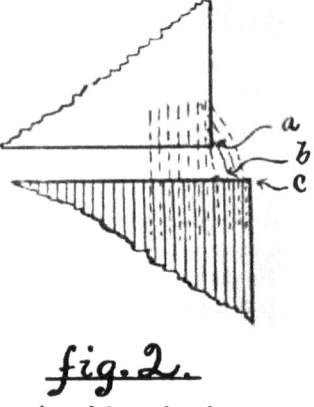

, and pole piece, but it would seem a rational solution of the question to make *ab* equal *cb*, see Fig. 2, or at least nearly so.

ARMATURE SPIDERS.

The mechanical design of armature spiders is usually considered to be identical with that of an ordinary fly wheel. Large factors of safety are used so as to have ample strength in case of a short circuit on the machine. The revolving armature is treated as the rim, and the spider as the hub and spokes, of the wheel.

The enormous kinetic energy which must be contended with in case of a heavy short circuit is given as the reason for the large factors of safety generally used in spider designs. A careful study of the conditions in cases of the above nature will show that this method of procedure in the mechanical design of armature spiders is radically wrong. If we assume that the armature is brought to a dead stop in a given time, we can compute the kinetic energy by the formula:

$$K = \tfrac{1}{2} M V^2,$$

where K equals foot pounds, M equals mass, and V equals the velocity in feet per second.

The first impulse is to use M as meaning the mass of the revolving core and conductors, but this is just where the mistake lies. The mass to be considered is *not that of the armature*, but that of the shaft, pulley, belting, and moving parts of the engine.

The major portion will of course be the fly wheels of the engine, and this will suggest that an engine with heavy fly wheels and moving parts will be more severe on a generator in case of a short-circuit than one where corresponding parts are light. Of course, so far as concerns the strength needed to transmit the strain from the shaft to the core in ordinary running at full load, or considerably beyond it, is concerned, the mechanics of the cantilever is of prime importance.

STRENGTH OF CONCRETE AND STEEL IN COMBINATION.

BY PROFESSOR FRANK H. CONSTANT.

Within the past few years the use of concrete and steel in combination has become quite common.

We see it first coming into extended use with the advent of the modern steel foundation for tall buildings; we find it applied to the solid floors of bridges and buildings; and later we see it appearing as the principal element of certain kinds of arch bridge constructions known as the Monier and Melan Systems.

By concrete-iron construction is meant such cases in which the two materials, by their combined resistance, carry certain specific loads. With the increased use of concrete in combination with steel or iron, a study of the elastic properties of the combination becomes correspondingly important.

Unfortunately, our knowledge of this subject is as yet extremely limited, owing to the few experiments that have been made to determine the elastic behavior of cement or concrete under varying conditions. The tests that have been made upon various types of concrete-iron construction, were conducted for the purpose of determining the ultimate strength of certain forms, rather than as a scientific interpretation of the mechanical laws governing such construction.

Generally, in large and important structures subjected to heavy loads, such as foundations and Melan arch bridges, the steel portion is made sufficiently strong to carry the entire load. In such cases, the concrete serves as an additional factor of safety. The writer would not advocate a change from this custom, for concrete, as it comes from the hands of the ordinary shift of laborers, is too uncertain a quantity upon which to place the safety of a structure. Nevertheless, with proper supervision, excellent concrete may be produced, which may aid materially in the strength of the structure. Especially is this true when the concrete is so placed in the combination as to develop its compressive strength simply. The compressive resistance of concrete being quite large compared with its tensile strength, considerable latitude might be permitted before its usefulness is destroyed.

The strength of two elastic materials in combination depends upon the relative elasticities of the two substances. The coefficient of elasticity of steel is about 30,000,000, while that for ordinary concrete varies from 750,000 to 2,000,000, but which, in what follows, we will take at 1,000,000.

Now the stresses developed in different materials under like conditions, are directly proportional to their coefficients of elasticity. Thus, suppose bars of cement and steel are placed side by side in a testing machine and subjected to the same pull and stretch. Then the actual stress developed in the cement will be but one-thirtieth of that developed in the steel. In other words, the cement will resist one-thirty-first part of the total pull. To make this illustration more clear, substitute for cement a bar of rubber.

Let us first consider the very common condition (occurring in foundation work, and in the Melan Arch) of a beam buried in concrete. (See Fig. 1.) As the formulæ which we are to obtain are to be working formulæ, we will assume that the steel is not strained beyond the elastic limit. Two cases must be considered: a.—When the concrete does not crack, and when the steel does not pass the elastic limit. b.—When the concrete cracks on the tension side, but the steel does not pass the elastic limit.

Should the concrete fail on the compression side also, the condition becomes that of the steel beam acting alone. In each of these cases the controlling fact is that the adhesion between the concrete and the steel is assumed to be sufficient to cause equal distortions in the two materials; that is, the radius of curvature under flexure is, at any point, the same for both materials. In order that this condition may be satisfied, the beams should not be spaced very far apart.

CASE I. (See Fig. 1.)

Let M'' = moment of resistance of the concrete,
M' = moment of resistance of the steel beam,
r = radius of curvature of both, under flexure,
E'' = coefficient of elasticity for concrete,
E' = coefficient of elasticity for beam,
I'' = moment of inertia of concrete,
I' = moment of inertia of beam.

b = width of concrete,
h = depth of concrete.

Then,
$$\frac{1}{r} = \frac{M'}{E'I'} = \frac{M''}{E''I''}$$

$$\frac{M'}{M''} = \frac{E'I'}{E''I''} = \frac{30\,I'}{I''}$$

$$\therefore M'' = \frac{M'I''}{30\,I'} \qquad (1)$$

M' and I' can be obtained from the handbooks of steel manufacturers, and $I'' = \frac{1}{12}bh^3$. The total moment of resistance of the combination is,
M = M' + M'' = external moment.
To find the extreme fibre stress in the concrete:

$$f'' = \frac{M''h}{2\,I''} \quad \text{and} \quad f' = \frac{M'h}{2\,I'}$$

$$f'' = \frac{M''I'f'}{M'I''} = \frac{1}{30}f', \text{ where}$$

f″ = extreme fibre stress in concrete,
f′ = extreme fibre stress in beam.

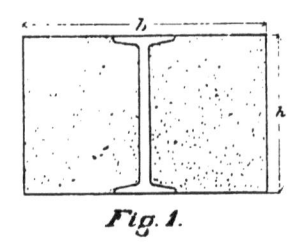

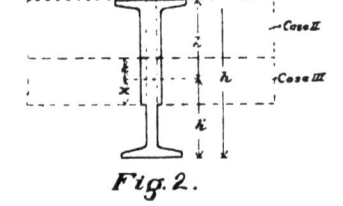

Fig. 1.   *Fig. 2.*

This last result is evident from the fact that fibres in the concrete, and in the steel, situated at equal distances from the the neutral axis, must have equal distortions, and hence the fibre stresses are proportional to the coefficients of elasticity of the two substances. Since the fibre stress in the concrete is at all points but one-thirtieth of the corresponding fibre stress in the steel beam, we might, without changing the total amount of stress in the concrete, have considered the width of the concrete reduced to one-thirtieth of its former width, while the fibre stress at each point is increased thirty-fold so as to be equal to that in the steel beam at the corresponding point. The concrete, which has thus been metamorphosed into a steel beam of one-hirtieth of its former width, may now be considered as part of the steel beam and its properties obtained in the ordinary manner. This method will be used in the solution of the next case.

If the ultimate tensile strength of concrete be taken at 200 pounds per square inch, the corresponding fibre stress in the beam should not exceed 6,000 pounds, in order that the concrete may not crack. Should the beam be subjected to the usual working fibre stress of about 15,000 pounds per square inch, the corresponding fibre stress in the concrete will be 500 pounds, which, while great enough to crack the concrete on the tension side, is well within the limit of its compressive strength.

This first case is of importance in the consideration of such forms of elastic arch construction as the Melan arch, where a reversal of stress is possible. In this case, should the tension, which may occur at different times in either flange, exceed 6000 pounds, cracks may start from both sides and pass entirely through the concrete, thus seriously impairing or destroying its value as one of the elements of strength in the structure.

CASE II. (See Fig. 2.)

In this case we shall assume that the concrete cracks on the tension side, and owing to vibration or other causes, the crack extends to the neutral axis. In other words, we shall assume that none of the concrete is in tension.

The special difficulty in this case is that the neutral axis is now no longer in the center of the beam, but has moved toward the compression side a small distance k. The value of k, however, can readily be found as follows:

As was shown in Case I, the concrete may be replaced by a strip of steel, having a width equal to one-thirtieth of the width of the concrete. This substitution does not affect the total amount of stress in each small horizontal layer and hence does not disturb the condition of equilibrium. This strip becomes part of the steel beam and we have now to deal simply with the modified steel beam shown in the figure. The neutral axis is in the center of gravity of the figure and may be found by dividing the moment of the area of the additional strip about the center of the beam by the area of the entire figure.

Let A = area of steel beam.
    I' = moment of inertia of steel beam about its own center.
    I'' =   "   "   " strip about the new neutral axis of beam.

I''' = moment of inertia of steel beam about the new neutral axis.

k = distance from neutral axis to the center of beam.

h' = ½h. of beam.

Other quantities have the same values as for Case I.

Then
$$k = \frac{b(h'-k)[\frac{1}{2}(h'-k)+k]}{30[A+\frac{1}{30}b(h'-k)]} \quad (2)$$

whence
$$k = \frac{bh^2}{4(60A+bh-bk)}$$

This equation can readily be solved by trial, as bk is a small quantity.

Thus, if b=24″; h=15″; A=12″; then, k=1.28″
b=24″; h=12″; A=9.4″; k=1.00″
b=24″; h=10″; A=7.5″; k=.90″

After k has been found, the moment of resistance of the modified beam is obtained from the equation

$$M = \frac{f(I''+I''')}{(h'+k)} = \frac{15000(I''+I''')}{(h'+k)} \quad (3)$$

Where $I''' = I' + Ak^2$, and $I'' = \frac{1}{90}b(h'-k)^3$.

The extreme compressive stress in the concrete is

$$f' = \frac{1}{30} \cdot \frac{15000(h'-k)}{(h'+k)} \quad (4)$$

Likewise, since $M'' = \frac{f'I}{(h'+k)}$, where $I = 30I''$

$$M' = \frac{f'I'''}{(h'+k)}$$

and $f' = 30f''$

$$M'' = \frac{M'I}{30I'''} \quad (5)$$

Which is identical with equation (1) of Case I.

CASE III. (See Fig. 2).

In this case we shall assume that the concrete cracks on the tension side, but that the crack does not extend beyond that point at which the tensile fibre stress in the concrete is just equal to its ultimate tensile strength: viz., about 200 pounds per square inch. In this case the lower portion of the sound concrete will be in tension.

Let x = distance from center of beam to the point in concrete at which the tensile fibre stress is 200 pounds. The corresponding fibre stress in the beam is 6000 pounds. The exact value of x in terms of h and k might be expressed, and likewise a second equation written, giving the value of k in terms of h and x; from which equations both k and x might be obtained. The resulting equations, however, would be very cumbrous. It will be sufficiently accurate to obtain the value of x on the supposition that the neutral axis remains in the center of the beam, and then from the result subtract about three-quarters of an inch, which is nearly the value of k.

$$\text{Hence, } x = \frac{6000}{15000} \times \frac{h}{2} - \frac{3}{4} = .2h - .75 \qquad (6)$$

$$k = \frac{\frac{1}{30}(h'+x)\,b\,[\frac{1}{2}(h'+x)-x]}{A + \frac{1}{30}b\,(h'+x)}$$

$$\text{whence } k = \frac{b\,(h'^2 - x^2)}{2\,[30A + b\,(h'+x)]} \qquad (7)$$

If $b = 24''$; $h = 15''$; $A = 12''$; $x = 3.75''$; $k = .80''$
$b = 24$; $h = 12$; $A = 9.4$; $x = 2.85$; $k = .68$
$b = 24$; $h = 10$; $A = 7.5$; $x = 2.25$; $k = .60$

The moment of resistance of the modified beam is, as before,

$$M = \frac{15000\,(I'' + I''')}{(h' + k)} \qquad (8)$$

where $I''' = I' + Ak^2$

$$I'' = \frac{b\,(h'+x)^3}{360} + \frac{b\,(h'+x)\,[\frac{1}{2}(h'+x)-(x+k)]^2}{30}$$

Equations (4) and (5) hold for Case III as well as for Case II.

Other cases can be solved by the same general method given above. For example, the concrete may have a greater depth than the beam.

CASE IV (See Fig. 3.)

In this case we shall assume that a square steel wire is imbedded near the lower portion of a concrete beam, and that the stress in the wire does not exceed 15,000 pounds per square inch.

Let t = length of side of steel wire.
b = breadth of concrete beam.
h = depth of concrete beam.
d = distance from center of beam to center of steel wire.
k = distance from center of concrete beam to neutral axis of the concrete-steel beam.

As before, let us consider the concrete beam as transformed into a steel strip having a breadth of one-thirtieth of its former width. We then have to deal with a modified shape similar to the preceding cases.

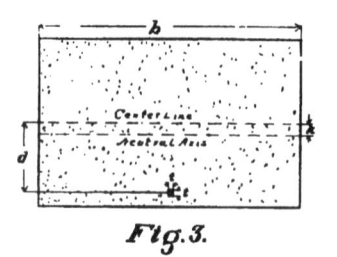

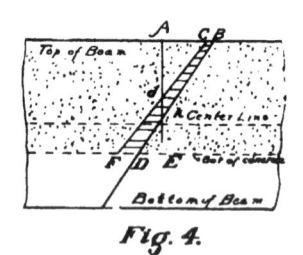

Fig. 3. *Fig. 4.*

Then, $$k = \frac{t^2 d}{\frac{1}{30}bh + t^2} \qquad (9)$$

Let I'' = moment of inertia of concrete strip about the new neutral axis.

Let I''' = moment of inertia of steel wire about the new neutral axis.

Then, $$M = \frac{f(I'' + I''')}{(d-k)} = \frac{15000(I'' + I''')}{(d-k)} \qquad (10)$$

where $$I'' = \frac{bh^3}{360} + \frac{bhk^2}{30}$$

$$I''' = t^2(d-k)^2.$$

The maximum tensile fibre stress in the concrete is,
$$f' = \frac{15,000(h'-k)}{30(d-k)} \qquad (11)$$

where $h' = \frac{1}{2}h$.

The maximum compressive fibre stress is,
$$f' = \frac{15,000(h'+k)}{30(d-k)} \qquad (12)$$

Equations (10), (11), and (12), are readily adapted to any other stress in the wire than 15,000, by making the proper substitution. The one condition is that the stress shall not exceed the elastic limit of steel.

It is evident that other cases might be solved by the method here outlined but it is believed that these few cases sufficiently illustrate the method.

That the strength of a beam is greatly increased by imbedding it in concrete is readily seen by solving Case I for one or two conditions. Thus suppose $b = 24''$; $h = 15''$; $A = 12.0''$; $I' = 424$. Then from Eq. (1);

$$M'' = \frac{6750}{12720} M'.$$

That is, the moment of resistance of the concrete is more than one-half of that of the steel beam, and the strength of the beam is therefore increased fifty per cent. by the addition of the concrete.

Again, let us take the case of a concrete beam having a wire imbedded one inch from its lower edge.

Let $b = 6''$; $h = 12''$; $d = 5''$; $t = .5''$.

From Eq. (9), $k = \frac{1}{2}''$.

$I'' = 30$; $I''' = 5$.

$M' = \frac{35}{30} M''$.

In this case the strength of the concrete beam is increased one-sixth by the presence of the steel wire, although the area of the wire is but four-tenths of one per cent. of the total area of the beam.

If the width of concrete is great enough, the concrete will bend about its own neutral axis, rather than that of the combined concrete-steel beam. The adhesion between the concrete and steel should be sufficient to prevent such independent rotation.

In the case of the wire imbedded in the concrete beam, if the adhesion between the materials is destroyed, the stress in the wire at once reduces to zero and the concrete beam carries the entire load. For this case, then, the adhesion between the wire and concrete must be equal to the total maximum stress which can occur in the wire. Practically, the stresses in the wire should be computed for adjacent sections near the end of the beam and the adhesion between the two materials for the length included between the sections should be equal to, or greater than the difference between these stresses. It is thus seen that the procedure is similar to that for the determination of the pitch of rivets in the flanges of a plate girder.

For a beam imbedded in concrete, the case is not so simple, since each of the two materials will carry a portion of the load after adhesion between them has been destroyed. It is necessary to determine, then, what portion of the bending moment each part is capable of resisting when rotating independently about its own neutral axis. From this the fibre stress in the concrete may be obtained and the strain line CF, Fig. IV. can be drawn.

From the formulæ given above the fibre stress in the concrete, where the two materials act as one beam, may be determined, and the strain line BD drawn for this condition. In the figure, k is the position of the neutral axis of the combined beam, and d the neutral axis of the concrete when acting alone. The depth of the concrete is AE. Then the effect of the adhesion is to change the stress in each fibre of the concrete from an amount proportional to the abscissas of the line CdF, to an amount proportional to the abscissas of the line BkD. The shaded area CBDF represents the amount of constraint caused by the adhesion. The area of the shaded portion, multiplied by the breadth of the beam, is equal to the total amount of adhesion, between the section and the end of the beam.

Consider two sections near the end of the beam. Determine the shaded area for each section. Then the difference in area of the two shaded portions included between similar horizontal lines, multiplied by the breadth of the beam, is equal to the amount of adhesion required for that part of the beam bounded by the two sections and the two horizontal layers. Thus any part of the beam may be tested.

In the above discussion the concrete was assumed to have a coefficient of elasticity of one million, and to otherwise act as an elastic material. It must be remembered, however, that cement is not a uniform product, such as steel or iron, but that it may vary in quality from utter worthlessness to that of the best cements upon the market. The elastic properties vary correspondingly, and hence actual tests may frequently give results widely different from those of this article. The writer hopes to conduct, in the near future, a series of tests, to determine experimentally the elastic behavior of cements and concretes under different conditions.

The writer is indebted to Mr. Frank B. Walker, '97, for the sketches accompanying this article.

CORN OIL.

BY HARRY W. ALLEN.

The investigation of the physical and chemical properties of corn oil, discussed in this article, was undertaken with the purpose of finding some economical use for which the oil might be adapted, either in its natural condition or after some cheap method of treatment. The oil is obtained from the refuse of corn distilleries, and from corn before it is employed in the manufacture of starch. Enormous amounts are now wasted which could be obtained at a nominal expense if some practical use for the product were known. The only manner in which the oil at present is utilized to any extent, is as an adulterant. Being much cheaper than linseed oil, it is used with the latter for mixing paints, and in the adulteration of lard it has been substituted for cotton-seed oil. Corn oil has been employed for burning, as a lubricant, and in soap-making, but for none of these uses does it seem specially applicable.

The experiments with the oil from this economical standpoint were rendered difficult from the beginning of the work by a scarcity of literature on the subject. No thorough examination of the oil seems ever to have been made, and only short references as to its general appearance, specific gravity, etc., are to be found. The latest works on the subject of oils contain a few generalizations in regard to its physical and chemical characteristics, but nothing as to its chemical composition. This lack of foundation for the work compelled first a systematic examination of the oil, and for this reason the efforts to find some practical use for the oil have not been completed.

When the oil is obtained from the residue of the fermentation vats, it is of a reddish brown color. The samples used in these experiments were evidently prepared by the second method, for the oil possessed a golden-yellow or straw color. The odor is very peculiar, smelling somewhat like a mixture of corn and beeswax, and is very sweet. Heated to a high temperature the oil smells like burnt lard, and under ordinary pressure cannot be distilled without decomposition. Its specific gravity at 15° C. is .9233.

The oil is extremely soluble in ether and benzol, but is very slightly affected by cold alcohol. According to Smith,[1] absolute alcohol at 16° C. dissolves two parts in 100; at 63° C., 13 parts in 100. In commercial acetone at 16° C. it dissolves 24 parts in 100, and at 16° and 63° C. it is soluble in glacial acetic acid 3 and 9 parts in 100.

Small quantities of the oil treated with various reagents of different strengths gave the following color tests, which, however, do not seem to be very distinctive:

Reagent.	Result.	Reagent.	Result.
NaOH 1.34	Dirty, yellowish white.	HNO$_3$ 1.33	Light brown, with yellow tint after heating.
H$_2$SO$_4$ 1.475	No action until heated. Brownish-red, with traces of dark purple. On further heating decomposed with fumes, leaving a black, sticky mass.	HNO$_3$ Con.	Without heat same color as above.
		H$_2$SO$_4$ + HNO$_3$	Light brown color.
		Aqua Regia	Slightly yellow.
H$_2$SO$_4$ 1.635	No action until heated. Brown color, about same as preceding, with similar traces of purple.	+ NaOH 1.34	First a light yellow. More NaOH—light brown, resembling orange. Fibrous.
H$_2$SO$_4$ Con.	Reaction without heat. Yellowish-brown. Heat changed this color to preceding brown.	HNO$_3$ 1.33	Light brown.
HNO$_3$ 1.18	Slightly yellow with a brownish tint after application of heat.	+ NaOH 1.34	Fibrous, light yellow.

A column of the oil examined through the spectroscope showed a marked absorption of the violet end of the spectrum beyond the line "8." In a polariscope with a 20 c. m. tube, the oil was found to have a dextro-rotatory power of 2.05°.

The oil is readily saponified by means of caustic soda or potash. It was found more expedient to use alkali of only moderate strength in order to thoroughly saponify the oil, as sodium hydroxide 1.34, at first employed, seemed to produce an unsaponifiable oil, which was changed into a sodium salt only after constant mixture with the alkali. An odor peculiar to "Johnny-cake" is very noticeable upon heating the oil with the sodium hydroxide. The sodium salt is a yellow, corn-colored soap readily soluble in water and hot alcohol, but not in ether to any extent. It would not crystallize out of alcohol,

[1] Jour. Soc. Chem. Ind., 1892, p. 505.

but was precipitated when cold as an amorphous mass. All efforts to get a good crystallized salt of the oil were without success, the barium, lead, and magnesium salts, also being almost structureless in appearance.

The free acids of the oil were readily obtained from the aqueous solution of the sodium salt as a yellowish colored oil, somewhat lighter in shade than the original corn oil, and having a specific gravity of .909. These acids are readily soluble in ether, almost insoluble in hot and cold alcohol, and not in the least soluble in water. The last fact is corroborated by the quantitative results obtained. According to these facts, the corn oil yields about 97% of its weight as free acids, which corresponds very closely to the Hehner value—the per cent. of insoluble fatty acids—accorded to the oil by numerous chemists. The acids solidified at about 14° C., becoming thick and jelly-like, and again became fluid at from 16-17° C. The oil cannot be distilled at ordinary pressure, but breaks down, partially at least, into acrolein, as it evidenced by the peculiar odor. The free acids were also separated from the corn oil directly, without conversion into the sodium or potassium salt, by passing a current of super-heated steam through the oil. From the different characteristics of the acids it is almost certain that the greatest proportion of the mixture is oleic acid, with a greater or less amount of oxy-oleic acid, depending upon the sample of oil taken.

The determination of the iodine absorption of the corn oil and the free fatty acids gave most surprising results. The corn oil gave iodine values from 129-131%, and the free acids a value from 114-115%. These large iodine values would place the corn oil in the class of drying oils, which is not in accordance with its other qualities. It does not dry readily and the ordinary process of "boiling" and the addition of lead oxide have but little effect upon it. It has been observed, however, that certain drying properties are imparted to it if heated in a current of air at 150° C., with the subsequent addition of manganese borate. It was with the view to convert the oil into a better "dryer" either by a reduction or oxidation of the fatty acids contained, that most of the experiments with the oil were planned. These investigations have not yet been finished, but have developed a number of interesting facts in regard to the oil.

The result of the oxidation of the oil was one of the best proofs obtained that oleic acid is a constituent. One method adopted for the purpose was to simply pass a current of dried air through the oil at a temperature of 200-260° C. for a number of hours. The oil became heavier, having a specific gravity of .96, and changed its color to a reddish shade, resembling that of "golden oil." It was thick and viscid and in odor resembled hot lard. The sodium salt from this oil also had a reddish color and differed from the soap from the original corn oil in that it was readily soluble in hot sodium or potassium hydroxides. The free acid from this salt was a reddish-yellow oil, thick and jelly-like up to 17° C. Its specific gravity at 30° was .93. Unlike the original acids from the corn oil, it was readily soluble in cold alcohol. From its general behavior and characteristics, it was undoubtedly oxy-oleic acid.

The reduction of the oil by means of metallic magnesium in a digester, under high pressure and temperature, was not successful in changing the composition of the fatty acids but resulted in a very good magnesium salt of these acids. The salt was fairly well crystallized, of a yellowish-white color, and was obtained after heating the oil with magnesium powder to 206° C. and a pressure of 20 atmospheres. This salt was insoluble in alcohol and water but readily so in ether. At a temperature of about 100° C., it conglutinated but did not melt. This presumably was magnesium oleate, and the method for its preparation seems to be an entirely new one, according to any literature on the subject which has been available.

LATITUDE.

BY PROFESSOR WILLIAM R. HOAG.

[By altitudes of the sun near the meridian, employing a graphical chart for increasing the accuracy.]

In the determination of terrestrial latitude the instruments, methods and plans of reduction employed are as varied as the demands placed upon such knowledge; the choice being governed largely by the degree of accuracy required. In exploration and reconnaissance, the nearest five minutes of arc frequently being sufficient, we find the pocket sextant well suited to such needs. A single observation on the sun or star at meridian passage with no correction for refraction will give the desired accuracy. For purposes of navigation and for use with the solar compass the latitude is gained within about a minute using the telescopic sextant on ship-board and the engineer's transit or solar compass on land.

In geodetic work as the inauguration of a system of triangulation, or the establishment of a boundary line between two countries the highest accuracy is desirable, and the nearest tenth of a second is secured by the zenith telescope, employing a large number of observations together with an elaborate plan of reduction.

The Civil Engineer is sometimes called upon, in taking up a detached piece of topographic work or in locating certain prominent political stations for purposes of mapping, to establish latitude within 5 or 10 seconds.

A small zenith telescope, with the Talcott method of observing, will enable the engineer to secure this accuracy. This instrument is rarely accessible and its cost, $300, forbids its addition to his instrumental outfit. It is proposed in the following article to show how the engineer, with his ordinary transit, can easily establish his latitude within 10 seconds and with the aid of a sextant, a comparatively common instrument costing about $75, he can determine it within 3 to 5 seconds.

The scheme in a somewhat less elaborated form is substantially what we developed several years ago and have since used with the classes in Geodesy and in the practical work of the State Topographical Survey, securing a degree of accuracy in most cases within the limits named above. The basis of accuracy in all physical measurements is repetition. When the function to be determined is constant, as a geodetic angle or the length of a standard bar, with the determining observations all made under equally favorable conditions, the simple arithmetic mean of all, *i. e.* the average, is the best value which can be given to the function based upon such observations. When we have a uniformly varying function, as the rod intercept for different distances from the transit, in stadia work, or velocity of water corresponding to varying angular velocity of the current meter exposed to its action, we can readily determine the equation between the two dependent variables by plotting the one along the axis of X and the second along the Y axis. These plotted observations will give a series of points agreeing with a straight line in so far as the observations are perfect, and a straight line adjusted to agree most nearly with all, giving to each its due weight, determines the relation sought.

In case the equation between the two variables is above the first degree, the plotted points give a curved line and we can no longer employ the usual methods for obtaining graphically a mean value of the function.

A glance at the conditions involved in our problem shows that we have this last form of relation present, and to avail ourselves of the desirable feature of multiple observations we must devise a plan to treat these observations giving a curved locus, after the manner of the right line with the simple direct variables or the arithmetic mean in case the desired constant is capable of direct measurement.

Figure 1 represents the sun at meridian passage or at apparent noon. By observing the altitude of the sun at this time and subtracting from it the sun's declination we readily obtain the co-latitude and hence the latitude of the place.

The usual plan of observation with this method is to follow the sun in altitude with the instrument to its highest point and accept this one determination as its altitude at meridian passage and from this deduce the latitude.

This method is not only wrong in theory, since the sun is not at its maximum altitude at meridian passage, owing to its

Latitude. 87

change in declination, but it violates the fundamental principle touching repetition of observations in permitting but one observation to be taken.

Now, if instead of idly watching the sun in its change of altitude during the ten or fifteen minutes preceding its passage and about an equal time after, to make sure that it has passed, let the observer make independent observations for altitude as frequently as careful work will permit, noting the watch time of each observation. With the sextant the interval need not be greater than one minute of time and not more than two and a half minutes when the transit is used. This extra time

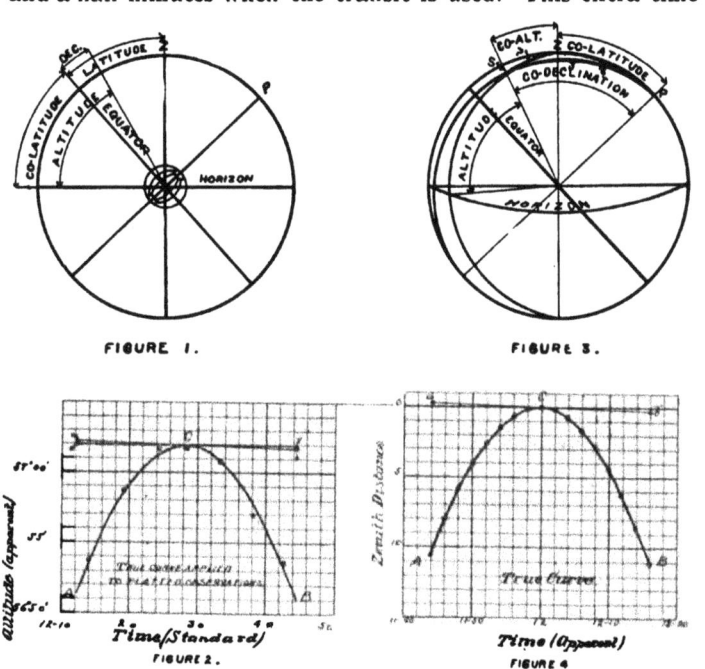

FIGURE 1.

FIGURE 3.

FIGURE 2.

FIGURE 4.

is required in using the transit in the direct and reversed position to eliminate instrumental errors.

When accuracy exceeding five seconds of arc in latitude is desired the sextant must be employed on two or three successive days, observing on alternate limbs of the sun each day to eliminate personal error of contact.

Of course when this degree of accuracy is desired the sextant must be in precise adjustment, with its index and eccentricity errors accurately known, and the refraction errors of its shade glasses equally well determined, should these be used instead of the eye-piece shade.

We now plot these observations for altitude using the mean of each direct and reverse measure with the transit as one observation, laying off time on the axis of X, and altitude as Y ordinates.

Figure 2 exhibits, according to this plan, the observations taken at Pipestone, August 20, 1895, with a Troughton & Simms transit reading to ten seconds. Each plotted observation is the mean of a direct and reverse observation.

Now if we had a curve plotted to the same scale representing the true altitude of the sun during the whole time covered by these observations, by applying it to the observation curve we have the same ability to determine its true position relative to the plotted observations as in the case of the straight line. The adjusted position of the true curve shows the true altitude of the sun during the time covered by the observations as a function of latitude and reading it for the instant of noon shown by the central Y axis of the curve we have the altitude and from it the latitude desired. We will now develope a plan for easily constructing this true curve.

Figure 3 shows the spherical triangle involved, SPZ, in which we wish to study the changes coming to ZS consequent upon changes to α (the hour angle), and SP, the co-declination, with ZP a constant.

The fundamental equation involving these parts written out for this spherical triangle becomes,

$$\cos.p = \cos.z \cos.s + \sin.z \sin.s \cos. \alpha$$

By substituting in this formula the s and different values of α and the corresponding values of z covering the desired time before and after noon we can determine p for the several values of α which passing to the co-functions enables us to construct the true curve.

This method, however, is seen at once to be laborious requiring the use of an equation not adapted to logarithmic computation and further by requiring the determination of the full function a 7-place table must be used and must be entered nine times for each substitution or about fifty times in all in determining nine points of the curve.

Latitude.

To construct this curve we do not need the whole function v, but only the changes coming to it. Hence if we differentiate the equation above with respect to p and α regarding z and s as constants, the latter is such as has been noted and z can be regarded so for the present since the maximum change which can come to it is less than 15 seconds, we shall obtain the relative change between these variables which we desire. Even the 15 seconds will not remain an outstanding error since in platting the final curve we shall employ oblique axes to take account of this declination change.

Differentiating equation (1) gives

$$-\sin. p\, dp = -\sin. z \sin. s \sin. \alpha \, d\alpha \qquad (2)$$

or $\quad \dfrac{dp}{d\alpha} = \dfrac{\sin. z \sin. s \sin. \alpha}{\sin. p} \qquad (3)$

Equation (3) expresses the rate at which p is changing relative to α for any given value of α and p. Calling this curve a circular arc which it is sensibly within the limits we use, i. e., four degrees from the center, we can make this equation express the full change coming to p for a given $d\alpha$ by writing the equation for $\dfrac{\alpha}{2}$ instead of α since at the middle point of the arc, the curve being circular, the ratio change between dp and $d\alpha$ is the half of what it is at α or for this point p and α are having their mean relative change.

A rigid substitution in Eq. (3) would of course require that p corresponding to each α be used. To call p uniformily ZS can introduce as a maximum error to p about 4 seconds of arc. This occurs only near the outer limits of the curve, and if thought necessary even this can be eliminated.

Equation (5) for our use then becomes

$$dp = \sin. z \dfrac{\sin. s \sin. \tfrac{1}{2}\alpha}{\sin. p} d\alpha \qquad (4)$$

an equation adapted to logarithenic computation and dealing with the correction only.

To make one substitution in this formula requires the use of the table six times. If we desire nine points in the curve, by allowing for no declination change, except in the plotting, the curve as computed is symmetrical, hence the vertex point and four on one side will give the whole curve. The vertex point falls on the axis of X since it is for $\alpha = 0$. This reduces the needed substitutions to four and as z, s and p remain constant throughout the four, the additional substitutions require but three logarithms each. This gives our curve complete with nine points fixed by using the table but fifteen times. One case of such reductions will show how this work can be further reduced one-half.

Let it be required to determine the true curve for the 24th of August, 1895, in latitude 43° 58'—declination on that date being 10°56' north.

This is the date on which the observations were taken at Pipestone platted in Fig. 2.

We then have $z = 79°.04'$, $s = 46°.01'$, $p = z-s = 33°.03'$ and we assume for α the successive values of 1°, 2°, 3° and 4° and $d\alpha$—letting the minute be the unit—will equal 60, 120, 180 and 240.

This substitution is best followed by the following tabular form:

Log. sin. 79°04'................................. 9.992044
Log. sin. 46°01'................................. 9.856934
Log. sin. (a.c.) 33°03'........................ 0.263308

0.112286(c)

½α log. sin. (a)	d α	log. (b)	(a)+(b)+(c)	Minutes	Seconds
30' 7.940842	60	1.778151	9.831279	.6781	40.68
1° 8.241855	120	2.079181	0.433322	2.712	162.72
1°30' 8.417919	180	2.255273	0.785478	6.101	366.11
2° 8.542819	240	2.380211	1.035316	10.847	650.62

We notice that these values are very nearly proportional to 1, 4, 9, 16, showing the curve to be closely parabolic. This gives the second order differences uniform and enables us to readily interpolate eight additional points on the curve, making 17 in all. Thus:

```
      10.2    10.2
40.7   30.5   20.3
91.5   50.8   20.3
162.7  71.2   20.4
254.2  91.5   20.4
366.1  111.9  20.3
498.3  132.2  20.3
650.9  152.5
```

Thus one substitution requiring the use of the table six times readily gives us seventeen points on the curve, quite sufficient for the most exact construction.

The following table has been made using equation (4) and

Latitude.

checked at intervals with (1) and gives this value of dp for $\alpha = 1°$ for 10° of north latitude, *i. e.*, from 40° to 50° and the full range of the sun's declination.

TABLE

Showing the difference between the altitude of the sun at apparent noon and at four minutes before or after noon, neglecting declination change.

LATITUDE.

N'th Decli-nat'n	40°	41°	42°	43°	44°	45°	46°	47°	48°	49°	50°
25	84.2	77.8	72.1	67.1	62.7	58.8	55.2	51.8	48.6	45.7	43.2
20	66.0	62.1	58.5	55.2	52.1	49.1	46.7	44.4	42.2	40.1	38.0
15	54.5	51.8	49.3	46.9	44.7	42.6	40.8	39.1	37.4	35.7	34.1
10	47.3	45.2	43.3	41.5	39.7	38.1	36.5	35.1	33.7	32.3	30.9
5	41.8	40.1	38.6	37.1	35.7	34.4	33.0	31.7	30.6	29.5	28.4
0	37.4	36.1	34.9	33.7	32.6	31.5	30.4	29.3	28.3	27.3	26.3
So'th											
5	33.9	32.7	31.6	30.7	29.8	28.9	27.9	27.0	26.1	25.3	24.5
10	30.9	29.9	29.0	28.2	27.4	26.7	25.8	25.0	24.2	23.5	22.9
15	28.3	27.5	26.8	26.1	25.4	24.8	24.1	23.4	22.7	22.1	21.5
20	26.0	25.3	24.7	24.1	23.5	23.0	22.4	21.8	21.2	20.7	20.2
25	24.0	23.4	22.8	22.3	21.8	21.4	20.8	20.3	19.8	19.4	19.0

The value taken from the table for the latitude of the place and the declination of the sun and multiplied by the following numbers—¼, 1, 2¼, 4, 6¼, 9, 12¼ and 16 gives the ordinates of the true curve. Entering the table for the case which has been analytically determined above—Lat 43° 59′, Dec. 10° 56—gives 40.66. Using the multipliers ¼. 1, 2¼, etc., gives for the ordinates of the curve 10.2, 40.7, 91.5, 162.6, 254.1, 366.0, 498.2, 650.6. This shows a sufficiently close agreement.

For the sun near the meridian we can call the change in declination so much change in altitude and apply it with the proper sign to the quantities above. Thus on the date of our observations the sun was passing south at the rate of 52″ per hour or 1.7″ per each 2 minutes of time to which our ordinates above correspond.

This gives the following corrections: 1.7, 3.4, 5.2, 6.9, 8.6, 10.4, 12.1, 13.9. These are to be applied with the minus sign to correct the forenoon ordinates and with the plus sign to the afternoon ordinates when they are to be laid off from the rectangular axis.

By drawing the oblique axis *ab* (Fig. 4) first, the original ordinates can be used and set off from this axis, thus using the same ordinate for equal time before and after meridian passage.

In constructing the true curve we have taken no account of differential refraction which would tend to slightly flatten the curve. In latitude 45° this error would be greatest when the sun was at the winter solstice and would then be 0.6", and this affects only the outer portions of the curve. For all work done during the summer months this error is less than 0.2".

Reading the plat shows the sun's apparent altitude
at meridian passage to be...................................57° 1'54"
Refraction (corrected)... 38"
True altitude...57° 1'16"
Declination ....11° 1'17"
Zenith distance...45°59'59"
Latitude ....44°00'01"

GROUND DETECTION ON ELECTRIC CIRCUITS.

BY H. M. WHEELER, '96·

1. *Lines in General.* Grounds may often be detected by the eye. When they are caused by swinging contacts or by broken or "down" wires, a very little observation is generally sufficient to locate the faulty place; while in many cases, faults in the insulation, blackening of woodwork, and fusing of metallic conductors are all immediately patent and require no special instruments for their detection.

2. The most common method of detecting grounds is by the use of the magneto. This is applicable to overhead circuits; but does not give reliable results on underground circuits, submarine cables, or any circuits which contain condensers or a large amount of distributed capacity. The line to be tested must be "idle" and disconnected from all other lines and the earth. The magneto leads are connected to the line and to the earth respectively. If on turning the handle the bell rings, the insulation resistance of the line is not >10,000 to 25,000 ohms and the line is probably grounded. The line is now broken at the joints and the sections tested separately. Then the grounded sections are examined foot by foot until the exact spot is located. If the line is very long, distributed capacity may cause the bell to ring when there is no ground. In such a case it is well to use an ordinary electric bell, or a galvanometer, with the direct current from a battery or small dynamo, the battery and bell being in series and the test carried out as before.

3. *Central Station Indicator for D. C. Constant Potential Circuit.*[1] The connections are shown in Fig. (1). L and L' are two incandescent lamps of equal candle power joined in series parallel to the mains, the sum of the normal voltages of L and L' being equal to that of the dynamo. Ggf is a ground connection through the galvanometer (or bell) g. When there is no ground on the mains, the galvanometer deflection is nil and L and L' have equal brilliancy. When either main is

---
[1] Abbott: Electrical Transmission of Energy, pp. 219, 220.

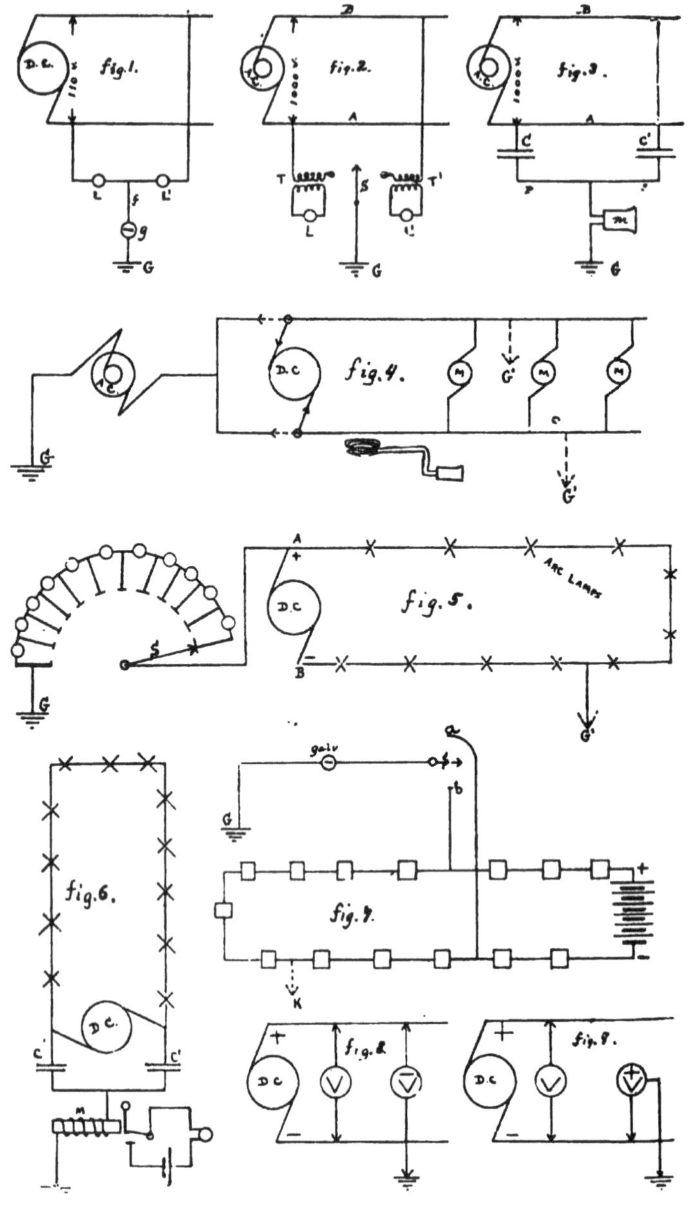
fig. 1.
D.C.
110 v
L
L'
f
g
G
fig. 2.
B
A
A.C.
1000 v
T
T'
S
L
L'
G
fig. 3.
B
A
A.C.
1000 v
C
C'
P
m
G
fig. 4.
A.C.
D.C
M
M
M
G'
G'
G
fig. 5.
A
B
+
−
D.C
S
ARC LAMPS
G
G'
fig. 6.
D.C.
C
C'
M
fig. 7.
galv
a
b
G
K
+
−
fig. 8
D.C
+
−
V
V
fig. 9.
D.C
+
−
V

grounded, the lamp on its side burns less brightly, being shunted around by the ground current, and the galvanometer needle is deflected. While this method gives continuous and automatic indication of grounds, it is objectionable from the fact that the permanent ground causes a constant and unnecessary strain on the insulation, and insures a ground current through anyone touching the line. These objections are obviated by using a switch at f and testing at stated intervals.

4. *Central Station Indicator for A.C. Constant Potential Circuits.*[1] In Fig. (2) the transformers T and T' have one end of their primary coils connected to the mains A and B respectively, while the other ends can be grounded through the switch S. Across the secondary coils are the lamps L and L'. If on grounding T, there is a ground on B, the ground current through the transformer will render L incandescent. A ground on A is indicated when, upon grounding T', L' becomes bright.

5. A still better method which gives continuous and automatic indication of grounds is shown in Fig. (3).

C and C' are small condensors, M is a telephone receiver (or an electric bell with polarized coil) When either main is grounded the telephone sets up a continuous hum. If on breaking the circuit at S the hum still continues, A is grounded. In tne same way break circuit at P to test B.

6. *Mr. E. M. Bentley's Ground Detector for Electric Railway and Power Circuits.*[2] In Fig. (4), full line switches show ordinary series double metallic power circuit. Move switches to dotted position and send a pulsating or alternating current through the mains (now in parallel) and the ground GG'. Carry a coil attached to a telephone receiver along the line and notice the humming. At grounded points G', the humming suddenly ceases owing to the disappearance of the current into the earth. Hence grounds are easily located by this method. The coil and telephone arrangement may be used in locating grounds in regular alternating circuits or in tracing the same where the wires are behind the plastering or under the floors as in a house.

7. *Mr. M. D. Law's Method of Locating Grounds on Arc Circuits,*[3] Fig. (5). At the left is a bank of incandescent lamps, in series, with one terminal grounded. The lamps are taken of such

---
[1] Ibid, pp. 220, 221.
[2] *Electrical World,* XIII; p. 214.
[3] *Electrical World,* XIX; p. 160.

voltage as to be a multiple of the average drop between contiguous arc lamps, while their total resistance is such that all just come up to normal candle power when connected directly across the mains. To test for grounds make connection as shown and move the arm S till the grounded lamps just come up to candle power. Next disconnect S from A and join to — terminal B and move S as before. If the sum of the lamps brought up to candle power in the two tests equals total number of lamps in bank, the arc circuit is "dead" grounded at a point G' distant from A such that

$$\frac{G'A}{G'B} = \frac{\text{no. lamps lighted in first test}}{\text{no. lamps lighted in second test}}$$

If the sum of the lamps lighted in the two tests does not equal total number in the bank, the ground is only partial; but the above proportion is still approximately true.

A modification and improvement upon this method has been made by Mr. E. E. Stark.[1]

8. *The Rudd Automatic Ground Alarm for Arc Light Stations*,[2] Fig. (6). Two condensers, C and C', are joined across the mains with a common ground connection through the magnet coil M. For normal conditions on the arc circuit no current passes through M and the coil is neutral; but when a ground occurs, the sudden change in voltage sends a momentary pulsating current through M. This energizes the coil which attracts a trigger, thus loosening a shutter S, which falls and completes a second circuit consisting of a bell and one cell. The bell rings and continues till the attendant breaks its circuit. The ground may now be located by Mr. Law's method or by the bridge method exploited by the Western Electric Company, Chicago, who manufacture the ground alarm.

9. *Ground Detection on the Minneapolis Fire Alarm System*, Fig. (7).[3] The alarm boxes are in series with the battery as shown. When there is no ground the galvanometer shows no deflection if the switch s is moved to a or b. If there is a ground at k, the galvanometer will be deflected when the switch is in contact with b or a, and the corresponding deflections will be proportional to the distances k+ and k—. Hence we have a rough method of determining where the ground is.

---
[1] *Electrical Engineer*, vol. XIV, p. 25.
[2] *Electrical World*, vol. XIV, p. 131.
[3] In Fig. (7), test wires a and b should be directly across terminals of battery.

Ground Detection. 97

10. Since the freedom of any circuit from grounds depends primarily upon the insulation, an easy method of detecting a ground, or leak, is to measure the insulation resistance of the line in question. If this insulation resistance is found to be some high figure, depending upon the requirements of the case, the line is considered all right; a certain minimum insulation resistance being deemed sufficient. The ground is "partial" below this limit, and becomes "dead" when the insulation resistance equals zero. The following methods deal with the detection of grounds through measurement of insulation resistance.

11. *The Voltmeter or "Drop" Method.*[1] This method may be used to measure the insulation resistance of a line to the earth, of armature or field coils to the frame or, in fact, of any conductor to another; it requires, however, for its operation, a direct current. A voltmeter of known resistance R is connected across the dynamo terminals (Fig. 8), and the voltage V is noted; next the — lead of the voltmeter is grounded and the reading −V is noted.

Since the same current passes through the voltmeter and the ground

$$C = \frac{-V}{R} = \frac{V - -V}{-R}$$, where −R is the insulation resistance of the — side.

Whence, $-R = \frac{V - -V}{-V} R$

In the same way (Fig. 9) the resistance of the + side is found, and $+R = \frac{V - +V}{+V} R$.

For a voltmeter of R = 20,000 ohms, a 110 volt circuit, and the voltmeter readable to $\frac{1}{10}$ volt, no deflection means that

$$\pm R > \frac{110 - .1}{.1} 20,000 \text{ or } 21,980,000 \text{ ohms.}$$

In the same way values of $\pm R$ may be calculated for different values of $\pm V$ and the results may be written in megohms opposite their corresponding scale readings. Of course, in further tests, the voltage V must be that on which the table was based.

12. *The Insulation Resistance of Storage Batteries.*[2] Suppose three storage cells in series to be grounded at A_1, A_2, A_3, etc. Connect any point A through a sensitive

[1] *Electrical Engineer*, Vol. X, p. 687.
[2] *London Electrician*, XXXV, p. 855.

high resistance galvanometer in series with an ammeter to the earth, choosing A such that there is a large readable deflection

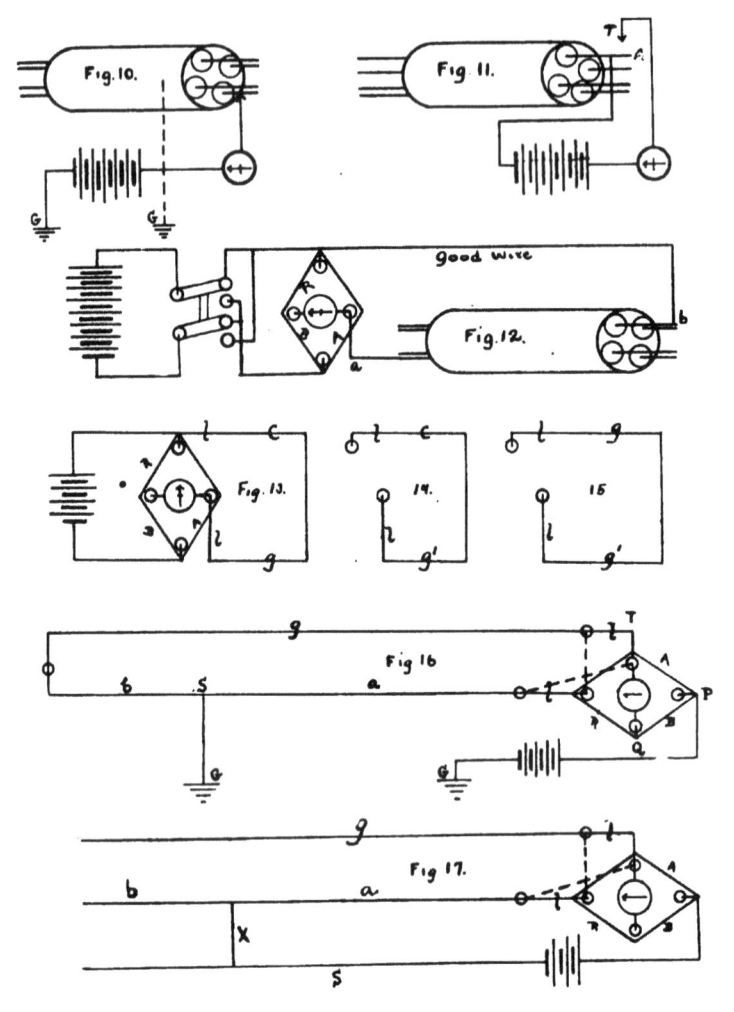

of the needle. Now add resistance in series to the galvanometer till the deflection is reduced to one-half its first value. If c = current for a galvanometer and ammeter resistance g, and c' is current in the second case for a total resistance g'; then the total insulation resistance of the battery is

$$R = \frac{cg - c'g'}{c' - c}.$$

For proof of this see original article. This method may also be used for measuring the total insulation resistance of a complicated network, such as a three or five wire system.

13. For those desirous of further information on the subject of insulation resistance, reference may be made to: Maj. Cardew's method of using the electrometer,[1] Mr. Carl Hering's[2] description of methods of testing wire cables before leaving the factory; The U. S. Government tests[3] and others.[4]

14. *To Identify Cables at Any Point in a Conduit.*[5] Ground the desired cables at one end, disconnect and keep insulated from each other all other terminals at both ends of the conduit. With a sharp needle point, connected to a galvanometer which is grounded through a battery, make connections successively with the cables at the required point by piercing the insulation. The cables which cause a deflection of the galvanometer when so touched, are those required, provided the cables were not already "grounded or crossed." To determine which cable in a conduit is grounded[6], make connections as in Fig. (10), the circuits being open at both ends. Ground the galvanometer through a battery as shown and touch its other terminals to the various cable ends. When the terminals of the grounded cables are touched, the needle will be deflected. In case two or more cables are "crossed,"[7] connect as in Fig. (11). Detach A join to T. If A is crossed with any other cable the needle will be deflected, if not mark A and try another. Those cables which cause deflections are crossed.

15. To identify the terminals of a cable when several of the cables in the conduit may be grounded—method of H. W. Fisher.[8] In Fig. (12) a and b are any two cable terminals at opposite ends, the rest of the terminals being open. A good wire whose terminals are known is connected as shown. Balance R on the

---

[1] *Electrical World*, Vol. XIX, p. 197.
[2] *Electrical World*, Vol. XVII, p. 188.
[3] *Electrical Review*, Vol. XXIV, p. 126.
[4] H. W. Fisher, *Electrical World*, Vol. XVII, p. 482.
[5] *Electrical World*, Vol. XVII, p. 481.   [6] Ibid.   [7] Ibid.
[8] *Electrical World*, Vol. XVIII. p. 6.

bridge till the galvanometer deflection is 0 and then reverse battery switch. If there is still no deflection, a and b are terminals of the same cable. If not b is disconnected and replaced by another terminal and the same thing tried again. This method is fairly accurate.

16. To locate the "ground" in a cable;[1] where two good wires whose terminals end near those of the cable are accessable—this being a very accurate method. The instruments necessary are: A Wheatstone Bridge, Thomson Reflecting Galvanometer, battery and two lead wires of equal resistance.

1st. Measure carefully the resistance of the lead wires.

2nd. Measure carefully the resistances in series as shown in Figs. (13), (14), (15).

Let Res. of Fig. (13) be K. l = resistance of one lead wire.
Let Res. of Fig. (14) be K' [c = resistance of the cable C].
Let Res. of Fig. (15) be K'' g = resistance of first good wire.
g' = resistance of 2nd good wire.

$$\text{Then } c = \frac{K + K' - K'' - 2l}{2}$$

3rd. Make connections as in Fig. (16), where B and A are bridge arms; g is first good wire, C is grounded at S, a and b being the respective resistances of the parts of the cable on either side of S.

It is now evident that we have a bridge PQST such that for no deflection QS : PQ = TS : PT, substituting values from the figure,

$(a+l+R) : B = (b+g+l) : A$, but $c = b+a$,

∴ $(a+l+R) : B = (c-a+g+l) : A$,

$$\text{and } a = \frac{B(c+g+l) - Ba}{A} - l - R$$

$$= \frac{B(c+g+l) - A(l+R)}{A} - \frac{Ba}{A}$$

$$= \frac{B(c+g+l) - A(l+R)}{A+B}$$

but from Fig. (13), $K = c+g+2l$,

∴ $a = \frac{B(K-l) - A(l+R)}{A+B}$, of which all quantities are now known. If D = length of cable C, then $\frac{a}{c}D$ is the distance which the ground S is from the measuring end.

---
[1] Ibid.

Ground Detection. 101

To check this value of a, reverse the leads 1, as shown by dotted lines (Fig. 16), when for no deflection, as before,
$$QS : PQ = TS : PT, \text{ or } (R'+1+g+b) : B = (a+1) : A$$
$$a+1 = \frac{A(R'+1+g+b)}{B}, \quad b = c-a$$
$$a = \frac{A(R'+c+g+1) - Aa - Bl}{B}$$
$$= \frac{A(R'+c+g+1) - Bl}{A+B},$$
but $K = c+g+2l$
$$\therefore \quad a = \frac{A(R'+K-1) - Bl}{A+B}$$

These values of a should check very closely.

The author gives also the method of procedure where only one good wire, or no wire at all, is accessible; also where the cable is composed of two spliced cables of different diameters.[1]

17. *Determination of the Terminals of Crossed Wires.*[2] Given two crossed wires in the same conduit and one good wire whose terminals are known.

Let 1 and 2 be crossed wires' terminals at testing end.

Let 3 and 4 be crossed wires' terminals at other end.

Let 5 and 5 be good wire's terminals at ends.

At testing end join bridge leads to 5 and 1; at other end join 5 and 3, find Res. $= R'$.

At testing end join bridge leads to 5 and 1; at other end join 5 and 4, find Res. $= R''$.

At testing end join bridge leads to 5 and 2; at other end join 5 and 4, find Res. $= R'''$.

At testing end join bridge leads to 5 and 2; at other end join 5 and 3, find Res. $= R''''$.

If $R' + R''' < R'' + R''''$ then 1 and 3, 2 and 4 belong to same wire.

If $R' + R''' > R'' + R''''$ then 1 and 4, 2 and 3 belong to same wire.

If the resistance of the cross is constant it $= \frac{1}{2}$ the difference between $R'+R'''$ and $R''+R''''$.

18. *To Locate the Cross.*[3] The same instruments are required as is §16, also two good wires whose terminals are near

---
[1] *Electrical World*, XVIII, pp. 6, 30.
[2] *Electrical World*, Vol. XVIII, p. 30.
[3] *Electrical World*, Vol. XVIII, p. 30.

the crossed wires. 1st. Find the exact resistance of one of the crossed wires as in Figs. (13), (14), (15), §16. Next make connections as in Fig. (17), where a + b = Res. of crossed wire which we have just found. S is second wire crossed with it at X. ll are lead wires of known resistance. g = good wire.

Then, as before,
$$a = \frac{B(c+g+1) - A(l+R)}{A+B} = \frac{B(K-1) - (l+R)}{A+B}$$
and on reversing leads as shown by dotted lines
$$a = \frac{A(c+g+l+R') - Bl}{A+B} = \frac{A(R'+K-1) - Bl}{A+B}$$

Hence the Res. to the cross from testing end is found and the distance $= \frac{a}{c} D$.

The author gives further methods for location of crosses, but this is the most accurate.

RAILWAY MECHANICAL ENGINEERING.

BY PROFESSOR H. WADE HIBBARD,
Member of American Railway Master Mechanics Association.

What man of mechanical tastes has forgotten the awe which a locomotive used to impress upon his boyish heart? He would go as near as he dared to the wonderful engine, gaze at the engineer as if at a higher being, admire the unconcerned manner with which he started the ponderous machinery, and when an express train thundered past the otherwise quiet village home he felt the thought that, if only it were possible, the summit of his ambition would be to become such an engineer. This feeling of the lad is perhaps but the early exhibition of the general respect that exists in humanity for what is great and powerful. Many of those boys have indeed become engine runners; others have risen slowly through the shops and in later life have attained the higher positions of master mechanics, in charge of many locomotives; some few have taken the royal road of technical education and, uniting with it a brief practical shop experience, have advanced with leaps and bounds past the merely so-called practical man to the influential superintendence of motive power while yet in the full vigor and energy of young manhood. It is the purpose of this writing to point out the means of success by the latter method, to mention some of the pleasures, duties and responsibilities of railway mechanical engineering, its study at the technical school, the personal qualifications needed, and the opportunities presented to the ambitious man in this comparatively new and unfilled branch of skilled engineering.

In this country the term "engineer" has been appropriated in mechanical lines by those who run a stationary or locomotive engine. Abroad the title Locomotive Engineer is reserved for those who design, build and superintend the management of locomotives. That reservation will perhaps obtain in America when technical graduates become more numerous in railway work. Railway Mechanical Engineering includes not only the design, construction, operation and maintenance of locomotives, but also of cars, both freight and passenger. It covers the use not only of steam, but of electric traction and of electricity for lighting. In shop management modern methods of arrangement and equipment must be understood and practiced; electric, pneumatic and hydraulic appliances used, each as most suitable; and the economics of labor must be studied. The skilled purchase of railway material and the testing of existing equipment has to be supervised. European methods of locomotive engineering must be watched, and particularly is this true if one is engaged in contract shop manufacture of locomotives to compete in the world's markets of Japan, Australia, South America and elsewhere.

What are the pleasures of railway work? Chief of these is its variety. Man has ever been a creature fond of change. The traveler, passing from crowded London Cheapside to the grandeur of silence on Alpine glacier, from the æsthetic pleasures of the Louvre to view the squalid hordes of Islam at Mecca, is but the ancestral nomad—civilized. The same spirit that actuates him makes the railway mechanical engineer delight in his profession. First among his creations and care is a thing of power, not tied down to place, but roaming over the land from beside the sea, along green valleys, through deep gorges, or skirting lofty cliffs amid the wildest scenery. In the less important lines of his work he is called out upon this machine, not enough to be irksome, but to quicken his heart, freshen his lungs, and give zest to his spirits for more vigorous work in drafting room and office. The problems, too, that come up in design or management are almost infinite in number. He does not sit at a microscope and study the eyes of a beetle half a year, though that may be permitted to a few to be pleasure, but in superintending the design of a new locomotive his entire faculties must for a brief period be concentrated upon a judgment of one draftsman's detail and then upon another's work entirely different, and so on. The decisions to be made in the many lines of management; the choices from among the great variety of railway supplies; the necessity for his frequent presence and direction at different points along his road; the visits to car and locomotive works, steel works and foundries, or to the shops of other railroads, either in supervision of work being executed for him or in a general inspection visit to gain information and keep abreast with the best methods; the regular attendance at frequent Railway Club meetings and the annual Master Mechanics' Association convention,—all these preclude the remotest possibility of stagnation. To this is perhaps due the marked joviality of railway men. Incidental to the official position is also the free transportation for the man and his family—free over the whole road for the least important worker; free over neighboring railway systems and free in Pullman's for those a little higher; passes everywhere throughout the country for the Superintendent of Motive Power.

The work of the Locomotive Engineer if successful is sure of universal admiration, it is illustrated and praised in the railway technical press, discussed in the engineering associations, the designer congratulated by better known locomotive engineers whose long years of experience make their comment prized, the traveling public patronizes the road and remarks upon its improved equipment, all of which is indissolubly connected in the railway world with the engineer in charge.

Thus far the study of Locomotive Engineering has not received sufficient attention in the technical schools. There may be several reasons for this. Graduates have gone more largely into other lines. The schools in their management have been influenced by, even largely made up of, these graduates returned. Experimental laboratory engines, large and small, are with one notable exception exclusively of the stationary type and their behavior has been most minutely studied by stationary specialists. Students have naturally gravitated therefore into lines with which they were most familiar. They have not been familiar with the locomotive which is boiler plant, carriage and double engine combined to give a thousand horse-power within most contracted limits and to run under most adverse concomitants of instability, dust, weather, forcing and rough usage. Railroad motive power departments, with one exception, were not open to such men, while practical shop and road knowledge of locomotives seemed of more worth than theoretical ignorance. The steam engineering training of our technical schools has thus been almost altogether for the stationary engine, although the census of 1890 rated the horse power of our locomotives at ten million as against the four million developed in all the stationary steam power plants.

It is not at all difficult to obtain a professor of steam engineering who, a technical graduate, has risen to be in responsible charge of work. Many such successful teachers might be mentioned. It is needless to say that a professor of Locomotive Engineering should have been a technical graduate with extended and official railway experience. As noted before, technical railroad men are few in the first place and the pleasures and rewards of railway engineering are so great that it has been rare that one has made the change to the quieter life of the university.

American locomotive practice has been preëminently commercial along the lines of the larger economies. In Europe economies are closer. Only recently the receiver of a prominent Western railroad told the writer that engineering is a science of getting the most cents of interest out of a dollar of investment. Engineering students get filled up with the theories of scientifically correct designing and management. This is proper if not to the exclusion of commercial common sense. Coal is not burned in a locomotive with eyes squinted on careful economy so as to get the most steam, and hence the most pulling power, out of a pound of coal. If the consumption can be increased from a ton to a ton and a half per hour, *i. e.* one-half more, and by so doing that locomotive can pull one-third more cars per train, the author of the improvement is a great engineer. The theoretical man would call it wasting coal, while the railroad business man would say that though the half more coal does not pull a half greater train, still the extra coal is almost the only cost of hauling the extra cars because no increase is made in the interest on locomotive, track and bonds, in the wages of that engine and train crew, or of the track and signal men, in salaries of officials, in taxes or insurance. Practical railroad men have made these large savings, and in the newness and vast expansion of American railroad business the general managers have not felt the need of the smaller economies, large though the aggregate might be, which the educated engineer alone could introduce.

In considering the place of Locomotive Engineering as a technical school study the question of time is at once confronted. The addition of a graduate year is the best solution. This additional year of careful preparation is worth so much towards the later rapid advancement. Some thoughts upon that arrangement will be followed by a consideration of what can be done in undergraduate time.

A course of lectures should be given, as briefly outlined in the catalogue of the University of Minnesota of date 1895-6. They are divided as follows:

Past and future development of the locomotive.

Materials of construction. Motive power specifications and standards.

Locomotive and train resistances. Ruling grade as affected by kinetic energy. The track from motive power point of view.

The locomotive boiler; types, proportions, details, grates and heating surfaces, lagging, smoke prevention, circulation, water, fuels, effect of temperature upon metals, testing, accessories and attachments, shop work.

The locomotive engine; details, piston speed, reciprocating parts, bearing surfaces, link and valve motions, steam distribution, heat insulation.

The locomotive as a carriage; limitations, frames, spring and equalizing rigging, running gear, journals, truck wheels, drivers and their counterbalancing, brakes, steam heat, cab.

The tender; tank and attachments, wood and iron frames, built up and solid trucks.

Locomotive management; engine loads, coal premiums, working crew systems, expert instruction, lubrication, performance sheets.

Compound locomotives; systems and types, requisites for economy, cost of building and repairs.

European locomotive engineering and conditions of competition with American locomotives.

The domain and outlook for electric traction. The involved problems from electrical, railway and business standpoints.

Drawing room practice; preparation, management and classification of work, preservation of records, relations with the shops.

The shops; their arrangement, tools, cost and subdivision of power, labor paying, apprentices, reduction of costs by specialized machinery, by replacing hand work with machine work, by standardizing and duplication of parts, and watchfulness for wastes.

The railway test room and test department; inspection and purchase of materials, service tests of equipment, relations with general store house.

The railway mechanical engineer and superintendent of motive power, their qualifications and duties.

Actual designing should be carried on along the following lines, carrying out the principles of the above lectures, but keeping always in sight the restrictions to theoretical design which railway experience has found financially and practically to exist:

Designing of locomotive parts by the best modern methods. Link and valve motion designing by the geometrical diagrams with practical modifications and working models. The indicator diagram and inertia in designing. Determination of drivers, cylinders, steam pressure, boiler and grate for a given power and service, for simple and compound locomotives.

The engineer is also to be trained along the special needs of railway service test department work and a suitable amount of the following should be included:

Testing of railway appliances and supplies, as safety valves, injectors, gauges, air pumps, brakes, springs, metals of construction, lubricants, fuels, feed waters and their purifiers. Locomotive testing in road service with and without dynamometer car; also on laboratory experimental plant to eliminate the variables and permit finer manipulation and closer inspection.

The work in car design should be short and cover a few of the leading types of freight cars and trucks, including couplers and air brake work. Passenger and sleeping car design is rather specialized work and to it not much time should be given.

The graduate year would be arranged as follows, the hours being credit hours per week, designing and testing requiring two hours work for each credit hour:

	1st Term.	2nd Term.	3rd Term.
Locomotive lectures...............	5	5	5
Locomotive designing...........	5	5	5
Testing..................................	3	1	3
Seminar work, railway journals, and thesis...............	2	2	5
Car lectures and designing....	2	4	0
Elective, subject to approval but preferably electrical engineering.....................	3	3	2
	20	20	20

In studying Locomotive Engineering during the undergraduate period something will be accomplished in the senior year by using the time allotted to electives and to designing. The required subjects could also, if so decided, be very easily arranged with special reference to the needs of the prospective locomotive engineers. The course in valve gears might omit many types of high speed stationary design and then treat thåt most important part of the locomotive exhaustively. Thermodynamics would avoid all principles not related to locomotive practice; and windmills, gas engines and water motors would be omitted. Problems in design could be solely upon locomotives. The senior year in Locomotive Engineering would then be made to take a place similar to the senior year in electrical engineering at Cornell University.

Having taken Locomotive Engineering at the technical school the question arises as to the best way to enter the railway service. Locomotive shop work is unquestionably necessary. The summer vacations should be so spent if possible. Of course the best equipped man for an engineer is one who has had a general college education before his professional course. If two or three years can then be spent in a locomotive shop at nominal wages so as to be transferred frequently from one class of machines and work to another, the technical course to follow will be better improved. If the technical course precedes the railroad shop the latter will be better understood. On the Pennsylvania Railroad promotions can be made only from their own special shop apprentices who have been technical graduates. There are however very many railroads having no routine system of admission for technical graduates which are glad to get hold of one with shop experience even though it has not been gained in their own shops. A graduate therefore having had two or three years' shop practice will do well to get into the test department or drawing room of a first-class railroad. His abilities will soon make promotion for him because railroad officers are always on the watch to advance bright men into the upper places or to get them from another road.

A brief description of a railroad drawing room may not be amiss. The hours are short, seven or eight per day. The salary for the beginner about $60 a month. Two weeks' vacation in the year with no deduction of salary. Passes when desired. In railway work it is not customary to employ cheap help to make tracings for blue printing, the work being usually of such a nature that the designer can more profitably make his own tracing from his incomplete drawing.

In a small office with six or eight draftsmen the chief draftsman or mechanical engineer usually deals directly with each subordinate; with fifteen or twenty draftsmen three or four are leading men, the chief directing often through them, all the designs finally coming to him for approval. To avoid errors a second draftsman checks up the work of the original designer and is held equally responsible with him for mistakes to an extent variable with the nature of the drawing. The chief draftsman then signs it as correct, and forwards it in case it is a drawing sufficiently extensive or standard as to require the approval of the superintendent. The shops are not permitted to make any changes from the drawings without the approval of the chief draftsman. Thus all work is correctly recorded in his office and for future designs his records only have to be consulted. Where a general drawing room is established on an old road, particularly where there has been a great variety in locomotive equipment, the correspondence and investigation before each standard can be adopted and changes made is exceedingly extensive.

Draftsmen are often sent out on the road for information and into the main shops or to the test department. There is thus opportunity to broaden one's knowledge and to get fully acquainted with the railroad's usages.

In handling a drawing room the methods of government should be different from shop rules. Draftsmen are a much higher class of workers and must be treated accordingly. The best work can be obtained when they are permitted as engineers to feel personally interested in their work and informed as to its success after leaving their hands. This may seem axiomatic, but the writer knows where that is not the custom and where draftsmen have been regarded as mere machines to turn out the thought of the superiors.

The routine work of a railway test department is in part the inspection of car wheels and axles before purchase, and the testing of flat and coiled springs, boiler steel and other material which is in constant and severe use. Special tests are often made when changing the place of purchase, especially when adopting new standards, as for piston rod packing, cylinder sight-feed lubricators, brake shoe material or journal bearings. Tests requiring greater engineering ability are made, as in regard to the size and height of exhaust nozzles, smoke stack shape, grate area, driver counterbalance, valve action, injector efficiency or throttle and steam pipe area. Still more extensive are competitive trials between different classes of locomotives, covering coal and water consumption, with and without the dynamometer car, and involving the use of all the expert instruments of the testing engineer.

The variety of work of the Railway Mechanical Engineer has been mentioned. He must be the road's encyclopedia of mechanical information, knowing all that is best of other engineers' designs and how to apply them usefully and cheaply to his own road. He must cultivate cordial relations with officers of other roads, and this will naturally result in exchange of information, but be discreet in imparting such information to his competitor as shall not hurt his own road. The great army of railway supply men will often consume much of his time in urging him to favor their specialty, but he remembers that he cannot possibly be as well posted about injectors as the energetic man who is posted about nothing else, and so the engineer always learns something from the supply man and should listen to him patiently. In his periodical visits to the different division shops he should be on the lookout for improving methods of work, the need of modern or alteration and relocation of old machines, be quick to notice anything going wrong and take measures to have it corrected. Some cases in the writer's experience will illustrate: A connecting rod finisher was observed to be using too fine emery, requiring too long to work down the surface and putting on an unnecessarily fine polish He acknowledged that he had asked for coarser emery, but it was not given him. That matter was corrected in a few days. A bolt heading machine was seen to be running much too slow for economical production; chilled car wheels, mounted on their axles, were unloaded with danger of cracking by dropping them off the end of the flat car; new cast iron eccentric straps were being bored out for passenger engines, though bronze straps had been made the standard. In this latter case the store-house was in fault for not properly filling the division master mechanic's order for bronze castings. In some shops and round-houses split cotters were absent from some of the connecting-rod keys, letting the keys be thrown out on the road when loose. A modern wet emery tool grinding machine was not being used to grind tools to the tabulated cutting and clearance angles found by the manufacturers to be the best for quick and durable cutting.

Reference has been made to the charge of design in the drawing room and of the test department. The mechanical engineer is also the consulting engineer to his superintendent of motive power. If his road does not design its own locomotives he is asked his professional opinion and reasons as to what type is needed. If an engine does not make enough steam he has to be its doctor. If crank pins start to breaking he must show a draftsman how to calculate them to see if strong enough under proper usage. If the brakes of a coal car are not holding he must diagnose the difficulty. Compound locomotives come up and he is called upon to advise which of the twenty-five types is the best for the service of his road, and if need be design one perhaps in the early days when all compound locomotive designers had only stationary and marine practice as guide. All the railway technical papers are taken in his office and all glanced through and digested in part. He must keep up with the progress in steel manufacture, particularly of boiler, axle, tire and rod steels, as also the adaptation of steel and malleable iron castings and pressed steel in latest locomotive and car practice. Specifications for various materials are to be prepared and to be kept abreast with improvements. Acetylene gas for car lighting, roller curtain fixtures that the most stupid or irascible passenger cannot get out of order (and there are such fixtures), wrecking frogs that unfailingly replace a derailed car, air power for locomotive bell-ringers, the sand blast to prevent slipping of drivers, ribbed boiler tubes, stationary boiler design, chimney work, repair shop artillery for driving out refractory bolts, all of these hint at the varied details and responsibilities of this position. In all the designs and the standardizing the position of the general storehouse and their effect upon its stock carried and upon the interest accounts of the company must be born in mind. Electricity is recognized as a motive power and competitor to be absorbed, and on short lines where passenger traffic is fast and frequent, with light trains and the line to itself, it will probably be necessary in the near future to have plans ready for its adoption. The general consensus of expert electrical and railroad opinion is that for long distances or infrequent trains its possible adoption is very remote. The present efficient locomotive is capable of greater speed than the public is willing to pay lor, than the signal system could safely permit, or more profitable traffic make way for. The cost of installation is a commercial hindrance, and even if installed the low efficiency of intermittent and long distance operation, interest on investment of costly power stations and large conductors, would give no saving over the present steam traction.

Success in this profession implies learning, ability, untiring energy, alertness of thought and quick, independent decisions. It cannot be attained by the incompetent or by the use of chicanery or the artifices which in some pursuits are substituted for worth and work. Yet it lacks the charm of oratory, the dazzle of publicity, the swaying of opinions, which are so dear to the politician or the lawyer, and which surround them in their influence over men. Of all the branches of mechanical engineering, however, this one is the most abounding in life and activity. It is no place, this high tension of railway service, for the lover of quiet and moderate living. The speed of motion permeates the very air of offices, and the typical American as pictured abroad finds there his original.

I am led to close with some observations concerning the opportunities in railroad life. America is the distinctively railroad country of the world. In 1893 the world had 419,000 miles of railroad, of which 179,000 were in the United States and 148,000 in all Europe. A single American railroad has 3,400 locomotives, with 97,000 men on its pay rolls. It has been its policy for many years to take technical graduates into its shops, test department and drafting room, and from these "special apprentices" alone, to promote up through the various grades. Some of its best division superintendents of motive power are very young men, who have shown themselves worthy of rapid promotion. Another road, two years ago, appointed as superintendent of motive power a young man who from school went into the shops, was soon promoted, became a division master mechanic, where his tact in managing men, and his good judgment in caring for equipment while still keeping down expenses, brought him into notice, and he was called from this road to his present position. This road had previously done without technical men, but the policy of the management is now absolutely reversed. Only such men are now chosen for the various openings and trained for the future. A year ago one of their division master mechanics, who had seen several decades of service as a "practical" man, was replaced by a technical graduate of successful experience on another road. Many other leading railroads are inaugurating the same system to get expert technical men into the motive power positions. The highest official of one of the largest lines in the Northwest told the writer recently, having been in conversation about a number of unsettled problems in locomotive engineering, that he was wanting to secure technical graduates, as they alone were fitted to handle such matters intelligently; that the average master mechanic who had risen through the shops and road service was entirely at sea in technical investigation; and that the scientific solution of some matters mentioned meant thousands of dollars on the monthly income sheet of his road. In conversation at the Railway Clubs and the American Railway Master Mechanics' Association the writer has heard the same sentiments reiterated. Railroads are to the mechanical graduate a vast field ready for harvest. The thousands of positions as shop foremen and master mechanics, round-house foremen, road foremen of engines, motive power engineers, division master mechanics and superintendents of motive power, engineers of tests, foremen of drawing rooms, chief draftsmen and mechanical engineers, general superintendents of motive power, chiefs of motive power,—these, in the immediate future, are to be filled by the technical graduate of today.

ORE DEPOSITS IN MINNESOTA.
BY ARTHUR H. ELFTMAN, M. S.

Minnesota, while largely an agricultural and lumbering state, has within her limits vast mineral deposits which, even in their early stages of development, have become an important factor of the state's resources. The only ores which have been developed to any extent are those of iron. Gold, silver, nickel, cobalt, copper, and manganese are known to exist in the various rock formations, but have not yet been developed.

It is intended to give only a very brief sketch of the present status of development of the above mentioned metals.

IRON.

The iron ores are found chiefly in the Vermilion and Mesabi iron ranges. On the former range they occur in the Keewatin or Lower Huronian formation. The only places where they are mined are Tower and Ely. Eastward from Ely, extending through the eastern part of Hunter's Island, are very favorable indications of immense bodies of ore which have not been explored to any great extent. The ore at Tower is hard hematite, quite free from phosphorous and sulphur. The Ely ore differs from the preceding only in the degree of hardness, it being a soft ore easily worked.

The ores of the Mesabi found in the Animike or Upper Huronian are: hematite on the western end; magnetite in the central and eastern Mesabi. Several attempts have been made to mine the magnetite, but pure ore has not yet been found in quantities large enough to cover the expenses of mining. The magnetite-bearing rock occurs in large quantities, but ore rich in iron is quite limited.

The workable ores are the hematite deposits of the western Mesabi. The ore here occurs in large irregular bodies of soft hematite, easily worked and accessible without difficulty. Like the Vermilion ore it is found at the surface, usually concealed by a slight covering of glacial drift.

Development and exploration have not gone far enough to show the entire extent of the ore bodies thus far discovered. It cannot be said that all of the existing ore bodies have been discovered. The relative quantity of ore on the Mesabi and Vermilion ranges cannot be computed, owing to incomplete explorations on both ranges.

Titaniferous magnetite is found in the Keweenawan gabbro overlying the Animike of the Mesabi. This magnetite is generally a lean ore and does not occur in such large quantities as are usually reported. Even if processes were known by which the metal could be profitably extracted from titaniferous ore, this part of the mining industry of the state would be an insignificant factor.

MANGANESE.

Manganese has been found in workable quantities on the Mesabi range associated with the hematite. Analyses of iron ores from other localities show traces or a low percentage of manganese.

GOLD AND SILVER.

During the last thirty years discoveries of gold and silver have been announced from every part of the state. Every county at some time seems to have had its gold excitement, whether based upon the unearthing of a brass kettle or the finding of a speck of gold in a mountain of granite. Gold occurs in small quantities in the oldest rock formations. It is still very doubtful whether deposits rich enough to pay for extracting the gold will ever be found within the limits of the state.

NICKEL AND COBALT.

Nickel and cobalt have been found only in very small quantities in bog manganese ores, pyrrhotite from the Animike, and the ferro-magnesian silicates of the Keweenawan basal gabbro. The most reliable analyses show that nickel runs as high as three per cent. and cobalt 0.71 per cent. Usually analyses show but a trace or less than one per cent. of the two combined. The two metals are associated with each other. The nickel, however, predominates.

Up to the present time nickel ore has not been found in quantities warranting an outlay for mining development. Since the nickel has been found in small quantities scattered throughout the great gabbro mass of northeastern Minnesota it is possible that it may be found in paying quantities in localities where the basic constituents of the gabbro have been collected by the differentiation of the original magma and a further concentration of the nickel effected through the decomposition of these segregated basic masses.

COPPER.

Copper is found in the Keweenawan rocks of the lake Superior region. Although this formation covers an extensive area in Minnesota the copper-bearing horizon is limited and carries but very little copper. Chalcopyrite is scattered throughout the basal gabbro in small quantities. Malachite is found as incrustations upon the rock. Native copper occurs in the amygdaloidal lava flows which largely compose the copper-bearing horizon. The copper fills the cavities of the vesicular layers of rock and is also found in veins.

Numerous specimens of float copper have been found in the glacial drift throughout the state, but these were presumably derived from the Keweenawan rocks. From the character of the rock and other mineralogical associations it is not expected that copper will ever be found in paying quantities. At present this is confirmed by explorations at numerous places along the north shore of lake Superior. Very little copper was found at Cascade River and the Stewart river "mines." These two localities the writer considers show the most favorable indications in the state.

FOUNDATIONS FOR A POWER HOUSE.

BY GEORGE J. LOY, B. C. E., '84

In the recent construction of the foundation of a power-house for the new water-works system of Spokane, Wash., a number of unexpected difficulties, which necessitated certain changes in the proposed plans, were met with. These difficulties and their remedies will be briefly described.

The plans for the power-house were prepared by the Consulting Engineers of the water-works system, and then submitted for approval to the well known Hydraulic Engineer, J. T. Fanning, of Minneapolis. The power-house was to be located on the river bank about 40 feet from low water mark; the subsoil consisted of a bed of gravel known to be of considerable depth. The gravel occurred in layers of different thicknesses, consisting of coarse sand alternating with coarse gravel and small boulders.

According to the plans the masonry was to be of granite, laid on concrete footing, three feet thick. The depth of the excavation for the river wall was to extend ten feet below low water mark, to allow a discharge under water from the five water wheels (each fifty-four inches in diameter), and also of the concrete footing. When excavating was first started, three centrifugal pumps, with a combined capacity of 22,000 gal. per minute, were put in operation. The size of the excavation was 120 x 40 feet. After working a short time with these pumps it was found that the best they could do was to lower the water two feet below the water level of the river. When this became apparent the excavation was divided into smaller areas, but still the pumps failed to fulfill the requirements. This state of things lasted for ten days, when a rise in the river necessitated an abandonment of all work.

The plans were now entirely changed, a grillage foundation on piles being substituted for the masonry and concrete. This foundation was to consist of piles, driven in rows, two feet eight inches apart, the rows being three feet apart. On top

Foundations for a Power House. 119

of the piles was placed ordinary grillage, consisting of timbers fourteen inches square, which in turn was covered with a solid flooring of timber twelve inches square, which was to support the wall.

The excavation was finally completed by finishing small portions at a time and driving sheet piling around the sides and carefully filling up all leaks. The sheet piling was driven to a depth of about ten feet so that it would stand after the excavation was finished. It was found by experiment that the best sheet piling in this case was made of small piles about ten inches in diameter. Other kinds, such as grooved lumber piling and piling of square timbers, crushed before it could be driven to the required depth. Even with this method the final work of excavating had to be done in three feet of water.

The bearing piles for the wall and piers were driven from eleven to sixteen feet below the point of cut-off. Each pile will have a load of seven tons, with a factor of safety of at least two. The bearing piles had to be cut off by hand under water with an ordinary cross-cut saw, which, however, was fastened in a frame somewhat resembling that of a buck-saw, but having the middle brace raised three feet above the saw blade. The piles were first all cut off at a point out of water just three feet above the required point, and were afterwards cut off at the required point, the saw being held in its true position under water by means of the brace, which slid across the top of the pile out of water. Four men were required to run the saw. Cap timbers were next drift-bolted to the piles. Owing to their buoyancy eight men were required to hold them down while they were being drift-bolted. To facilitate drift-bolting a follower, composed of a piece of gas pipe and a round iron bar, was used. The 12-inch timber floor was next drift-bolted to the caps in a similar manner.

It was now thought advisable to bring the flooring up to the surface of low-water mark, which was done. This would enable the masonry to be laid out of water with more care.

At this juncture another change in the plans was made; this was due to the desire on the part of the city officials to rush the construction work. A concrete wall was substituted for the proposed masonry wall, the dimensions of the larger wall being 120 feet long by 30 feet high with an average thickness of 7 feet. The construction of this wall took ten days, this being about one third the time necessary to construct a similar masonry wall. During one day of nine hours this wall was built up six feet. The composition of the concrete was as follows: Portland cement, 1 part; sand, 2 parts; gravel, 1 part; broken stone, 3 parts.

A wooden frame-work for holding the concrete in place until set, was constructed as follows: 6" x 8" posts, spaced eight feet apart, were set up on either side of the wall. Each pair of these posts was tied together every seven feet in height by ¾" circular iron rods. A wall inside of these posts was now constructed by laying up 2" x 12" planks surfaced on the inside. The wall was braced on the outside by light braces to prevent bending. Light mouldings were placed on the inside of these walls, so as to form joints on the concrete, thus giving it the appearance of masonry. These mouldings were ⅜" x ⅞", three being used for each joint; two were laid flat, and the third edgewise between them, thus giving the joint an appearance of having a chisel draft on either side. Arch stones were shown in like manner. The inner surfaces of these walls were coated with soap-suds to prevent the concrete from sticking to the wood. Openings, anchor-bolts, etc., are easily placed in such a wall.

In laying the concrete, special care was taken that the material which was placed next the mouldings should contain no coarse gravel or rocks. This of course gives a smooth and neat appearance to the surface and joints. After this wall was completed it was carefully watched for a month; but no settling or cracks could be detected.

The principal changes in the plans, that of substituting grillage for concrete footings and a concrete for a masonry wall, gain several advantages; first, it reduced the depth of excavation ten inches and avoided the necessity of laying concrete or masonry under water; second, it allowed of a change in the 72" circular draft tube, to a smaller oval one; third, it saved at least fifteen days in the time of construction of the work. This last was an important item, inasmuch as the rainy season, bringing with it high water, was close at hand.

DESCRIPTIVE GEOMETRY AND WORKING DRAWINGS.

BY PROFESSOR W. H. KIRCHNER.

For technical purposes, it is of the utmost importance to represent solids and other figures in three dimensions by a drawing in one plane. A variety of methods have been introduced for this purpose. All, however, are systems of projection.

Descriptive (practical solid, darstellende) geometry, is the theory of making projections of any accurately defined figure, such that from them can be deduced the figure itself and all its metrical properties.

Under the name of "Geometrie Descriptive" Monge (1746-1818) about 1794 invented a method of drawing in plan and elevation or orthographic projection.

Plans and elevations, especially of buildings, were in use before his time, and rules had been developed to determine by construction from drawings the shape of stones in vaults and arches. These rules were reduced to a consistent method of projection by Monge.

Thus descriptive geometry dates from Monge, whose treatise appeared in 1800.

Since then purely geometrical methods have been continuously extended, especially by Poncelet, Steiner, Chasles, Von Staudt and Cremona. The beginnings of perspective date from the time of the Greek mathematicians; at present it is generally treated as a special case of projection. The theory of geometry in general, treated by the means of projection, is now considered as descriptive geometry.

Descriptive geometry was first taught at West Point, in 1817, but did not find its way into other institutions for a number of years. Today it is considered an important element and is found in the curriculum of all schools of engineering.

A working drawing is simply the application of practical geometry to the representation of any object and conveys to the mind of a skilled workman clear and exact information of form and magnitude. Drawing has become such an important factor in constructive engineering that it is well described as "the language of the work-shop." The use of working drawings in engineering has become so general and so extensive that all persons engaged in manufacturing industries are expected to understand a projection drawing, whether contractors, users, buyers or sellers of machinery.

This continued and ever increasing use of projection in the works and the office of the commercial engineer has demonstrated that for practical purposes it is expedient to make some changes in the nomenclature and the preliminary notions of descriptive geometry, such as the names of the projections or views, and the relative position of the object and the planes of reference.

The *rabatting* of the planes and consequently the arrangement of the views on the drawing depend upon this preliminary relation.

Descriptive geometry is a science when it shows a mathematical basis for its methods, and an art when it deals with the execution of its methods, hence any change we make in its preliminary notions of arrangement, execution and nomenclature, will not necessitate any change in its mathematical demonstrations.

Practical experience in making working drawings has shown that a few changes in this direction are for the better. The subject is taught more and more with reference to its practical application, and is pursued mainly by students who expect to follow some of the engineering professions as a livelihood.

In passing from his text book to the practical problem the student experiences some trouble in becoming familiar with the new arrangement of the projections. The difficulty is not great, but there is no occasion for its existence, and it should be removed.

It may be said that this is merely passing from the first to the third quadrant in descriptive geometry, and that no difficulty should arise in the transition. However, such is not the case, and the record of the class room shows that the average student cannot in the limited time work with equal facility in all four quadrants, and manipulate a problem involving more than one quadrant with ease. Nor is it advisable to devote a considerable portion of his time to acquire facility in passing from one quadrant to another at the expense of his knowledge of principles and methods.

In practice he will have no occassion to pass from one arrangement of views to another, and whatever facility he may have acquired in that direction is discipline gained at the expense of practical knowledge.

The time saved by avoiding unnecessary transformations is used in extending the scope of the subject matter, and nothing is lost in the disciplining of the geometric imagination, in fact there is a decided gain. All English and Continental descriptive geometries use the first quadrant. The first chapters generally contain problems illustrating the projections for points and lines placed in all four quadrants, but the larger part of the treatise confines the problem to the first quadrant.

Nearly all American texts treat the subject in the same manner, not even excepting those published within the last ten years. In works on mechanical drawing we find that many of the books written within the last decade have adopted the shop method of plan above the elevation, and a few have discarded the terms plan and elevation and used top view, front view and the like.

The line of intersection of the horizontal and vertical planes of reference is called both the ground line and the axis. As an illustration of the different notations we find the following: For this line of intersection, GL, G, XY, X, AB, etc. For the horizontal and vertical projection of a point (A) we find, $a^h a^v$, aa', $A_1 A_2$, $a^T a^F$, $a'a''$, and others. For lines, traces and planes we find the same diversity, and in addition the use of heavy and light lines, broken and dotted lines, dash and dot lines. In passing from one text to another we are quite often compelled to study the notation. Fortunately, all of this notation and the ground line or axis do not appear in the practical problem and no difficulty arises.

Foreign drafstmen use the first quadrant method of arrangement of views. The third quadrant is used in American practice. Many teachers prefer to dispense with the question of quadrants, and obtain the projection by assuming the object to be placed within a transparent cube, the projections being called the top view, front view, right view, and so on. The unfolding or *rabatting* of the faces of the cube brings the views in their proper relation. This avoids the old method of having the right view or elevation on the left side of the front elevation.

The third quadrant or shop method is generally followed in our manual training and mechanic arts high schools, where the study of projection is pursued without taking up the more complicated and general constructions of descriptive geometry. In colleges of engineering the practical or shop method is used in the drawing room, but not nearly as much as one would expect.

Treatises on machine drawing and design and kindred subjects, if English, are illustrated chiefly by first quadrant methods, if American we find that nearly all of the new illustrations of details of engineering practice are drawings in the third quadrant and in some occasionally an old cut introduced showing the first quadrant method. If simple details are shown no difficulty is experienced, but when we have right views in addition to the front, it is annoying to say the least.

The use of photography in making illustrations has done much to bring the drawings of the practical draftsman into engineering periodicals. It is to be hoped that it will not be long before we can say there is no essential difference between the method of the recitation room and the office of the engineer.

SOCIETY OF ENGINEERS

IN THE

UNIVERSITY OF MINNESOTA

LIST OF MEMBERS
MAY, 1896

HONORARY MEMBERS.

C. W. Hall, M. A.
Wm. R. Hoag, C. E.
J. E. Wadsworth, C. E.
Geo. D. Shepardson, M. E.
Wm. R. Appelby, B. A.
Frank H. Constant, C. E.
H. Wade Hibbard, M. E.
C. H. Kendall, C. E.

H. E. Smith, M. E.
W. H. Kirchner, B. S.
J. H. Gill, M. E.
B. E. Trask, C. E.
G. H. Morse, B. E. E.
H. T. Eddy, Ph. D.
F. W. Denton, E. M.

ACTIVE MEMBERS.

Beyer, A. C.
Blake, R. P.
Burch, A. M.
Coleman, L. M.
Craig, Robt.
Cross, Chas. H.
Dahl, H. F. M.
Donaldson, Tom.
Erickson, H. A.
Garvey, J. J.
Gilchrist, C. C.
Glass, C. A.
Hastings, Clive.

Herzog, J. S.
Hildebrandt, H.
Hilferty, C. D.
Irwin, J. B.
Jones, C. P.
Lee, E. A.
Linton, J. H.
Long, F. W.
Lonie, J. H.
Magnusson, C. E.
McKellip, F. W.
McKinstry, Wm.
Miller, W. L.
Neil, V. A.

Pratt, G. A.
Pratt, Sid.
Roberts, J.
Silliman, H. D.
Shumway, E. J.
Taylor, E. W. D.
Wagner, A. W.
Wales, R. T.
Walker, F. B.
Wheeler, H. M.
Woodford, G. B.
Wright, R. V.
Zeleny, F.

CORRESPONDING MEMBERS.

1875.

NAME AND RESIDENCE.	DEGREE.	OCCUPATION.
LEONARD, HENRY C. (B. S., '78) Minneapolis, Minn.	B. C. E.	Physician.
RANK, SAMUEL A. Central City, Colo.	B. C. E.	Civil and Mining Engineer.
STEWART, J. CLARK Minneapolis, Minn.	B. C. E.	Physician: Prof. of Pathology at University of Minnesota.

1876.

GILLETTE, LEWIS S. Minneapolis, Minn.	B. C. E.	President of the Gillette-Herzog Manufacturing Co.
HENDRICKSON, EUGENE A. St. Paul, Minn.	B. C. E.	Lawyer.
THAYER, CHARLES E. Minneapolis, Minn.	B. C. E.	Grain Dealer.

1877.

PARDEE, WALTER S. Minneapolis, Minn.	B. Arch.	Architect.

1878.

BUSHNELL, CHARLES S. Minneapolis, Minn.	B. M. E.	Mf'r. Stoves, Ranges and Furnaces.

1879.

DAWLEY, WILLIAM S. Chicago, Ill.	B. C. E.	Chief Engineer C. & E. I. Railway Company.
*FURBER, PIERCE P.	B. C. E.	Died April 6, 1893.

1883.

BARR, JOHN H. Ithaca, N. Y.	B. M. E.	Ass't Prof. Mechanical Engineering, Cornell University.
PETERS, WILLIAM G. Tacoma, Wash.	B. C. E.	Vice-President Columbia National Bank.
SMITH, LOUIS O. Le Sueur, Minn.	B. C. E.	Civil Engineer.

1884.

LOY, GEORGE J. Spokane, Wash.	B. C. E.	Civil Engineer.
MATTHEWS, IRVING W. Waterville, Wash.	B. C. E.	Real Estate, Insurance and Abstracts of Titles and Civil Engineer.

1885.

BUSHNELL, ELBERT E. New York City.	B. M. E.	Proprietor of the Hooper Typewriter Prisms.
*FITZGERALD, PATRICK T.	B. C. E.	Died, April 2, 1887.
REED, ALBERT I. Hastings, Minn.	B. C. E.	Civil Engineer.

1886.

NAME AND RESIDENCE.	DEGREE.	OCCUPATION.
WOODMANSEE, CHARLES C. Midway Park, St. Paul.	B. ARCH.	Book-keeper.

1887.

ANDREWS, GEORGE C. Minneapolis, Minn.	B. M. E.	Heating Manufacturer and Contractor.
CRANE, FREMONT (B. S. '86) Prescott, Arizona.	B. C. E.	Civil Engineer.

1888.

ANDERSON, CHRISTIAN Portland, Ore.	B. C. E.	Civil Engineer.
LOE, ERIC H. Minneapolis, Minn.	B. M. E.	Mechanical Engineer, with Nordyke & Marman.
MORRIS, JOHN West Pullman, Ill.	B. M. E.	Assistant Superintendent and Mechanical Engineer of Plano Manufacturing Co.
HOAG, WILLIAM R. (B. C. E. '84) Minneapolis, Minn.	C. E.	Prof. of Civil Engineering, University of Minnesota.

1889.

COE, CLARENCE S. Wenatschee, Wash.	B. C. E.	Civil Eng. in Engineering Dept. of C., M. & St. P. Ry. Co.

1890.

BURT, JOHN L. Minneapolis, Minn.	B. C. E.	Commission Merchant.
DANN, WILBER W. Minneapolis, Minn.	B. C. E.	Civil Eng., Supt. of Water Works Con'st'n at St. James, Minn.
GILMAN, FRED. H. Minneapolis, Minn.	B. C. E.	Editor Mississippi Lumberman.
GREENWOOD, WILLISTON Minneapolis, Minn.	B. C. E.	Civil Engineer.
HAYDEN, JOHN F. Minneapolis, Minn.	B. C. E.	Newspaper work.
HIGGINS, JOHN T. St. Paul, Minn.	B. C. E.	Physician.
HOYT, WILLIAM H. Duluth, Minn.	B. C. E.	Assistant Engineer, Duluth & Iron Range R. R.
NILSON, THORWALD E. Minneapolis, Minn.	B. M. E.	Manager of the Norgren Distilling Co.
SMITH, WILLIAM C. St. Paul, Minn.	B. C. E.	Assistant Engineer, N. P. R. R.
WOODWARD, HERBERT M. Boston, Mass.	B. M E.	Instructor of Wood Working, Mechanic Arts High School.

1891.

ASLAKSON, BAXTER M. Dayton, Ohio.	B. M. E.	With the Stillwell-Bierce & Smith-Vaile Co.
CARROLL, JAMES E. Minneapolis, Minn.	B. C. E.	City Engineer's Office.

NAME AND RESIDENCE.	DEGREE.	OCCUPATION.
CHOWEN, WALTER A. St. Croix Falls, Wis.	B. C. E.	On Survey of Minneapolis, St. Paul & Ashland R. R.
DOUGLAS, FRED L. New York City.	B. C. E.	Civil Engineer.
GERRY, MARTIN H., Jr.(B.M.E.'90) Chicago, Ill.	B. E. E.	Superintendent of Motive Power of the Metropolitan West Side Elevated Railroad Co.
HUHN, GEORGE P. Minneapolis, Minn.	B. E. E.	Flour City National Bank.

1892.

BURCH, EDWARD P. Minneapolis, Minn.	B. E. E.	Electrical Engineer, Twin City Rapid Transit Company.
BURTIS, WILLIAM H. Minneapolis, Minn.	B. E. E.	Electrical Engineer and Contractor.
FELTON, RALPH P. Minneapolis, Minn.	B. M. E.	Fire Insurance.
GOODKIND, LEO. St. Paul, Minn.	B. ARCH.	City Supt. of Schoolhouse Construction.
GRAY, WILLIAM I. Minneapolis, Minn.	B. E. E.	Electrical Engineer and Contractor.
HANKENSON, JOHN J. Minneapolis, Minn.	B. C. E.	Bridge and Sanitary Engineer.
HIGGINS, ELVIN L. Hutchinson, Minn.	B. C. E.	Teacher.
HOWARD MONROE S. Minneapolis, Minn.	B. E. E.	Electrical Engineer and Contractor.
MANN, FRED M. Philadelphia, Pa.	B. C. E.	Instructor Architectural Design, U. of Penn.
PLOWMAN, GEORGE T. Minneapolis, Minn.	B. ARCH.	Draughtsman.

1893.

ANDERSON, OLE J. Nicollet, Minn.	B. C. E.	Nicollet County Surveyor.
AVERY, HENRY B. Minneapolis, Minn.	B. M. E.	With Gillette-Herzog Manufacturing Co.
BATCHELDER, FRANK L. St. Paul, Minn.	B. C. E.	With C. F. Loweth.
CHASE, ARTHUR W. Hastings, Minn.	B. E. E.	Electrician.
COUPER, GEO. B. Northfield, Minn.	B. M. E.	Manager Northfield Electric Light Company.
DEWEY, WILLIAM H. New York City.	B. E. E.	Assistant Engineer for the American Boiler Co.
ERF, JOHN W. Minneapolis, Minn.	B. C. E.	With Gillette-Herzog Mfg. Co.
GUTHRIE, JOHN D. Minneapolis, Minn.	B. E. E.	Medical Student, University of Minnesota.
HOYT, HIRAM P. Minneapolis, Minn.	B. C. E.	With Gillette-Herzog Mfg. Co.

NAME AND RESIDENCE.	DEGREE.	OCCUPATION.
MORSE, GEORGE H. Minneapolis, Minn.	B. E. E.	Instructor in National School of Electricity.
REIDHEAD, FRANK E. Minneapolis, Minn.	B. E. E.	Electrician, Minneapolis General Electric Co.
SPRINGER, FRANK W. Minneapolis, Minn.	B. E. E.	Scholar in Electrical Engineering.
WASHBURN, DELOS C. Minneapolis, Minn.	B. ARCH.	With J. T. Fanning.

1894.

BRAY, GEO. E. Virginia City, Minn.	B. M. E.	Electrical Superintendent of Mining Company.
CHALMERS, CHARLES H. Minneapolis, Minn.	B. E. E.	Electrician and Assistant Manager, D. & D. Electric Mfg. Co.
CUNNINGHAM, ANDREW O. New Orleans, La.	B. E. E.	With Gillette-Herzog Mfg. Co.
GILL, J. H. (B. M. E., '92) Minneapolis, Minn.	M. E.	Instructor in Shop Work, University of Minnesota.
GILMAN, JAS. B. Minneapolis, Minn.	B. C. E.	With Gillette-Herzog Mfg. Co.
JOHNSON, NOAH. St. Paul, Minn.	B. C. E.	General Office G. W. R. R.
TRASK, BIRNEY E. (B. C. E., '90) Highland Park, Ill.	C. E.	Professor of Mathematics, Northwestern Military Academy.
WEEKS, WILLIAM C.	B. C. E.	Venezuela Survey.

1895.

ADAMS, GEO. F. Minneapolis, Minn.	B. E. E.	Electrical Engineering Co.
BISHMAN, A. E. Willmar, Minn.	B. E. E.	Supt. Electric Light Plant.
BOHLAND, JOHN A. St. Paul, Minn.	B. C. E.	Draughtsman, Bridge Dept. G. N. R. R.
CASSEDY, GEO. E. St. Paul, Minn.	B. C. E.	Draughtsman, Bridge Dept. G. N. R. R.
CHAPMAN, L. H. St. Paul, Minn.	B. C. E.	Engineering Dept. G. N. R. R.
CUTLER, HARRY C. (B. E. M. '94) Camp Golden, Whitehall, Mont.	E. M.	Assayer, A. D. & M. Co.
CHRISTIANSON, PETER. (B. E. M. '94) Minneapolis, Minn.	E. M	Instructor in Mining at University of Minnesota.
EDDY, HORACE T. Minneapolis, Minn.	B. E. E.	Graduate Student, University of Minnesota.
ROUNDS, FRED M. Minneapolis, Minn.	B. E. E.	Electrician, with Standard Telephone & Electric Co.
SHEPHERD, B. P. Minneapolis, Minn.	B. M. E.	Draughtsman, with Paul & Hawley.
TANNER, H. L. Minneapolis, Minn.	B. E. E.	Electrician, with Burtis & Howard.

NAME AND RESIDENCE.	DEGREE.	OCCUPATION.
TILDERQUIST, WM. Vasa, Minn.	B. M. E.	
VON SCHLEGELL, F.	B. E. E.	With St. Anthony Water Power Company
WEAVER, A. C. Minneapolis, Minn.	B. M. E.	Draughtsman.
WILKINSON, CHAS. DEAN, Gibbonsville, Idaho.	B. E. M.	In Chlorination Plant, A. D. & M. Co.

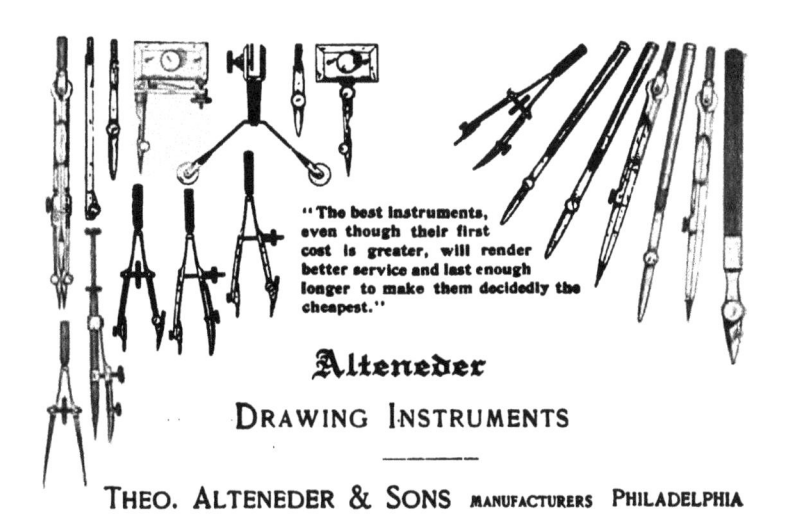

"The best instruments, even though their first cost is greater, will render better service and last enough longer to make them decidedly the cheapest."

Alteneder
DRAWING INSTRUMENTS

THEO. ALTENEDER & SONS MANUFACTURERS PHILADELPHIA

Send five cents in postage stamps for new Catalogue

Work of Compressed Air in and about Chicago, by the Rand Drill Co. Machinery:

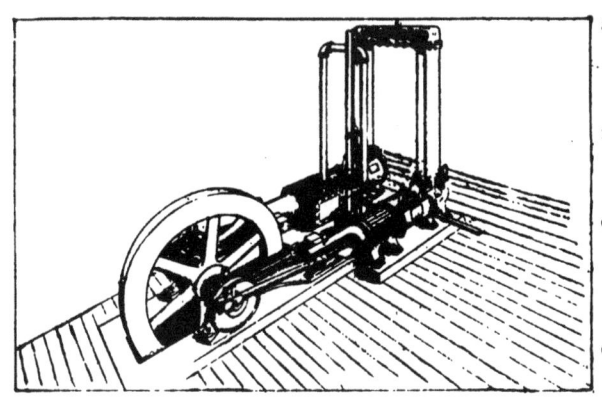

3,600,000 cubic ft. of air furnished to Machine, Boiler and Railroad Shops per day.

15,000,000 gallons of water pumped from deep wells daily.

Over 9,000,000 cu. yds. of rock excavated from the Chicago Drainage Canal in three years.

Over six miles of Intake Tunnel being driven.

Rock Drills.   **RAND DRILL CO.**   Air Compressors.

1328 Monadnock Block, CHICAGO, ILL.

100 Broadway, NEW YORK CITY.

MEACHAM & WRIGHT

Manufacturers' Agents for

Utica Hydraulic Cement

and Dealers in

PORTLAND and LOUISVILLE CEMENTS,

MICHIGAN and NEW YORK STUCCO.

98 MARKET ST., CHICAGO.

Telephone Main Express 59.

Engineers......

expecting to enter railroad service, or to do any work connected with railroading, will find invaluable assistance in reading the

RAILROAD GAZETTE......

It is published weekly and contains practical articles, many of them illustrated with accurate drawings, in all departments both engineering and operating. It is the largest, oldest and by far the most complete weekly railroad publication in the world. It is published at

32 Park Place, New York,

and the subscription price is $4.20 a year. Specimen copies free. Catalogue of The Railroad Gazette publications free.

LOUISVILLE CEMENT.

The undersigned is agent for the following works:

Hulme Mills,	producing	Star Brand.
Speed Mills,	"	" "
Queen City Mills,		" "
Black Diamond Mills (River) "		Diamond "
Black Diamond Mills (R. R.), "		" "
Falls City Mills,	"	Anchor "
Silver Creek Mills,		Acorn "
Eagle Mills,		Eagle "
Fern Leaf Mills,		Fern Leaf "
Peerless Mills,		Crown "
Lion Mills,		Lion "
Mason's Choice Mills,	"	Hammer & Trowel "
United States Mills,		Flag

These works are the largest and best equipped in the United States.

Orders for shipment to any part of the country will receive prompt attention. Sales in 1892, 2,145,568 barrels.

Western Cement Company,

247 West Main St., - - - - Louisville, Ky.

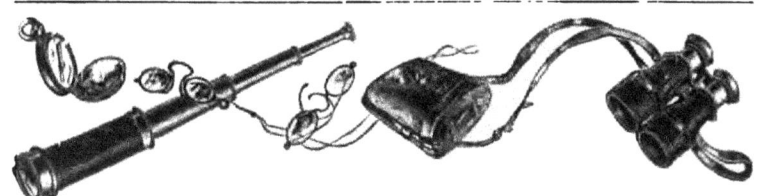

BOERINGER & SON,
OPTICIANS,

Engineers', Architects' and Surveyors' Instruments,

TECHNICAL DRAWING MATERIAL.

54 East Third St., ST. PAUL, MINN.

Pioneer Electrical Journal of America.

Most Popular of Technical Periodicals.

THE ELECTRICAL WORLD, weekly, is the largest, most handsomely illustrated and widest circulated journal of its kind in the world.

It is **ably edited** and **handsomely illustrated**, and is noted for **popular treatment** of subjects in simple and easy language, devoid of technicalities. No other technical journal has as many general readers.

THE ELECTRICAL WORLD devotes a large part of its space to **alternating and multiphased currents**—subjects that no student can afford to neglect, and which no other electrical journal in the world treats so fully—while the **Weekly Digest of Current Technical Electrical Literature** gives a complete resumé of current **progress** in electrical science and its application, **both in this country and abroad.**

SAMPLE COPIES FREE.   AGENTS WANTED.

BOOKS ON ELECTRICAL SUBJECTS.

There is no work relating to Electricity, Street Railways or kindred subjects that is not either published or for sale at the office of THE ELECTRICAL WORLD, from which is also issued weekly the ELECTRIC RAILWAY GAZETTE (subscription $3.00 a year) and annually JOHNSTON'S ELECTRICAL AND STREET RAILWAY DIRECTORY, price $5.00.

Books promptly mailed, POSTAGE PREPAID, on receipt of price. Catalogue and information free.

THE W. J. JOHNSTON COMPANY,

253 Broadway, New York.

Minneapolis Radiators

for

Minnesota People.

MINNEAPOLIS RADIATOR AND IRON COMPANY

829 Ninth Street S. E.,
Minneapolis, Minn.

GIVES PROMINENCE TO
MUNICIPAL AND BUILDING ENGINEERING
WHICH INCLUDES

Water-Works (Construction and Operation), Sewerage, Bridges, Metal Construction, Pavements, Subways, Road Making, Docks, River and Harbor Work, Tunneling, Foundations, Building Construction, Industrial Steam and Power Plants, Ventilation, Steam and Hot-Water Heating, Plumbing, Lighting, Elevator and Pneumatic Service.

"The success of this publication has been marked in many ways; not only has it become a source of profit to its projector but it has been of incalculable value to the general public whose interests it has always served."—*Cincinnati Gazette.*

"It stands as a fine example of clean and able journalism."—*Railroad Gazette.*

Published Saturdays at 277 Pearl St., NEW YORK.

$5.00 PER YEAR. SINGLE COPY, 12 CENTS.

THE ENGINEERING RECORD is the recognized medium for advertisements inviting proposals for all Municipal and U. S. Government Engineering and Public Building Work. Its subscribers include the experienced and reliable Contractors and Manufacturers of Engineering and Building Supplies in all sections of the United States and Canada. The RECORD'S value to secure competition in bids is therefore obvious.

KUHLO & ELLERBE

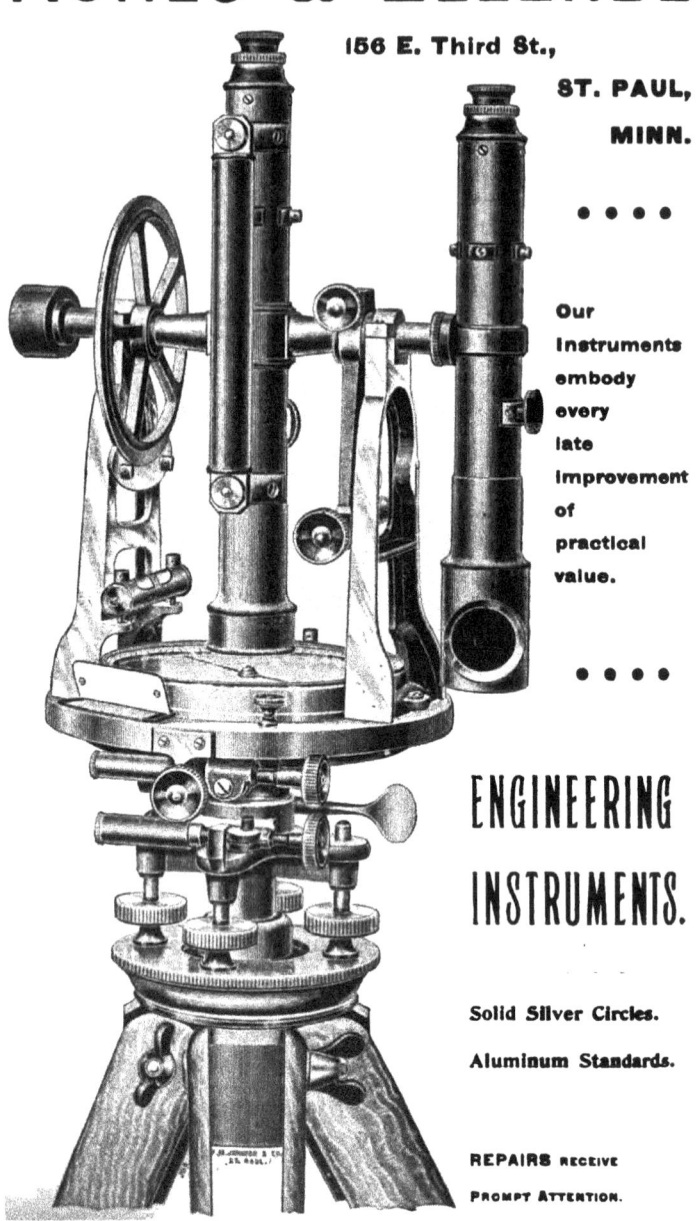

156 E. Third St.,

ST. PAUL,
MINN.

• • • •

Our
Instruments
embody
every
late
improvement
of
practical
value.

• • • •

ENGINEERING
INSTRUMENTS.

Solid Silver Circles.

Aluminum Standards.

REPAIRS RECEIVE
PROMPT ATTENTION.

THE UNIVERSITY OF MINNESOTA
COLLEGE OF
ENGINEERING AND THE MECHANIC ARTS

CYRUS NORTHROP, LL. D., President.

THE COLLEGE OF ENGINEERING AND THE MECHANIC ARTS offers four years' courses in *Civil Engineering, Mechanical Engineering, Electrical Engineering*, with Special Courses in the above-named departments and in *Industrial Art*.

The large and increasing equipment of the shops and laboratories gives *a modern and practical character* to the work.

The more recent lines of work opened to students are the following:

Structural Engineering, with especial emphasis on bridge building and the designing of steel and stone structures.

Railway Mechanical Engineering which has been expanded to form a complete senior year for those wishing to prepare for official positions in railway motive power departments. Courses of lectures are offered in the design, construction, operation and maintenance of railway equipment, with drawing, locomotive testing and electric railway engineering. The demand from the railways for graduates who have, during the past two years, taken partial work in this specialty, has justified this expansion.

Upon completion of a full course of study, the degree of Civil, Mechanical or Electrical Engineer is conferred.

The situation of the College in the manufacturing and commercial center of the upper Mississippi region is particularly favorable. Railway shops, manufactories, electric light and power stations are open not only for inspection but for study and research work. Thus the student can become acquainted with large engineering enterprises conducted under business methods and in a scientific way.

Requirements for admission: English grammar and composition, with essay; algebra, plane and solid geometry; history or civil government; natural sciences; free-hand drawing; German or French. While four years' work in Latin may be offered, students are urged to present their preparation in German. Persons of mature years may be admitted to special lines of study provided they give evidence of ability to pursue the same with profit.

Tuition is free. Shop and laboratory fees are very low.

For Catalogue and Further Information, address

CYRUS NORTHROP, President, Minneapolis, Minn.

THE BEST PLACE

For an Engineer to buy

Books, Instruments, Cameras,

Or anything else in his line
——————IS——————

The University Book Store.

SCHAEFFER & BUDENBERG,

Works: Brooklyn, N. Y.

Sales Offices:

66 John St., New York.
22 West Lake St., Chicago, Ill.

Manufacturers of the Improved "Thompson" Indicator and "Lyne" Indicator for Steam and Ammonia.

Prof. R. C. Carpenter's Calorimeters, for determining the percentage of moisture in steam.

Also of Tachometers, (indicating and recording), Revolution Counters, Pyrometers, Pressure and Vacuum Gauges for all purposes, &c.

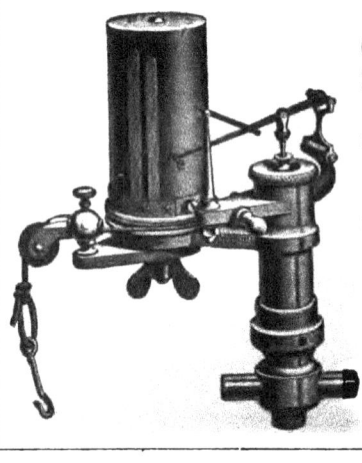

THE
YEAR BOOK
OF THE
SOCIETY OF ENGINEERS.

The Year Book is a scientific publication issued by the Society of Engineers in the College of Engineering and the Mechanic Arts and the School of Mines and Metallurgy in the University of Minnesota.

It is essentially technical in its scope and contains articles contributed by active and honorary members of the society.

This is the fifth publication of the Year Book, and, as in previous years, is in charge of an editorial committee from the departments of Civil, Electrical, Mechanical, Mining and Chemical Engineering.

The price of the Year Book is fifty cents per copy, or by mail (post paid), sixty cents. Copies of '95 and '96 Year Book may be had at same price. Address

The Year Book of the Society of Engineers,

UNIVERSITY OF MINNESOTA,

MINNEAPOLIS, MINN.

1897.

THE SOCIETY OF ENGINEERS

IN THE

UNIVERSITY OF MINNESOTA.

OFFICERS, 1896-97.

ROBT. CRAIG, *President.*
FRANK W. McKELLIP, *Vice-President.*
R. P. BLAKE, *Business Manager.*
CHAS. H. CROSS, *Secretary.*
CLIFTON A. GLASS, *Treasurer.*

COMMITTEE ON PROGRAM:

F. B. WALKER.   J. B. McINTOSH.
TRUMAN HIBBARD.

THE
YEAR BOOK,
1897.

BOARD OF EDITORS.
OFFICERS:
JAS. H. LONIE. *Managing Editor.*
HENRY D. SILLIMAN, *Business Manager.*
ROBERT CRAIG, *President of the Society.*
ADOLPH W. WAGNER, *Assistant Business Manager*

DEPARTMENT EDITORS:

E. A. LEE,	*Civil Engineering*
W. L. MILLER,	*Electrical Engineering*
J. H. LONIE,	*Mechanical Engineering*
GEO. BECKER,	*Mining Engineering*
H. C. HAMILTON,	*Chemical Engineering*

MINNEAPOLIS:
HOUSEKEEPER CORPORATION.
1897.

TABLE OF CONTENTS.

Some Suggestions Concerning Strikes....Frank L. McVey 1
Engineering in South America..............W. C. Weeks 8
Chemistry as an Elective for Engineers......F. F. Sharpless 17
A Novel Three-Phase Alternator.......Geo. D. Shepardson 26
Inspection—Its Place in Modern Engineering..A. C. Beyer 30
The Heat Treatment of Iron and Steel......Harry E. Smith 37
Improvements in the Three-Wire System..Truman Hibbard 45
Recent Railway Economies..............E. E. Woodman 58
A Multifunctional Dynamo............Geo. D. Shepardson 74
The Modern Explosion Engine..............E. S. Savage 79
The Structural Laboratory............Frank H. Constant 86
The New Dam at Minneapolis............H. M. F. Dahl 93
Recent Railway Economies..............E. E. Woodman 58
Copper Minerals in Hematite Ore..J. H. Eby, C. P. Berkey 108
The New Distributing Reservoir..........C. H. Kendall 118
A Convenient Ammeter Plugboard....Geo. D. Shepardson 124
The Science of Photography............H. C. Hamilton 128
A Cut-Off for Steam Compression......Dr. Henry T. Eddy 136
The Gillette-Herzog Competition........................ 138
Editorial 140
Obituary 142
Members of the Society of Engineers and Alumni......... 143

MECHANIC ARTS BUILDING.

THE
YEAR BOOK
OF THE
SOCIETY OF ENGINEERS.

MAY, 1897.

SOME SUGGESTIONS CONCERNING STRIKES.

FRANK L. McVEY, Ph. D.

When the average man looks at some great building towering many stories above him, or crosses a river on an ingenious arrangement called a bridge, he is likely to consider these structures as the products of the brain of an engineer or an architect. The work and labor behind the engineer and architect are forgotten in his admiration. Possibly even the contractor is overlooked, while the hundreds of men whose daily toil has made the accomplishment of the plans possible are not given a thought. The last two factors are largely neglected in our technical and mechanical education. The problems which arise from association with contractors are being met in some degree, in our technical schools, by law lectures on contracts; but very seldom is any attention paid to the constantly increasing difficulties of labor employment. And in that human labor is a great essential in engineering enterprises the young engineer is far from being equipped for his work unless he understands something of labor organizations, their purpose, and methods of work; and still more, knows what human nature is in all its phases. The first he can get by study and observation; the second can be acquired if the young engineer has a native ability for leading and understanding men.

Recognizing this as a valuable opportunity to present to engineers the necessity of understanding industrial conditions, the writer has readily undertaken to say something about the great labor problem, i. e., strikes. In the limited space assigned it will be his intention to show the dangerous and expensive character of strikes, the reason for their existence, the elements which make them difficult to deal with, and, finally, some general suggestions as to the treatment of men with grievances.

A strike can be defined as an attempt on the part of employees to secure higher wages or a settlement of a grievance by quitting work.

In the beginning of this century strikes were a necessity in England. The laboring class at that time was largely destitute of political franchise. As a class they were burdened with taxes and were uneducated. Society, at the same time, had a fixed way of considering the laborer, regarding him as a kind of machine without human sympathies and wants. There was need in consequence, of a series of fierce revolts that would break up the customary and established ways of looking at the workingman. And it was only by means of the revolts that he was able to secure new liberties. The damage done was small in comparison to the good accomplished. The competition between manufacturers at the time was limited by lack of communication and facilities for transportation, so that the effect of such an insurrection was confined to the locality in which it took place.

All of this is changed at the present time. The influence of a great strike is now felt almost everywhere. Railroad traffic is stopped and the supplies which are necessary to the existence of people are held back. Society is so highly organized that a blow at one part is transmitted to every other part. Both the public and the contestants have found this to be true. The expense of such a strike to those immediately concerned seldom justifies the results, and the disastrous effect on the general welfare of society can hardly be estimated. Under these circumstances we should look for a decrease in the number of strikes; but this has not been the case. In the year 1881 there were 471 strikes, involving 2,928 establishments, causing 130,000 to be thrown out of employment. In 1894 there were 896 strikes, involving 5,154 establishments; 482,066 men were thrown out of employment in consequence thereof. During the period between the years just mentioned

SOME SUGGESTIONS CONCERNING STRIKES.

14,389 strikes occurred. Some 69,166 establishments were involved and 3,700,000 men were made idle. One-half of these strikes occurred in the last five years of the period mentioned and involved 32,780 establishments and caused 1,600,000 men to be thrown out of employment. Naturally enough the greater number of these strikes occurred in the larger cities. From June, 1887, to June 30, 1894, there were 5,909 strikes in the cities of the United States, 2,614 happening in New York City alone.

The statistics given above do not tell one-half the story. The loss in wages to employees from strikes and lockouts in the period 1881 to 1894 amounts to $190,493,173. The employers, on their part, have lost from the same causes $94,825,000. In addition to these sums the labor organizations advanced $13,438,000 to help carry on the conflict. The loss to all immediately concerned is the grand total of $298,756,000. Nor does this sum of nearly three hundred millions include the indirect injury to society. In order to get complete returns it would be necessary to know the damage done to business of all kinds by delays in shipments, increased rates of insurance and the thousand and one inconveniences and privations which third parties have been occasioned. Just what sum of money could be said to cover this loss to the entire country is difficult to say, but it certainly is very large. It all goes to demonstrate the fact however, that this method of settling differences is expensive and unjustifiable.

The causes which occasion these industrial conflicts are not seen in their full importance at first glance. The one which appears most prominent is the failure on the part of both employer and employee to understand their mutual relation and interest. This cause can be separated into arbitrary acts of employers, refusals to hear grievances, and the constant misunderstandings concerning wages, time and discharges. There are other causes, such as false economic theories, like the following: Profits fall when wages rise; limited numbers mean higher wages; and reduced products will increase piece prices. As a general thing strikes are caused by misunderstandings. This is due to the refusal of one side or the other to explain the situation. Then follows a pitched battle, lasting an indefinite period, in which the public is compelled to take one side. The employer sends out circulars warning other employers against certain laborers, and setting forth the cause of the trouble from his standpoint. The strikers do the same, only giving a much broader circulation to their manifesto. Possibly the strike spreads to other industries, simply because there has been no intelligent effort to get at the cause of the trouble in the beginning.

A further analysis of causes enables us to group them as follows: (1) Wages, (2) hours of labor, (3) men to be employed (often nothing more than a mere discrimination between union and non-union men), and (4) sympathetic strikes. Of the 46,862 establishments that were connected with strikes or lockouts in the period January 1, 1887, to June 30, 1894, forty-three per cent or 20,394 of them had trouble over the question of wages. Some 6,199 were disturbed by the single question of hours of labor, and 3,095 over a combination of the questions of wages and time. Reduced to a percentage, we may say that 13 per cent of the total number of establishments were engaged in strikes because of attempts to increase or reduce the time of a working day. Under the third group of causes 5,090 establishments or 11 per cent, in the same period, were engaged in strikes over the employment or discharge of men. There were 3,620 establishments, or 7.75 per cent of all the concerns, dragged into strikes because of the sympathy of the employees with strikes elsewhere. Besides the causes mentioned there were minor ones which involved such a small number of employers and employees that it is not worth while considering them in a short article of this character.

The principals in a strike can, for simplicity's sake, be confined to employer and employee. Both of these are influenced by outside forces. The position of the employee cannot be understood without a thorough consideration of his trade union relations, and the policy and leaders of the union to which he belongs. In his turn the employer cannot be separated from his business, his success in it, his policy of treating his men and the attitude of the public toward enterprise. In examining the case of the laborer we find that in the act of selling his commodity he seldom comes in contact with the buyer. He (the laborer) is nothing else than the object whose use he sells. The purchaser takes possession of the employee's person and controls it during certain hours of the day. The purchaser thus determines residence, place ol working and employment of time. Naturally, combinations of workingmen followed in an endeavor to modify these conditions

SOME SUGGESTIONS CONCERNING STRIKES.

to at least some degree. Hence the trade union is always an element to be considered in a strike. In this connection, however, it must not be thought that trade unions were established for the purpose of striking, for such is not the case. It is true that the union of laborers is a combination of employees to meet the employer, but that combination is the result of the factory system. The first and primary object of such an organization is to benefit its members generally in connection with some insurance plan. But being so organized, the union becomes a factor in strike troubles.

The policy of a union is too often opposed to reason. The members are blinded by prejudice in many instances and fail to see the results which would follow from their course. The reason for such policy on their part is largely due to the fact that it is guided in many cases by the man who can talk the loudest and frame the best sounding phrases about the tyranny of capital and the corruption of government. The conservative thinking man, and there are many in the labor organizations, is silenced and his counsel refused. Many mistakes naturally follow. If the union holds together under the pressure of the burden it has brought upon itself by listening to unwise counsel, the future actions of that union are apt to be guided by reason, and because of this to be an aid rather than a hindrance to the employer.

On the other hand the employer is by no means free from responsibility. The old relation of man to man before the growth of corporations has been superseded by the manager representing a corporation whose stockholders want dividends. In this connection the workingman becomes a tool that can be laid aside when necessary. When this is done no explanation is offered. The employees are forced to take up their own case in self-defense, refusing to accept the kindly offices of a third party for fear that the decision would not be accepted by the employer unless favorable to him. In all probability there are good reasons for the action of the employer in laying off men; but he refuses to make them public because his competitors in the same line of business may come to know his condition. But such a view is a false one, as the men are sure to accept a reasonable explanation, and the advantage to the competitor is small in comparison to the good will of the men. A refusal to explain, on the other hand, is regarded as a "causa belli."

FRANK L. McVEY.

There is another consideration. The business man of to-day is a money maker. That is his sole object and nothing is allowed to stand in his way. Sentiment and conscience are laid aside, for he finds he cannot afford to consult them. With him it is a question of success. His standing in the community depends very largely upon his ability to acquire wealth. So long as he can maintain such a position the public receives him with open arms. There is, in consequence, a temptation to drive hard bargains with employees and to take advantage of their necessities in order to maintain this position. The public overlooks the means and only regards the end as worthy of attention. Strike conditions prevail, therefore, without any restraint from the public which is deeply interested in the maintenance of industrial peace. The principals in a strike have been pointed out as employer and employee. The responsibility for strikes is wider, and rests not alone upon the factors mentioned, but also upon a third, the public. If strikes can be avoided without interference by the public, a great advance will have been made toward the true relation that should exist between employer and employee. How trouble can be avoided, at least to some degree, will now be considered.

A strike is not a spontaneous thing. It takes place after much agitation and planning. A few men, more intelligent than the others, are always consulted in such an event. To know these men is the first duty of the young engineer. Unless he does become acquainted he can hardly understand their point of view in questions involving their rights and comforts. He must also avoid any tendency toward arbitrary actions, for men are quick to take offense at any inclination to be overbearing. In fact, they are to be considered as allies, not as mere hirelings. Nor is such talk mere theory. Firmness, coupled with consideration and kindness, will be a true preventative of strikes. In case trouble should arise every step should be taken to prevent misunderstanding and the development of hostile feelings. This can be brought about by asking the men to send a committee to meet the employer and his representatives for the purpose of talking over the situation. The one great object of such a meeting is to secure an understanding, and if possible, an agreement, to which both sides pledge their support. A spirit of concession and fair play will produce valuable results toward this object. If the trouble has not gone too far it is likely to be settled at once. Sometimes the trouble may have advanced to such a stage that it is impossible to secure reconciliation; then the employer or his representatives should endeavor to secure a submittance of the case to some third person or persons, who are known to be upright and honorable. Although the decision rendered may not be entirely satisfactory, it is far more economical in the long run to abide by the decision than it is to undertake a long and expensive fight whose result is by no means certain.

In conclusion it can be said that the relation of men to employer and vice versa rests upon judgment and common sense. It is easy to start a quarrel which may develop into a strike. The young engineer may find in such an event that he has cost his employer far more than all of his mistakes in construction. The one way to avoid labor complications is to be firm, kind, just, and to act as good, hard common sense dictates.

W. C. WEEKS.

ENGINEERING IN SOUTH AMERICA.

W. C. WEEKS, '94·
U. S. Assistant Engineer.

It is with considerable hesitation that I write under this heading in a technical publication, of my experiences during several months of last year while engaged in exploring a portion of the Orinoco Delta in the Republic of Venezuela. Owing to the peculiar character of the country and a lack of knowledge of

Starting on an Exploring Trip.

existing conditions, very little work of an engineering character was done, and it will be the object of this article to give the reader an outline of the trip, a description of the conditions under which engineering or surveying must be performed, and make a few suggestions as to the most feasible and economical methods of prosecuting a survey in the tropics.

The Orinoco Company, limited, has obtained from the Venezuelan government a ninety-nine year lease of a valuable tract of nearly ten million acres of mixed low and hilly country lying in and south of the Delta of the Orinoco. It was with a view of exploring this vast tract of practically virgin territory that our party of sixteen men, for the most part, residents of Minnesota, set sail from New York on the 5th of March, 1896, my connection with the party being in the capacity of engineer. Our route was the usual one taken by steamers plying between New York and Port of Spain, a city of about seventy thousand inhabitants, situated on the island of Trinidad, British West Indies. We were equipped with supplies and instruments for conducting a tour of exploration, including plenty of arms and ammunition, and a steam launch complete. Six days after starting we passed through the treacherous Boca de Dragos and dropped anchor in the harbor of Port of Spain. Here a delay of two weeks occurred, caused by the destruction of the hull of our launch during the process of unloading, and it was not until the 25th of March that we embarked on the Orinoco trader "Bolivar" for Cuidad Bolivar, whence all who enter Venezuela from the west must repair, in order to make their peace with the Venezuelan custom officers. Our route was across the Gulf of Paria, through the "Serpent's Mouth" and up the Macareo channel of the delta to Banancas, and thence by way of the Orinoco to Bolivar, making a few stops on the way for the delivery of mail—no one save the officials being permitted to land. Friday morning, March 27th, the boat was laid alongside the sandy bank of the river in front of the City of Bolivar, the thermometer registering 102° in the shade, and not a breath of air stirring. Cuidad Bolivar is situated on the south bank of the Orinoco river, in the state of Bolivar, 273 miles from the sea, and has a population estimated at 11,000 souls. The city is built on a side hill in the usual style of Spanish architecture, the large buildings being for the most part of heavy masonry, the roofs of red tiles and protected against rifle fire by an extension of the side walls. These roofs offer a very secure refuge to the inhabitants in times of trouble, for from their advantageous position they can command the streets with comparatively little danger to themselves. The dwelling houses of the poorer classes are made of pole frames, to which bamboo poles are lashed, and plastered with mud, the roof being thatched with an incombustible palm leaf. This latter style of architecture prevails throughout the delta.

Two days were spent arranging our affairs and in trans-shipping our goods to a light draft, stern-wheel steamboat, and we re-embarked for the concession. The following night we stopped at the iron mines of Manoa, where an English company had for a short time been mining, or rather quarrying, iron ore. A delay of a day occurred here, after which we went back up the Orinoco to the Island of Poloma, opposite the mouth of the unknown Toro river. At this time the dangerous illness of one

River Bank—Manoa.

of our party made it imperative that he be returned to the hospital at Port of Spain with the least possible delay, and our goods were hastily placed on shore at a native's hut and the steamboat moved away. No communication with us was possible for at least two weeks, and in the meantime we were to make camp comfortable and push the exploration as the means at our command permitted.

The two great drawbacks to any extended exploration at this time were lack of boats and motive power. The machinery of our wrecked launch was being transferred to another hull at Port of Spain, and not until the completion of this reconstruction could we hope for any other motive power than by the usual means of oars and paddles—and the ability of a white man to wield a paddle is strictly limited on those lower Orinoco tributaries, winding about through the low, alluvial delta lands walled in by a mass of impenetrable tropical vegetation, through which not a breath of air percolates to relieve the intolerable heat. This vegetation extends down and into the water of the streams, making it difficult to land from a boat, and almost an impossibility to gain the shore from the hazardous position of a swimmer. No footing can be had on the banks except in well chosen places, and then considerable work with the machete is necessary. When once on the bank, the explorer is surrounded by a vegetation so dense that any idea of making a survey from the bank is at once abandoned.

Our party remained two weeks in camp on the Island of Polona. Tent flies were hung to keep off the rain that now fell daily, and we were protected in our cots and hammocks at night from innumerable flies and mosquitoes by the "toldetta," an absolute necessity with the tropical explorer. The air was hot, moist, and produced a feeling of lassitude quite difficult to overcome, and in no way contributing to the successful prosecution of our work. Several short excursions were made along watercourses, and our routes "surveyed" as well as might be under the circumstances. By utilizing the clearing in front of a native's hut and that made by ourselves for the camp, a base line of 300 feet showed a point on the opposite bank to be 1,700 feet away, and this was the initial point on the rough survey of the Toro. Our only available boat was a steel lifeboat, well made, and incapable of being sunk. A couple of ranges were established on shore leading across the river, and the direction of these ranges carefully determined by compass while standing on the bank. Then by testing these courses from the boat when on the river, the error due to the boat's influence on the needle was determined and allowed for on all courses so taken. Rifles and other articles which might of themselves disturb the direction of the needle were kept at a distance—the magnetic declination was taken from a marine chart. Distances were all estimated, with the exception of our short base lines. Seated in the middle of a boat, crowded with men and piled high with camp equipage, covered with tarpaulins to protect it from the frequent showers; compass and book in one hand, pencil in the other, the notes were taken and sketches of important points made, while the attention of the observer was being continually taken from his work by the screaming of strange birds, the stealthy movements of monkeys and aligators, the snort of the water dog or

Changing Camp—Paloma.

splash of the iguana from his leafy perch. A condition of mind bordering on distraction was continually maintained by the laziness and deliberation of our native oarsmen, ever ready to stop rowing at the smallest excuse, and often without any excuse at all, and in utter disregard of the fact that at such times our progress entirely ceased.

These explorations showed the country south of the Orinoco

river to consist of dense forests growing from a sticky, alluvial soil near the margin of the river, changing into irregular savannahs further inland with occasional ridges running down from the broken mountainous country still further south. Walking in these savannahs is attended with great fatigue. The grass is high, strong and as fire never runs here, the last season's growth forms an unstable, insecure footing having a great tendency to entagle one's feet. The total absence of any protection from the fierce heat of the tropical sun and the torture of the "bete rouge" renders travel across these grassy seas anything but pleasant. These savannahs are inundated to a depth varying between two and ten feet from May to September of each year.

At a later date, further exploration along this river showed an abrupt step from the delta to a higher and mountainous country. The river descends from the high lands in a long series of cascades and rapids with vertical falls of six to eighty-four feet, this latter waterfall marking the end of our exploration along this river. After the first ten miles from the Orinoco, there was nothing to indicate that human beings had ever preceded us, and we were without doubt the first white men to gaze upon those splendid waterfalls which have for untold ages broken the stillness of the tropical forests with their thunder. These falls were surveyed, vertical distances measured by barometer, horizontal distances carefully paced wherever practicable and checked by numerous triangulations to stations marked by poles. As far as we penetrated, the rivers abounded with alligators, electric eels and dangerous stinging rays, and it was fortunate that none of our party ever received serious injury.

Two attempts were made to conduct an exploration by land and these were carried on much the same as a preliminary railway survey. The vast amount of physical labor necessary to clear a path along which a passage may be made can only be appreciated by one who has been in the tropics, and the progress of an exploring expedition by land is most exasperatingly slow. Considerable difficulty would be found in leaving permanent marks of a survey except possibly by means of reference trees. I never saw a dead tree or piece of rotten wood in the tropical forest, so quickly is the destruction completed. The Orinoco river has never been surveyed except for an occasional "running survey" by the officers of some ascending steamboat. Any attempt to survey the delta portion after the manner of the Mississippi river commission would necessitate a great expenditure, for there are no roads or trails except the "canoe path" and no elevated points upon which to establish stations. Perhaps the most feasible method of surveying the river would be by transit and stadia, running a carefully checked stadia line along each bank with tall signal poles at long intervals upon which to check up. Azimuth could be carried along this line of signals by the usual means of triangulation, and certain of these triangulation stations located

Ceropias Sarcupana.

astronomically by means of sextant and horizon. The soundings could be located by transits in the usual way from those stations giving the best intersections. For near by soundings, a stadia board fixed rigidly in the bow of a boat and observed upon by a single transit would be found practicable. The season for the prosecution of this work would be from November to May, the low water season. It is then possible to obtain footing outside the line of dense vegetation, and the exposed sand bars could be utilized for temporary stations. These could be referenced to living trees on the banks. Even at this season, the river and its multitude of canos varies from a thousand feet to five miles in width; hence the necessity of a line on each bank. A steam launch and canopied barge—the larger the better—would be indispensable, for it is utterly impracticable to camp on shore. Three or four fair sized covered rowboats, manned by acclimated oarsmen, would be necessary for the transportation of the transitmen and their assistants; the signal poles could well be set by the launch. No trouble would be experienced from the river's current for at this season it flows one way as much as the other. The tide is said to extend its influence a hundred miles above Bolivar and I have myself seen boats at anchor at that city with their bows pointing down stream. Only instruments of moderate power, inverting images and equipped with platinum wires should be used. The latter is essential, as in that hot, moist climate, spider lines soon become useless. The instruments should receive a most careful examination under precisely the same conditions in which they are to be used, before the survey begins, and reductions made by tables calculated from the results of this examination. Aside from the measurement of base lines for testing purposes, I would not advise the use of tape or chain at all as being far too costly. Stadia boards may be of unusual length and size, say twenty or twenty-five feet long and six inches wide, with proper stiffening. The characters should be large and distinct. The mean of a number of observations with varying positions of the stadia wires should be taken as the most probable distance between stations, a single reading for "point shots" would be sufficient. The atmosphere is remarkably steady considering its moist, heated condition, and in the morning and evening there is no inconvenience from heat waves. Sunshades for the instruments must be provided, for the heat of his direct rays is almost incredible and on the exposed sand bars is terrible to endure. Light rain might be expected each day, for the distinction between the dry and rainy seasons in the delta is not very marked. The temperature in the shade throughout the delta would rarely exceed ninety degrees, and the difference between the wet and dry bulb thermometers rarely more than one degree—this latter showing the excessive humidity of the atmosphere. Its effect is very marked on the unacclimated, and little physical labor should be attempted by white men.

All surveying should cease in April, for this is the beginning of the rainy season, when the whole face of the country is changed. The sun is obscured by clouds, rain falls frequently in violent showers, and where a short time before the white sand bars glistened in the sun, now runs the mighty current ten to forty feet deep, the dark muddy waters rushing onward to the sea regardless of the tide. For miles inland the banks are inundated, and the alligators swim unimpeded over the grazing ground of a month ago. Fog fills the forest, moisture drips from the trees, the whole region is abandoned to birds and monkeys and presents a most gloomy spectacle. Not until September do the rains cease; then the sun shines, interrupted only by occasional showers; the rushing flood of three months subsides gradually, and by November the banks and bars once more appear and make it practicable for the surveyor on the Orinoco to proceed.

CHEMISTRY AS AN ELECTIVE FOR ENGINEERS.

F. F. SHARPLESS.

The proper selection of electives is one of the most difficult duties of the undergraduate; difficult, because so many are offered while so few can be taken, and because of the fact that the undergraduate is lacking in the one element essential for intelligent selection, namely, experience. It is owing to this lack of experience that few electives are offered to engineering students and that the greater part of their subjects are required. Sooner or later, nevertheless, the time does come when he has an opportunity for making some choice of studies, and, naturally, among other questions, asks himself whether chemistry shall be one of these studies, whether chemistry is a luxury or a necessity, so far as it concerns his future welfare. If he decides that it is a necessity, as he certainly should,. he then asks himself what amount of time he can afford to devote to the subject.

The subject of chemistry, and the proper amount for engineers, has been the cause of many controversies in the technical journals, and in the faculty meetings of many colleges; controversies presenting two good sides and not always ending in perfect harmony of the views of the disputants. If professors, broad-minded and of ample experience, are unable to agree upon these points after years of observation, how much less able to decide must the student of engineering be in the comparative embryotic state of development in which he finds himself at the time the questions present themselves?

When the student knows the line of work that it will be his fortune to follow, the task of selecting subjects is not a difficult one. It is a very easy matter to pick out those studies that have a direct bearing on his work and easy to avoid wasting time on subjects irrelevant to the known line that he is to pursue.

Unfortunately, it will be found that a very small percentage of undergraduates have any definite plans for their future; they are studying mining, civil, mechanical or electrical engineering, expecting, of course, to obtain a position in one branch or the other soon after graduation. It is advisable for such students to obtain as broad an education as possible, thus preparing themselves for whatever positions may be offered, and avoid specializing until after the character of their work has been definitely ascertained. It is to this class of students, seeking as broad an engineering education as can be obtained in four short years, that the following remarks are addressed:

The expediency of the required courses in chemistry as now given in the curriculum for engineering degrees can hardly be questioned. They are the result of careful study and experience and may well be considered, by the undergraduate, as essential portions of his course. Taking for granted, then, that the required courses are essential, the question naturally presenting itself to a young man after they have been finished is, "shall I take any of my electives in chemistry or are there other subjects that will be of more benefit to me?" This is a question demanding serious thought and due deliberation. There are numerous factors which must influence one in making a choice of electives, personal and impersonal factors that are so different in individual cases that it is impossible to treat them collectively.

In making a selection from electives one naturally chooses the courses that appeal to him as being valuable and interesting and omits those courses which are void of apparent benefit. Early associations frequently develop or warp the mind in one direction; one young man has grown up in an iron manufacturing town, a second has been the office boy of an hydraulic engineer, while a third has seen nothing of engineering except on the location of some small mine where his parents lived. Each of these students will probably make his choice of electives in different directions; each knows what subjects are of interest to him; because his early associations have developed in him certain lines of thought to the exclusion of others, and in making his selection of studies, he frequently selects unwisely, merely because of lack of rudimentary knowledge of the subjects that are offered him. It can be easily seen that the case of each student is largely a personal matter. No broad statements can be made to cover all engineering branches, therefore, rather than point out a path for the engineering student to follow, this article will suggest only some of the uses to which the engineer may put his chemical knowledge, and leave him the option of making his own selections of the courses offered.

To the chemical engineer but few suggestions are to be made. He should prepare himself for the endless variety of work that is to come under his supervision; he should carry out carefully

CHEMISTRY AS AN ELECTIVE FOR ENGINEERS. 19

some one line of work, but it is not advisable to limit himself outside of this one line.

The experience of several years as a public analyst has shown the writer the great diversity of material coming to the hands of the chemical engineer. During a year scarcely two samples, with the exception of gold, silver and iron ores, will be similar or will be subjected to the same treatment; from the examination of horse shoes, grindstones or cement rocks, he may be called to work upon a patent medicine or shoe polish, or he may be subpoenaed as a witness in a poison case. So broad is the field of this subject that one can only say to a chemical engineer who is not specializing, take all the chemistry you can get and vary your work in chemistry as much as possible.

As little advice need be given to the metallurgical engineer as to the chemical. It is obvious that the problems submitted to this engineer will fall within two lines—chemical and mechanical. The range of subjects handled by the metallurgical engineer is generally narrow; he adheres very closely to one subject; one will follow the working of lead, another of copper, while others will adhere closely to the metallurgy of iron and steel, but few have the time or opportunity to follow more than one course. While the lines of work are narrow, the details are elaborate and require hard, unremitting study for their mastery. An ideal metallurgical engineer is Sir Henry Bessemer, whose work has been confined to the chemistry of steel and to the study of economical methods for carrying out on a commercial scale the results of his laboratory experiments. In other words, his chemical and mechanical knowledge have been a happy combination, and its value to mankind can scarcely be estimated. It is evident that two courses are open to the student of metallurgical engineering, namely; that of specializing and the broader course treating in a general way of all the metals. Which of these courses should have preference? Generally speaking, the broader one is to be preferred, even though the line of one's future work is known. For example, if one's future work is to be confined to steel works, the writer would still advise the general course for the undergraduate.

Before one has had actual practice in his profession, before he has served his time at the furnace, he is unable to carry on judicious experiments and original work with the comprehensive understanding that should accompany such courses.

The undergraduate, when pursuing special lines, will naturally be influenced by his instructor in the selection of work and methods of investigation, and in fact is not only influenced, but is entirely dependent upon him and helpless without his continued assistance. The graduate who returns to his college after he has worked at the furnace knows the lines that will particularly interest him, understands the nature of the information that he seeks, and makes the legitimate use of his professor, namely, in obtaining advice as to methods and means for securing certain results and information.

To the metallurgical engineer the suggestion to be made is, take all the chemistry you can relating to the metals, but unless you have some definite plan for the future do not specialize.

Closely allied to the work of the metallurgist is that of the mining engineer, but his field is broader; it is seldom confined to one particular line, as that of the metallurgist, but includes nearly all of the branches that may come under the subject of metallurgy, civil and mechanical engineering. One of our well-known mining engineers of this country says: "To be a modern mining engineer one must know everything—all the laws of nature and all the laws of the land." This is very nearly true and is particularly true in regard to the laws of chemistry. The mining engineer is very frequently called upon for solution of chemical problems identical to those of the metallurgist. He must be familiar with the ores of all the metals, with their properties and composition; he must know how they are reduced and brought into the form of the pure metal. When he undertakes the examination of a mining property and its ore he must be able to report on a satisfactory method of working the deposit and for treating the ore. This demands an intimate chemical knowledge of the ore, fuel, fluxes and reactions of reduction furnaces.

A science of which the mining engineer makes continual use is geology, and particularly chemical geology. He uses this in his study of the formation of ore deposits, their shape, character and probable extent. In short, the mining engineer uses chemistry constantly; he uses it for all those purposes that the metallurgist uses it, and it serves him continually as it does the civil and mechanical engineer, to whom, as will be shown presently, it is a most valuable assistant.

At first thought, one might suppose that the civil and

CHEMISTRY AS AN ELECTIVE FOR ENGINEERS. 21

mechanical engineer might well dispense with chemistry, especially as an elective, if not from among their required studies. A close examination of the work that they do and the materials that they handle reveals the fact, however, that both are continually confronted by problems which, for their solution, require a knowledge of chemistry. Think for a moment of the frequency with which problems are presented to these engineers in which iron, steel and all classes of structural material enter as factors. A thorough knowledge of the metallurgy and chemistry of iron becomes a necessity as does also a knowledge of building stones, limes, mortars and cements.

It is customary for the engineer of any structural work to require that the material furnished shall fulfill certain physical requirements, and generally that it shall fulfill certain chemical requirements. The purpose for which any structure is used must be purely physical and it would seem natural that if the material fulfilled all physical requirements it would be unnecessary to burden the manufacturer with useless chemical requirements, but this is by no means the case. In the construction of a metal bridge, for example, the engineer may call for a metal of tensile strength and elasticity falling within certain limits, for material with certain minimum compression strength, etc., but he cannot call for material that will enable him to give an indefinite life to the structure. A metal fulfilling all of the physical requirements could be furnished and yet be entirely unsuited to the purpose for which it is intended. It must be remembered that the material is to be used for a structure carrying heavy or light loads, a structure that may be subjected to all variations in temperature, to continued vibrations and shocks, but above all, it is to stand for an indefinite time. If the time clause could be introduced so that the engineer could demand that at the end of 100 years the metals and materials used should possess the properties they had when the construction took place, then he could disregard the question of chemical composition. The engineer must strive to obtain material that will give the best physical tests to-day and will stand the ravages of time as long as possible. In order to obtain such proper material he may use the specifications of structures that have been standing for a number of years, or he may draw up new requirements based upon the present knowledge of the laws of chemistry. As a matter of fact, specifications are frequently but mere copies of some structure built years ago which has worn well and given perfect satisfaction; and one might suppose that a duplication would, when used again, give perfect satisfaction, but the "rule of thumb" is no more applicable to engineering work than it is to the more common walks of life.

The progressive engineer must be alive to the discoveries that have recently been made in the properties of alloys and other structural material. He must be able to take advantage of the chemical knowledge that is daily being offered to him. If certain elements are now being used that will increase the strength of the metal and thus lighten members of the structure, the engineer should know that fact and take advantage of it. He must understand the effect of varying the percentage of carbon, silicon, phosphorus, nickel, etc., in the material he is using. When a new alloy of apparently particularly desirable qualities is produced he should be ready to examine and adapt it to his uses if applicable, or reject it if inapplicable.

In short, it is just as important that the engineer understand the capabilities of the elements as it is that he understand the capabilities of his mathematics.

Again, if chemical knowledge is lacking, the engineer may demand a material that it is impossible or impracticable to produce. Last year, specifications written for steel rails for one of the Northwestern railroads, on which manufacturers were asked to bid, called for a steel so low in phosphorus that the bids came in at $32.00 to $35.00 per ton when it was expected that they would run from $22.00 to $25.00. The rail ordered would have been an excellent one, but would have required a very low phosphorus ore to make it, also a special run of the blast furnace and steel mill, hence the high bids. The engineer who drew up those specifications simply did not realize that he was calling for something impracticable to produce.

And still further, the chemical and physical properties may both be met in steel or iron manufactured by different processes. For the tension members of a bridge, acid Bessemer or basic Bessemer, acid open hearth or basic open hearth, or wrought iron might be used. If specifications are written so that any two or more of these metals may be used, the engineer should understand sufficient of the metallurgy and chemistry of the metal that he may know which metal to select. The metal will be offered at

CHEMISTRY AS AN ELECTIVE FOR ENGINEERS. 23

different figures and he must have sufficient knowledge of it to know whether the cheaper one will answer his purposes or whether he must use the more expensive.

Chemistry also plays an important part in the masonry of the structure. Of what material is the masonry to be built? Suppose it is decided to use stone, then the question arises as to the most suitable variety. The specifications require that the material to be used must have a certain crushing strength, and a number of stones from different localities and at different costs, which will fulfill the requirements, are offered. In order to make a proper selection it will be necessary that the engineer submit the samples to a careful physical and chemical analysis and thus ascertain whether or not decomposition will probably take place while standing in the structure, under water or above, and whether time will improve or deteriorate the rock. Then as to the mortar and cement that are to be used. It is known that if the best quality of Portland cement is used it will probably be all right; but one may not be justified in doing this when several other brands are to be obtained at one-half its cost. The selection must again turn upon the chemical and physical properties of the various brands and this necessitates that the engineer must himself understand their chemical properties.

Turning from structural engineering to railroad engineering, we find here also that the knowledge of chemistry is a most valuable assistant. If the railroad engineer confines himself to the rod and transit, the less time spent on chemistry the better, but if he is a progressive man he will strive for something more and just as soon as he is through with grades and curves, then chemical questions begin to confront him. Without any knowledge of chemistry he does not appreciate that he is dealing with chemical problems, but it cannot be said in this case that "where ignorance is bliss 'tis folly to be wise."

That the chemist can have anything to do with the movement of trains, carrying of passengers or freight, may at first seem strange, but, nevertheless, he does play an important part in obtaining the final result. The most forcible way of illustrating the value of chemical knowledge to a railroad engineer is by giving a list of some of the questions that the engineers of one of our railroads have referred to their chemist.

The valves of certain steam cylinders are wearing rapidly.

What is the cause and remedy?

The varnish on passenger coaches becomes dull very rapidly. Can this be remedied in any way?

Certain boilers are forming scale very rapidly. How may this be prevented?

The lubricating oil used on the road is costing as much as ever but is not giving the satisfaction formerly given. What is the matter?

It is noticed that tires on certain engines are wearing well while those on other engines must frequently be renewed. What is the difference between the metals giving good and giving poor results?

We are paying twenty-five cents per pound for a patent scale remover. What is its composition, and what should it cost?

Such questions might be continued indefinitely. It is not contended that the railroad engineer should be able to answer them; this can only be done by the close study of an experienced chemist; but it is contended that the railroad engineer should know enough about chemistry that he may know how and when to use it for the advancement of his interests. He should understand enough of the laws of chemistry to know that in certain cases a chemist can solve the problem, can turn failure to success, and loss into profit.

The electrical engineer is a physicist and a chemist. The science that treats of electricity brings together and holds inseparably the subjects heat, light, magnetism and chemistry, and a knowledge of one is incomplete or cannot be had without an understanding of its associates. The science of chemistry is one of the foundation stones of the science of electricity, indeed it is an inseparable part of that science. The actions of wet and dry cells and storage batteries are all based upon the laws of chemistry. One of the most important and fruitful fields of electrical engineering is in the line of electric smelting and refining; here again are found the two subjects so interlaced and dependent upon each other that unless the engineer is well informed on both sciences he is absolutely helpless.

The goal that many an electrician has been striving for during the past decade and is still striving for is the economical production of electricity direct from fuel. Some results have already been obtained, but much more remains to be done, and when success finally crowns the efforts of some student it will be found that his discoveries have been accomplished through the study of electro-chemistry.

The attention of the student of electrical engineering is earnestly called to this field of electro-chemistry. Much has been done, but the possibilities for original and profitable investigation are scarcely equaled in any branch of applied science.

Did time and space permit, examples illustrating the value of chemistry to the engineer could be continued indefinitely.

Before making his selection of electives the student should consult some of those numerous works dealing with engineering chemistry alone, in order that he may know what valuable assistance is to be derived from a study of this science.

The writer does not wish to be misunderstood, he does not suggest that all electives of the engineers should be given to chemical subjects; but he does wish to impress upon the engineering student the great importance of chemistry in the daily practice of his profession. He would suggest the selection of those courses that give the broadest and most general knowledge, a knowledge showing the usefulness of the science, a knowledge that will inform him of the possibility of solving many problems through the assistance of the chemist, and, though lacking in skill of chemical manipulation himself, his mind may ever be in a state of chemical activity and prepared to take advantage of every assistance that the science may offer.

A NOVEL THREE-PHASE ALTERNATOR.

GEO. D. SHEPARDSON, M. E.

The peculiar construction of the Thomson-Houston arc light dynamo makes its transformation into a three-phase alternator an easy matter. The armature has three coils which are equally spaced and whose inside terminals are connected at a common point. The electromotive forces in the three coils may be represented approximately by sine curves whose corresponding values follow one another by a constant lag of one hundred and twenty degrees. If therefore the outside terminals of the three coils are connected to suitable collecting rings, three-phase alternating currents may be obtained when the field is separately excited.

A Novel Three-Phase Alternator.

A three-ring collecting device was designed for the T-H arc machine in the electrical engineering laboratory, but it was decided preferable to modify the regular commutator so that the machine might be changed more readily from one form of generator to the other.

A NOVEL THREE-PHASE ALTERNATOR. 27

The regular commutator has three copper segments screwed to brass bridges supported by brass rods passing through insulating hardwood bushings in two end flanges which are fastened to the armature shaft by set screws. Each end flange consists of a central brass disc nine thirty-seconds inch thick and three and three-fourths inches in diameter, on each side of which is a hard rubber disc four inches in diameter and one-eighth inch thick. Shellacked paper is wound over the edge of the central brass disc for insulation. Copper wire was wound in the grooves thus formed between the rubber discs of each flange, making two collecting rings which were connected to two special binding posts set in rubber bushings through the flange at the outer end of the commutator. The terminal wires of two of the armature coils were removed from the regular binding screws connecting with the commutator segments and were fastened into the special binding posts connected with the rings. The connection between the third armature terminal and its commutator segment was not disturbed. The three commutator segments were then coupled by a wire passed under the heads of the segment screws. In this way each armature terminal was connected with a continuous ring.

In order to make contact with the rings, the brushes were easily modified. One pair of the regular brushes was left undisturbed and served to make continuous contact with the commutator. The other two brushes and the cable connecting the two brush holders were removed. In their stead were placed two special brushes made of spring copper one thirty-second inch thick and three sixteenths inch wide, each one being bent so as to make contact with either of the new collecting rings on the flanges.

When thus arranged, the fields are separately excited and three-phase alternating currents may be obtained through the cross-connected commutator and the two special rings. It is a matter of only a few minutes to replace the regular brushes, remove the jumper from the commutator, replace the two armature leads in the regular cups, and change the field connections, when the machine is again ready for operation as a direct current arc light dynamo.

The machine modified as above described is style "E 12," rated as a nine light 1200 c. p. machine, but found capable of delivering 6.8 amperes at 650 volts at normal speed of 1,000 revolutions. Arranged as a three-phase alternator, the machine gives 800 volts between any two brushes when delivering 0.5 ampere and when excited by a field current of 6.8 amperes, the armature reaction being negligible with so small current.

This machine is useful for operating induction or synchronous motors and will itself operate as a synchronous motor when suitably excited and brought up to speed. The low frequency makes it unsatisfactory for operating incandescent lamps, as the pulsations of current candle power are apparent. A third collecting ring at the pulley end of the armature and connected with the common junction of the inside terminals of the three armature coils, allows experimental study of unbalanced circuits. Also graphical and analytical determinations of the resultant of alternating electromotive forces in series may be checked by the aid of this contact with the common center.

A few details of construction may be of interest. The brass disc inside the commutator end flanges was found to be slightly out of true and a light cut was taken off in the lathe. The original paper insulation covering this brass disc was found to be too thin to stand the strain and the rings first made became grounded on the frame when the external armature circuits were suddenly opened. Increasing this insulation by the superposition of five layers of dry paper seems to have made this point safe.

In making the new collecting rings it was desirable to avoid the use of acid soldering flux and also to solder the ring quickly so as not to heat the rubber discs unduly. The ring was therefore made of No. 16 B. & S. copper wire which had been previously tinned. A small hole was drilled through the two hard rubber discs just outside of the paper insulation above the brass central disc. One end of the wire was threaded through the hole in the outer disc and was temporarily anchored to one of the binding posts. The wire was then held taut while the commutator was turned slowly winding a layer of the tinned wire across the groove. After fastening the band by a drop of solder, the last end of the wire was pulled through the hole in the inner flange and lightly hammered into place. After soldering the wire together into a solid band, the outside end of the wire was cut off close to the flange, and the inside end was brought down and soldered to the inner end of one of the new binding posts. When both bands were built up in this way while still in the lathe, a light cut was taken off each band so as to give smooth surfaces for the brushes. The half tone engraving shows the general appearance of the modified commutator as arranged for a three-ring collector.

INSPECTION—ITS PLACE IN MODERN ENGINEERING PRACTICE.

A. C. BEYER, '96.

It is now only about fifteen years since the first inspection bureau was established in this country, which advertised itself prepared to assume the responsibility of seeing that the ideas and wishes of engineers, and purchasers of engineering materials, were carried out by the contractors who had undertaken the work. This bureau has been followed by many others, as the field, at first profitable, has opened up and developed.

Inspection was not, however, a new line of engineering work, for the advisability, economy and ofttimes necessity for it had long been recognized by many prominent engineers who were desirous of feeling assured that no defects in materials, or careless or dishonest workmanship should vitiate the results of their designs, or endanger the stability and safety of their structures. This work had usually been performed by some member of the engineer's force who was sent out for the purpose, and almost anyone was considered qualified to do it, the number of men who spent all their time at this being comparatively insignificant. But with the advent of the inspection bureau new methods were employed, more efficient and systematic service was rendered, and the inspecting engineer has now entered the field claiming recognition as another specialist in the engineering profession.

All the bureaus have a corps of men who do this work and nothing else; men who have been trained for it, who have made a special study of it, and whose judgment and ability are the results of long experience. The inspector of engineering materials has become an important and almost an essential factor in the professional practice of to-day, and his field is continually broadening. To a large extent inspection outside of government work was originally confined to materials used for structural purposes, principally to iron and steel, and at the present time its most general application is in connection with the materials for bridges, buildings and similar structures. With but few exceptions all the important structures completed during the past few years, or now being built, have been given a careful and thorough inspection in all stages of the work. And many others of less

INSPECTION—ITS PLACE IN MODERN ENGINEERING. 81

importance are examined and passed upon by inspectors in one or more stages, usually either in the mill or shop. Many railroad companies have men stationed at the shops where their bridges are being turned out to see that both in quality and quantity the work comes up to the desired standard. New York City appoints its large force of building inspectors through the municipal civil service commission, and while they cannot detect careless or dishonest work as easily or to such an extent as it could have been done at an earlier stage, yet their reports show that they are able to correct a great deal of it, and in a city where steel-frame and cast-iron-column construction is so common, the importance of having this examination, when private inspection is often omitted, is readily seen. Other large cities have followed New York's example.

Inspection in connection with structural work naturally divides itself into three stages, technically known as mill, shop and field inspection, following each other in the order given and all considered essential in the best class of work. Inspectors have not yet specialized to any large degree as to these three classes, but do more or less of all. By this is not meant to imply that one man follows a given job through all of its stages, for this is quite unusual. He who approves the material at the rolling mill is done with it there, as is also the man at the bridge shop when the built-up work is shipped away to the site where it is to be erected under the supervision of a third party. This method is certainly as economical as it is efficient; it furnishes a means of checking the work, since what is overlooked by one is quite likely to be seen by the next, especially when, as is usually the case, the parties for whom the work is being done do their own field inspecting. Again, under this method the inspection bureaus can have men stationed continuously for considerable periods of time at the various mills and shops in a certain territory, thus enabling them to more easily secure work that comes there, and to do this at a greater profit in that one or two contracts will pay expenses while all others that may be obtained at the same time are almost clear gain. There are, of course, many men who never leave the mills, for a great deal of material, such as eye-bars, beams and rails, needs no other inspection. As a result the number of mill inspectors is larger than that of any other class.

It is easy to see that the work of an inspector at foundries and rolling mills is of the utmost importance. The engineer who designs the structure has counted on using material of a definite strength and possessing certain specified physical and chemical qualities which he regards as more or less essential, as shown by the past experience and observation of himself and others. He knows that he can obtain such material, and, being ready to pay, has a right to expect it. The inspector, as his authorized representative, makes tests and examinations and accepts or rejects the material according as, in his opinion, it meets the requirements of the specifications or not.

Without going into detail as to the great variety of tests which are made, depending on the kind of material, the purpose for which it is intended, and the ideas of the engineer who draws up the specifications, it can be said in general that for structural work an inspection of the surface of the material when finished and the examination in a machine of test pieces of specified shape, size and number are all that are considered necessary to determine whether the metal is suitable or not. From the surface the inspector, by simply looking it over and using a file or chisel, endeavors to ascertain whether there are seams, cracks, pipes, laps or other defects resulting from the melting or rolling which tend to injure the soundness of the material in any way. And an experienced man can in this way detect faults which another would not be able to see with an ordinary magnifying glass. Hot and cold bending, nicking, drifting punched holes and other tests are often specified and used as a means of revealing any unsoundness that may exist. The testing machine gives data from which the tensile strength, ductility, elastic limit and reduction of area are obtained. Through their own laboratories, or others to which they have access, the bureaus are able to determine the chemical properties of the materials from drillings and specimens sent them by their inspectors. But this is not very often done, as all the large mills have their own chemist who makes analyses, and these are in most cases accepted as correct.

After being accepted the material is ready to be shipped to the bridge shops, where the structure is to be built in accordance with shop drawings made from general sketches furnished by the engineer. The construction work is now watched by the shop inspector, who in general is expected to see that it is made as called for on the drawings, and that the workmanship comes up to the specifications as regards neatness and finish. His first duty is by far the easier and can be performed by almost anyone who can read a shop drawing. The second involves a great variety of details, including the matter of painting, the tightness and appearance of rivets, the accuracy of connections, the condition of surfaces in contact and the innumerable other small points which go to make up the difference between first-class and careless work. The shop inspector, if not a practical machinist, must have a large acquaintance with the methods and tools of the latter, and the kind of work he can do. To him are referred all mistakes and difficulties which so often arise during the construction of the work, and the good judgment which he must exercise on such questions can come only with experience and intimate knowledge of such matters. He needs something more than a technical education and the ability to understand a detail drawing.

Of the three classes of inspection on structural work the one which, if the others were to be omitted, would best insure good results, is undoubtedly that in the shop. As has been suggested, the inspection at the mills is, even with the best men, largely superficial, being usually but a more or less hurried surface examination with the pulling of a few tests, which must often fail to reveal all defects. That this is so is shown by the fact that at the shops during the various processes and handling to which the material is subjected, flaws are discovered whose presence was not even suspected by the man at the mill. It is claimed that iron which will stand the treatment it gets in a shop, where it is punched and sheared and drilled and pounded, and shows no flaws, is good material. . And it certainly is a trying ordeal. Again, the custom of engineers of using a large factor of safety—usually four or five—makes it possible to allow for a certain amount of unsoundness not shown on the surface, even if it were not the general practice of all reputable mills, for the sake of their business character and standing, to endeavor to turn out good material. Thus by giving the shop inspector power to reject everything that shows signs of weakness, one can be reasonably certain of securing good work with shop inspection alone, if it is necessary to omit one or the other.

On the other hand, however good the material is, it may be so maltreated in the shop as to entirely undo the care taken in producing it at the mills, and this contingency must be guarded against. Pursuing this argument further, it may be claimed that mistakes and carelessness in erecting the structure can undo all the care and good work that the shops may have contributed and therefore a man to watch against this is very essential in the field.

And so he is. But with good connections made in the shop and care taken before shipping to plainly and correctly mark all pieces, his duties become very light and consist principally in seeing that tight field rivets and good connections are made. His is, of course, a very responsible position, and wise judgment and good common sense are qualities that he can always find use for. As an example of some of the points over which he may have control, the case of a bridge in a large city may be mentioned. Here the field inspector had to pass judgment on not only the erection of the structure itself, but on the masonry substructure, the concrete foundation and asphalt surfacing of the pavement, and the laying of the rails and stringing of the wires of an electric railway which ran across the bridge.

As the inspector is expected to see that the engineer's ideas, as they are expressed in the specifications, are carried out, it needs no argument to show how important it is that these latter should be plain and explicit and as nearly as possible capable of but one interpretation, namely, that intended by the engineer. It is in this matter of indefinite specifications that inspectors often find their duties made hard and unpleasant. It does not suffice to say in a general way that all workmanship shall be first-class, but definite clauses should exist stating in what particulars the work should be first-class. The specifications of the New York Central & Hudson River Railroad can be named as good examples of explicit statements on many points, enabling the inspector to secure a very high grade of work which cannot but prove satisfactory to the company.

The relations existing between the inspectors and contractors are as a rule, and indeed should be, characterized by a spirit of courtesy and mutual helpfulness, with a desire to work together for the best results. The inspector is not usually of the opinion that he is the only honest man in the transaction, and his unceasing watchfulness is not because he thinks someone is trying to hoodwink him, but because he understands that human nature is

INSPECTION—ITS PLACE IN MODERN ENGINEERING. 35

not infallible and the common laborer in a bridge shop or mill is no better than other men.

Having now considered at such length the inspection of structural materials, a brief reference, at least, should be made to that given to others. Its application to these has not, however, become as general as to the former. Building stones, paving materials, limes and cement, wood and other materials may be mentioned. The Pennsylvania Railroad is an example of railway companies which have an organized corps of inspectors as one branch of the engineering department and their duties are performed in connection with ties, rails, splices, bolts, spikes and all manner of railway supplies.

While preparing this paper letters were sent out to a half-dozen or so of the prominent inspection bureaus in the country asking them for information regarding the preliminary training or experience that their inspectors had before entering upon the work. From the replies received the following figures have been compiled:

Number of inspectors reported...................... 80
Number of these who were technical graduates or college men .... 49, or 61 per cent.
Number of these who were skilled mechanics, but not college men .... 21, or 27 per cent.
Number of these who were experienced shop men, but not skilled mechanics .... 10, or 12 per cent.

From these figures it would seem that a very large proportion of the men engaged in inspection have had a college training. They are not all civil engineers, electrical and mechanical engineers being found among the number. As has been stated before, an inspector must know a great deal that is familiar to a skilled machinist, and if as a college man he does not have this knowledge he has to pick it up as best he can, for it is absolutely essential.

May we, in closing, be pardoned for letting the imagination carry our thoughts forward into the future when the time will come in which, with all the other varied specializations of the engineering profession, the inspecting engineer shall be recognized as such by the engineering fraternity; when technical schools shall have chairs of inspecting as they now have of hydraulic or sanitary engineering; when after the preparatory years, with their curriculum of general studies, special attention will be paid in the upper classes to such work in practical botany and geology, the chemistry and testing of engineering materials, the study of processes of manufacture, and shop, foundry and machine practice as will give the inspector some of the many things he has to become acquainted with when he engages in his work. Nothing can, of course, take the place of actual experience, but special training can prepare a man so that he will receive the maximum benefit from this experience. Does this seem to be a fanciful picture? Time will show.

NOTES ON THE HEAT TREATMENT OF IRON AND STEEL.

HARRY E. SMITH, M. E.

To the engineer who is brought into intimate contact with the uses of iron and steel, the processes of their manufacture are of interest. Their varying characteristics during and after casting, rolling, forging, annealing or tempering, form a fruitful source of study and investigation, while some knowledge of the changes that these materials undergo in chemical and physical qualities, during these operations, is essential to their successful use. Enormous advantages may be anticipated from the systematic study of "heat treatment" when applied to iron and steel, of which we now know but little.

Let us consider some of the characteristics which a piece of tool or cast steel can be made to assume by the application of heat. If a bar of tool steel be nicked at intervals, then heated from the temperature of the air at one end, to that temperature at which the corners just begin to melt, at the other end, then quenched and the bar broken at the successive nicks, the structure of the bar will be found to have undergone varying changes.

Beginning at the end which was not heated above the temperature of the air, and passing to each successive piece we will find the grain or structure of the steel to be the same as that of the original bar until we reach the portion which was heated to a full red. Here the grain suddenly changes to an extremely fine and nearly amorphous form, and beyond this point the crystalline structure becomes more and more coarse.

There appears to be a maximum degree of coarseness, or size of grain, for each temperature, varying with the composition of the metal and rising with the sectional area and with the temperature and time of exposure.

The rate of cooling also appears to have some influence on the size of the crystals, especially in large forgings.

A bar of steel 1.5 inches square, when quenched from a very high temperature shows on the outside coarse crystals. They become rigid so quickly that they preserve the form acquired at the high temperature. The interior of the bar is flaky and might even be called fine grained, much finer, indeed, than if the bar had been cooled slowly. This is probably due to the internal stresses, which tend to break up the crystalline structure of the center of the bar. It appears to be a general rule that the finer the grain, other things being equal, the better the condition of the steel to resist shock. Thus the means of acquiring and preserving this grain are of great importance, and not less important are those of avoiding the coarse grain. This fine grain may be obtained in two ways, or under two sets of conditions, one, by suddenly cooling the steel from the molten state, and the other, by producing a change in the carbon from the nonhardening to the hardening form at the full red heat. The first method probably breaks up the coarse structure due to the high temperature by the intense internal strains induced, while the second is due to the sudden rearrangment of the crystals, which seems to be so violent as to efface all previous crystallization.

Forging and rolling iron or steel oppose and break up crystallization, if they are performed while the iron is at or above the red heat.*

But if forging or rolling be completed at a high temperature and the piece be then slowly cooled, the higher the temperature at which the forging ceases, the coarser the grain and the worse the steel, hence the importance of a low finishing temperature in this work.

The superiority of thin over thick forgings, which is usually attributed to the extra work performed on them, is probably due in large part to the usual low finishing temperature of thin or small pieces.

An experiment performed by a competent metallurgist serves to illustrate this fact. Three uniformly heated test bars, cut from a single ingot, were carefully rolled, one very fast, one normally, one very slowly. Their merit was found to be inversely as their finishing temperature.

As power, hence cost, is saved by rolling or forging at high temperatures, means have been proposed of accelerating cooling after forging has ceased. Such a method is used in the Coffin Rail Process, where the rails are immersed in water or have streams of water play on them immediately after coming from the rolls, until their temperature falls to the dark red, from which

---

* Rivets are said to fail at the head which is not struck, rather than at the struck head. The blows of the riveter prevent and overcome crystallization and consequent brittleness.

point the rails are cooled slowly to the temperature of the air.

It will be found that the pieces of our experimental bar, having the finest crystalline structure, exhibit the greatest ductility, toughness and shock resisting power.

This fine grain is then very desirable in the finished product, whether it be rolled or forged iron or steel, or steel castings.

This structure cannot be obtained by any system of cooling the pieces but can be produced by reheating after cooling, to the full red, and then cooling quickly. The more or less complete restoration of overheated and even burned steel may be accomplished by reheating to the red heat, followed by forging and then quickly "quenching" or slowly cooling.

The proper annealing of steel castings or forgings relieves internal stress, effaces columnar structure, gives fine grain, increases strength and ductility and raises the elastic limit.

On account of the tendency of steel to crystallize above the full red heat and also while the temperature is falling to the dark red, the proper heat for annealing is the full red, followed by quickly cooling to the dark red and then slowly cooling through the remaining range of temperature.

The above treatment is applied to car axles and other steel forgings, by the Cambria Steel Co., of Johnstown, Pa.

Cherrioff gives the following as the results from three samples cut from the same bar of steel. After forging the samples were treated as follows:

A was slowly cooled.

B was reheated to a full red and then slowly cooled.

C was reheated to a full red, quenched to a dark red and then slowly cooled.

A broke under a single hammer blow.

B required five such blows.

C could be broken only by a blow of a five-ton steam hammer.

It will also be found, on applying tests for hardness to the pieces of our experimental bar, that beyond the point where the crystalline structure changes suddenly, the steel has become extremely hard and especially very brittle toward the end that was exposed to the highest heat.

Chemical as well as physical change has also taken place in the steel. The carbon it contained exists, in combination, in at least two perfectly distinct modifications called cement and hardening carbon.

The chemical evidence of these two forms was first demonstrated by Faraday in 1822, when he found that steel, when suddenly cooled, dissolved completely in dilute, cool, hydrochloric acid, and when the steel was annealed it left a carboniferous residue. In the first case, the carbon is said to be in the hardening form, in the latter, in the non hardening or "cement" form.

The exact difference between the two forms of carbon is not definitely known, but their effects upon the qualities of the steel are quite evident.

The carbon in our samples will be found in the cement form in all the pieces that were not heated above the full red heat, and in the hardening form in all pieces heated above this temperature.

In annealed steel practically all the carbon is in the cement form, while in hardened steel scarcely any of it is, from which we infer that it is in the hardening state.

In tempered steel an intermediate proportion is in the cement state.

In molten iron and steel the carbon is probably in a condition closely related to the hardening carbon, since on suddenly cooling from fusion we find it chiefly in this form.

According to Brinnell's experiments the change from cement to hardening carbon does not take place below the red heat but occurs suddenly and completely at the red heat. This can be proven by heating a piece of steel, in which the carbon is initially in the cement state, up to nearly the red heat, and then quenching, when the carbon will be found still in the cement state, while if quenched from the red heat or above, the carbon is found wholly hardening. On the other hand, if we heat a bar above the red heat and slowly cool it, the carbon remains in the hardening form until the temperature at which red just shows in the dark, when the carbon begins to change slowly to the cement form, and so continues until about the temperature of the air is reached.

That the transfer of carbon from the hardening to the cement form may occur at very low temperatures is suggested by the reported fact that table knives gradually lose their hardness if habitually washed in hot water, or that razors if used cold retain their temper longer than those that are warmed before using, though for other reasons the razor may cut better while hot than while cold.

Other phenomena indicate an important chemical change at a temperature somewhere between the full red and the dark red heats.

The temporary magnetism, whether induced by an electric current or by another magnet, suddenly vanishes at this temperature, and its thermo-electric behavior is abnormal.

Here a remarkable evolution of heat takes place, known as the "after glow." This is accompanied by a marked rise in temperature, a sudden expansion, and a great temporary increase of flexibility.

The late John Coffin, of the Cambria Steel Co., found that the expansion of a bar of iron when cooling through this range increased greatly with the proportion of carbon. A four-foot bar with 0.90 per cent of carbon, in cooling from an orange heat, contracted 1-8 in., re-expanded 1-32 in., then again contracted 9-16 in. A similar bar with 0.17 per cent of carbon contracted regularly during 45 seconds, then ceased to contract for 20 seconds, then again contracted. In iron containing a very small per cent of carbon, no perceptible decrease in the regular contraction was noted.

An experiment which gives striking illustration of the violent rearrangement of the molecules of steel when heated to the full red heat was performed before the Philadelphia meeting of the American Society of Mechanical Engineers, November, 1887, by Mr. Coffin.

A piece of one-fourth inch square tool steel was broken, the fresh fractures placed in close contact, the pieces inclosed in platinum foil to exclude the air and heated to the full red in the flame of a Bunsen burner. After cooling it was found that the pieces were very perfectly joined. A number of larger pieces, both soft steel and tool steel, were shown at the same meeting, that had been united in a similar manner, which, by their color, sharpness and freedom from scale indicated that they were united either much below the ordinary welding temperature or with the nearly perfect exclusion of oxygen.

Another critical temperature for both iron and steel is at the so-called "blue heat," which is from 450° to 600° F. Here iron and steel are much more brittle than when cold or at redness. This heat, however, does not seem to leave any bad effect on the iron so heated. But if the piece be worked in this range of temperature, it will retain the brittleness after cooling and show a great loss of ductility as measured by the bending test, although it has not been conclusively proven that there is a loss of ductility when the piece is pulled apart by static tensile stress.

The loss of ductility and shock resisting power is not due to any incipient cracks, as can be proven by the restoration of the former qualities of the metal by reheating to redness or by annealing. Several experimenters have called attention to the increase of "blue shortness," or the effects of "blue heat," with the increase in percentage of non ferrous elements present in the iron, but these reported facts have failed to be fully sustained by recent experiments. There seems to be some close relation between the effects of "blue shortness" and "coldworking," but the injurious effects of the former are more intense than those of the latter.

Soft steel plates which have cooled to the blue heat, while in the process of rolling have been reported by Bessemer and others to have been shattered in the last passage through the rolls.

While blue-working lessens ductility, it is not always fatal to the metal. Examples may be cited in proof of this. In hand riveting the operator usually continues hammering while the rivet cools past this critical temperature, and yet few rivets have been known to fail from this cause alone. Shafting is often heated to blueness by a dry bearing while under stress and shock, but is apparently uninjured.

Blue-working without subsequent annealing is prohibited in boiler work for the United States navy, but not for hull work. Many good American and British engineers do not forbid blue working in their practice.

There is another phenomenon exhibited by iron and steel when cooling from the molten state, that is, the evolution of large volumes of gas, which assist in the formation of blowholes and other imperfections against which the designing engineer has to contend. These gases have been found to consist principally of nitrogen, hydrogen and carbon oxide gases, and may be held in the iron in several conditions:

1. In chemical combinations. 2. In solution. 3. In adhesion. 4. Mechanically retained.

In the first three conditions the gases are no longer gaseous but rather in a condensed form. This condition of the gas can

NOTES ON THE HEAT TREATMENT OF IRON AND STEEL. 43

be illustrated by some examples. A piece of charcoal will, at atmospheric pressure, absorb ninety times its own volume of ammonia gas. The ammonia could not possibly be still in the state of a gas, for if it were, its pressure would burst the charcoal. Similarly, electro-deposited iron may hold 248 times its own volume of hydrogen. These gases are evidently in a liquid or solid form.

It appears that iron, when molten, absorbs nitrogen, hydrogen and carbonic oxide gases freely, and in solidifying evolves them, since it is no longer able to retain them. The escape of these gases is what gives the steel maker and foundry man their trouble, as gas which is caught in the pasty mass of iron forms minute bubbles which enlarge, by accretions from the adjacent metal, into openings called blowholes. The gases in these openings, if unable to swim to the top of the casting, cause the metal to swell or rise, or, if they escape, cause boiling and scattering, which produces unsound pieces.

A grey cast iron which passes quickly from the liquid to the solid state is less likely to form blowholes than white cast iron, which passes through a pasty condition, unfavorable to the escape of gases.

The more free iron is from carbon, silicon and manganese, the more does it tend to form blowholes. High carbon steel is usually comparatively free from blowholes, but is more liable to form openings at the center of the ingot, called "piping," due to its greater contraction in cooling.

Irons containing a large percentage of phosphorus, however, disengage gas much more violently and tend to produce unsound castings. The escape of gas continues for some time after the iron has solidified, and can be obtained from the minute cavities of all irons by subsequent borings which penetrate these openings.

Sir William Bessemer proved that the escape of gas from molten steel was largely governed by the existing pressure. When a partial vacuum was formed over the solidifying ingot the evolution of gas was much more violent. If a pressure were produced, the solubility of the gases in the iron was increased, and any gas which was evolved during solidification would be more promptly expelled. This process, which he called "liquid compression," was described by him in 1856. It remained, however, for Sir Joseph Whitworth to put this knowledge into practical use.

In his process the steel is cast in a special mold or flask of great strength and so arranged that the gases can escape. The metal while solidifying is subjected to a pressure, occasionally reaching 20 tons per square inch of horizontal cross section. This pressure hastens the discharge of the evolved gases and prevents to a considerable extent the formation of blowholes, and by producing a flow of metal to the contracting center of the ingot, partially overcomes piping, the ingot being actually compressed about one-eighth of its longitudinal dimensions. The Bethlehem Iron Co. uses this process in the steel castings from which large shafts and other structures are forged.

It is not the intention of the writer to do more, in this brief article, than to outline some of the physical and chemical changes which may take place in iron and steel while being formed for the many purposes of the engineer.

There are many phenomena in this field, which should be investigated and the possibilities which may be derived from a more complete knowledge of this most valuable metal, iron, are far reaching.

IMPROVEMENTS IN THE THREE-WIRE SYSTEM.
TRUMAN HIBBARD, '97.

In the year 1880, after costly and tedious experimenting, Edison introduced his incandescent electric light on his low pressure constant potential system of distribution, as a competitor for commercial lighting. The difficulty of making a high voltage lamp required the use of low voltage lamps, 50 to 125 volts, and therefore low voltage and heavy currents on the distributing lines, causing a heavy loss therein.

In order to diminish this loss, Edison devised his three-wire system for which the United States patent No. 274,290 was granted to him March 20, 1883. A similar patent was granted to Dr. John Hopkinson in England July 27, 1882. By this system the pressure on the mains is doubled while the same low voltage lamp is used, allowing the carrying capacity of the line to be decreased to less than 40 per cent of its original capacity and still have the same per cent lost on the line.

Where a single generator is used it must generate the whole pressure. Fig. 1 shows the original method proposed and employed by Edison for two machines operating on the three-wire system. When more than two machines are desired they are connected two in series, and as many pairs of these series as desired are operated in parallel, as shown in Fig. 2.

It will be readily seen that when the current consumed on each side of the middle wire is the same, the fall of potential between it and either main will be the same, and it will have a potential midway between the two, as it also has at the generators. Therefore the potential is the same at both ends of the middle wire and there is no tendency for current to flow. It is therefore called the "neutral wire."

The increase of pressure to twice its former value allows a decrease of current to one-half its former value, and as the line loss is proportional to the square of the current, this decreases the line loss to one-fourth its former value, or allows a conductor of one-fourth the carrying capacity. The neutral wire is usually one-half the size of the outside mains and this makes the total weight of the conductors of a three-wire system five-sixteenths of that for a two-wire system furnishing the same number of lamps with current. This enormous saving in one of the largest items of expense has led to an almost universal use of the three-wire system for low pressure distribution to distances over a few hundred feet and under a mile. Of the 207 miles of underground mains of the Edison stations in New York City in 1896, all but seventeen one hundreths of one mile are of the three-wire system.

This economy of line copper is the sole advantage of this system. Opposed to this are the following disadvantages:

1. The necessity for two dynamos or a complicated compensating apparatus.

2. The wiring is slightly more complicated involving the installation and maintenance of three instead of two conductors, and more complicated switchboard apparatus.

3. Sometimes the system may become badly unbalanced and cause some of the lamps to have a higher pressure than either of the generators ([1]). When the system is badly unbalanced, if by any means the neutral conductor should be broken it would mean sure death to the lamps burning on the light side. But destructive unbalancing can usually be guarded against by judicious planning at the beginning and care in later extensions.

The one advantage remains so much greater than all the disadvantages, that as stated it is almost universally used for low potential distribution, where the plant is of sufficient size to warrant the use of two machines.

From the three-wire system five-wire systems using four machines and even seven-wire systems using six machines were easily developed ([2]). But the greatly increased complication of these systems prevents their use only for very dense supply. Edison also proposed four- and six-wire systems using three and five machines, but neither of these systems can be divided into other systems with a smaller number of wires, and still properly maintain the balance, hence they are rarely used. A great deal of effort has been directed toward improvements and simplifications of the system. Edison suggested using a third brush for the neutral wire, on a single bipolar dynamo, but the brush caused vicious sparking on account of the conductors being in a strong field when passing under the brush, and we find no accounts of its successful use.

Edison also proposed the use of a storage battery, with the

IMPROVEMENTS IN THE THREE-WIRE SYSTEM. 47

neutral wire connected at the middle and one main at each end, the batteries being in series, and an adjustible resistance being put in series with each main for regulating the pressure of each side, which is manifestly a wasteful method.

Rankine Kennedy in England patented a new machine in 1882 which he called a "multiple current dynamo"([2]). His scheme was to use a two-pole armature in a four-pole field (Fig. 3). The

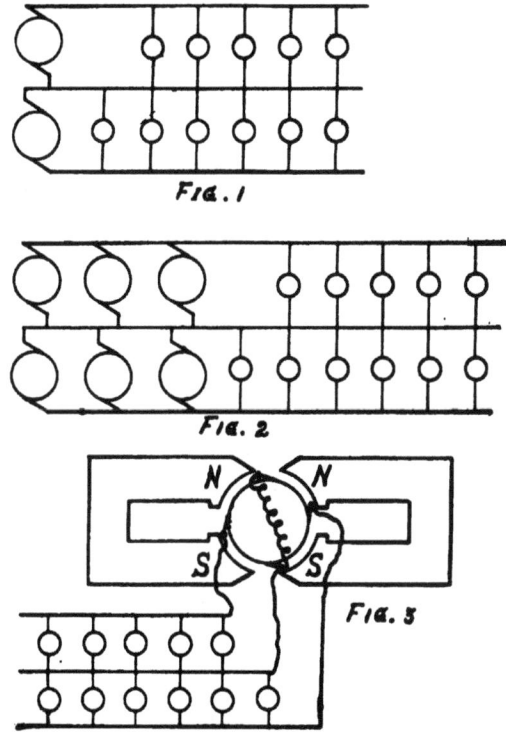

FIG. 1

FIG. 2

FIG. 3

windings of the fields were so arranged as to bring two north poles together and two south poles together. The main brushes collected current from the usual neutral points between north and south poles. Two other brushes were placed on the commutator in the gap between the two parts of each north and south pole, and connected together, and to the neutral wire of a three-wire system. These auxiliary brushes being at a point where the armature conductors were generating no pressure, on account of the gap in the field, gave very little trouble from sparking when the load was fairly well balanced between the two sides.

It is interesting to note that this machine was patented before Edison's three-wire system was brought out, and originally built for another purpose. In 1895 Kennedy proposed using the machine at 460 volts on the outside mains and 230 volt lamps on each side. Motors at 460 volts could be used directly from the two outside mains as also arc circuits at constant potential. He also suggested using a six-pole machine of similar design on a five-wire system.

In 1880 Herr Hermann Muller([4]), of Germany, proposed to use only a single generator with two brushes, generating the pressure on the outside lines, transmitting the current on two wires to the point of distribution and there using an auxiliary small machine of special design to convert it into a three-wire system and effect a balance, distributing thence, on the three-wire system. Fig. 4 shows the arrangement. The balancing machine had one pole split and the other solid, and the two main brushes, 1 and 2, bear on the commutator on the usual neutral points between north and south poles. These brushes are connected directly to the mains from the generator. A third brush, 3, connected with the neutral wire bears on the commutator in the neutral point formed by the gap in the north pole, and is half way between the main brushes. When the load is unbalanced as in Fig. 4, the difference of the two currents flows from brush 3 to brush 1 through the armature conductors, acting as a motor to drive the armature. This motor action develops a pressure in 3 2, of the armature which is added to the pressure already existing there, thus raising the pressure on the heavily loaded side of the system. One-half of the armature always acts as a motor; one remaining quarter acts as a dynamo, and the other quarter as a motor, these quarters changing functions as the overload varies from one side to the other. When the system is balanced the machine runs as a motor without load. This particular method does not seem to be much used, though methods quite similar are often used.

A set of storage cells([5]) has been used in place of the equalizing machine of Muller's scheme, by connecting the ends of the set of cells to the mains and the neutral wire to the center of the

IMPROVEMENTS IN THE THREE-WIRE SYSTEM.

battery as in Fig. 5. When the system is unbalanced, the heavily loaded side of the circuit receives current from that half of the battery while the other half of the battery is charged. This method affords no economy in first cost but at times of light load the batteries may furnish all the current that is required and allow the machinery to be shut down.

Mr. R. E. Crompton[6] was the first to use storage batteries in connection with the three-wire system in Europe. He employed them in a plant at Vienna in 1886 and the system was still in successful operation in 1895. He used them in many subsequent installations in Great Britain with good success. The use of accumulators also gives a more uniform load for the generating machinery, being charged by the lines at times of light load and discharging to the lines in times of heavy load, thus allowing the machinery to be operated at a higher efficiency. By slightly shifting the point of connection of the neutral wire, the pressure for the two sides may be very closely regulated for almost any degree of unbalance. Storage batteries may also be used in connection with one generator and a five-wire system. A large plant using accumulators on a three-wire system was installed in Mulhausen, Germany, in 1888 and is giving excellent economy and regulation[7].

In 1887 Elihu Thomson patented a method of using storage batteries[8] in connection with a three-wire system, in which he used a battery giving only one-half the pressure of the outside mains. By means of a revolving commutator he rapidly connected the battery first to one side and then to the other so that it would receive a charge from the side of light load and instantly be switched over and deliver it to the heavily loaded side. This does not seem to be of any commercial value.

Thomson also proposed several other methods. One was to use a small motor[9] mechanically connected to the single generator, and an automatic switch to keep the motor always connected into the light side of the system and always running as a motor, using the surplus current and converting it into mechanical work to assist the engine to drive the generator. The field of the motor had a shunt winding across the outside mains and was differentially compounded by a series winding of the neutral wire, so that as the load became more unbalanced the motor tended to speed up. The connections are shown in Fig. 6. The automatic switching arrangement for throwing the motor connections from one side to the other depended on the differential action of two sets of coils on an iron core. Fig. 7 shows the arrangement of the switch in which coils N N' carry the current in the neutral wire, coil A the current in main A and coil B the current in main B. When any current flows in the neutral wire one of the sets of coils will act together and the other set act in op-

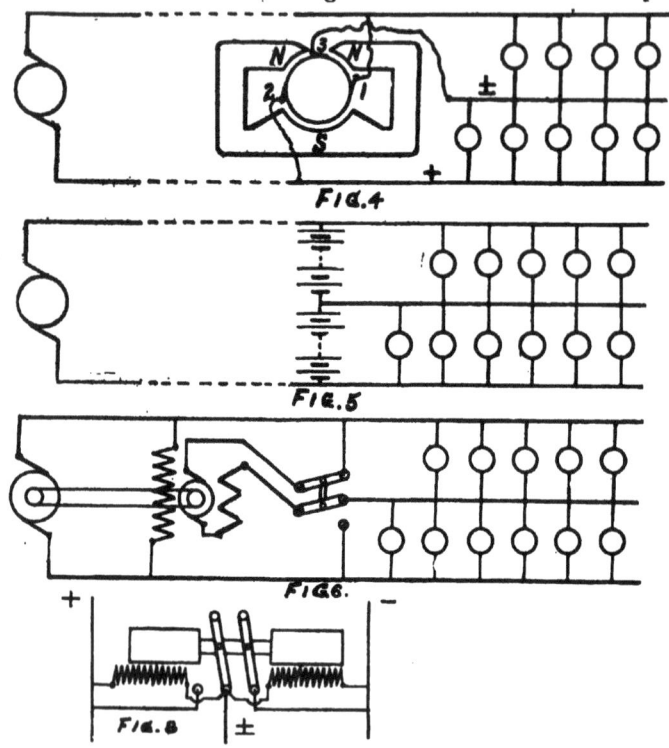

FIG. 4

FIG. 5

FIG. 6.

FIG. 8

position, so that the core is pulled in the direction of the two coils acting together, in the figure, N' and B. When the current in the neutral changes direction, due to a change of balance, the other two coils act together and draw the core in their direction throwing the switch and changing the connections of the motor to the lightly loaded side.

Another switch actuator for controlling the same switch, for

IMPROVEMENTS IN THE THREE-WIRE SYSTEM. 51

the same purpose was proposed, and is shown in Fig. 8. It depended upon the difference of potential of the two sides of the system but was found not to be sensitive to less than about four volts difference between the two sides, and thus caused a bad flicker when it did throw, and it was abandoned.

Mr. E. W. Rice, Jr., patented in 1891 a method([10]) of using a single auxiliary machine permanently connected to one side of the circuit, and mechanically connected to the generator as in Thomson's scheme. The electrical connections were never broken, but on the other hand it changed its function as the balance changed, running as a motor assisting the engine when the load was light on its side, and as a dynamo helping to supply the current when its side was heavily loaded. To cause it to change its functions automatically the field was wound with a shunt across the outside lines and a series winding of the neutral wire. Then as the neutral carried current one way the series winding assisted the shunt, making a strong field and causing the machine to generate current. When the neutral current reversed the series opposed the shunt and the machine tended to speed up in the weak field and run as motor.

Mr. H. Ward Leonard modified this system by using a small generator and motor connected together near the center of distribution([11]). (Fig. 9.) One machine was connected to the outside mains and the other to the neutral and one of the mains permanently. When that side was lightly loaded machine A ran as a motor driving B as a dynamo, and in parallel with the main generator, supplying current to the mains. When the balance changed A would become a dynamo driven by B, supplying current to its own side. The two small machines could be placed some distance from the central station and required little care, and gave quite satisfactory results.

Another method proposed by Thomson([12]) was to use a motor generator, in which two armatures were connected rigidly together. They were connected in series across the outside mains and the neutral connected in between them (Fig. 10). Then the armature on the light side drives the other as a dynamo at a slightly increased speed, thus developing a slightly higher pressure on the heavily loaded side, and thus maintaining the lamp pressure. As the balance changes from one side to the other, the armatures exchange functions automatically. By means of a rheostat in the field circuit of one of these small machines, the regulation may be made quite close by constant attention.

In this method of regulation the C R drop and the C^2 R losses in the armatures are detrimental to close regulation and should be reduced to a minimum in the design.

Under balanced load both armatures run as motors, if the fields are of the same strength, dividing the friction load between them. This method of compensation is perhaps the one most commonly used, aside from the original two dynamo system. The windings of both armatures may be on a common iron core, placed in a single field, and provided with two commutators. By this method armature reactions are neutralized and the commutation is sparkless at all loads, and the friction loss is reduced, but the advantage of independent field regulation is lost.

A combination of the motor-generator and accumulators for compensating is used with good success by the Burnley station, at Burnley, England[13]. (Fig. 11.) The motor generator is built in one machine and connected between the two armature windings to the neutral wire. The neutral is also connected into the middle of a large storage battery giving something more than the generator pressure. The end of one armature winding is then connected to a battery regulating switch B, and the end of the other armature to another similar switch, D, at the other end of the battery. Two similar switches on the other side of the batteries as shown, A and C, connect to the outside mains. These switches are so arranged that a few more or a few less cells may be put in circuit at will, and by this means the bad effect of armature losses are overcome. For instance, if the plus side were loaded heaviest and the pressure low, moving the switches A, and C, to the left, raises the pressure between plus or minus by putting more cells in circuit, and at the same time lowers the pressure between plus or minus by cutting out some cells, thus maintaining the pressure at the lamps the same in each. By moving B and D to the right the armature on the light side speeds up as a motor, and drives the other armature as a dynamo, delivering current to the heavily loaded side. When the overload changes to the other side, the switches must all be moved in the opposite direction, though not necessarily the same amount for each, and the armatures exchange functions. The motor generator tends to

IMPROVEMENTS IN THE THREE-WIRE SYSTEM. 53

keep the state of charge the same in both sides of the battery, which is quite desirable. The accumulators furnish all the required current for hours of light load, so that the engine need only be run during hours of heavy load. This method is somewhat complicated and perhaps expensive to install, but it is claimed that the operating force is largely reduced and the economy of operation increased.

Mr. A. Churchward has perfected a motor generator([14]) of the single machine type, to be used as an equalizer on three-wire systems, which has met with good success in New York during the last three years. Experience has shown that the capacity of the equalizer need be only ten per cent of the load it is to equalize.

A "double voltage" dynamo([15]) for operating the three-wire system was brought out in 1894 by Mr. Fuller of Detroit. It was simply two generators built in one having two commutators and operated very much as two dynamos, except that independent field regulation was lost. It has not come into general use.

In 1893 Dobrowolsky ([16]) devised an ingenious compensating device to be used with a single generator, which was exhibited at the Lyons Exhibition in 1894 by the Five-Lilles Co. (Fig. 12.) The method was to connect the two ends of a choking coil to diametrically opposite points of an ordinary Gramme armature, used as the generator armature. Then the neutral wire was connected in to the middle point of the choke coil. When the machine is running there is an alternating pressure set up between the points where the choke coil is connected to the armature, which would send a large current through the coil were it not for the choking action due to the high selfinduction of the choke coil. By making the selfinduction quite large, the ohmic resistance of the wire coil may be made quite small and still only a very small current will get through, due to the alternating pressure. The middle point O, of the choke coil winding is always at a potential midway between that of the ends A, B, and at the same time the difference of pressure between O and brush N is equal to that between O and brush P. Now by connecting the neutral wire to the point O and the brushes P, N, to the outside mains, if the load is unbalanced as shown, the neutral current will come back to O and there divide and pass through the windings of the choke coil, to the armature since it is continuous and causes no selfinduction. Since the difference of pressure between O and A, and O and B, is always maintained the same

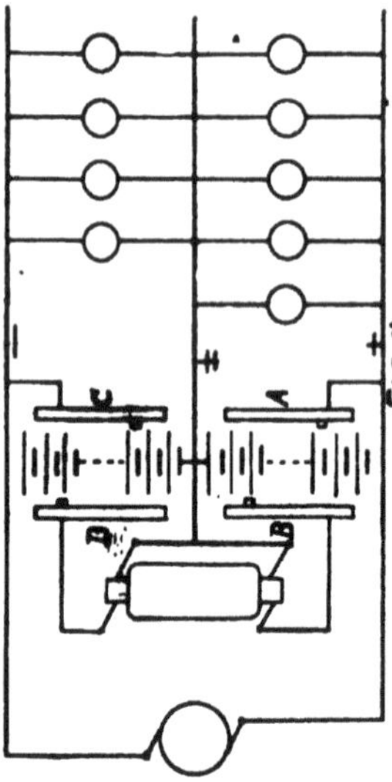

Fig. 11

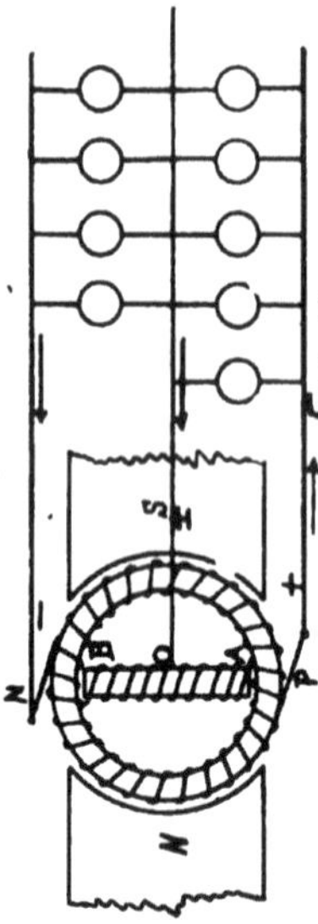

Fig. 12

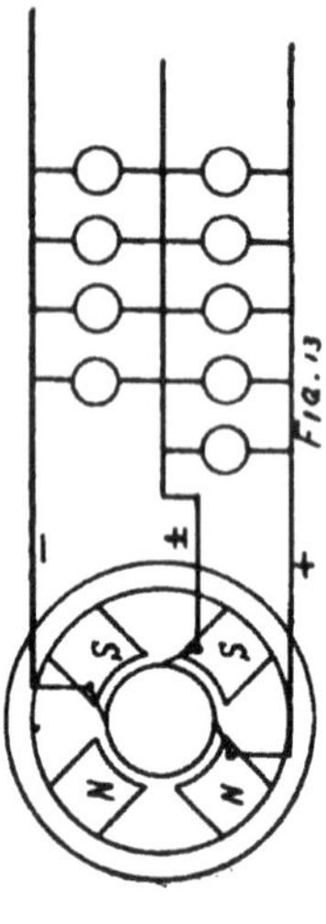

Fig. 13

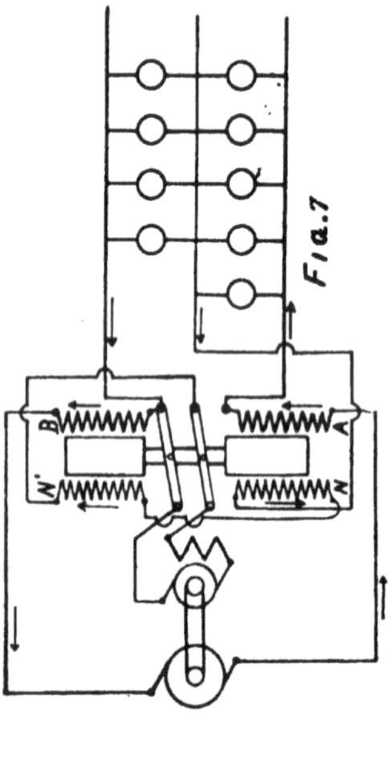

Fig. 7

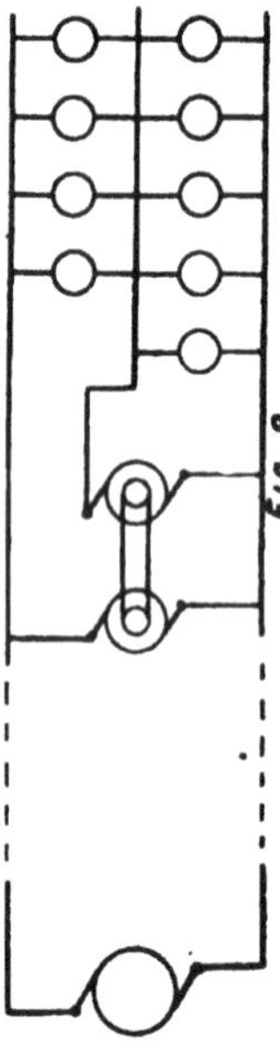

Fig. 9

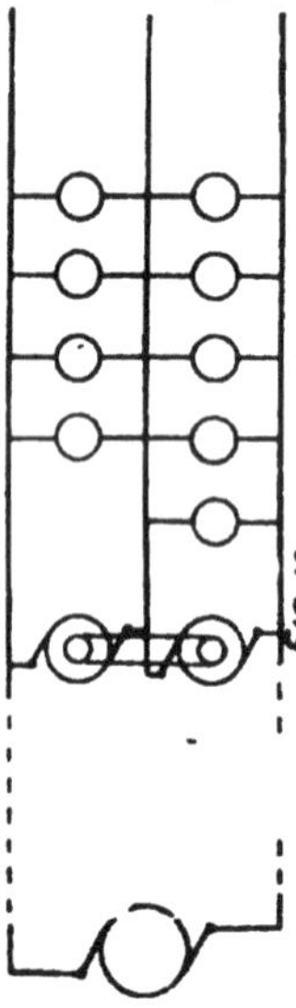

Fig. 10

IMPROVEMENTS IN THE THREE-WIRE SYSTEM. 55

by this choke coil, for all conditions of balance we will have practically the same pressure between the neutral and each outside main, at the station. As the load on one side may quite largely exceed that on the other it may be desirable to raise the pressure on the heavily loaded side above that of the other to take care of the drop on the neutral line, which is manifestly impossible with this arrangement. It was found to work well only on pretty evenly balanced systems.

A small machine of this type may be used as a tension regulator at any distance from the generator, just as a motor generator would be used. Usually the choke coil is placed outside the armature and the connections made through collector rings.

Pilcher suggested a modification[17] of this machine in which a Gramme armature is wound in the usual manner, and in addition, an auxiliary winding of smaller wire, of sufficient size to carry the neutral current, and of exactly one-half the number of active turns of the regular winding, is evenly distributed over the core. The two ends of this auxiliary winding are connected to two points of the main winding diametrically opposite, in a two-pole machine, and the middle point of the auxiliary winding is connected to the neutral wire through a contact ring and brush. Thus the two halves of the auxiliary winding are connected in series, and the pressures generated in each half added together. This pressure is alternating and its maximum value is equal to that of the main winding. By this arrangement the pressure generated in the auxiliary is at all times equal and opposed to that impressed at the two points of the main winding, to which it is connected. Thus a counter E. M. F. generated by the fields and auxiliary winding checks the flow of alternating current, instead of the selfinduction as in Dobrowolsky's machine. However, this method does not afford any means of independently regulating the pressure in each branch of the system.

Apparently the most satisfactory method proposed, up to this time, is that used by the A. E. G. Co., formerly Lahmeyer & Co., of Frankfurt on the main[18]. The machine was invented in 1894 but did not reach the market until late in 1895. Several of the machines are now running successfully.

To divide the total pressure in halves they use a third brush between the positive and negative brushes, and suppress sparking by the use of a four-pole field in which two adjacent poles are north and two are south, as shown in Fig. 13. The third brush bears between the two north poles at the neutral point formed by the gap between them.

To obtain independent regulation for each branch it was found only necessary to strengthen one north and one south pole, those being the ones diametrically opposite, and weaken the other two, giving any required difference between the pressures of the two sides. This is easily done by exciting the opposite north and south poles by one circuit with a regulator in it, and the other two by a separate circuit with another regulator. Some difficulty was experienced at first by the tendency of a strong pole to reverse a neighboring one of the same polarity. This was overcome by crossing the exciting circuit so that one pair was excited from the circuit in which the pressure was generated by the other pair and vice versa. This made it impossible for one set to come up to strength without both coming up. A 75 K. W. machine of this type was run for some time at the Berlin Exposition in 1896, with the frequent occurrence of 200 amperes on one side and 400 amperes on the other, without any noticeable change in the working of the machine.

This type of machine requires a little different design from the usual multipolar on account of the change in the magnetic path. The armature requires about twice the iron for the usual type and the yoke between N. and S. poles needs twice the cross section, as it now has to carry all the lines of force due to one pole piece. But there is no use for any iron between the two N. or between the two S. poles, except for appearance and mechanical strength. In this machine it is essential that only one neutral brush be used, for if another were placed opposite the first and connected to the neutral, the independent regulation of the two sides would be lost.

For general discussion of the three-wire system see the following:

W. J. Jenks, Elec. World, vol. 14, p. 39, 1889.
F. B. Crocker, Elec. World, vol 29, p. 126, et seq., 1897.
A. V. Abbott's Electrical Transmission of Energy, p. 423.

---

(1). Crocker Elec. World, Vol. XXIX., p. 126, 1897.
(2). Crocker Elec. World, Vol. XXIX., p. 237, 1897.
(3). London Elec. Rev., Vol. XXXVI., p. 805, 1895.
(4). London Electrician, Feb. 28, 1800. Elektroteck. Zeit., Vol. XI., p 57, 1890.
(5). Electricity, Vol. VIII., p. 310, 1895.
(6). Proc. Institute of Electrical Engineers (England), Vol. XVII., Nos. 72 and 73.

(7). London Electrician, Vol. XXXIII., p. 282, 1893.
(8). Patent No. 360,125.
(9). Patent No. 468,122.
(10.) Patent No. 460,364.
(11). Elec. Engineer, N. Y., Vol. XIII., p. 572, 1892.
(12). London Elec. Rev., Vol. XXXIV., p. 146.
(13). Elec. World, Vol. XXIX., p. 32, 1897. Am. Electrician, Vol. IX., p. 21, 1897.
(14). Elec. World, Vol. XXIV., p. 505, 1894.
(15). London Elec. Rev., Vol. XXXIV., p. 624, 1894. La Lumiere Electrique, Vol. LI., p. 30, 1894.
(16). Zeit. f. Elek., Feb. 1, 1897. Elec. World, Vol. XXIX., p. 265, 1897.
(17). Elec. World, Vol. XXVIII., p. 569, 1896. Elektrotecknischer Zeitschrift, Vol. XVII., p. 611, 1896.
(18). Elec. World, Vol. XXVIII., p. 569, 1896. Elektrotecknischer Zeitschrift. Vol. XVII., p. 611, 1896.

RECENT RAILWAY ECONOMIES.*

E. E. WOODMAN, C. E.

A late news item recounted the discovery of a burning coal bed in Wyoming, and there is little doubt that the incident was only the most recent of many re-discoveries of the same thing that I saw at the head of Little Powder river thirteen years ago, when examining the Black Hills country for a railway line and to report on its natural resources. At that time the geological insularity of the Hills, as an outlier of the Rocky Mountains, was matched by the isolation of the people, who were two hundred miles from the nearest railroad, possessed of a Sancho Panza's Island in the midst of the Great Plains. I went a hundred miles west of Deadwood with a guide, riding in a light buckboard, to examine an exposure of lignite. We found the half-formed coal showing twenty feet of thickness, where a creek had made a shallow valley, and at this deepest place it was on fire and had been burning an indeterminable time. We fried our bacon and boiled our coffee on the glowing edge of this Inferno, and as I lay at night on the ground, with feet to the heat, in the midst of a vast solitude, and looked up to a few friendly stars, whose acquaintance always relieves me from the sense of loneliness when in uninhabited places, and somehow co-ordinates the universe, it was not impossible to guess at the grand scale of those operations of nature that resulted in this enormous bed of fuel. The old cosmogony, of which man and his home were the center, would have made all this for the very purpose of comforting the vagrant engineer. But Galileo, with the first feeble telescope, pierced that fog of error and proved that finite relations and adaptations do not really make clear any designs of the Infinite. But here was this exposure of lignite, five hundred miles intermediate between two similar beds known on the Northern Pacific and Union Pacific Railways. It was then a probable inference that it extended across the entire one thousand miles, and this opened broadly to the mental view the state of that young world in which peculiar and adequate causes were effective to such vast results.

The geologist and the chemist have delighted us with the natural history of a lump of coal. Igniting it as it were by the fire

---
*Lecture before the Society of Engineers, March 26, 1897.

of the imagination, they have made it glow for us with the light and the heat of a sun "before time was," of ages long antecedent to man, the clock-maker. They have clothed that ancient globe with a luxuriance of vegetation far surpassing the tangles of the Amazon, peopled it with corresponding animal monsters; and when I broke out pieces of the lignite in which the bole, the knots, the bark, the grain of the wood, were almost as palpable evidences of the tree as like parts would identify the allied tamarack of to-day, the vast swamps in which this immense deposit of half carbonized wood accumulated were conceived with the same strength as the ancient civilization of the Nile Valley is now apprehended, when we look upon the photograph of Rameses, the very Pharaoh who persecuted the Hebrews. The imagination has indeed notably played its part in science. It is a kind of sublimated reason, by which truths now beyond the range of demonstration are gained by a leap and held till the slower faculty can slide a cantilever over the gap. It has also been a powerful instrument of invention, and the chief aid in effecting those ameliorations of the physical conditions of life in our world on which all other forms of progress depend. Doubtless the cave dweller brooded over the problem of how he could subdue the mammoth and make it carry wood for his fire. For in after ages he accomplished it in India, meantime improving his weapons and tools from flint to copper, to bronze, to iron, to steel, ever further subjugating the hostile forces of the world. With the rise of intellect brute forces appear more and more to have declined, as if man was ever the integration of all that had gone before; though far weaker physically, yet able to cope with the same obstacles by a higher method, which we may typify by the lever. In due time he thought out his steam engine, by which he brought back the mighty saurians, harnessed them to tread-mills the world over, and trained pterodactyls to fly in lines at fifty miles an hour, pulling the limited. And so we come to our theme of the railroad.

Next to the invention of the steam engine, that of the railway has done most for our material welfare; and next again the Bessemer process, by which both of the others have been so vastly amplified in their utilities, may be given rank. Using this term to cover all the improvements by which steel has been made so much better and cheaper, we may affirm that the enormous increase of wealth shown by many countries, including our own, is primarily due to the success of this invention, of which it was long since said: "It has been of more value than all the gold of California." In forty years our constructive arts have been revolutionized by it. Whereas, forty years ago, iron was too costly and unreliable to be much used by the engineer, and steel was entirely beyond his reach, twenty years since the new metal had not only superseded iron for most constructive purposes, but at a saving to mankind of five thousand millions of dollars, computed on the total product and the diminished cost. To the date of this invention we refer the introduction of improved heavy machinery, steel vessels, heavy ordnance, armor plates and modern projectiles, the widespread use of rolled forms in architecture and bridge construction; while it would be well-nigh impossible to enumerate all the present uses of steel, which from its superior strength and ductility, comparative lightness, uniformity of quality, and adaptability to special requirements, has finally almost driven iron from the engineering field. The most admirable and useful point of difference between this metal and iron is, that the fibrous structure of iron has in steel given place to a homogeneous molecular arrangement, so that, as Sir William Siemens described it, "it has its full strength in every direction."

The most extensive and useful application of the Bessemer metal has been in the improvement of railways, chiefly in the track, though otherwise largely in the related structures of bridges and buildings, and in machinery and rolling stock. The locomotive was the first essential of a steam railroad, the track the second; but soon these conditions were reversed, the track having become the more difficult to maintain with the requisite economy. The Rocket weighed 2½ tons, our present engines weigh 60 tons. The former carried 2,500 pounds on each driver, the latter carry over 20,000, and these differences are a fair index of the economy of using steel. Since the introduction of this material the improvement of our railways, the great increase in their mileage and the corresponding development of our national wealth have severally been very great, and may be traced to the possession of this metal. The tendency of railway economies along mechanical lines during the last twenty years, by which these results have largely been obtained, has been controlled by the Bessemer process.

RECENT RAILWAY ECONOMIES.

The first Bessemer rails introduced into this country from England cost one hundred and fifty dollars a ton. Like rails of our own make have recently been sold at seventeen dollars a ton. Sir Henry himself, while in the first flush of his success, and amassing a fortune, immediately reduced the cost of tires in England from four hundred and fifty dollars a ton to one hundred. A Bessemer rail will last from six to nine times as long as an iron one under the same traffic. Here is a clew to many economies. With the life of rails so greatly prolonged, the cost of renewal of rails and repairs of track, a heavy item of expense to every company, is of course greatly reduced. From being able to obtain rails at once cheaper and far more durable, other important consequences have flowed. Heavier rails have been placed within reach. This in turn has made possible the use of heavier engines, stronger cars and heavier loads, resulting in such economies of operation as to be a principal cause of the continual decline in freight rates in this country, which to-day are only a third of what they were when the new metal had become fairly introduced.

If an engine of new model proves to be twice as serviceable in tonnage hauled as one of the kind previously used, then half as many trains, drawn by such new engines, will do the same business. This will chiefly apply to through traffic, because local traffic usually does not make long trains, or at all events the same degree of economy can not be worked in it, since heavy engines can not so frequently be loaded to their full capacity in local business as in through traffic. The most obvious reduction in expenses due to the new engines will be, a certain number of train crews dispensed with. Also, there may be economy in fuel, the general rule being that large boilers and engines use fuel to better advantage than small ones. I may illustrate this by recent experience. The freight traffic of the C., St. P., M. & O. Railway Company during October, 1896, was the heaviest in the history of the road. As a consequence of this, engines were loaded to their full capacity, and the engine performance became especially interesting. Following are the mileage and coal consumption of three classes of engines:

Cylinders.	Miles run.	Tons coal used.	Pounds coal. per train mile.
17 by 24	133,750	6,245	93.4
18 by 24	149,400	6,853	91.7
19 by 24	54,664	3,014	110.3

It has not been practicable for me to obtain the tons hauled by these engines. But assuming that they were fully loaded, which we may do in consideration of the deficiency of cars under the unusual demand, a comparison of these engines in respect to consumption of fuel may be reached by dividing the quantity of coal consumed per train mile by the adhesive weight, or weight on the drivers, since this weight is closely proportional to the hauling capacity, and in the present case to the number of tons hauled. Applying this test, we find that the relative consumption of coal per ton mile is, for the 17 by 24 cylinders, 20; for the 18 by 24 cylinders, 16; for the 19 by 24 cylinders, 11; or, the largest shows an economy of 45 per cent over the smallest in the item of fuel consumed per ton mile.

Another illustration may be drawn from passenger service in the same month. The new express locomotives used on the Chicago limited have cylinders 19 by 24, drivers 73 inches, and weigh 126,000 pounds each. The engine and tender and supplies of fuel and water weigh 113 tons. If we add to the last mentioned weight the weight of the cars that compose one of these trains, and of the several items comprised in the load, we shall find that the average number of tons moved per train mile is 456. In October, 1896, twin engines in this service together ran 17,015 miles and consumed 746 tons of coal. From these data it follows that these engines moved a ton a mile by the consumption of 19-100 of a pound, or 3 ounces, of coal. The card time of this train, covering a distance of 200 miles, is 31 miles per hour, and the actual speed between stations often 40 to 50 miles.

In comparison with this we take the Stillwater run, with engine and tender weighing 58 tons, and cars 57 tons. These trains made 5,840 miles in October, with the consumption of 170 tons of coal, showing that it required half a pound of coal to move a ton a mile, or 8 ounces as against 3 in the case of the limited, and at lower speed. Part of this difference is undoubtedly due to the relative preponderance of engine friction in the second case. It is no doubt true that the lighter engine could take a heavier load and yield a better result per ton mile than it did in this case.

Three ounces of coal seems a small quantity to move a ton a mile by any kind of mechanism, yet it is well known that the locomotive is a form of engine comparatively wasteful in the use of fuel. Bituminous coal is used in operating the trains under consideration, Spring Valley, from Illinois, in running north, and No. 8 Pittsburgh, from Pennsylvania, in running south. For the purpose of illustrating fuel economy,I have selected a distance of 4.55 miles in which no stops are made by these trains, the time allowed being seven minutes with the grade and eight against it. The average speed is 36½ miles per hour, and the average of the rise and fall 66 feet, corresponding to a grade of 14.5 feet per mile for the whole distance. I neglect the effect of a little light curvature. Using Clark's formula for train resistance, and comparing the foot-pounds exerted by the engine in moving a ton a mile under these conditions with the stored energy in the 19-100 of a pound of coal consumed in that service, it appears that 7 per cent of the stored energy of the fuel is converted into useful work. By the Searles' formula a very much higher economy is indicated. But the 7 per cent is itself an improbable value. Working tests skillfully planned and carefully made, of equally good non-compound engines, showed only 5½ per cent of the theoretic value of the fuel realized. Moreover, the 7 per cent comes too near the performance of the best stationary engines. I am indebted to Mr. J. D. Estabrook, C. E., for the result of a seven days' test of the triple expansion Schichau engine of 1,100 horse power now in the Washburn-Crosby Company's flouring mill at Minneapolis. This fine engine will be remembered by many who saw it in operation in Machinery Hall at the Columbian Exposition. The test was conducted by the late Prof. Wm. A. Pike. and Prof. H. E. Smith, both of this college, and showed one indicated horse power for 1.928 pounds of Youghiogheny coal consumed, the coal having a determined calorific power of 13,222 B. T. U. Thus the engine developed in the cylinders 10 per cent of the theoretic value of the fuel. Deducting one-tenth from this result for engine friction, we see that no single expansion locomotive can be held to realize 7 per cent. economy without supposing that it comes well up to a triple expansion stationary engine in efficiency. I think the principal source of error here is not in the calorific value assumed for the fuel, but in the formula for train resistance. For high speeds both those named appear to give excessive results. Consequently, when we use either for high speed, the engine appears to be doing an amount of work that in reality it can not get out of the coal. So the above approximation of 7 per cent net out of the fuel is only to be considered as suggestive, or as possessing a negative value in that it impugns the traction formulae. In order to reach a result of greater value, conditions that I have assumed, such as the calorific power of the fuel, and the traction force exerted, would have to be determined by experiment.

However, as a rough check upon the foregoing, I cite the performance of English freight engines, as given by Mr. D. K. Clark, in his elaborate treatise on the steam engine, one of which moved a ton a mile by the consumption of 16-100 of a pound of coal and another 14-100. Of course the speed was here much lower, and the coal almost certainly better, and we have no account of the grades. Also, the Scotch Express, which makes about fifty miles an hour on the average, and mounts some long and heavy grades, is drawn by a compound engine that moves a ton a mile with two ounces of coal estimated to be fifty per cent better than ours, though it can not be so much better than the coal above described. Again, the freight train statistics of the C., St. P., M. & O. Railway Company for the years 1885 and 1895 show that the average performance of engines in each year was to move a ton a mile with 19-100 of a pound of coal. These results are based on the consumption of coal per mile run of engines in hauling 796,000 cars in the earlier year, and 978,000 in the later one, and therefore are deductions in which all kinds of train resistance are averaged.

But cost of fuel is only one item of engine expense, and heavy engines are costly in other ways. The repairs upon them are usually very expensive. Their fire-box plates are so severely tried by the fierce combustion, and by expansion and contraction, as to require frequent renewal. Strenuous endeavors are made to secure the best material for this purpose, yet a sheet has been known to show more than a hundred and fifty cracks after a short service. Also, the great weight of the reciprocating parts aggravates the destructive effect of a lack of balance in those parts, and consequently these monsters soon pound flat places in the tires of drivers, and must be sent to the shop to have those defects turned off. Their effect upon the rail will be mentioned later.

An increase in the capacity of freight cars has kept pace with the growth of engines, the two are correlatives of a single development. This increase has worked an economy in heavy traffic because less dead weight proportionally is hauled. A car weighing 20,000 pounds carries 20,000 pounds; one weighing 22,000 carries 40,000; one weighing 29,000 carries 60,000. From 1865 to about 1876 nearly all freight cars were of the capacity of 20,000 pounds. The increase then began. On the road of which I have been speaking the average capacity of the freight car has increased twenty-five per cent in the last eleven years. But the Lake Shore and Michigan Southern, by the fullness of its reports, offers one of the most striking and useful illustrations of this economic tendency. This is a trunk line, so called, one of the main routes of traffic between the East and the West, on which the pressure of competition would naturally give a powerful stimulus to mechanical and engineering ingenuity. Also the movement indicated has been so long in operation with this company that it may be measured by results. Accordingly we find that the freight traffic of this road shows a growth in ton miles, between 1870 and 1895, of 318 per cent, due to an increase in train mileage of only 80 per cent, but in train loads of 132 per cent. The car equipment of this road also throws light on these changed conditions. Out of 18,700 freight cars, 5,700 carry 40,000 pounds each, 3,200 have 45,000 pounds capacity, 3,200 have 50,000 pounds capacity and the remaining 6,600 are of 60,000 capacity. Many roads have special cars designed to carry exceptional loads. The Pennsylvania and the Northwestern have a number whose capacity is from 70,000 to 150,000, but these are not sufficiently numerous to further emphasize the tendency towards heavier loading. Nor is it likely that the movement will go further in the near future. Indeed, a reaction has already set in. Some managers have discovered the limited use of cars of great carrying capacity. Such cars are in fact only suited to heavy through traffic, and a company may make the mistake, and some have made the mistake of putting too much money into this kind of rolling stock. A car of 50,000 pounds capacity, found in a local train, strutting as it were with a few boxes of soap, is an example of pride going before a fall, it is losing money for the company. The equipment should be adapted to the traffic. One source of loss in these cars of large capacity is that, especially on foreign roads, they stand idle a great part of the time because no suitable loads are offered.

In connection with the subject of dead weight, the steel car is likely to work an economy. One of these cars, that weighs 25,900 pounds, carries 80,000 pounds, as compared with a wooden car of the same weight that carries but 50,000 pounds. The principal objection made to this car is the conservative one that it is apt to prove costly to handle and repair after it is wrecked. But this implies that the accident is the rule and not the exception in car service. The objection may be overcome, and doubtless will be overcome, by improvements in design, whereby such cars can readily be taken to pieces. Then they may easily prove to be more economical in repairs than wooden cars, which so often are destroyed as to all but the iron in them.

With fewer trains passing over the road by reason of using more powerful engines, renewals of track repairs may be less costly. In short, out of a total of forty items of operating expenses, sixteen will be favorably affected by such a change to heavier engines, and many more will not be affected either way. On the other hand there is to be charged against the improvement the increased cost of engine repairs and the interest cost of making the change in engines, track and bridges. Though the last two would by some companies be deemed betterments, and their money cost would be added to the cost of the road, yet interest on the amount must be earned if the assumed economy is justified. This is the general principle by which all such experiments are to be measured. For all our railroads have been built with borrowed money. The main end of the operating department must be, to earn interest on the bonds and a reasonable dividend on the stock. It is primarily the volume of traffic that does this, supposing, of course, that the business is conducted at a profit. Any specific economy in operation will bear usefully on this result, but is of secondary importance compared with the volume of business. That this is true is shown by the fact that any particular economy, such for instance as the introduction of more powerful engines, will usually not be obvious on examining the general published statistics of most companies. Thus, it might be expected that the effect of introducing heavier rails, with heavier engines and cars of greater carrying capacity upon them, would immediately be seen in the items of renewal of rails, cost of repairs of track, of repairs of engines, and in the percentage of operating expenses to gross earnings. But with a successful road fifteen hundred miles long such anticipated results will not often be realized; that is, the general statistics will not disclose the particular economy. And it is for the reason that the relatively small results are lost in the magnitude of the totals. They have, so to speak, only local bearing and significance. The accounts of the division on which heavy engines have displaced light ones may show that expenses have been reduced by the change in a degree justifying the experiment. But after all it is a complicated matter to balance the account between certain visible and incontestable savings, such as come from employing fewer engineers and firemen, and from a smaller consumption of coal in moving a given tonnage, on the one side, and on the other not only interest on the cost of additions to the property, but some expenses not at first readily determinable; as, for instance, whether the big engines will not pound themselves to pieces much faster than lighter ones, and whether they will not file away the rails and their own tires much faster than the others did. However, during the past twenty years the movement has been in the direction indicated, and so far as cost of rails is concerned is sustained by some general considerations.

If the weight of rails were to be increased from sixty-five pounds per yard to eighty, with corresponding fastenings, the increased cost of track at the present time would be about $1,400 a mile. If, on a division two hundred miles long we add $20,000 to this for alterations in bridges, to enable them to carry the heavier loads, we shall have a total of $300,000, the yearly interest on which, at five per cent, would be fifteen thousand dollars. Now, such a division may be earning $3,000,000 annually, on which the interest charge for account of improved track, as above given, would not be a burden, even in the extreme case that the heavier metal did not prove economical in connection with the heavier rolling stock. But on the other hand it may almost be decided without going into details, that an economy of $15,000 a year can be worked in operating expenses amounting to $1,800,000 by the purely administrative reforms continually being introduced. Thus there is a margin of opportunity for the experiment; besides which are the possibly longer life of the rail, fewer track repairs, the saving in fuel on the large engines, and in the wages of trainmen, leaving some other items doubtful. So far then as this movement has proceeded, it seems to be justified by general prudential considerations.

The further question arises, what is the rational limit of this economy? How long can we go on, increasing the weight of engines and cars and loads on the present quality of metal? The chief limiting fact would seem to be the ductility of the rail. It will not do to use a brittle rail, but on the other hand the soft steel now used flows under the cold rolling given it by the heavy wheel loads. Rails give out partly by wear and partly by deformation. On the outside of curves they are rendered unserviceable as much by the latter cause as by the former. In some of the new engines each seventy-three inch driver presses on the rail with a weight of over 20,000 pounds, and it was shown by Mr. Chanute's experiments that with 14,000 pounds on a five-feet driver the small area of contact indicated a pressure in excess of the elastic limit of the steel. So also these great machines, with the larger wheels but heavier loads, are certainly very destructive of the rails, first by grinding them, and again by squeezing them out of shape much in the manner that they were originally shaped. These results may easily be observed in large yards, where the inside of many railheads will be found worn away to the splices, the outside of them showing at the same time a projecting lip formed by excessive pressure. It is obvious that increasing the depth and weight of the rail has its principal effect in strengthening it as a girder, to carry the load from tie to tie, and does not prevent the head from yielding outwardly under pressure. Therefore we must proceed cautiously if we further increase the load on the rail, carefully balancing the account between maintenance of way and movement of traffic. This may involve a more minute distribution of expenses than has been customary heretofore. On the Lake Shore road, before mentioned, the item of renewal of rails is about one per cent of the income. Expressed in dollars it is a large sum, but the small percentage shows that wear of rails is not a place in which large economy can now be hoped for. It is the place where expense is most certainly proportional to use, since it depends almost wholly on the number of trains run over the rails. There is therefore less chance here for waste or extravagance than in any other branch of operating expenses.

Supposing the limit to have been now nearly reached in adapting the weight of engines to Bessemer rails, I think we may look to see future economies take one of two directions. The first is in the line of a better rail, either of steel possessing greater durability in this kind of service, or of some alloy offering higher resistance to wear and distortion than this steel, yet sufficiently ductile to be rolled and to prevent brittleness. Our ordnance officers are dealing with a similar problem, namely, to increase the power of guns without increasing their weight, for which they need a metal for the inner barrel that will resist the erosive action of gases at high pressure better than the steel now used is found to do. The invention of Bessemer metal was due to Sir Henry Bessemer's desire to obtain a better metal for ordnance. He knew little or nothing of metallurgy when he began his experiments, which resulted in revolutionizing the steel industry of the world. It is therefore not unreasonable to expect that a better rail, for which there will be, perhaps for ages, an ever increasing demand, will be forthcoming. Bessemer rails, superior as they are compared to iron ones, may yet be the first windfall of a tree that is to yield better fruit. As a mere hint, I may allude to the possibility of cold-rolling the rails in manufacture.

The other direction in which I think we may look to see economy operate, on the supposition that wheel loads are now at a profitable maximum and that we must continue to rely on our present rail material, is in the use of more frequent but lighter and faster trains. The problem always is and will be to administer a railway most profitably. Accordingly, we must keep in view the fullest utilization of the property and not lose sight of the fact that an idle railroad not only is earning nothing, but is decaying and disintegrating as a physical structure. When we reflect on how small a part of each day any particular rail in the line of this vastly expensive railroad actually is used in transportation, and earning something, the possibilities of traffic enlarge upon us. We then see the bearing of the idea of trying to work economy by more and faster trains in place of fewer and slower ones; of thus securing a greater average daily movement of freight cars, now surprisingly low; in short, of making time more an element of profit throughout the service, as now it only is in respect to passenger trains, to some few articles of perishable freight, and to the carrying of cattle. A street railway in a populous city illustrates the opposite tendency to that observed in recent steam railway management, for there we see cars or trains in sight one from another, the rails kept almost continually warm by the friction of use, thus realizing the foregoing ideal of a productive property. Circumstances have made our practice very unlike the European in respect to train service, but I think we may look to see changed social conditions, due to denser population and a fall in wages, bring in a similar economy.

Among the most remarkable facts of our subject is the constancy of the ratio of operating expenses to income on railroads in general. On the Lake Shore and Michigan Southern the average for the last twenty-five years has been 65 per cent, with but one very marked fluctuation. On the Chicago, St. Paul, Minneapolis & Omaha Railway it has been practically stationary at 64 per cent for the last fifteen years. On the first mentioned road the economies that we have outlined were long ago in full operation, on the latter they are not yet so completely developed. These facts seem to show two things of interest: First, that the proportional cost of operating a particular road is remarkably constant under all circumstances; and, second, that all railway improvements, not only betterments, but also those which make for economy of operation, inure to the benefit of the public. The most prosperous roads have only paid interest on their bonded debt and a fair dividend to stockholders. All the rest of their earnings have gone to the public, largely in the way of improved and cheapened transportation. I think we may infer from this stability of the relation of expenses to earnings that fast, light freight trains may be substituted for slow heavy ones without loss to the companies, but with a growing profit due to more rapid circulation of commodities; and that this will prove true, notwithstanding the theoretical, and indeed actual cost of increasing speed.

Another direction which economy of operation has recently taken is toward loading engines to their full hauling capacity. Until within a short time the rule was to load them with the number of cars that the engineer said he could take up the limiting grade. This made the rule sometimes coincide with the caprice of the engineer. If on approaching the ruling grade he told the conductor that two cars must be dropped, they were accordingly set on the siding, to wait there till some other engineer would not rail at the suggestion of picking them up. While this system was in operation, an examination of way-bills often disclosed that the number of cars in the train was sufficient fully to load the engine if the cars had been fully loaded, but that in reality the train load was far too light for the engine because of cars very lightly loaded. All this has now been changed to great advantage by adopting what is called tonnage rating. Under this system the traction of locomotives has been determined experimentally and the proper loading has then been derived from the results. To this end a freight engine is tested on the controlling grade of each of the sections into which the division is naturally divided by the existence of such grades. This engine is run up the grade at a proper speed, say 12 miles per hour, with a sufficient number of loaded cars of various known weights to stall it. Then an assistant engine goes behind and helps the train to start up again and acquire a speed of five or six miles an hour, when the helper drops off with one or two cars, and thus another trial is made, and so on till the load in tons that the engine can surely take up the grade is practically determined. On each hard spot in the division, caused by grades or by grades and curves combined, the engine is similarly tried; and then small deductions are made from the maximum loads so found to allow for specially long grades, such as tax the steaming capacity of engines.

An engine having been thus tested on the governing grades, and its performance recorded, one engine of each of the remaining classes in service is then tried similarly on one or two of the same grades and its relative power thus determined. This relative performance affords the rule for loading engines of these classes proportionally to their developed capacity. Thereafter all loading is by tons instead of cars.

In preparing the rating sheet to govern engineers under this system, some superintendents make three ratings, corresponding to three assumed conditions of weather and rail, while others rely on the discretion of the engineer to make deviations from the printed single ratings based on the trial conditions, at the same time holding him accountable for satisfactory engine performance. The tons hauled and coal consumed by each engine are bulletined monthly, and the engineers are found anxious to make good records by hauling full train loads. The economy thus effected is considerable. It has increased the train load from ten to twenty per cent on several roads whose officers have kindly given me their results.

Still another way in which economy has been worked in recent years is by reducing grades and curvatures. Most of our Western roads were originally located with an eye to facility and economy of construction, by the eye of the promoter rather than that of the engineer. They were what is called surface roads, conforming closely to the slopes of the country with their grades, and to the plan of the hills with their curves. Heavy grades and sharp curves and combinations of both marked this era. Many companies have recently spent large sums in improving these defective pieces of alignment, which were originally due to poverty of resources, and did not reflect on the intelligence of the engineer. To reduce a maximum grade is equivalent to increasing the capacity of engines. The equation is between the cost of operating the line as it is, and interest on the cost of the improvement, and as to all very bad places there can be no doubt that the betterment will be justified by events. It is largely by such works as these that lines geographically well placed are making their future prosperity secure against the competition of inferior lines, while by the same stroke they are reducing the cost of service to the public.

But of all causes that have contributed to economy of operation, the consolidation of short lines into large systems has been the most effective. Twenty years ago a thousand miles under one management was a giant railway. Now we see from five thousand to eight thousand miles in one system, often practically controlled by a single administrative officer. Such a combination of interest and of utilities has reduced order out of chaos, established long periods of peace where there was continual war, and has given greater stability to private enterprise in the districts so served, because such a great system possesses an inertia comparable to that of a planet in its orbit, which hardly any force can disturb. The result is a development peculiar to this form of human activity under freedom. Under free competition the result is what we see, what no man could have foreseen or forestalled: it is that we have the best and cheapest service in the world.

I omit many minor economies, only intending to present a broad outline. It is for you who are soon to take possession of the field of engineering to be alive to the opportunities and requirements of this so learned profession. We have an enormous country, nearly the whole of it still undeveloped as compared with the wealthiest parts of Europe. We have one mile of railroad to eighteen of area, as compared with one to less than four in Belgium. Our mileage is almost sure to be doubled in your lifetime. Yet this is only one branch of engineering and each of the others presents as wide a prospect. There is therefore every incentive to your ambition. You have but to read the story of Aladdin as a parable. Every man is born with an enchanted lamp of mental aptitudes. His fortune depends on his keeping it burnished.

As a last word, I should like to lodge in your memory the counsel of one of the great men whose names adorn the roll of the profession. Tyndall was a practicing engineer before he shone out as a philosopher. Speaking of his having failed in tens of thousands of experiments to wring a certain truth from nature (a struggle in which, however, he was finally successful), he gives this advice to young men:

"The experimenter, and particularly the young experimenter, ought to know, that as regards his own moral manhood, he cannot but win if he contend aright. Even with a negative result, the consciousness that he has gone to the bottom of his subject, as far as his means allowed—the feeling that he has not shunned labor, though that labor may have resulted in laying bare the nakedness of his case—reacts upon his own mind, and gives it firmness for future work."

A MULTIFUNCTIONAL DYNAMO.

GEO D. SHEPARDSON, M. E.

The Edison dynamo in the electrical engineering laboratory of the University of Minnesota is an interesting example both of the expedients to which one may resort in teaching with somewhat limited appliances and also of the manifold value and flexibility of electrical machinery.

This machine was purchased in 1887 for the purpose of lighting the Mechanic Arts Building and for supplying current for the

physical laboratory, which was then located in that building. In 1891 it was moved to its present location. It has been in operation for an average of probably two hours each day during nine months of each year and has cost nothing for repairs except one set of brushes, two turnings of the commutator and one re-babbetting of the boxes, costing in all about eight dollars, or less than one dollar per annum. The machine seems to be as good as new and more valuable than at first, its attachments overbalancing the great reduction in prices of recent electrical machinery. This machine, part of which is shown in the accompanying halftone engraving, has been equipped with eight series coils having individual terminals for coupling in various combinations, with a series resistance to ease starting as a motor, with two switches for short-circuiting either the series resistance or the series magnet coils, with six collecting rings and brushes for obtaining single-phase, two-phase and three-phase currents, with a field exploring wire and instantaneous contact device, and with divided segments for voltage distribution measurement by single or two-brush methods.

By various combinations of switches on the machine and at the switchboard, this machine may be operated as a shunt, cumulatively compound or differentially compound motor, or as a dynamo either singly or in multiple with the Mather machine in the Mechanic Arts engine room over four hundred feet away. When the Armory is lighted, these two dynamos are operated on the three-wire system, the lamps being nine hundred feet from one dynamo and thirteen hundred feet from the other. A similar condition exists when the Library and Chapel are lighted simultaneously. It thus happens occasionally that on a single day the Edison machine is driven as a motor by the Mather machine to operate the line shaft of the laboratory, later is driven by the laboratory engine and operated as a dynamo in parallel with the Mather machine to help carry the afternoon lighting load in the shops and library, and then in the evening of the same day is operated as a dynamo in series with the Mather on the three-wire system of lighting the Armory or the Library Building. The installation of a larger direct connected unit in the lighting plant during the coming summer will obviate the necessity for this varied use except for demonstration purposes. By the recent addition of six collecting rings, the machine is available as a rotary transformer, changing the direct current from the Mather machine into single, two or three-phase alternating currents. When driven by belt, it will operate as a self excited alternator for giving any of the three alternating currents or for any combination of them, while delivering continuous currents at the same time. By taking account of the various possible arrangements of the series and shunt coils, a list of several hundred arrangements and uses of this truly multi-functional dynamo might be enumerated.

A few points in the mechanical construction of some of the attachments may be of interest. As seen in the figure, the collecting rings are mounted upon the armature between the commutator and the polepieces. In order to obtain a substantial support for the rings, the canvas cover was removed and the uneven space over the armature leads was partially filled with hard cord wound on as tightly and evenly as possible and then coated with shellac. After this had become thoroughly dry, it was covered with a number of layers of paper. Heavy manilla paper was cut into strips about an inch wide and five or six feet long. A number of these strips were soaked in water until quite soft and were then painted on one side with shellac. The strips were then wound over the cord back and forth until the whole surface was covered two or three layers

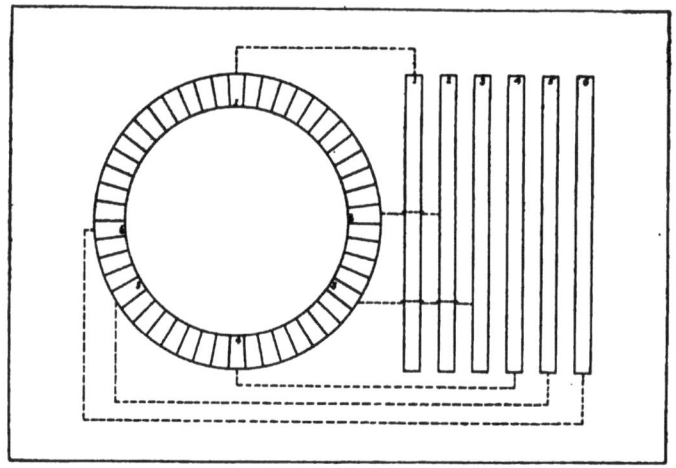

deep. This was then allowed to dry for twenty-four hours, during which time the paper shrunk and dried so as to be very hard and dry. Three layers more were then wound on and allowed to dry. In this manner the space was filled up to a size nearly equal to the outside diameter of the commutator. A tool rest was then made from a bar of iron one inch wide, one-quarter inch thick and about a foot long, by bending about three inches at one end at right angles and drilling two holes through the short end which was screwed to one of the polepieces in place of the fuse block. By the help of this temporary tool rest and a sharp diamond pointed hand turning tool, the hardened paper was turned down to an approximately true cylinder. A sheet of red fiber one-sixteenth inch thick was then cut wide enough to extend from the back of commutator to the armature head and about six inches longer than the circumference of the paper surface. After being well soaked in water, this was bent around the paper surface and allowed to dry while held in place by twisted wire bands. When dried, the overlapping ends were cut obliquely at angles of about forty-five degrees and to such a length that when drawn tightly together the edges just met. By making the joint oblique rather than square, any tendency for the fiber to open out between the rings was overcome. Before the fiber ring was finally placed, strips of sheet brass one-sixteenth inch thick and three-eighths inch wide were prepared for connecting the collecting rings with the desired commutator segments. These were fastened by flat head countersunk screws tapped into the radial projections of the commutator segments. Grooves for the brass strips were cut in the paper so that the strips should be flush with the surface of the paper. Slits were then cut at proper places in the fiber ring, and the inner ends of the brass strips bent out so that when the fiber ring was drawn together, the ends of the connecting leads projected through the fiber. These were so spaced that the strip for each ring projected on the side furthest from the commutator, in order that in case of breakage the strip would still be long enough to fasten to the nearer side of the same ring.

The surface was then ready for the collecting rings, which were made from spring brush copper about three sixty-fourths inch thick and cut into strips one-half inch wide. These were cut from a wider strip in the squaring shears and were formed to approximately the right curvature in the rolls of a convenient tin-shop. Each strip was cut to the right length by trial and the inner side of each end was filed to a taper. A shim about an inch long was then cut from the same stock and filed to a corresponding bevel on each end. The shim and both beveled edges of the curved strip were then thinned and put in place, the shim being laid under the joint. The whole was then drawn tightly together by means of a No. 8 copper wire passed once around the ring and then drawn taut by block and tackle. When all was in place, the ring just touching one of the brass leads, a hot soldering bolt quickly made a strong splice an inch long, and made connection between the ring and the brass lead. After the six rings were made and put on in this way, they were turned off to a true surface by means of the rest previously used and a round nosed hand tool.

Brushholder studs were then provided by drilling and tapping a hole in each polepiece and screwing in a Bessemer steel rod eight inches long and three-eighths inch in diameter. The outer end of each rod was filed flat, to afford a grip for screwing it in by a wrench. Each brush-holder was made from a block of vulcanized fiber three-quarters of an inch thick, turned to a cylinder one inch in diameter, with two flat sides. A strip of spring copper was fastened to this by brass machine screws, to serve as a brush. A second strip of thinner copper was soldered to the first strip, so as to give two points of contact with each ring to avoid sparking. Longer machine screws serve to hold the brush-holders in place on the stud, also to raise and lower the brushes and to adjust the tension on the brushes. Copper washers, soldered into the slots in the screw heads, convert them into convenient thumbscrews. Short·lengths of lampcord soldered to the brushes at one end and clamped in double-pole fuse blocks screwed to the iron name plates on the yoke, furnish convenient means for connecting with external circuits.

In order to obtain the proper spacing of the commutator to give the desired angular distance between various derived alternating currents, the commutator bars were numbered consecutively from one to fifty, as shown in the line sketch. From the rings connecting with bars numbered one and twenty-six, a simple alternating current may be taken. From bars thirteen and thirty-eight a second alternating current may be taken at right angles to the first. As fifty is not exactly divisible by three, it is not possible to obtain three currents exactly one hundred and twenty degrees apart, but this angle may be closely approximated by taking bars, one, eighteen and thirty-four.

THE MODERN EXPLOSION ENGINE.
E. S. SAVAGE, '97.

The great interest which is now being taken in this youngest of motive powers is strikingly shown by the large number of new makes of gas engines which have recently been placed on the market. In the United States alone over one hundred manufacturing establishments are building these engines, and thousands are sold each year.

The modern gas engine may in a legal sense be considered to have reached its maturity, as it is just twenty-one years since Dr. N. A. Otto first brought out his "Otto Silent" engine, which is the prototype of the majority of recent explosion engines. But though just outgrowing its infancy this young motor has made a name for itself which may well arouse jealousy on the part of our old friend "Steam Power."

It is to be understood that the term "gas engine" is generally used to denote any form of explosion engine whether the fuel is coal gas, gasoline or any hydrocarbon vapor. The uses and applications to which the gas engine can be put are numerous and varied in the extreme, and it would be difficult to suggest any application of medium power where it could not be profitably employed. These engines are now generally used for units of from ¼ to 25 H. P.; the 4 or 5 H. P. being the size for which the demand is greatest. Gas engines of very much larger power are also in demand. Several manufacturers advertise single cylinder engines of 200 H. P., and there is now in operation at the Pantin flour mills in France a single cylinder gas engine which gives 450 indicated H. P., with illuminating gas as fuel, or 320 H. P. with producer gas.

Since gas engines have now become so numerous it is suprising to note the total lack of knowledge which exists among the great majority of people, regarding the principle of their operation. This is no doubt principally accounted for by the youth of the industry and also by the scarcity of books on the subject. When the time comes, which many believe is close at hand, when the horseless carriage will be a common sight on our streets, then the explosion engine will be more universally understood than the steam engine is now. Then it will be possible for a party to step into their carriage after lunch, take a little pleasure trip of a hundred miles, and be home again in time for dinner.

The power exerted by an explosion engine is derived from a mixture of combustible gas or hydrocarbon vapor, and air, which is confined in the clearance space of the cylinder and exploded by the electric spark or otherwise. The combustion of the mixture producing a very high temperature, the expansive pressure of the heated gases drives the piston forward. A circulation of water is necessarily sustained around the working cylinder to prevent its overheating.

Beau De Rocha in 1862 first set forth the requisite conditions of an economical explosion engine as follows:

1. The greatest possible cylinder volume with the least possible cooling surface.
2. The greatest possible rapidity of expansion.
3. The greatest possible expansion and the greatest possible pressure at the beginning of expansion.

He proposed to obtain these conditions by means of the following cycle of operations: 1. Suction of the explosive charge. 2. Compression of the charge during the following in-stroke. 3. Ignition at the dead point and expansion during the third stroke. 4. Exhaust of burned gases from the cylinder on the fourth and last stroke.

Thus with a single acting cylinder one impulse is given for two revolutions of the crank. This cycle is commonly known as the four-stroke cycle or the four-cycle, often called the Otto cycle.

The compression of the charge is exceedingly important as the efficiency increases with the compression, and without compression no economical gas engine has ever been constructed. The majority of those now made employ the four-stroke cycle.

Another cycle used in some small engines not exceeding 15 H. P., is known as the two-stroke cycle or two-cycle. This cycle gives one impulse for each revolution of the crank. These engines have an inclosed crank case into which the charge is drawn during the out-stroke of the piston. During the in-stroke the charge is slightly compressed and when the piston reaches the end of the in-stroke ports are opened which allow the fresh charge to rush into the cylinder driving the burned gases through the exhaust port. During the out-stroke the charge is com-

THE MODERN EXPLOSION ENGINE. 81

pressed and ignition occurs near the dead center. These engines are usually made in the upright form with one or two cylinders and are well adapted for boat propulsion.

The fuels commonly used with the explosion engine are: Illuminating gas, producer gas, natural gas, gasoline, kerosene and crude petroleum. Acetylene gas gives a very powerful explosion and will undoubtedly be used extensively when the calcium carbide is sold in quantities. Gasoline and illuminating gas are most generally used in America. Gasoline has the advantage that it can be used at any place and is very easily vaporized. In Europe oil engines are coming widely into use. Objections to the use of oil are: the necessity of some sort of heated vaporizer; the low mean effective pressure given during the stroke, and the liability of premature explosions in the cylinder. It is a curious fact that the vapor of heavy oils in contact with heated metal will ignite at a much lower temperature than light vapors and gases. This fact is utilized in the Hornsby Akroyd oil engine which ignites the charge in a chamber projecting from the cylinder head and is kept hot by the heat of the explosions.

To obtain an effective explosive mixture it is necessary to mix with the gas or vapor a volume of air containing sufficient oxygen for complete combustion. Owing to the burned gases contained in the clearance space, the ratio of air to gas is usually greater than that theoretically necessary. The ratio of the volume of gas or vapor, to air, giving the best results are: with coal gas, 1:6; gasoline vapor, 1:9; oil gas, 1:8. If the proportion of air used is too great or too small, ignition of the charge will not occur. The limits within which ignition is possible are: with coal gas 1:14 and 1:3, with gasoline vapor 1:17 and 1:2, with oil vapor 1:17 and 1:5.6. The ratio of air to gas necessary in any case depends of course upon the number of heat units which the gas contains. The average calorific values of the above fuels are: illuminating gas of 15 candle power, 620 B. T. U. per cubic foot; producer gas, 111 B. T. U. per cubic foot; gasoline vapor, 690 B. T. U. per cubic foot, or 11,000 B. T. U. per pound; light petroleum (sp. gr. .84), 18,400 B. T. U. per pound.

Nearly all gas engines are single acting of the trunk piston type with no piston rod. Objections to the use of a piston rod are: the difficulty of keeping the rod and the stuffing box sufficiently cool and the loss of energy due to the cooling effect of the rod upon the gases surrounding it. Each engine is usually provided with two fly wheels which are necessarily large and heavy in comparison with the size of the engine, the inertia of the fly wheels being depended upon to carry the engine through three of the four strokes of the cycle.

Two valves, usually of the poppet type, are used; one for the admission of the charge, and the other for the exhaust of the burned gases.

The governing of a gas engine can be effected by one of two principal methods, namely: 1. Entirely suppressing explosions during one or more revolutions of the crank. 2. By varying the quantity of the gas or vapor used. The first method is the one which is most commonly adopted and if properly obtained is the most economical method where the variation of load is great. By this method the governing is accomplished either by entirely cutting off the supply of the hydrocarbon element, or by holding the exhaust valve open during one or more strokes, or by arresting the operation of the igniter, the last being obviously very wasteful.

The second method in which the supply of gas or vapor is graduated by the governor is both economical and satisfactory if the variation of load does not carry the charge beyond the limit of combustion. A combination of the two methods would seem to be the most satisfactory; the graduating device operating to one-fourth load for instance and the "hit or miss" method operating with less load.

Ignition of the charge can be effected by four distinct methods, namely: 1. Flame ignition. 2. Incandescent tube heated externally. 3. Electric spark. 4. Ignition internally by heat from the explosion.

No engine now manufactured in the United States has a flame igniter. The incandescent tube and the electric spark are the methods almost universally employed; in fact most American engines are furnished with both the electric and tube igniters so that either can be used as preferred.

The incandescent tube igniter consists of a small tube usually made of one-eighth inch or one-fourth inch wrought iron pipe about four inches in length closed at one end, the other end being screwed into an opening in the cylinder head and connected to the explosion chamber. The tube is heated to a red heat by a flame directed against it. The length of the tube and position of the

THE MODERN EXPLOSION ENGINE.

burner are so adjusted that as the compression nears a maximum some of the fresh charge is forced back into the hot tube and causes ignition near the dead center. The advantages of this igniter are that it is easily managed by any one and not liable to get out of order. Its disadvantages are: the objection to the use of an external flame; the time required to heat the tube at starting, which takes from five to ten minutes, and the frequent renewal of the tube. Several English engines employ a timing valve, in connection with the tube igniter, which admits the charge into the tube just at the time for igniting.

The electric igniter, which is the one most generally employed in America, requires a battery, a self-induction coil and the igniter proper which consists of some form of make and break device in the engine cylinder. The battery must be long lived and easily renewed when exhausted. Eight or ten Salamoniac cells are often employed. The igniting spark is the spark of selfinduction, caused by breaking the electric circuit. No secondary wire is used, the spark coil consisting of a coil of insulated copper wire (No. 16) having a soft iron wire core. The igniter which can be made in a great variety of forms, is designed to make electrical connections within the engine cylinder and break the connection when explosion is desired. The contact points are often tipped with platinum to prevent corrosion. The igniter, battery, spark coil and a switch are all connected in series. The best form of electric igniter is that which gives a slow contact and a quick break, thus producing a long spark and avoids battering the contact points. The time of contact during which the current is flowing should be about one-twentieth of a second at the normal speed of the engine.

The temperature of the explosion of a mixture of gas and air is about 2,800 degrees F., or above the melting point of cast iron, which is near 2,000 degrees F. The maximum pressures commonly obtained range from 150 lbs. per square inch to 350 lbs. above the atmosphere. The maximum pressure with different mixtures of gas and air can be readily calculated, which gives the pressures obtained by explosion from atmospheric pressure, and also the time of the explosion.

1 volume of gas to 13 vols. of air gives 63 lbs. in .18 sec.
1 volume of gas to 9 vols. of air gives 69 lbs. in .13 sec.
1 volume of gas to 7 vols. of air gives 89 lbs. in .07 sec.

1 volume of gas to 5 vols. of air gives 96 lbs. in .05 sec. If the compression is increased the maximum pressure will be increased in the same proportion. For example, if the mixture is 1 to 7 and the compression is 15 lbs. above the atmosphere, then the maximum pressure of explosion will be $(89+15)(30 \div 15) = 208$ lbs. per sq. in. absolute, or 193 lbs. above the atmosphere.

It is customary to compress the charge to about 45 lbs. above the atmosphere, causing a rise of temperature of about 280 degrees F. Higher compression is theoretically advantageous, but practical difficulties prevent the general use of compressions above 45 to 60 lbs. in most single cylinder engines. The mean pressure for the working stroke varies from 45 to 100 lbs. per square inch, depending upon the fuel, the mixture and the compression employed.

The pressures in gasoline engines are usually about 20 per cent greater than in gas engines. In oil engines the pressures are low, being about 30 per cent below those in gas engines.

The horse power obtainable from a single cylinder gas engine may be roughly approximated by dividing the volume of the stroke in cubic inches by 65, considering the speed as 200 revolutions per minute.

The power given with producer gas is about 75 per cent of that with illuminating gas, and the volume necessary is four times as great. An engine using producer gas requires very large gas ports.

The speed of gas engines ranges all the way from 100 to 1,000 revolutions per minute. The ordinary speed is 250 R. P. M. It is found that increasing the piston speed increases the pressure, quickness of the explosion and the efficiency.

In point of economy the gas engine is ahead of any other fuel consuming source of power. A consumption of 1½ lbs. of coal per I. H. P. is regarded as a very good performance for a steam engine even of large power. This means an actual thermal efficiency of 12 per cent, while 6 lbs. per I. H. P., or an efficiency of 3 per cent, is often considered economical. With gas engines an actual efficiency of 20 to 25 per cent, or a consumption of coal of 1.2 to 1.4 lbs per I. H. P., is a common performance.

D. Clerk states in a recent paper that the actual efficiency varies from 45 to 58 per cent of the theoretical, the theoretical ranging from 37 to 45 per cent. Gasoline engines consume about one pint of gasoline per B. H. P. hour. A small gasoline engine using gasoline, bought at retail, will equal in economy a steam engine using coal at $3.50 per ton and six lbs. per B. H. P. besides other savings as attendance, handling fuel, etc.

The advantages of a gas engine for electric lighting have often been pointed out. It is an interesting fact to note that if illuminating gas is used in a gas engine to run a dynamo, twice the quantity of light can be obtained from the lamps that would be given by burning the same quantity of gas in gas burners.

In closing it may be said that the present form of gas engine, notwithstanding all its advantages barely touched upon here, has many imperfections and there are yet many unsolved problems on which hundreds of engineers and inventors are now working. The gas engine has certainly a bright future before it and will ultimately displace steam in nearly all if not all small and medium power plants.

THE STRUCTURAL LABORATORY IN AN ENGINEERING COLLEGE.

FRANK H. CONSTANT, C. E.

Science, from its very foundation, has to do with the properties of matter. Its methods are inductive; its processes of reasoning, based upon observed phenomena of matter, lead to generalizations which, when sufficiently verified by physical tests, become part of the common stock of discovered law. The study of matter is the only accessible path to a knowledge of the laws of matter.

A first knowledge of the properties of matter was obtained accidentally by daily contact with the material world, and doubtless many crude scientists were thus developed. But at length scientists were no longer content to wait for accidental disclosures of the hidden things in nature. In the judgment chamber of the laboratory they have studied matter under many conditions and have learned some of its secrets. The marvelous advance in scientific knowledge during this century is mainly due to the development of the scientist's workshop—the laboratory. The laboratory has now become universally recognized as the only medium through which scientific investigation can be conducted.

Structural engineering is simply applied science. If the scientist is concerned with the laws of matter, to a greater extent is the engineer whose life is spent in battling with the forces which control matter. If the laboratory is necessary for the acquisition of scientific knowledge, its need is no less urgent to the engineer whose skill and success depend upon his intimacy with the properties of the materials with which he builds.

As in the case of science, engineering knowledge was at first acquired accidentally. If a structure, built largely by guess or rule of thumb methods, exhibited a disinclination to fail under a load, other structures of a similar character were built in like manner. This method for the advance of knowledge was costly, dangerous and slow, and under it structural engineering made but little progress.

With the advent of the iron age, mere experience was forced to give way to the more scientific methods of the laboratory. Timber, formerly the material almost universally used in structural work, generally gives some suggestion of weakness when overstrained. On the other hand, if the material has been used too abundantly, an excess does not greatly increase the cost of the structure. With iron, however, the reverse is true. It fails with little warning, while excessive weight is prohibitive in cost. The engineer now studies his structure part by part, knowing that if the individual members are strong enough the entire structure will be safe. Thus the superior material has developed the laboratory, which, in turn, has reacted upon structural engineering to give it an impetus that could scarcely have been anticipated. Iron was adopted, at first timidly, then as its properties became better known, with greater confidence, only to be finally cast aside for a still better material.

But if the laboratory was necessary for the development of iron, its need is infinitely more real in the case of steel, whose properties may be wholly altered by the presence or absence of a fraction of one per cent of some impurity, or by a slight change in the mode of manufacture.

At the present time the testing laboratory is indispensable to the practicing engineer. Every large steel mill and bridge company, as well as many railroads, manufacturing plants, engineering firms and engineering colleges, possess well endowed testing laboratories. When engineers do not have convenient access to testing laboratories, they employ special engineering or inspection firms who have such equipment and are organized for this purpose, to test and inspect the material in the various processes of manufacture. Thus inspection has become a necessary part of structural engineering.

As familiarity with the properties of materials is as essential to the structural engineer as a knowledge of mechanical laws, so a training in the use, methods and results of the testing laboratory should form as vital a part of the course in an engineering college as the theoretical work of the class room.

The function of the laboratory in a course in structural engineering is four-fold.

First—To supplement the work of theoretical instruction in the class room.

Second—To acquaint the student with the common properties of the materials of engineering and with laboratory methods.

Third—To instill in him the spirit of original research.

FRANK H. CONSTANT.

Fourth—To form a medium for valuable contributions to the common stock of engineering knowledge by the publication of results in engineering journals.

In the class room the student is presented with theoretical principles and taught to use the knowledge of mechanical law thus gained in the designing of practical structures. In the laboratory he develops and verifies this theory for himself. Thus the primary function of the laboratory is to co-operate with the work of the class room and render this latter more fruitful. In this capacity the laboratory serves as an inspiration both to student and instructor.

In the laboratory the student also acquires a knowledge of the characteristic properties of the materials with which, in after practice, he will build his structures. This is by no means old ground, although the methods used may be well established. Two pieces of steel will not exhibit just the same qualities; yet from the generality of tests made under various conditions trustworthy deductions may be drawn. These tests, to be of any value, should be as nearly like those made by practicing engineers as possible.

But the engineer should not only know what his material ought to be, but what it really is. No engineer will allow material to enter a structure of any importance until a sufficient number of tests has been made to satisfy him that all of the material in that structure possesses certain approved qualities. Even the general properties of such a material as steel are constantly changing with the introduction of new and better modes of manufacture. The engineer who would seek the text books when writing a specification for his material, will find himself directing the use of iron after it has been supplanted by steel, or Bessemer steel when other engineers, who have kept in closer touch with the more recent progress of the mills, specify open hearth steel. Structural engineering is advancing with such rapid strides that he who waits for the carefully digested results of the experiments of others will always find himself just a little behind the times.

It is a necessary part of the training of the structural engineer to be able to enter the testing laboratory and make or direct a series of tests upon material that he proposes to use in his structure. Any laboratory instruction in a college, which does not give the student a real working knowledge of the laboratory methods of actual practice is inefficient.

Only a small amount of original research work can be conducted wholly by under graduate students, this being, perhaps, more profitably reserved for graduate students. Yet sufficient work of this nature should be prosecuted, under the direction of the instructors, to instill in the student the spirit of scientific research. This work should be chosen with reference to its practical character, and it should be conducted with such accuracy that the results may be a real contribution to engineering knowledge.

There is no more enticing field in engineering science than this one; and as yet it has not been much more than entered. Heretofore the efforts of experimenters have been directed to the examination of materials. But of late years more attention has been directed to tests of material after it has passed through one or more processes of manufacture into a finished structure.

These tests are of importance, as they determine into what forms and by what modes the parts of the structure may be shaped. From these tests the engineer learns to what extent material is injured by punching, what is the effect of annealing, how the strength of an eyebar compares with the sample piece, what is the strength of many kinds of riveted joints, what is the most efficient form into which a column may be built. An almost endless list of fit subjects for experimental examination could be named, some of which have already been studied, most of which have not, and nearly all of which will, from time to time, have to be studied anew, as the modes of construction and of metal manufacture change.

The present tendency is in the direction of tests upon a larger scale, upon full size parts of structures. Only a few of these problems can be mentioned within the scope of this article.

Perhaps as interesting a field for investigation as any is a fuller study of plate girder construction. The plate girder has now become the almost universal form of bridge for spans from twenty feet up to one hundred feet. It thus constitutes a large percentage of the total mileage of railroad bridges built each year. Yet the conditions involved in its design are uncertain and complex, and beyond the scope of theoretical treatment. Some of the problems to be investigated are: The real distribution of stresses in flange and web, the function, size and spacing of stiffeners, and the spacing of rivets in flanges, splices and stiffeners. In fact, nearly the whole subject of plate girder designing is wrapped in uncertainty. No doubt a clearer insight into the action of molecular and internal forces in matter will be of great assistance in the solution of these problems.

Another problem upon which so much has been done and so little has been accomplished is the determination of the proper working strength of long columns. The difficulty is that most of the past tests upon long columns have been made beyond the elastic limit of the material, and hence are valueless to the practical designer.

The subject of working stresses per unit of area is one upon which there is no close agreement among engineers. It is, however, one of the most important within the province of structural engineering. Some valuable results have been obtained by German experimenters upon the effects of impact and reversal of stress. Further experiments should be made in this direction and within limits which conform more nearly to actual cond'.ions.

The strength of riveted joints and a study of sec' · dary stresses induced in such joints and in the members cc¯ ::ected will bear further investigation.

This very partial list will serve to indicate how fundamental are some of the problems which still require more satisfactory solution.

Passing from steel to other materials, the field for the investigator is almost untrodden. The extended use of hydraulic cement and the necessity of avoiding the use of a spurious or inferior article in important work have made the cement laboratory necessary. The fact that a well-equipped laboratory of this nature may be had at a cost of a few hundred dollars has also largely encouraged its rapid multiplication among private engineering firms. The result of placing the cement laboratory at the command of a large number of individuals is to induce a more careful and widespread study of the properties of cement, to raise the standard of specifications and to force the manufacturers to produce a superior article.

Steel and cement in combination are now abundantly used for floors of buildings and bridges, for certain classes of masonry arches, and in foundations of steel buildings. The elastic properties of this combination have received little or no attention.

Cast steel has already proven its vast superiority over cast iron, and in many cases may profitably supplant rolled steel. More familiarity with its properties and an improvement in its mode of manufacture, which will lessen its cost, will widen its field of usefulness immeasurably.

No material has been more popular or more universally useful to the engineer than timber. Yet it has only been within recent years that a scientific study of this material has been attempted. At the present time the government is conducting an elaborate series of tests which will furnish valuable information to the engineering profession.

The proper equipment of a structural laboratory is a problem which must largely depend for its solution upon local conditions, such as the individuality of him who will most habitually direct its operations, upon the number of students, and upon the funds at hand. The apparatus should be the best obtainable. It should be large enough to produce genuine results, for nothing is more disheartening to a student than to feel that his work is not real.

While the individual proclivities of the experimenter will largely determine the choice of apparatus, there are certain fundamental pieces which form the basis of every laboratory. For example, each laboratory should contain a tension and compression machine of sufficient capacity to take the largest pieces that the engineer will have occasion to test. A machine of one hundred thousand to two hundred thousand pounds capacity, with extensions, will take full size columns and beams and is sufficiently large for most purposes. A smaller machine, of about fifty thousand pounds capacity, will be found convenient for the bulk of small tests. Several pieces of minor apparatus are needed to make the service of the larger machines most efficient. A complete outfit for cement testing is now considered indispensable in a structural laboratory. In addition, the experimenter will desire to add other pieces of apparatus to his equipment, the choice depending largely upon the nature of his work.

The college has a two-fold duty to perform. The first is one which should never for a moment be forgotten—the training of its own students for future usefulness in their chosen profession. In truly performing the first task the second is partially accomplished—the raising of the standard of engineering education. But no less essential is it that the college should aim, from time to time, to impart directly to the engineering profession such new truths as may be of value to the practical interests of the profession. And this is best accomplished by the intelligent use of the engineering laboratory.

THE NEW DAM AT MINNEAPOLIS.

H. M. F. DAHL, '98.

Below the historic falls of St. Anthony the Mississippi river rushes swiftly over its rocky bottom in a series of rapids, extending about one mile below the falls. The descent is the greatest in the upper portion and for years the argus eye of industry had jealously watched this lavish waste of energy, with a desire of obtaining control of some of that unused power. But no particular move was made to utilize this great force until within very recent years, when it was finally decided to build a dam across the river below the falls, and to transform the head obtained into electric power. The work was commenced May 15, 1895, and a general description of the structure follows.

The dam is built across the rapids about a quarter of a mile below the falls. It extends 430 feet from the west bank, nearly at right angles, and about two-thirds the distance across the river to a massive pier rising five or six feet above its crest. Here it makes an obtuse angle and runs parallel to the east bank for a distance of 575 feet to the power house, which extends the remaining 200 feet to the bank.

A retaining wall extends along the east bank 575 feet up-stream. It is thirteen feet high, three feet wide at the top and five feet at the bottom and is built of blue limestone. The headrace, between this wall and the dam, was excavated to a uniform depth of thirteen feet below the mean water level. Another retaining wall extends along the west bank for some distance above and below the point where the dam is built into the foot of the bluff. Alongside of it is a log sluice, six feet wide, whose uniformly sloping floor is faced with steel rails imbedded in concrete. A similar sluice is built at the east end of the dam alongside the power house.

The bear-traps are three folding gates, 50 feet long, one at the west end next the log sluice, the other two, side by side, near the east end, 56 feet from the power house. They work between vertical piers built at right angles to the dam. These piers are five to eight feet thick and about six feet higher than the dam.

The gates are raised and lowered by hydraulic pressure controlled by two sets of valves; one set, on the up-stream side of the dam, serves for raising the gates, the other, on the down-stream side, for lowering them. They consist of steel frames, covered with planking from three to six inches thick, and weigh about 300,000 pounds each. A four-foot girder spans each gateway, and serves for raising the gate by hand in case any portion of it should get out of order and need repairs.

Referring to the cut of its cross-section, the foundation of the dam proper is seen to consist of a layer of concrete, 33 feet wide and nearly three feet thick. Underneath this concrete are two dikes of rubble rock laid in cement; the one under the crest of the dam being about 12 feet deep and six feet thick, the other, four to five feet deep.

The dam proper is 19 feet high with a base 18 feet in width. The up-stream face is vertical. The down-stream face, consisting of St. Cloud granite, is stepped in such a manner as to change the direction of the falling water from the vertical to the horizontal with a minimum shock. Steel rails are imbedded in the concrete at the base of the dam. These rails are laid with their outer ends six inches higher than the end next to the dam; thus a pool of water is formed which acts as a cushion, and protects the con-

THE NEW DAM AT MINNEAPOLIS. 95

crete from being eroded by the sheet of water falling over the dam. The coping is a steel skeleton laid on a course of concrete. The section of the dam west of the angle is two feet four inches lower than the east portion, and is similar to it, excepting that one of the middle courses is omitted. By placing "flash" boards on its coping the water level may be raised and the head, which is normally from 16 to 19, may be increased to 21 feet. The backing of the dam is protected by rubble rock piled along the up-stream side.

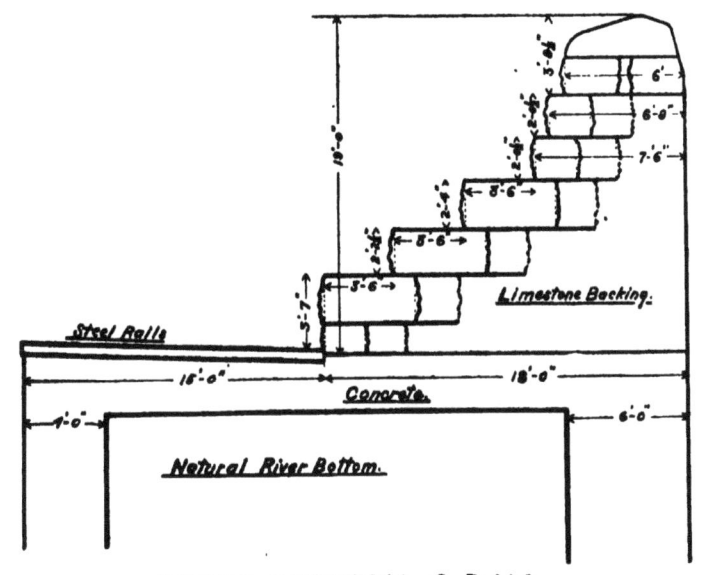

— CROSS-SECTION of DAM.—
From east bear trap to angle at middle
of dam. Other portion 2'-4½" lower.

The power house is built on a concrete foundation, 200 feet square and two and one-half feet thick. This also forms the floor of the tailrace, extending from the culverts to the piers of the Minneapolis and Western railway bridge, 80 feet distant. The tailrace is lined along the east side with a limestone retaining wall capped with Kettle river sandstone. Another similar wall separates it from the log sluice at the west end of the power house.

The ten tailrace culverts form the sub-structure of the power house. Their partition walls are of limestone, and the arches and skew backs of Kettle river sandstone, as is also the lower portion of the south and west walls. The latter, which separates the dynamo room from the log sluice, is five feet thick.

Since the culverts are each 20 feet wide, center to center, and there are ten 1,000 H. P. units above them, one to each culvert, it may be said that 50 horse power are developed per frontage foot, neglecting the exciters.

A short culvert, built into the bank, takes care of the tail water from the exciter turbines; it extends a little in front of the building in the form of a quarter circular bay in which the water gauge is placed.

The partition and end walls of the ten turbine chambers are of limestone. A wooden gate of heavy timbers may be lowered before the openings of the chambers and thus shut out the water of the headrace. The partition wall between the chambers and the dynamo room consists of three divisions; the middle portion of concrete—not shown in the cut—the outer portions of limestone. The circular, eight foot aperture, is lined with a riveted, steel plate cylinder, flared at the inner end to better grip the masonry. The hydraulic pressure is resisted by means of a heavy cast iron plate bolted to the cylinder. The turbo-generator shaft passes through a stuffing box in the center of this plate.

The following description of the working of the turbines will be evident by referring to the sectional view of the power house. The hollow, cast iron casing which looks like an inverted, saddle backed U, forms the central part of each twin turbine outfit. It contains inside only a horizontal cross-arm, carrying a bearing in which the shaft rotates. On either side of it are shown the guide vanes. When the gate at the entrance to the chambers is lifted, the water from the headrace fills the chamber to about seven feet above the shaft, and, rushing swiftly between the guide vanes, strikes the buckets of the turbine wheel, which the moving water drives around in passing through it into the casing, where it is deflected downward through the suction tube and into the tailrace. This suction tube is the hydraulic feature. Without it, the turbine would have to be placed in or near the tail-water to utilize the full head. The tailrace is usually a disagreeable and difficult place in which to repair a turbine. In this plant, however, the work-

THE NEW DAM AT MINNEAPOLIS. 97

man shuts the gate of the chamber, empties it completely through the two-foot pipe in the center of the floor, and then descends through the six-foot circular opening in the roof, there being one directly above each turbine.

How can the suction tube utilize the full head of water? The lower edge of the suction tube is under the level of the tail-water. No air can enter and the nine or ten feet of water in the tube sustains a steady suction. Therefore the pressure on the turbines is due to the height of the water above it plus the column of water in the tube, whose sum is equal to the entire head. Two courses

of brick above Kettle river sandstone skew backs form the arch-roof of the chamber. This is covered with concrete forming the horizontal roof. A frame work of structural steel supports the rack, which extend in front of the chamber gates the whole length of the power house. The rack is a steel screen to prevent the entrance into the chambers of floating debris, which would foul the turbines. It consists of flat bars, one-fourth of an inch thick, three and one-half inches wide, and spaced one and one-third inches apart. In front of each chamber is a four-foot iron pipe

98

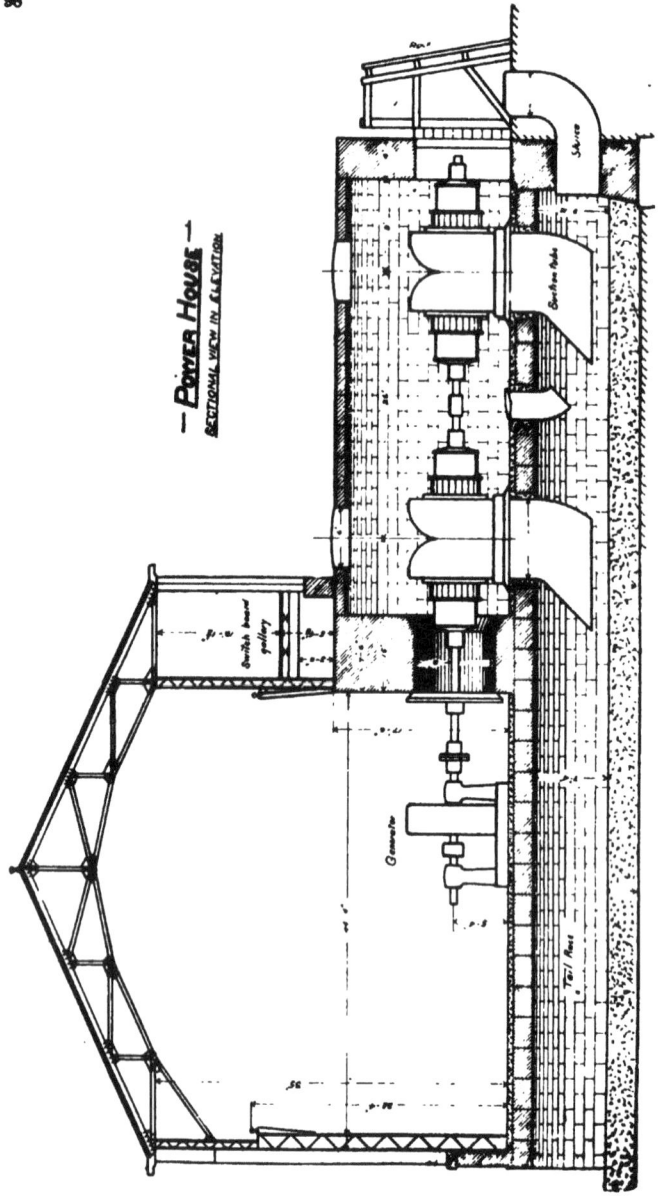

Power House
Sectional View in Elevation

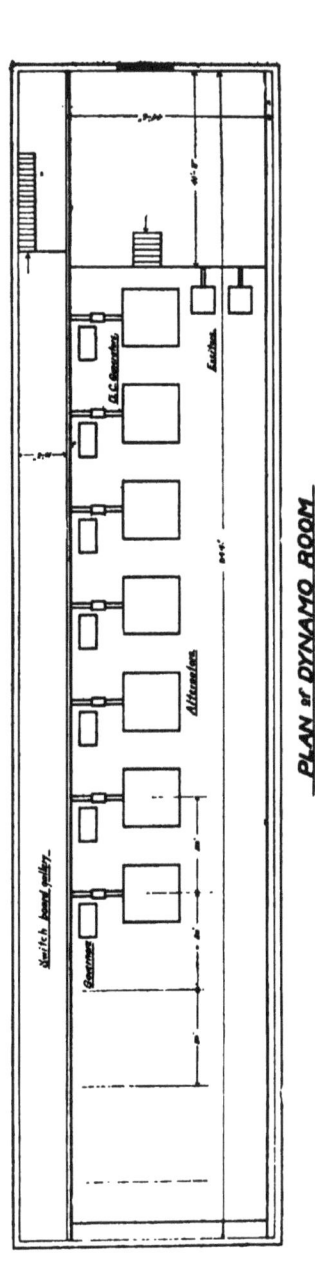

elbow making a direct connection between the headrace and tailrace. These are the waste sluices by means of which the head of water is nicely regulated, and which dispose of surplus water morning and evening.

The dynamo room is one large hall, fire and water-proof and excellently lighted. The floor, like those of the turbine chambers, is paved with concrete, on which the generators are located. The surface is dressed with two inches of asphalt. The roof is covered with Brook tile under an outer layer of slate. The walls are built of salt-glazed tile. At various points along the switch board gallery, tiles are built in the wall with their perforations horizontal furnishing eight holes for the exit of electric wires. This part of the wall is veneered inside with brick. The gallery, which is 206 feet long and ten feet wide, commands a view of every part of the room. The floor is of hard wood on three-inch pine planking. A twenty-ton traveling crane runs the entire length of the room twenty-five feet, four inches above the asphalt floor. The floor of the east end of the room is of pine, and is about four feet above the rest. Two direct connected, 100 K. W. exciters, running at 280 revolutions per minute, are each driven by a pair of twenty-four inch horizontal turbines. The water driving these turbines is carried from the headrace through a six-foot underground pipe. Each 1,000 H. P. generator is also direct connected to four 250 H. P., 24 inch, horizontal cylinder-gate turbines. These run at 130 revolutions per minute.

Being a private undertaking, the work was pushed energetically from the start. The river was first deflected to the west half of its channel by a dike of loose rock and gravel faced with sheet

piling. Then the floor of the headrace, the site of the power house, and the dikes of the east section of the dam were excavated. When the stone masonry of the power house and the east section of the dam had been completed, work was commenced from the west end; dikes were thrown up and soon the massive walls were reared. The beginning of the year 1897 found the section of the dam between the pier at the angle and the west bear-trap still unbuilt, where the waters were then rushing between their narrowing confines. Would it be done before the coming of spring freshets?

According to the plans of Chief Engineer Wm. De La Barre, everything was to be completed by June, 1897. Superintendent of construction C. A. J. Morris, and his picked force of men did the work of Titans. The dike at the upper end of the now finished headrace was torn away, and the river turned into its new channel. It was barred from its former bed by a cofferdam consisting of cribs of 12x12 inch timber filled with stone and faced with sheet piling. The foundation was laid. The wall rose rapidly from the river bed, and scarcely had the last stone been laid and the coping placed, when the spring flood came.

RACE TRACKS.
HOW TO BUILD AND REPAIR THEM.
W. R. HOAG, C. E.

Having been called upon last season to take charge of the preparation of a trotting course, and finding almost no literature relative to their construction or repair, the writer offers to the profession the following, suggested by his personal experience, with the hope that it may prove helpful, in some small measure, to the profession.

A mile race course must be 5,280 feet in length, measured three feet from the pole line. The above condition, though seemingly uncalled for, appears to be the only one imposed by the racing fraternity upon track associations which desire to have their records accepted by any race association. They must be of their nominal length at that distance from the inside fence line which a horse, in a sulky, must be to safely clear the fence.

We have mile tracks with a quarter of a mile in each end curve and a quarter in each stretch; we have them on the square plan, with an eighth of a mile in each quarter turn and an eighth in each of the stretches; then the kite-shaped track has its advocates. We have tracks of various widths, from 20 to 120 feet. Some are given the amount of superelevation on the curves demanded by theoretical considerations, while others use half this amount or even less. Some are made perfectly level along the course of the track, while others allow as much as a one-foot grade to the hundred. Some will have six inches slope across the track on the straight, for drainage, and others double this amount.

This wide range touching seemingly important details would appear to argue, at first thought, the absence of proper regulation. It might seem that, where an acceptable record made on one track is to be considered as having been made on every track of the circuit and is comparable with every other similar record on any track, these tracks should be quite similar. The practical horseman, however, understands that the quantity and time of the last feed of grain or the last drink before a race, to say nothing of the previous month's training, has much more to do with shortening the record time the last second or fraction than the matters of grade, elevation at the curves or general shape of the track.

To attempt to bring these unimportant details under a rigid set of specifications would bring a large extra expense upon many associations, and thus discourage the sport without increasing the interest in it. There are well founded mechanical considerations which would indicate that a straight away mile is undoubtedly the fastest track, but the interest at the start being second only to that at the finish, as well as the matter of expense, have ruled that some form of track bringing the start and finish in front of the grand stand shall be provided. Overshadowing all these considerations affecting the time is that of elasticity of the course. While we can control this in a measure by providing a suitable foundation and carefully selected surface material, yet the amount of moisture in the track, gratuitously furnished by Jupiter Pluvius, too frequently just before or during our great races, will change our records from three to eight seconds, do what we may. So it seems that no regulation is attempted by the racing world of these things which are modified by economic considerations or governed by local conditions over which we can exercise no control. We understand that the track with semi-circular ends and an equal division of its length between the curved and straight parts has become very general and is spoken of as the "standard track," but inquiry among several associations has failed to discover any standard specifications or any knowledge of the existence of any.

Allowing that these circumstances do, to any considerable degree, affect the speed of the horse, it would undoubtedly act differently upon different horses, and thus increase the pure chance element in racing, which must increase the interest in it for all lovers of sportsmanship.

THE SELECTION OF THE SITE.

Exclusive of building, a mile track requires about fifty acres and a half mile track about fifteen acres. We assume that the association has determined upon the general locality and sum available for the purchase of land and construction of track and necessary buildings. This would, of course, determine the general shape and size, whether oval or kite-shaped, mile or half mile. At this point the services of the civil engineer should be called in, as the problems beginning with the economic selection of the site and ending when the track and buildings are fully completed are purely engineering problems. The more nearly level the ground is, the less, of course, the expense of grading.

In this the engineer can bring his special knowledge into requisition, and by taking a few readings with a level and rod can readily compare the merits of the different sites, and, perhaps, without making a formal preliminary estimate can narrow down the choice to one, or possibly to the shifting of the track in a certain piece of land whose size permits such range of choice. An eighty-acre tract and a mile track might easily bring these conditions.

That suitable surface material shall be everywhere present over the tract is scarcely to be expected and not essentially desirable. For, except the ground be unusually level, very little, if any, of the natural surface can be adopted as a "grade" for the finished track surface, and so the engineer will find it to his advantage, except perhaps a few small patches at the grade points, to work the whole track to a sub-grade, and after this has been suitably prepared the surfacing material can be added from the side, or he can obtain it from any place within or near the track with very slight additional expense.

THE BEST MATERIAL FOR THE TRACK.

As might be expected, there is a wide range of opinion as to what is the best material for the track. Since in the construction of a dozen tracks an engineer might be compelled to adopt essentially different soils for the surface materials, we will not attempt to give the exact constituents of the best sub and surface materials, but rather call attention to the conditions which these materials are desired to furnish and ask the engineer to study the different materials available in his particular case as found in adjoining roads, and determine thereby what soils or mixtures of soils will most nearly bring these desired conditions. A solid, easily drained foundation and an elastic surface are the conditions sought.

Sand or gravel will thus be very suitable sub-material, but can form only a small part of a good surface material. Its inelasticity and the ease with which it is carried from the track by the wind, alike render it unsuited for surfacing. The clays are to be avoided as far as possible, as their impervious character renders them unfit for sub-material, and in the surface material the presence of clay in considerable quantity will render the track unfit for use for some time after a rain. The fine dust into which clay readily works when dry becomes a nuisance on a track.

The sandy loams, with sufficient clay to insure a packing

when moistened, seem to meet with general favor among trackmen. The natural soils are to be preferred. The Lexington track is surfaced with natural soil. We would advise the engineer, except he has unusual opportunities of study and experience in this matter of surface material, to leave this part of the work to the judgment of some officer of the association, or, if obliged to care for these matters, let him consult freely with horsemen, whom he will find as a rule very observant of the slightest circumstance that can in any possible way affect the speed of their horses. I would recommend that such advice be followed rather than that of the expert who, with his balances and his sieves, might be able to tell the desirable constituency of the surface metal even to the number and size of the grains of sand in a given quantity of the material.

LAYING OUT THE TRACK.

Having determined the approximate location of the track, the exact determination will be made by each engineer according to his peculiar methods. One might make a topographic map of that part of the field affecting in any way such final location, while another, of a less scientific turn of mind, would make a trial location and with the profile of this seek to have the next location prove final.

The first method will be likely to result in a more economic location of the track, but might cost a day or two more work with the engineering party than the "fit and try" method.

Since the question of final location will be governed somewhat by our allowable maximum grade, we will here briefly refer to this. There seems to be no sentiment among horsemen against easy grades. Two to three feet in a quarter of a mile cannot be detected, and much more than this is permitted on first-class tracks. There should be a slight descent on the last quarter, since this not only aids the horse in getting up speed at the start, but it accelerates his speed at the finish.

The location having been fixed, by whatever methods and from whatever considerations, the staking out of the track becomes a simple problem of running two parallel tangents, 1,320 feet long and 840.34 feet apart, and connecting these by circular curves of 420.17 feet radius, or by thirteen degrees thirty-seven minutes and thirty seconds curves.

The pole line, being run out, serves the same purpose that the center line in railroad work does from which to take "distance out" in cross-sectioning the course for determining the earthwork and setting the slope stakes.

With the grades established, the work of cross-sectioning differs in one respect only from that involved in common railroad work, and needs small comment. The roadbed, instead of being level across the track, will have a slope on the stretch of about one in sixty and on the curve one foot in twelve is much used, though this is only about half as much as demanded by theoretical considerations. The present tendency, since the introduction of the pneumatic tired sulky, is to even lessen this one in twelve superelevation. This rise should begin on the stretch about 150 feet from the beginning of the curve, ascending gradually and reaching full elevation about the same distance along the curve. The side slopes of one to one in cut, and one and one-half to one with embankment, is common.

The nature of the ground will rule, as in railroad work, as to the frequency of the cross-section stakes. For light work and common rolling ground at each hundred feet on the straight parts and at each fifty feet on curves would be sufficient for an experienced contractor's use. For purposes of determining the amount of earthwork involved, more frequent readings might be needed at certain sudden changes in the surface. All of these slope stakes will be set to the final surface as grade, then, as the work progresses, sub-grades will be set to a surface as much below final surface as it is desired to put into the surface material. Eight to ten inches makes a good depth for this and can be counted on to wear away about half an inch a year.

THE DRAINAGE OF THE TRACK.

Reference has been made to the slope of the track to secure a prompt removal of storm water from the surface of the track. It is unfortunate that this drainage must be toward and across the pole line. The necessity for superelevation at the curves makes this plan imperative. From the danger of making wash holes at the pole line at points where the storm water passes from the track to the ditch, the drains should be made about twenty-five feet apart along the track and should be very broad and shallow. A ditch about three feet inside of the pole line should be made, and of a depth necessary to insure a certain and immediate disposal of all storm water to points well remote from the track.

THE RESURFACING OF A TRACK.

The combined action of rain washing the lighter parts into the ditches and the winds blowing away these, together with the sandy portions, gradually reduces the surface material. To meet these losses it becomes necessary to renew the surface every ten or fifteen years, depending upon the nature of the material, the amount of service on the track as well as the atmospheric conditions of rain and wind. In case the resurfacing is to be done by contract based upon volume of material on the track, sections at one hundred feet apart along the track are sufficient, but on account of the concave surface levels should be taken at each ten feet across the track at the section. By platting these on cross-sections paper the areas can be easily determined by use of the planimeter, and the work of computing very much shortened and greater accuracy insured.

The material should be put on in two layers where over six inches is to be put on, the first being leveled and thoroughly rolled before the second is added.

The foregoing is not offered as an attempt to treat all of the problems incident to the building of a race track, but it is hoped that some of the more important considerations may have received such notice as to be of service to track associations desiring to build or to an engineer called upon to assist in the construction of a race track.

COPPER MINERALS IN HEMATITE ORE.

J. H. EBY, E. M., and CHAS. P. BER KEY, M. S.

THE OCCURRENCE OF COPPER MINERALS IN HEMATITE ORE, MONTANA MINE, SOUDAN, MINNESOTA.

Reprinted from Proceedings of the Lake Superior Mining Institute.

The two parts of this paper have been prepared independently. The description of the occurrence was written for the Minnesota Iron Company's records by Mr. J. H. Eby, engineer, and is here published by permission of President D. H. Bacon.

The mineralogical examination has been carried on in the Mineralogical Laboratory of the University of Minnesota by Mr. Chales P. Berkey, and is based upon a complete series of specimens illustrating this remarkable association. Mr. Berkey has kindly consented to the publication of his results.

In reply to a letter of inquiry Prof. C. W. Hall writes:

"I think Engineer Eby has given a very clear and complete statement of the occurrence—no further points occur to me.

"The question of the genesis and chemical composition of the material found in the several ores of copper are entirely distinct questions, and should afford the subject of further study and discussion."

Prof. Thos. Egleston writes of the occurrence as follows:

"I have heard of the occurrence of native copper in this association, but I have never seen it, and I have never seen any explanation of it. The precipitation of native copper, of copper oxide and oxide of iron, is not an unfrequent occurrence in metallurgical establishments, but it is never seen in this particular condition, as in these establishments it is never subjected to any very high degree of pressure. I have several such specimens from Europe in my own collection, and I believe it is not uncommon in European mines, but I have no specific information in relation to such specimens in this country."—Secretary.

PART I.
DESCRIPTION OF THE OCCURRENCE, BY J. H. EBY.

About January, 1896, the presence of copper was indicated by the appearance of fine coatings of Malachite (Cu_2, CO_4 H_2O, hydrous copper carbonate) upon the hard hematite ore extracted from the fourth level of the Montana Mine at a point 850 feet from the shaft. These indications of copper kept growing stronger, and a few pieces of native copper were said to have been found.

The Montana vein is a body of hard hematite, of bessemer quality, and high in iron; and is really a group of veins, the partitions between which become thinner as they extend westward, where the group apparently forms one large vein over eighty feet in width. The walls and partitions are composed of the very softest kind of soap rock and do not contain, nor do they show that they ever contained, metalliferous minerals.

The ore-body dips to the north at about 78 to 80 degrees, and pitches to the west at an angle of between 25 to 30 degrees. The veins of ore as they extend to the east pinch down to thin seams, and finally end entirely in the body of soap rock; while to the west on the second level (96 feet above the fourth, the third being dropped) it terminates against a body of jasper which forms a capping.

The specimens of copper appeared in a flat seam in the back of the breast stope of the most southern vein, fifteen feet above the floor of the fourth level, and extending west, keeping at the same height, but dipping 11 to 12 degrees towards the south. The extreme southern side of the seam was from three to ten feet north of the foot-wall, and extended 20 to 25 feet across the vein. The point where the signs of copper first appeared is about sixty feet below the jasper cap, and 265 feet below the surface of the ground. The thickness of the seam was from one-fourth to one-half inch. While no traces of malachite were to be observed above it, the ore was coated with it from three to five feet below, and often good specimens were to be found at that distance.

Until the breast had advanced eighty feet after the presence of copper was first noted the seam was filled with malachite, and not until then was there any native copper of importance found. The copper occurred in sheets one-fourth to one-half inches in thickness, and it was not entirely confined to the seam, but frequently extended through small crevices in the adjoining ore. At the point of contact with the ore the copper was altered, first to culprite (Cu_2O, red oxide of copper), then to malachite, and in some cases to azurite ($Cu_3[OH_2][CO_3]_2$ copper carbonate).

That the copper was altered to its oxide and carbonate was shown by the occurrence of an octahedron, the center of which was copper, the next layer oxide, and the outside carbonate.

Specimens of sheet copper were found as large as six by ten inches, and some were remarkably well crystallized and free from alteration products.

The seam has continued thirty feet farther than where the first important specimen occurred, and native copper is still to be found in varying quantities. One or two specimens of chalcopyrite (Cu Fe S_2, copper and iron sulphide), the size of a pea, were found, but as no importance was attached to them at the time, their position with relation to the seam was not noted.

No record can be found of any previous occurrence of copper ore in any of the workings, either on this or any other vein in the vicinity.

The accompanying sketch will show the position of the seam with relation to the foot-wall and ore-breast.

PART II.

STUDY OF THE MINERALS, BY CHARLES P. BERKEY.

In the series of specimens presented to the University of Minnesota, all phases of the associated minerals appear, with the one exception of the chalcopyrite mentioned by Mr. Eby. No specimen at hand shows this mineral. In September of the present year (1896) Mr. Eby added two specimens illustrating unusual phases of the "soap rock" enclosing the ore-body.

The specimens are grouped and entered in the catalogue of the Department of Geology and Mineralogy under the serial numbers 1591-1598 of the mineral collection, and 3901-3902 of the rock collection.

No. 1591—Chiefly hematite with copper, cuprite and malachite; one specimen.

No. 1592—Largely metallic copper, coated with cuprite and malachite; three specimens.

No. 1593—Chiefly hematite with cuprite, malachite, copper, silica and ochre; three specimens.

No. 1594—Chiefly metallic copper originally crystallized, coated with cuprite and malachite; three specimens.

No. 1595—Mostly hematite, with cuprite and malachite crystals and azurite; five specimens.

COPPER MINERALS IN HEMATITE ORE. 111

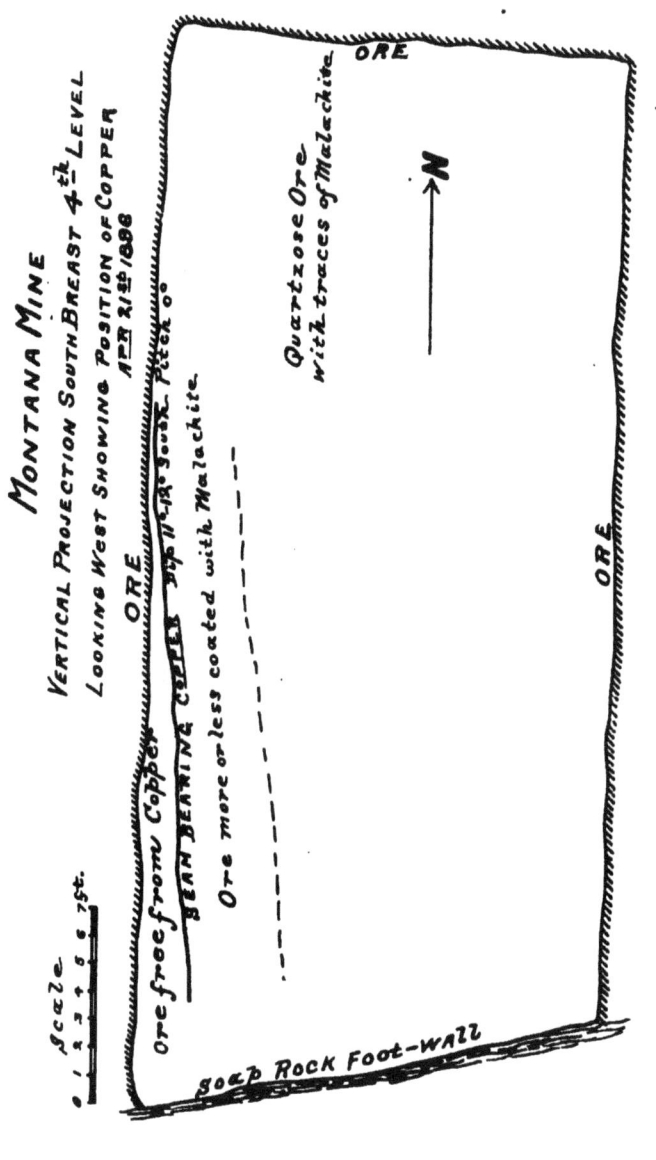

No. 1596—Largely cuprite with malachite, hematite and copper; ten specimens.

No. 1597—Largely hematite with copper, cuprite and malachite and ochre; one specimen.

No. 1598—Chiefly hematite with cuprite, malachite, azurite, copper and silica; five specimens.

No. 3901—"Carbonaceous shale," twelfth level cross-cut, Minnesota Mine; one specimen.

No. 3902—The "soap rock" of the mine from a horse rock, between north and south veins, fourth level, Montana shaft. The chief point of interest in the specimen is a small quantity of metallic copper which appears in the filling of a narrow vein; one specimen.

Hematite—The original mineral of the specimens, so far as the present discussion is concerned, is hematite. The fact that the various copper compounds are not scattered promiscuously through the ore, but are wholly confined to the lines of fracture in the brecciated mass and are without exception either cavity fillings or surface coatings on the hematite seems to place this conclusion beyond question. There is, however, a later development of red ochre and minute hematite crystals. This will be referred to later.

Copper—(A). In the ore-body—The metallic copper seems to be the earliest accessory mineral. It has been largely altered since the original precipitation, to the secondary products next to be mentioned; but such specimens as still bear the metallic copper present many points to support this view. Chemical tests disclose no impurity in sufficient quantity to warrant analysis, either of the metal or indeed of any of the secondary products. The native copper occurs in sheets from the seam in the ore (No. 1592), and in partially obliterated crystals coated with cuprite and malachite (No. 1594), and also in isolated threads and grains (Nos. 1591, 1593, 1596, 1597, 1598 and 3902). The evidence that copper is the primary mineral is stated below:

1. The original crystals and crystalline grains of copper remain, even with all surfaces coated by alteration products.

2. Cuprite, malachite and azurite all occur below the seam carrying the chief amount of metallic copper.

3. There is a constant occurrence of copper within an area of cuprite, with no noticeable development of porous zones. If

the metallic copper were a result of the reduction of cuprite, there would be a shrinkage in volume of 39.96 per cent. This process is always attended with the development of porous pseudomorphs. But, on the other hand, when cuprite alters to malachite, an increase in volume of 235.17 per cent results,[*] which is in entire accord with the bulging faces of the original crystals of copper now coated with cuprite and malachite.

4. The copper specimens freest from secondary products come from parts of the seam which were filled with sheets of copper, and show only thin coats of other minerals.

The foregoing points seem sufficient to establish metallic copper as the primary mineral of the group. None of the specimens furnish any dissenting evidence. The record of the discovery suggests no other view, unless the report of chalcopyrite, noted by Mr. Eby, can be so construed. But, on account of the apparent lack of reliable data and the total absence of any trace of this mineral in the specimens, it cannot receive serious consideration.

(B.) *In the adjacent rock.*—Small amounts of native copper occur quite free from alteration products in specimen 3902. It is confined, however, to the narow vein, and tests applied to other parts of the specimen do not show any trace of copper.

Cuprite—The red copper oxide, cuprite, Cu_2O, is one of the most abundant minerals present in the specimens. It is very pure; is clear red in color to very dark red; is thoroughly crystalline, and at some points is fairly well crystallized. The forms recognized are the cube, octahedron and the dodecahedron. All are small and rather imperfectly developed (No. 1592). It is always found on the specimens at hand and is a constant enclosing matrix for the metallic copper. There is no specimen of the metal without at least a thin coating of this oxide. It always forms an intermediate film between the native copper and the malachite. In many places it is quite compact and completely fills the fissures in the brecciated ore.

Malachite—The green carbonate, malachite, $Cu_2(OH)_2CO_3$, is as plentiful as the oxide. Its position is always next outside the coating of cuprite, when the two are in contact. But it extends further as a surface coating on the fragments of the ore

---

[*] Elements of Chemical and Physical Geology, Gustav Bischof, translation by Benjamin H Paul, Vol. III., p. 533.

than has been reported for any of the other minerals. There is no impurity in the better specimens. Many show fine groups of acicular prisms of very small size lining a few of the cavities. Others have the usual velvety or earthy characters (1595 and 1598).

Azurite—The blue carbonate, $Cu_2 (OH)_2 (CO_3)_2$, occurs in small amount associated directly with the malachite or hematite. A few specimens show crystallized azurite (1598). It is not at any point in direct contact with cuprite or metallic copper. Its most usual position is directly in contact with the hematite in small cavities.

Crystallized Hematite—Many areas of the ore are covered with a sparkling coat of hematite crystals, which are at least older than the malachite (Nos. 1593, 1598).

Quartz—A white and sometimes porous substance, reacting chiefly for silica, occurs sparingly as a coating and fissure filling in a few specimens. It is most closely associated with earthy hematite, and portions are colored yellow, etc., by oxide of iron (Nos. 1593, 1598).

Earthy Hematite—In a few limited areas there is a very bright red ochre. It was at first thought to be "tile ore," but examination proves it to be iron oxide (Nos. 1593, 1597).

Later than any of the minerals mentioned above is a reddish yellow, waxy film which covers some surfaces of the malachite. It is so exceedingly small in amount and so difficult to obtain in a pure state, that all attempts to determine it were unsuccessful. It contains silica, alumina, iron calcium and traces of magnesia.

PARAGENESIS.

The relationships of the different minerals may be seen by the diagram on page 111.

Origin—It is difficult to give a satisfactory explanation of the origin of this remarkable deposit. In the first place, however, it is narrowed down to a question of the origin of the metallic copper. The origin of the hematite is beyond our discussion. The oxide and two carbonates of copper are clearly secondary or derived products from the metallic copper by the action of water carrying oxygen and carbon dioxide. The cuprite developed first by simple oxidation. Malachite formed next by the addition of

COPPER MINERALS IN HEMATITE ORE. 115

water and carbonic acid, and azurite was formed last by the further action of the same process.*

Thus far the explanation seems to be borne out by the facts of occurrence. But the original source of metallic copper is not reached. It is reasonable to suppose, however, that the immediate source was a copper solution, either a sulphate, carbonate or silicate, from which the metal was deposited as a precipitate through the agency of ferrous oxide.**

This supposition seems to be so suitable in the case before us that the theories*** sometimes brought forward in explanation of the occurrence of copper will not be noted.

As to the ultimate source of the copper, data for satisfactory conclusions are not at hand. The three or four points which have any bearing on the problem are:

First—The limited amount of copper present;
Second—The local character of the occurrence;
Third—The comparatively recent development of the cupriferous minerals; and,
Fourth—The occurrence of metallic copper in the inclosing rock.

In accordance with these facts the following explanations are chosen as the most reasonable:

First—Small amounts of copper are scattered through the strata adjacent to the ore bodies, of which specimen No. 3902 is a fair example. By some change in the underground currents, this deposit was developed as a segregation. In such case, we should expect similar deposits at other points and we should detect a wider distribution in a district where so many mines are similarly situated. Specimen No. 3902 is the chief support for this explanation. But it comes from the immediate vicinity of the copper minerals, and instead of indicating a source of supply, the country rock copper may indicate the limit of penetration of the copper bearing waters.

Second—The original Keweenawan rocks were at one time more extensive than now. They doubtless have covered this very district in some period of preglacial time. They are noted for their copper producing minerals. If we assume that solutions

---
* Elements of Chemical and Physical Geology, Gustav Bischof, translated by Benj. H. Paul, Vol. III., p. 509, 1854.
** Geology of Michigan, Vol. I., pt. II., pp. 43-46.
** The Copper-Bearing Rocks of Lake Superior, R. D. Irving, U. S. G. S. Monograph V. p. 425.
*** Geology of Wisconsin, Vol. I., 1873-1879, pp. 110-114.

derived from the decay of these rocks had ready access to the locality under discussion, it would form the basis of a suitable explanation.

Third—A surface erratic or local accumulation in the glacial drift would furnish conditions apparently consistent with the facts at hand. The chief objections are: The depth of the copper and cupriferous minerals below the surface, and the very recent origin of the glacial drift itself.

Of these three possible sources no one is perfectly satisfactory to the writer. Work among the iron deposits is not far advanced, and in all probability additional data will soon be forthcoming that will be of more service in this question.

So far as the writer is aware there has been no previous report of any considerable segregation of copper minerals in workable deposits of iron ores. The only analysis at hand, furnishing anything definite from an American locality, is that of the Cornwall iron ore deposit in Lebanon county, Pa. In this ore there is reported[*] an average of about one per cent of copper oxide, melaconite, and about six-tenths of one per cent of copper pyrites.

Oxide of copper with some carbonate occurs with decomposed specular ore[**] near St. Clair, Franklin county, Missouri; and metallic sulphides occur sparingly in the Ozark region.[***]

Cuban ores also carry an appreciable amount of copper, estimated at about 0.05 per cent,[*] but in this case I am not sure of the mineral or chemical combination in which it occurs. Judging, however, from the amount of sulphur reported as an impurity, one would expect to find it as a pyrite. This is apparently the most common copper mineral found in iron ores.

It is not an unusual thing to encounter copper as an impurity in iron and steel, and as a furnace product in the metallurgy of iron and steel. But it is commonly in such minute quantities that no detrimental effect is produced, although two per cent of metallic copper, and usually very much less, will seriously interfere with the working of iron,[**] making it red short. A brief discussion of this subject does not necessarily include a treatment of

---

[*] Mineral Resources of the United States, 1883-1884, p. 270.
[**] Geology of Missouri, Vol. II., Report on Iron Ores, 1892, p. 95.
[***] Geology of Missouri, Vol. II., Report on Iron Ores, 1895, p. 153.

[*] Notes on Copper in Iron and Steel, by R. W. Raymond, Trans. Am. Inst. of Mining Engineers, Colorado Meeting, 1896.
[**] Notes on Copper in Iron and Steel, by R. W. Raymond, Trans. Am. Inst. of Mining Engineers, Colorado Meeting, 1896.

the effect of copper as an impurity in iron, but has been referred to in order to furnish a basis for whatever is new in the discovery of copper in the Montana mine at Soudan. Even after taking account of the occurrences noted above the following points still distinguish this discovery and claim for it an addition to our knowledge of the association of these minerals:

1. A seam of metallic copper occurs in workable hematite ore,

2. In which alteration has produced the red oxide, cuprite, and the carbonates, malachite and azurite;

3. This within a limited zone in a mine not showing any other like development, and,

4. In one of the greatest iron districts of the world where no other mine has yet shown a similar occurence.

THE NEW DISTRIBUTING RESERVOIR FOR THE WATER SUPPLY SYSTEM OF MINNEAPOLIS.

C. H. KENDALL, C. E.

For thirty years Minneapolis has been supplied with water pumped from the Mississippi river directly into the mains. During this period the population has increased from a few thousand to one hundred and ninety-five thousand, and the annual consumption of water has increased to over six and one-half billion gallons —a little more than eighteen million gallons daily average. This rapid growth of the city and the proportional increased demand for water, both for consumption and fire protection. has necessitated various changes and improvements, at frequent intervals, in its system of water supply. These have terminated in the present extensive improvement by which the supply is changed from a direct pumping system, using objectionable, unfiltered river water, to that of gravity distribution of filtered water from an extensive storage reservoir, into which the water is pumped from the river.

All previous improvements have been in the way of enlarging the pumping capacity, changing the location of intake pipes, and extending the system, but nothing has been done toward improving the quality of the supply, which for several years has been very unsatisfactory. The river is not only subject to contamination from the city sewerage and from the cities and towns above, but the water also contains an excessive amount of organic matter of vegetable origin, principally from the millions of logs floated in it yearly, polluting it at times to such an extent as to render it even dangerous.

From records, extending back twenty years, of careful examinations and analyses of the water by such competent men as Professors Peckham, Hewitt, Dodge, Leeds, Drew, Sidener, and E. G. Smith, we find them all concurring in the opinion that the supply is suspicious and not of proper purity for public use. Prof. Drew says: "That with water so heavily charged with vegetable matter, and perhaps more or less animal waste, as well as the discharge from the sewers of Anoka and the towns above us. there is continually a possibility of the development of an epidemic of disease as a result of contamination of the impure water with the germs of disease." Prof. Smith, of Beloit, in reporting analysis made for the city in July '94, says: "I can hardly see how, in their present condition, the waters can be considered healthful and desirable, and it seems to me that if it be necessary to use the river as a source of public supply, you should make every effort to determine whether its quality can be improved by some system of storage or rapid treatment."

With the consideration of purity as paramount, Mr. Cappelen, as city engineer, has given his attention to the problem, and in the new system, has sought to give the city an abundant supply of unquestionable quality.

After investigating the different possible sources and systems of supply suggested, and after studying the supply of other cities by personal examination, Mr. Cappelen, is his report to the city council, recommended the continued use of the river supply and the construction of a one hundred million gallon distributing reservoir, with a thirty-five million gallon filter plant. And provision to be made for another reservoir of the same size for future use when it became necessary.

From surveys made, a site, admirably adapted for such an improvement, was found just outside the city limits, located between Forty-fifth and Forty-eighth avenues northeast. The land belonged to Mr. Lowry, and he generously gave the city a tract of forty-five acres—sufficient for the two reservoirs—and also in addition about twelve acres for a hundred-foot boulevard across his land from the city limits to the reservoir.

The total estimated cost of the improvement was, in round numbers, $1,150,000, and was adopted by the council in November, 1894. Work was commenced the following May. The general design of the reservoir and its operation is best obtained from the following description, given by Mr. Cappelen in his paper before the American Water Works Association in August, 1894.

"The entire land required, including that for another reservoir in the future, is bounded by Forty-fifth and Forty-eighth avenues and 'B' and 'E' streets northeast. The south end of the reservoir is located about 150 feet from Forty-fifth avenue and runs up to Forty-eighth avenue, leaving a driveway of twenty feet all around at the bottom slope. The reservoir is divided into two chambers to facilitate cleaning when necessary. The dimensions of each chamber are:

"Inside top length, 877.5 feet; inside top width, 413.5 feet;

maximum depth of water, 20 feet; the top of the embankment being 3 feet higher, or at an elevation of 317 feet above city datum. The highest part of the bottom is at an elevation of 294 feet above city datum. The capacity of each chamber is 46,500,000 gallons.

"The south chamber is partly in excavation, the north chamber is formed entirely by embankments. The design of the reservoir contemplates earthen embankments with outside slopes 1½ to 1, and inside 2 to 1. The inside slopes are first covered with two feet of puddle, then eighteen inches of crushed rock, and then to within 10 feet of the coping with six inches of Portland concrete, put down in squares of about fifty feet each, the joints being filled with a composition of tar and lime. The upper parts of the slopes are covered with Kettle River stone, 14 inches thick, not less than 10 inches in width and not less than two feet in length. The top is surmounted on the water side with a four-foot coping, one foot thick, and the entire basin enclosed by a handsome railing. The bottom of the chamber will be puddled with two feet of puddle, and then covered with six inches of concrete blocks, as on the slopes. In all embankments, a puddle wall will be constructed through the center to such a depth as may be required.

"The top of the embankments will be 20 feet wide and can be utilized as driveways. The outside slope will be sodded. In the southeast corner of the land, at the top of the reservoir, which at that point is entirely incut, the filter plant will be located. This building will be of brick, with stone trimmings, about 275 feet long and 84 feet wide, with iron trusses and roof, ceiled up on the inside with oiled ceiling. In this building the 42-inch delivery and discharge mains will be located above the concrete floor, between the filtering tanks; these tanks will be open, so that the operation of filtration will be in clear sight. This building will also contain a couple of small pumps and boilers for washing filters. Also electric light plant, all duplicate.

"The water will leave the filters through a 42-inch main running along the north side of the reservoir. From this main two 36-inch mains will run below the embankments, one to each chamber, and enter into an ornamental gate tower located on the inside of the embankment. In this tower the water will be controlled by a series of gates so that the water can be made to empty into the reservoir either at the bottom, middle, or top, as conditions may make it advisable. This tower is called the 'inlet tower.' The outlet will be opposite in another tower, and the water can there enter, as at the inlet, at various heights.*

"The bottom of the reservoir slopes from inlet to outlet tower, two feet, and correspondingly from all sides, so that the chamber can be cleaned and flushed, and the dirty water be drawn off at the outlet tower through a drain pipe put in the same trench as the outlet pipe which delivers the water to the city main. The outlet pipes are 36-inch, running into a 42-inch pipe, which is located on 'B' street. From here the pipe runs to the river, where it is, by means of a submerged pipe, connected with the 36-inch and 30-inch mains, now leading from the North Side station to the city. The supply pipe to the reservoir will run from the North Side station by a submerged pipe and take the same route as the main from the reservoir, and also be 42 inches in diameter. At the intersection of Forty-fifth avenue northeast, the two mains will be provided with a gate each, so as to shut the reservoir off entirely, and the two mains will be connected up so that delivery can be made directly into the main going to the city."

The following is the estimated cost:

Reservoir	$327,000
Reservoir pipe lines	60,000
Main pipe lines	260,000
Drain pipe lines	40,000
Submerged pipes	49,000
Filter plant	260,000
New pumps and boilers	150,000
Total	$1,146,000

In order that the enterprise might directly benefit the unemployed of the city, the following special requirements were incorporated in the specifications:

"1st. That no laborer shall be paid less than fifteen cents per hour.

2d. That team and teamster be paid not less than thirty cents per hour, all teams to be owned by Minneapolis citizens.

3. That no men be employed except bona fide residents of the city of Minneapolis for at least one year immediately preceding

---

*As constructed, this has been slightly changed from the proposed plans. The towers being placed on top of the embankments, outside of the basins into which the water enters through a single pipe.

such employment, and that the men so employed shall also be heads of families, or have families depending upon them for support."

Work was successfully carried on from May to November during '95 and '96, and the amounts paid for labor during the two years was $116,735.45 and $84,724.72 respectively. The highest number of men employed in '95 was 550, and 150 teams. The highest number in '96 was 390 men and 86 teams.

A synopsis of all the work done in '95-'96, together with the financial statement, shows:

Main boulevard, from city limits to reservoir, completed. Cost .... $ 20,515.81
Construction of reservoir proper, complete except as noted below. Cost........................... 372,512.66

Total expenditures........................... $393,028.47
There still remain to complete the work:
Remaining coping and sheathing on the north basin, reservoir boulevards, railing. some pipe lines and sundries, estimated at....................... $ 49,300.00
To build pipe lines, steel....................... 315,957.00
To change old pump........................... 3,500.00
To electric pumping machinery................. 133,500.00
Suction work, etc., at pump house.............. 5,000.00
Filter plant estimated at....................... 250,000.00

Making a total cost of...................... $1,150,285.47

If cast iron pipes are used the cost would be about $27,000 more. No decisive action has been taken regarding the filter plant, but undoubtedly some form of mechanical filter using coagulents will be adopted, built on the gravity principle, from whence the filtered water will run directly into the storage reservoir. There are filter companies which guarantee to furnish water according to the following standard:

First. All odor, color, and impurities in suspension shall be removed.

Second. The free ammonia shall not exceed 0.05 parts in one million.

Third. The albuminoid ammonia shall not exceed 0.1 parts per million.

Fourth. No measurable amount of coagulant or other purifying agent used shall be left in the filtered water.

Fifth. The microbes in the filtered water shall not exceed 100 colonies per cubic centimeter.

For the material for this article, the writer is greatly indebted to papers and reports by the city engineer, Mr. Cappelen.

A CONVENIENT AMMETER PLUGBOARD.

GEO. D. SHEPARDSON, M. E.

Station ammeters like those of the Weston Electrical Instrument Company, which measure current by the fall of potential method, may be arranged without difficulty to serve a variety of purposes. These well known instruments are essentially d'Arsonal galvanometers. When designed for use as a voltmeter, the movable coil is wound with a number of turns of fine wire and is connected in series with an auxiliary coil of high resistance, the fall of potential through the movable coil being but a small fraction of the total voltage between the terminals of the instrument. When designed for use as an ammeter, the movable coil is wound with fewer turns of larger wire so as to have low resistance, and is connected in parallel with a "shunt" of much lower resistance. The current through the galvanometer coil, and consequently the deflection of the pointer, is therefore directly proportional to the fall of potential through the shunt. Since fall of potential equals the product of current by resistance, a given current may be sent through the galvanometer coil and a given scale reading obtained while the current through the shunt may have one of several different values, if the various steps in the resistance of the shunt increase proportionally as the main current decreases. The same galvanometer coil may therefore be used for measuring various ranges of current by taking suitable shunts for the galvanometer, or, in other words, by shunting the galvanometer around suitable shunts.

The determination of the resistance to be taken for a shunt to measure a desired range of current is a simple matter. Since the shunt and galvanometer are connected in multiple, the fall of potential in each circuit is the same when a given current is passing. We may write therefore,

$$e = rc = RC,$$

in which e is the fall of potential through each circuit, r is the resistance of the galvanometer circuit including coil and connecting wires, c is the current through the galvanometer, R is the resistance of the shunt and C is the current through the shunt.

A CONVENIENT AMMETER PLUGBOARD. 125

We have then as the desired resistance for the shunt,

$$R = \frac{rc}{C}.$$

If the current through the galvanometer is very small compared to that in the shunt, the latter may be considered as sensibly equal to the current in the main line, and we may write without serious error,

$$C = c\frac{r}{R},$$

or the current in the main line equals the current in the galvano-

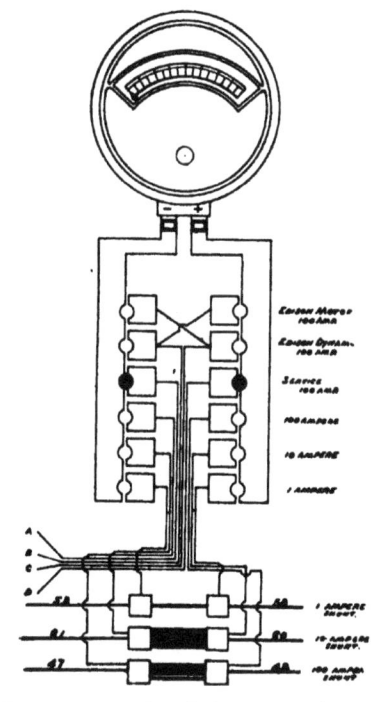

meter multiplied by the ratio of the two resistances. When the galvanometer requires a sensible current for a full scale deflection, the range may be multiplied by a series of shunts like those used with certain forms of Siemens torsion galvanometer. If a certain current passing through the galvanometer gives a full scale deflection, ten times as much current is required in the main line when the galvanometer is shunted by a resistance one ninth that of the galvanometer. Likewise for measuring one hundred times the current, the shunt should have one ninety-ninth the resistance of the galvanometer.

For multiplying the usefulness of a Weston round pattern ammeter, a series of shunts was arranged as indicated in the line cut. Two brass bars were arranged for fastening directly to the terminals of the galvanometer without introducing appreciable resistance. A series of smaller blocks were arranged in pairs so that by the use of tapered plugs any pair may be connected with the galvanometer bars. To one pair of blocks were screwed the galvanometer wires leading to the special alloy shunt sent with the instrument. By heavy wires leading to terminals 47 and 48 on the switchboard, this shunt may be connected in series with any desired circuit for measuring current of one hundred amperes or less.

A second pair of blocks forms terminals for other wires, A and D, which are attached to two points a few feet apart on one of the service wires supplying the laboratory with electric power from the Mechanic Arts Building. By taking a suitable length of main conductor and known length and size of lead wires to the galvanometer plugboard, calculated as above indicated, the ammeter reads directly to one hundred amperes.

The third and fourth pairs of blocks form terminals for a third pair of wires, B and C, which are connected to suitable points on the leads of the Edison dynamo in the laboratory referred to in another paper in this volume. Since the machine operates sometimes as a dynamo and sometimes as a motor, the direction of current in the leads changes. It is therefore necessary to have a means of reversing the current through the galvanometer. This is secured by having two pairs of blocks which are cross-connected as shown in the cut.

With each of the three shunts above noted, the galvanometer needle indicates one hundred amperes for full scale deflection. It was desired also to arrange the instrument for other ranges, and shunts of German silver wire were arranged so that by plugging the galvanometer around one, a full scale deflection means ten amperes. Still another shunt of higher resistance makes one

A CONVENIENT AMMETER PLUGBOARD.

ampere give a full scale deflection. Suitable conductors connect these shunts with terminals 58, 59, 60 and 61 at the switchboard so that either shunt may be plugged into any desired circuit.

Since the galvanometer gives a full scale deflection with considerably less than one-tenth ampere, another shunt might easily be arranged so that the instrument would read milliamperes. Furthermore, since a difference of potential of only about three one-hundredths volt gives a full deflection, a small resistance connected in series would convert the galvanometer into a convenient milli-volt-meter.

The instrument as outlined in the cut has been in constant use for two years and has given great satisfaction, being reliable within less than one per cent.

THE SCIENCE OF PHOTOGRAPHY.

H. C. HAMILTON, '97.

Until recently photographic work has been regarded as very closely allied to the "black arts," but with the introduction of so much prepared material for manipulation the secrets of the art are no longer confined to the few.

The processes involved in making photographs are so few and so simple that it is surprising how many people are entirely ignorant of them. Even a great many of the camera fiends who snap us unaware are ones who "push the button," leaving "the rest" for someone else to do.

Everyone at some time of his life would find a practical knowledge of the art to be a means of both pleasure and profit. Scientists have increased facilities for study and for imparting their learning to others; professional men in all departments have found it a valuable aid in their work. Its application to astronomy, its use in surgery since the discovery of the Roentgen rays, by the draughtsman to reproduce drawings and designs, by surveyors to reduce their plots to convenient size, prove its value.

The photographic lens is an eye capable of seeing much that we cannot. In astronomical work a photograph of a star group has revealed, in some cases, the presence of stars too small to be seen by the eye. The lens resembles the eye in that nothing in its direct line of sight is clear and distinct except objects lying at nearly the same distance as the point for which it is focused; while it differs in that any object at the same distance from the lens as the principal one and lying within the field, forms a distinct image. So that a photograph of some object of interest will usually contain additional objects, which often mar the effect.

There is a great variety of lenses, but all may be grouped under three heads: the single lens, the double combination, and the wide angle lens. The single achromatic lens—that having only one combination—is the best yet introduced for landscape work. This lens is plano-convex or concavo-convex, the convex side being turned toward the image formed. It is provided with a diaphragm or stop, which allows only those rays to enter which can come to a focus at approximately the same distance from the lens. This diaphragm is used also to correct spherical aberration. Rays of light from an object, falling on different parts of a lens, are differently refracted, and therefore do not come to the same focus. If all the rays were to come through a part of the lens so small as to have practically the same radius of curvature, they would be equally refracted and have a common focus.

The achromatic feature of the lens is to correct for color. Just as a ray of light in passing through a prism is decomposed into its constituent colors, so it is in passing through a lens. Besides having the image colored, the greater refraction of violet rays over those of the red places the actinic image—that which affects the sensitive plate—and the visual image some distance apart. This is overcome by having a combination of two lenses, one of dense flint glass and one of crown glass, so cut that they fit into each other. They have different refractive indices, the effect being to recombine the dispersed colors into white light.

For portrait work a lens is required which does not distort the image and which will represent straight or curved lines in their true character. This can be accomplished only by having two combinations, which will make every ray emerge after transmission in a direction parallel to that at which it entered. Its spherical aberration is corrected in manufacture, so that the image is sharply defined and at the same time it can be used with a large aperture. This allows an exposure to be made with very short time, on account of the amount of light admitted.

For this work the field is small and the aperture large, while for architectural and interior views the field is large and the aperture small. These lenses are very convex, almost spherical, allowing light from a wide angle to come to a focus at a short distance from the lens. Hence the name "wide angle."

In some of the cheaper forms of hand cameras there is no such thing as focusing; the lens is so constructed that rays of light from objects over eight feet distant are refracted the same as parallel rays, that is, their focus is at the same distance from the lens. But this is at the expense of distinctness and of the relative proportion of objects photographed.

There is a difference of opinion as to the change effected by the action of light, whether it is chemical—the silver bromide being partly reduced—or physical, the molecule being dissociated. Whatever it is, microscopic examination fails to show;

only the action of chemical agents make it apparent that a change has taken place.

Mercury vapors are attracted to those places in the film where an image has been received. Reducing agents, that is, compounds that easily take up oxygen, act only on those affected parts. There are two reasons for believing it to be a physical change. An exposed plate that has not been developed shows no signs of the image when the soluble silver salts are dissolved in the fixing bath. An exposed plate may have the latent image entirely removed by the action of vapors of acetic acid and the plate be as good as ever. However, it seems scarcely possible that the change is not chemical. It is probable that further research will show that a combination has taken place between the organic film and the bromine, leaving the silver salt partly reduced.

The sensitive medium in which the image is received may be on glass, celluloid or waxed paper, anything that is at the same time tough or firm and transparent.

In the earlier days the film was of collodion, made sensitive by being placed in a solution of silver nitrate.

Collodion is a solution of nitrate of cellulose—pyroxylene—commonly called gun cotton, in a mixture of alcohol and ether; to this is added iodide of cadmium or ammonium and bromide of cadmium, the combined effect being considered better than either alone. When this solution is poured on the plate, the volatile liquid evaporates, leaving a tough film. The plate is then put into a bath of silver nitrate to be sensitized. The silver combines with the iodine and bromin in the film, forming compounds sensitive to light; so this process must take place in a dark room.

In developing a negative the reducing agent acts on the silver iodide that has been affected, reducing it to metallic silver, and at the same time depositing silver from the excess of silver nitrate on the plate.

When the image is formed in all its details and with sufficient intensity, the unchanged silver iodide is dissolved with potassium cyanide or sodium hyposulphite, leaving the image formed of metallic silver in a very finely divided state, giving it a black appearance.

This process is the wet plate or collodion process and is no longer in use, except for tintypes and in reproductions for engravings.

Modern methods depend on the use of dry plates, using a film of gelatine instead of collodion, and being prepared for wholesale by large manufacturers. Their advantage lies in their convenience, their keeping indefinitely, and the greater sensitiveness and density of the film.

Gelatine is liquified in a small amount of water containing ammonium bromide, and is then well mixed with a solution of silver nitrate. The mixture is heated on a water bath at 100°C for three-quarters of an hour, and then cooled to 38° and more gelatine in liquid form is added. By varying the time of cooking, or by heating at a lower temperature for a longer time, different degrees of sensitiveness can be obtained.

In the development of dry plates, since there is no free silver nitrate present, from which silver can be deposited, the gradual increase of density of the image is due to the gradual reduction of the silver bromide deeper in the film, so that one can judge of the density of the silver image by noting when the reduction has penetrated the film and the outline can be seen through the glass on the back.

The salts of silver vary much in their sensitiveness to light. Silver nitrate is not sensitive except when combined with organic substances, and in that form it changes rapidly. Silver chloride is the least sensitive of the haloid salts, but undergoes the greatest visible change. Silver iodide is ten times and silver bromide sixty times as sensitive as the chloride but they change little in the light. The amount of change in the molecular condition of the sensitive silver salts does not seem to be directly proportional to the length of exposure, nor to the intensity of the light; but a definite amount of light is necessary to form the latent image. Too long an exposure is often as slow in development as one too short, but the usual result of an over exposure is a negative without that contrast between lights and shadows which gives life to a picture.

The negative differs from the finished picture in that when viewed by transmitted light the blackest parts represent the strongest lights in the objects photographed, while the finished picture or positive is reversed showing the lights and shadows naturally. With the exception of tintypes, daugerreotypes, and pictures of that class, all positives are made by the action of light passing through a negative. Where the light passes through the lightest parts the greatest effect is produced in the sensitive film of the positive, forming the shadows, and where less light acts through denser parts of the negative less change is produced and the paper is left only slightly tinted.

Printing paper for positives must be of good quality to resist the rough treatment it gets. Rives paper, made in France, is that most used. It is usually sized with a coating of starch or gelatine to prevent the sensitizing solutions from sinking into the paper. It is then salted by floating on a solution of sodium chloride containing liquid albumen with some organic salts; as, the citrate and tartrate of soda. After this it is floated on a solution of silver nitrate, by which process there are left in the film several sensitive silver compounds. The different silver compounds are of use to give to the paper different colors while printing. The gelatine and albumen compounds give a red color, the citrate a purple, and the chloride a violet blue. The chloride increases the sensitiveness, while the metallic silver, that gives intensity to the print, all comes from the excess of free silver nitrate. When printed dark enough, all the silver not changed to the metallic form is dissolved by putting the printed pictures into a solution of hyposulphite of sodium. This leaves an image, formed almost entirely of metallic silver.

When an image has been printed and is put into the fixing bath, as just described, the thin layer of silver has a red color which is anything but agreeable. This would, in time, become black from the formation of silver sulphide, the sulphur being taken from the hyposulphite; but the action is too slow and the results not satisfactory.

Instead of trying to obtain a good color with silver alone, gold is used, and by this means more agreeable tones are produced. A solution of gold chloride is used; its action seems to be to give up the chlorine to the silver forming silver chloride and precipitating free gold in a very finely divided state. The thin films of gold take various colors, but with a tendency toward the blue. The toning may be carried so far that blue is the only color left; but it is usual to stop the process while there is still some of the red silver film left in the shadows, which gives somewhat of a flesh color to the picture. The changing of most of the silver to silver chloride and the removal of this by action of the fixing both bleaches the print to a very great extent.

THE SCIENCE OF PHOTOGRAPHY. 188

Bleaching may be partly prevented by making the toning bath alkaline with sodium carbonate, which forms an oxide or oxychloride of the gold. This compound decomposes precipitating gold on the silver image while the effect of the chlorine on the silver is neutralized.

The use of albumen in the film is to give brilliancy to the picture by imparting a glossy surface. This in time loses its lustre and spoils the effect of the pictures.

There is a paper coming into general use which is not glazed and on which the image appears only in black and white when finished. The effect is obtained by means of platinum. The sensitizing solution contains both silver and platinum in the form of chlorides and silver nitrate; only the silver is affected in printing and the image appears the same as on plain paper.

It is toned a very little in gold to clear the high lights and the process completed in a solution of potassium chloro-platinite. From this solution metallic platinum entirely replaces the silver image, and the result is a picture in black and white and in the most permanent form of finely divided platinum black.

Carbon printing is an interesting method of reproduction. . The sensitive medium is gelatine and potassium bi-chromate. The action of light is to liberate oxygen from the bi-chromate, and this unites with the gelatine, forming a compound insoluble in water in proportion to the intensity of the light. The gelatine may be given any color, but lamp black is mostly used, and from this the method obtained its name.

On account of the short exposure necessary for the bromide of silver, it is sometimes used for positive printing, principally enlarging. The rays of light pass through a lens and the negative and are focused on the paper, which must be so placed that no other light acts on it. It is exposed and developed just like a negative. This method is used to make the outline for the cheaper grades of crayon portraits.

The principles of photography are made use of in almost all forms of reproduction of pictures.

Even since the discovery by Daguerre, attempts have been made to reproduce objects in their natural colors. One of the methods successfully tried is that discovered by Lippman, a prominent French physicist. He reasoned that, since color can be obtained by interference of light waves, strata of silver could be formed in the negative film so as to give color, and that these strata could be formed if a suitable reflecting surface were placed behind the plate.

When a ray of light meets a reflecting surface and is sent back on itself, the waves with vibration frequencies giving the different colors will meet incoming waves of the same amplitude of vibration in the opposite phase. This forms nodes or points of no light, and, therefore, no action in the sensitive film, and corresponding points having the double effect. Then when the fixing bath removes the unchanged silver bromide there are left particles of silver alternating with transparent film. These strata cause interference in transmitted light, taking out some colors and consequently showing a colored image. The film must be about one-tenth millimeter thick in order to be transparent, and the exposure made three minutes long. The reflecting surface behind the glass is pure mercury.

Another method of making colored prints based on photographic methods is that used in photo-lithography and similar processes. Colored printing inks are applied to surfaces which are in relief only where a certain color is wanted. These surfaces in relief are prepared from negatives on an asphalt or bi-chromated gelatine film covering lithographic stone, zinc or copper. The action of light makes the film insoluble while the parts not affected are dissolved in water, leaving the exposed surface to be etched by some corroding agent. The protected parts are sufficiently in relief so that when ink is applied by a roller they alone are touched and the color can be transferred to paper by the printing press.

Negatives for this work must be clear glass in those places where a color is to come out in relief. This can be done by screening out the light of that color; green will absorb red rays, violet screens the yellow and orange the blue. Where the red and yellow rays are intended to act on the film it must be made sensitive to that kind of light by being bathed in a solution of one of the eosin group of dyes.

It is not known definitely what the nature of the combination of these dyes with the silver salts is, but it is probable that a chemical compound is formed, sensitive to an entirely different part of the spectrum.

Another method makes use of the fact that when fine lines in different colors are ruled very closely together, or silk fibres dyed in those colors are placed side by side, the result is a mixture which acts similarly to the color screen, as in the first method. Red, green and violet colors are used 300 or 400 to the inch. A negative is made through this screen and from this a transparency which will be divided into parallel lines, according to the light transmitted. Another colored screen, made like the first, must be used with this in order to excite the three primary color-sensations. When this is viewed by transmitted light the result is an image in its natural colors.

Of all these methods none is very practical excepting the second, this being used for nearly all work in colored reproductions.

136   DR. HENRY T. EDDY.

THE MOST ECONOMICAL CUT-OFF FOR STEAM COMPRESSION.

DR. HENRY T. EDDY.

The result of an analytical investigation of this question was stated in the Year Book for 1895, the result being, that a given quantity of steam per stroke is used most economically in case the ratios of compression and of expansion are so adjusted to each other as to be equal.

This result is of such interest and importance that the author has thought it desirable to give, if possible, some simple and direct proof of it, instead of the rather complex analysis by which it was originally established. This seems especially desirable in order that the clearness of the demonstration may set at rest doubts that seem to have risen in the minds of some engineers as to the validity of the result. The following proof has been devised with this end in view:

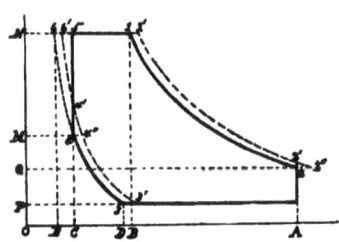

Let the area, 12345, be that of an ideal indicator card, in which the curve of expansion 12, and of compression 34, are both hyperbolic, while its other bounding lines are straight and not assumed to be rounded off at the corners, as in an actual card.

Assume the steam to be admitted at a constant pressure, p_1=ON, and let the admission cease at the point B so that a volume v_1=OB is contained in the cylinder and clearance together.

Let the release occur at a pressure $p_2 = OQ$, and at a volume $v_2 = OA$, which includes both cylinder and clearance. Let $v_4 = OC$ be the volume of the clearance, then $v_2-v_4 = CA$ the volume swept through by the piston per stroke. Let the exhaust cease at pressure $p_3 = OP$, and volume $v_3 = OD$. The cushion steam is then compressed by the piston to pressure $p_4 = OM$ and volume $v_4=OC$, and finally it is further compressed by the steam admitted from the boiler to pressure $p_1=ON$ and volume $v_6=OE$.

Then $v_1-v_6=EB$ is the volume of the steam admitted per stroke at pressure $p_1=ON$. For convenience let the ratio of expansion be $v_2\div v_1=r_1$, and the ratio of compression $v_3\div v_4=r_2$. It is our object to show that for a given volume (v_1-v_6) of steam admitted per stroke, the work done per stroke with given cylinder and clearance is the greatest possible when $r_1=r_2$.

Since the curves given are hyperbolas, it is known that

Area $N12Q = p_1v_1\log_e r_1 = A_1$, say;

Area $M43P = p_1v_6\log_e r_2 = A_2$, say;

Now suppose the admission to cease at a point slightly later in the stroke, so that the corner 1 of the card is in a new position 1', and 6 at 6', yet these changes are to be so adjusted that the volume of steam admitted at boiler pressure shall be unchanged, then 16=1'6' and 11'=66', or $dv_1=dv_6=dv$, say.

By this change the work per stroke has been increased by an amount represented on the indicator card by the area 122'1'= 122"1' ultimately; and it has been decreased by 344'3'=344"3' ultimately.

But area $N1'2"Q = p_1(v_1+dv_1)\log_e r_1 = A_1+dA_1$
and area $M4"3'P = p_1(v_6+dv_6)\log_e r_2 = A_2+dA_2$.
$\therefore dA_1 = p_1 dv \log_e r_1$, $dA_2 = p_1 dv \log_e r_2$.

Hence in the change, the total gain per stroke is represented ultimately by

$$dA_1 - dA_2 = p_1 dv(\log_e r_1 - \log_e r_2).$$

It is known from the mathematical laws of maxima and minima that this quantity must vanish in case the area of the card, and so the work done, are to be as great as possible, under the given conditions. The expression we have obtained vanishes when, and only when, $r_1=r_2$.—Q. E. D.

The particular advantage of this method of proof consists in making the algebraic difficulties as few and as small as possible.

THE GILLETTE-HERZOG COMPETITION.

In 1892 the Gillette-Herzog Manufacturing Company established the custom of offering annually, for competition by the students of the Engineering College of the University of Minnesota, a first cash prize of $50 accompanied by a gold medal, and a second cash prize of $30, accompanied by a gold medal. The conditions of the competition for the present year are:
I. The following subjects will be admitted:
1. Mechanical engineering in such branches as machine design or processes of construction; elevators and hoists, stationary and traveling cranes; motor wagons; heating system for building or for large plant; smoke prevention; power generation, transmission or distribution; plan of manufacturing plant, including arrangement and types of machinery, to give best results in plant cost and product cost.
2. Structural (architectural) engineering, as seen in the construction of fire-proof buildings, bridges and iron and steel structures generally. Designs of fire-proof buildings should include the construction of iron and steel roofs and trusses, girders, etc.; bridge designs should give a discussion of the whole question of strains; iron and steel structures should include a comparison of the efficiency of cast iron, wrought iron and steel columns.
3. Municipal engineering, as exhibited in the problem of water supply, e. g., the design of a reservoir or pumping station, with a discussion of the attendant questions of supply and distribution of water.
4. Electrical engineering, particularly in the electric lighting of manufacturing plants and the use of electric motors in such plants.
II. While the competition is open primarily to seniors in mechanical, structural, municipal and electrical engineering, special circumstances may make it advisable to admit graduate students to the competition.
III. The names of ten (10) students selecting suitable subjects shall be presented in good faith as signifying their intention

to compete for the prizes before the Gillette-Herzog Manufacturing Company shall be bound to declare the prizes open for competition.

IV. The Gillette-Herzog Manufacturing Company and the president of the University shall name the examining board to adjudge the prizes in this competition.

V. The company assumes the right to withhold all prizes if, in the opinion of the judges the theses and designs are not of sufficient merit to deserve prizes.

VI. The judges will be expected to decide on the merits of the theses in view of their practical usefulness.

VII. Honorable mention of any theses and designs of special merit, not awarded prizes, may be made by the judges.

VIII. All theses and designs in competition shall be in the hands of the judges on or before May 25th, 1897, and shall be presented in the following manner: Each thesis, accompanied by its designs, shall be handed, without the name of the writer, or any designating mark whatever, to the Dean of the College and by him shall be delivered to one of the judges.

IX. All theses and designs shall be presented in duplicate; one copy shall become the property of the University of Minnesota and the other shall become the property of the Gillette-Herzog Manufacturing Company.

The prizes awarded under each annual offer were as follows:

1892.

First prize to Leo Goodkind for a design of a fire-proof building with steel skeleton frame.

Second prize to James H. Gill for a design of a high speed Corliss engine.

1893.

First prize to Delos C. Washburn for a design of a steel frame of a machine shop or iron foundry.

Second prize to Frank E. Reidhead for a design of a series electric motor.

1894.

First prize to Andrew O. Cunningham for a design of a steel arch bridge.

Second prize to Harriet E. Wells for a design of a wrought iron gate.

THE GILLETTE-HERZOG COMPETITION.

Recommended for honorable mention:
Charles H. Chalmers for a design of a 15 h. p. electric motor.
George E. Bray for a design of a steel boiler.

1895.

First prize to Leslie H. Chapman for a design of a swing bridge.

Second prize to Harry L. Tanner for a design of an induction motor and a polyphase generator.

1896.

First prize to C. Paul Jones for an original design for the steel frame of a 10-story building.

Second prize to C. Edw. Magnusson for specifications for an electric light plant for the Gillette-Herzog Mfg. Co., Minneapolis.

EDITORIAL.

The advertisements in this issue cover every branch of the engineering profession. The firms represented are among the best and most reliable in this country, and it is their patronage that enables the Year Book to be published.

A favor will be conferred on the Society by its friends and members if on writing to any of these firms they will mention that they saw their advertisement in the Year Book.

The Engineers' Society wishes to acknowledge the courtesy of an exchange with the following colleges and societies.

GENERAL EXCHANGES.

University of Michigan, "The Technic"; Purdue University. "Proceedings of the Purdue Society of Civil Engineers"; University of Tennessee, "The University Scientific Magazine."

SINGLE EXCHANGES.

University of Illinois, "The Technograph"; Bulletin of the University of Wisconsin; Transactions of the American Society of Irrigation Engineers; Rensselaer Polytechnic Institute; "The Polytechnic"; Cornell University Quarterly; Pratt Institute Monthly.

EDITORIAL.

The lecture of Mr. Woodman, secretary of the C., St. P., M. & O. R. R., was especially appreciated, and is printed in full in this issue.

The Board of Editors wish to acknowledge their indebtedness to Miss Margaretta A. Stevens, of the School of Design, for the artistic cover page design of this issue.

A Suggestion.—If all lecturers before the Society would have the kindness to have their lectures written out, so that if desired they could be printed in the publication of the Society, it would greatly accommodate the Board of Editors of the Year Book.

The different methods of setting forth electrical phenomena to the student are quite often highly interesting, especially when making use of apparatus originally designed for altogether different purposes. Until the present year the electrical laboratory had been somewhat lacking in polyphase machinery, but owing to the inventive genius of Prof. Geo. D. Shepardson, this can no longer be said of it.

The T. H. arc machine and the Edison constant potential dynamo, mentioned in his articles, are now very successfully taking the place of expensive machinery, the purchase of which would otherwise have been necessitated. Although they are not so efficient as apparatus designed for that particular use, yet they are so thoroughly interchangeable that in the eyes of the student they far surpass any apparatus that could have been purchased for practically illustrating direct alternating and polyphase currents.

The exploring engineer is the pioneer in his profession. His work, preceding the tide of immigration, forms the basis of that of the geographer and geologist, and gives an invaluable aid to the naturalist. Among the graduates from our college, Mr. W. C. Weeks may well be ranked among the foremost and as having had the most varied experience in this line of engineering. To the embryo engineer who longs for adventure and imagines that the day of the explorer is past his article on engineering in South America in this issue of the Year Book will be a pleasant surprise, showing, as it does, that there are yet vast territories, even on our own continent, entirely unexplored.

LIST OF ADVERTISERS.

ASSAYERS, AND CHEMISTS' SUPPLIES.
The Denver Fire Clay Co., - - - - - Denver, Colo.
CEMENT.
Meacham & Wright, - - - - - - - Chicago, Ill.
Western Cement Co., - - - - - - Louisville, Ky.
DRAWING INSTRUMENTS AND SUPPLIES.
Theo. Alteneder & Sons, - - - - - Philadelphia, Pa.
University Book Store, - - - - - Minneapolis, Minn.
ELECTRICAL SUPPLIES.
Interior Conduit & Insulation Co., - - - New York, N. Y.
MECHANICAL SUPPLIES.
Schaeffer & Budenberg, - - - - New York and Chicago
L. S. Starrett Co., - - - - - - - Athol, Mass.
PUMPS.
Battle Creek Steam Pump Co., - - - - Battle Creek, Mich.
Deane Steam Pump Co., - - - - - - Holyoke, Mass.
SURVEYING INSTRUMENTS.
Buff & Berger, - - - - - - - - Boston, Mass.
Kuhlo & Ellerbe, - - - - - - - St. Paul Minn.
SEWER AND CULVERT PIPE.
Evens & Howard, - - - - - - - St. Louis, Mo.
MISCELLANEOUS.
Cutler & Wilkinson, - - - - - - Minneapolis, Minn.
Engineering Record, - - - - - - New York, N. Y.
Hartford Steam Boiler Inspection & Insurance Co., - Hartford, Conn.
Minneapolis Engraving Co., - - - - Minneapolis, Minn.
Sharpless & Winchell, - - - - - Minneapolis, Minn
Railroad Gazette, - - - - - - - New York, N. Y.
University of Minnesota, - - - - - Minneapolis, Minn.
W. J. Johnston & Co., - - - - - - New York, N. Y.

THE SOCIETY OF ENGINEERS OF THE UNIVERSITY OF MINNESOTA.

PROGRAMS FOR 1896-97.

October 10, 1896:
Bridge Building - - - - - - - - - - E. A. Lee, '97
October 23, 1896:
Wiring of Buildings - - - - - - - O. G. F. Markhus, '97
November 6, 1896:
Lighting System of Armory - - - - - W. L. Miller, '97
November 20, 1896:
Notes from Students' Trip to the Mines - - R. T. Wales, '97
January 15, 1897:
Gas Engines - - - - - - - - - - E. S. Savage, '97
January 28, 1897:
Railroad Construction Work - - - - F. B. Walker, '97
February 6, 1897:
Good Roads - - - - - - - - - - - A. B. Choate
February 11, 1897:
Electric Street Railways - - - - - F. W. Springer, E. E.
February 19, 1897:
Sanitary House Drainage - - - - Geo. L. Wilson, C. E.
March 12, 1897:
Safety in Electrical Installations - - - - C. L. Pillsbury
March 19, 1897:
Water Supplies for Small Towns - - - C. E. Loweth, C. E.
March 26, 1897:
Economic Limitations of Electric Lighting - Morgan Brooks
April 2, 1897:
Recent Railway Economics - - - E. E. Woodman, C. E.
April 14, 1897:
Electric Elevators - - - - - - G. Rosenbusch, B. E. E.
May 14, 1897:
Technical Education - - - - - - L. S. Gillette, C. E.

CHARLES A. LARSON

Died March 28, 1897

WHEREAS, It has pleased God, in His infinite wisdom, to remove from our midst our esteemed friend and classmate, CHARLES A. LARSON;

RESOLVED, That we, the members of the Engineers' Society of the University of Minnesota, hereby express our sincere regret and sorrow, and extend to his bereaved family our heartfelt sympathy in their great affliction; and be it further

RESOLVED, That a copy of these resolutions be spread upon the minutes of this Society, and that a copy be sent to his parents.

> ROY V. WRIGHT.
> ADOLF WAGNER.
> C. A. GLASS.

SOCIETY OF ENGINEERS

IN THE

UNIVERSITY OF MINNESOTA.

LIST OF MEMBERS,

MAY, 1897.

HONORARY MEMBERS OF ENGINEERS' SOCIETY.

A. B. Choate.
F. W. Springer, E. E.
Geo. L. Wilson, C. E
Chas. L. Pillsbury.
C. E. Loweth, C. E.
Morgan Brooks.
E. E. Woodman, C. E.
G. Rosenbusch, B. E. E.
C. W. Hall, M. A.
Wm. R. Hoag, C. E

Geo. D. Shepardson, M. E.
Wm. R. Appelby, B. A.
Frank H. Constant, C. E.
H. Wade Hibbard, M. E.
C. H. Kendall, C. E.
H. E. Smith, M. E.
W. H. Kirchner, B. S.
J. H. Gill, M. E.
H. T. Eddy, Ph. D.
F. W. Denton, E. M.

LIST OF ACTIVE MEMBERS OF ENGINEERS' SOCIETY.

Anderson, J. G.
Artz, E. A.
Becker, Geo.
Blake, R. P.
Blake, H. B.
Brush, C. F.
Campbell, W. L.
Chestnut, G. L.
Craig, Robt.
Crane, F.
Cross, Chas. H.
Currier, H. L.
Dahl, H. F. M.
Davis, S. E.
Donaldson, W. T.
Dow, J. C.
Garvey, J. J.
Gilchrest, C. C.
Glass, C. A.
Graling, V.
Hamilton, H. C.
Hewett, F. M.
Hibbard, T.
Hildebrandt, H. A. G.
Johnson, F. E.
Jones, E. B.
Kellar, F. H.

Miller, W. L.
Mills, E. C.
Mooney, F. X.
Morris, C. R.
Myers, M. A.
Nye, J. A.
Ober, W. M.
O'Brien, J. E.
Pease, Levi B.
Roberts, W. H.
Savage, E. S.
Schofield, F. M.
Silliman, H. D.
Shumnway, E. J.
Smith, H. A.
Sperry, T. A.
Stussy, W. M.
Taylor, E. D. W.
Thaler, J. A.
Thompson, R. E.
Towne, B. A.
Van Duzee, C. M.
Wagner, A. W.
Wales, R. T.
Walker, F. B.
Wennerlund, E. K.
Whitman, E. A.

1886.

Woodmunsee, Charles C...B. Arch.   Bookkeeper.
Midway Park, St. Paul.

1887.

Andrews, George C........B. M. E.   Heating Manufacturer and Con-
Minneapolis, Minn.   tractor.
Crane, Fremont............B. C. E.   Civil Engineer.

1888.

Anderson, Christian........B. C. E.   Civil Engineer.
Portland, Ore.
Loe, Eric H................B. M. E.   Mechanical Engineer with Nor-
Minneapolis, Minn.   dyke & Marman.
Morris, John..............B. M. E.   Mechanical engineer and assist-
West Pullman, Chicago, Ill.   ant supt. of the Plano Mfg. Co.
Hoag Wm. R. (B. C. E. '84)..C. E.   Prof. Civil Engineering U. of M.
Minneapolis, Minn.

1889.

Coe, Clarence S............B. C. E.   Civil Engineer in Eng. Dept. of
Wenatschee, Wash.   C., M. & St. P. R'y Co.

1890.

Burt, John L...............B. C. E.   Commission merchant.
Minneapolis, Minn.
Dann, Wilbur W..........B. C. E.   Asst. Engineer Lake Superior
Sault Ste. Marie, Mich.   Power Co.
Gilman, Fred H............B. C. E.   Associate editor Mississippi Val-
Minneapolis, Minn.   ley Lumberman.
Greenwood, W. W........B. C. E.   Editor and publisher of "The
Minneapolis, Minn.   Christian News."
.Hayden, John F...........B. C. E.   Editorial dept. Mississippi Valley
Minneapolis, Minn.   Lumberman.
Higgins, John T..........B. C. E.   Physician.
St. Paul, Minn.
Hoyt, Wm. H..............B. C. E.   Asst. engineer Duluth & Iron
Duluth, Minn.   Range R. R.
Nilson, Thorwald E......B. M. E.   Reporter.
Minneapolis, Minn.
Smith, William C..........B. C. E.   Asst. engineer N. P. R. R.
St. Paul, Minn.
Woodward, Herbert M....B. M. E. Teacher in Mechanic Arts High
Boston, Mass.   School.

1891.

Aslakson, Baxter M.......B. M. E.   With the Stillwell-Bierce and
Dayton, Ohio.   Smith-Vaile Co.
Carrol, James E...........B. C. E.   Draughtsman city engineer's
Minneapolis, Minn.   office.
Chowen, Walter A........B. C. E.   Engineer in charge of G. N. ele-
Buffalo, N. Y.   vator.
Douglas, Fred L..........B. C. E.   Civil Engineer.
New York City.
Gerry Martin H., Jr., (B. M. E. '90)   Supt. motive power of the Metro-
Chicago, Ill.   B. E. E.   politan West Side Elevated R.
     R. Co.
Huhn, George P..........B. E. E.   Flour City Nat'l Bank.
Minneapolis, Minn.

1892.

Burch, Edward P......... B. E. E.   Electrical engineer Twin City
Minneapolis, Minn.   Rapid Transit Co.

Kinsell, W. L.
Larson, C. A. (deceased.)
Latham, W. H.
Lee, E. A.
Leedy, J. W.
Lonie, J. H.
McIntosh, J. B.
McKellip, F. W.
Magnusen, C. E.
Markhus, O. G. F.

Wilson, M. F.
Woodford, G. B.
Woodman, H. H.
Woodward, E. E.
Wright, R. V.
Wright, W. H.
Yager, Lewis.
Zeleny, F.
Zintheo, C. J.

1875.

CORRESPONDING MEMBERS.

NAME AND RESIDENCE.	DEGREE.	OCCUPATION.
Leonard, Henry C. (B. S. '78.) Minneapolis, Minn.	B. C. E.	Physician.
Rank, Samuel A. Central City, Col.	B. C. E.	Civil and Mining Engineer.
Stewart, J. Clark Minneapolis, Minn.	B. C. E.	Physician; Prof. of Pathology at University of Minnesota.

1876.

Gillette, Lewis S. Minneapolis, Minn.	B. C. E.	President of the Gillette-Herzog Manufacturing Co.
Hendrikson, Eugene A. St. Paul, Minn.	B. C. E.	Lawyer.
Thayer, Chas. E. Minneapolis, Minn.	B. C. E.	Grain Dealer.

1877.

Pardee, Walter S. Minneapolis, Minn.	B. Arch.	Architect.

1878.

Bushnell, Charles S. Minneapolis, Minn.	B. M. E.	M'f'r Stoves, Ranges and Furnaces.

1879.

Dawley, William S. Chicago, Ill.	B. C. E.	Chief Engineer C. & E. I. R. R.
*Furber, Pierce P.	B. C. E.	Died April 6, 1893.

1883.

Barr, John H. Ithaca, N. Y.	B. M. E.	Associate Prof. of Machine Design, Cornell University.
Peters, William G. Tacoma, Wash.	B. C. E.	Vice-president Columbia Nat'l Bank.
Smith, Louis O. Le Sueur, Minn.	B. C. E.	Civil Engineer.

1884.

Loy, George J. Spokane, Wash.	B. C. E.	Bridge Contractor.
Matthews, Irving W. Waterville, Wash.	B. C. E.	Real Estate and Abstract of Titles.

1885.

Bushnell, Elbert E. New York City.	B. M. E.	Dealer in Typewriters.
*Fitzgerald, Patrick T.	B. C. E.	Died April 2, 1887.
Reid, Albert I. Racine, Wis.	B. C. E.	U. S. Inspector of Harbor Improvements.

Burtis, William H..........B. E. E. Electrical engineer and contract'r
  Decorah, Iowa.
Felton, Ralph P..........B. M. E. Fire insurance.
  Minneapolis, Minn.
Goodkind, Leo............B. Arch. City Supt. of schoolhouse con-
  St. Paul, Minn.   struction.
Gray, William I.......... B. E. E. Electrical contractor.
  Minneapolis, Minn.
Hankenson, John L........B. C. E. Asst. Engr. C. M. & St. P. Ry.
  Minneapolis, Minn.
Higgins, Elvin L..........B. C. E. County surveyor of McLeod Co.
  Hutchinson, Minn.
Howard, Monroe S........B. E. E. Electrical engineer and contract'r
  Waukan, Wis.
Mann, Frederick Maynard..B. C. E. Instructor in Architectural De-
  Philadelphia, Pa.   sign, Univ. of Penn.
Plowman, George T.......B. Arch. Draughtsman.
  New York City.

1893.

Anderson, Ole J............B. C. E. Civil Engineer.
  St. Peter, Minn.
Avery, Henry B..........B. M. E. With the Gillette-Herzog Mfg.
  Minneapolis, Minn.   Co. (Engineering Dept.)
Bachelder, Frank L........B. C. E. Head draughtsman St. Paul
  St. Paul, Minn.   Foundry Co.
Chase, Arthur W..........B. E. E. Shoe merchant.
  Hastings, Minn.
Couper, Geo. B...........B. M. E. Manager Northfield Electric
  Northfield, Minn.   Light Company.
Dewey, William H........B. E. E. Engineering.
  New York City.
Erf, John W...............B. C. E. With the Gillette-Herzog Mfg.
  Minneapolis, Minn.   Co.
Guthrie, J. De Mott........B. E. E. Medical Student, University of
  Minneapolis, Minn.   Minn.
Hoyt, Hiram P............B. C. E. Draughtsman with American
  Chicago, Ill.   Bridge Works.
Morse, George H..........B. E. E. Instructor in National School of
  Minneapolis, Minn.   Electricity.
Reidhead, Frank E.........B. E. E. Supt. East Side Station of Minne-
  Minneapolis, Minn.   apolis General Electric Co.
Springer, Frank W.........B. E. E. Instructor in University of Minn.
  Minneapolis, Minn.
Washburn, D. Cuyler......B. Arch. Draughtsman for J. T. Fanning,
  Minneapolis, Minn.   C. E., Specialists on Water
                                     Powers.

1894.

Bray, Geo. E...............B. M. E. Electrician for Mountain Iron
  Mountain Iron, Minn.   Mining Company.
Chalmers, Chas. H.........B. E. E. Manager of the Electric Machin-
  Minneapolis, Minn.   ery Company.
Cunningham, Andrew O....B. E. E. With Gillette-Herzog Mfg. Co.
  New Orleans, La.
Gill, Jas. H. (B. M. E., '92)....M. E. Instructor in Metal Work, U. of
                                     Minn.
Gilman, Jas. B..............B. C. E. Draughtsman, Gillette - Herzog
  Minneapolis, Minn.   Mfg. Co.
Johnson, Noah............B. C. E. C. G. W. Ry. Engineering Dept.
  St. Paul, Minn.
Trask, Birney E. (B. C. E., '90).C. E. Professor of Mathematics and
  Highland Park, Ill.   Drawing, Northwestern Mili-
                                     tary Academy.

Weeks, William Charles....B. C. E. U. S. Assistant Engineer.
Minneapolis, Minn.

1895.

Adams, Geo. F.............B. E. E. Electrical Engineering Co.
Minneapolis, Minn.
Bishman, A. E.............B. E. E. Chief Engineer of Lighting Plant.
Willmar, Minn.
Bohland, John A...........B. C. E. Draughtsman, Bridge Dept. G.
St. Paul, Minn.   N. R. R.
Casseday, Geo. A..........B. C. E. Draughtsman, Bridge Dept. G.
St. Paul, Minn.   N. R. R.
Chapman, Leslie H........B. C. E.   Bridge Dept. G. N. R. R.
St. Paul, Minn.
Cutler, Harry C. (B. E. M., '94).   Mining Engineer and Metallur-
Whitehall, Montana.   E. M.   gist with American Developing
                                    and Mining Co., and Gold Hill
                                    Mining Co.
Christianson, Peter..(B. E. M. '94). Instructor, U. of Minn.
Minneapolis, Minn.   E. M.
Eddy, Horace T............B. E. E. With General Electric Co.
Schenectady, N. Y.
Rounds, Fred M............B. E. E. Engineering Dept. Southwestern
Dallas, Tex.   Tel. and Tel. Co.
Shepherd, B. P............B. M. E. Draughtsman.
Minneapolis, Minn.
Tanner, H. L..............B. E. E. Sect. Decorah Electric Light Co.
Decorah, Iowa.
Tilderquist, William.......B. M. E.
Vasa, Minn.
Von Schlegell, F...........B. E. E. Bridge Dept. G. N. R. R.
Minneapolis, Minn.
Weaver, A. C..............B. M. E. Track Recorder, C. G. W. Ry.
Minneapolis, Minn.
Wilkinson, Chas. Dean....B. E. M.   Manager Gold Hill Mining Co.
Gaylord, Mont.

1896.

Beyer, Adam C............B. C. E.   Ass't Inspector for The Osborn
Elmira, N. Y.   Company, Cleveland, Ohio.
Burch, Albert M...........B. C. E. Draughtsman, Bridge Dept. G.
St. Paul, Minn.   N. R. R.
Erikson, Henry A.........B. E. E.   Teacher.
Rochester, Minn.
Hastings, Clive............B. M. E. Special Machinist's Apprentice,
Brainerd, Minn.   N. P. R. R.
Hilferty, Chas. D..........B. M. E. Draughtsman, D. Clint Prescott
West Duluth, Minn.   Co.
Hughes, Thos. M..........B. E. M.   Mining Engineer.
Congress, Arizona.
Hugo, Victor..............B. M. E. Mechanical Engineer.
Duluth, Minn.
Jones, C. P................B. C. E. Draughtsman, G. N. R. R. Bridge
St. Paul, Minn.   Dept.
Lang, James S.............M. E.   With West End Electric R. R.
Boston, Mass.   Co.
Long .... B. C. E. Ass't Engineer Georgia & Florida
Hayland, Georgia.   R. R.
Magnusen, C. Edw.........B. E. E.   Instructor in U. of Minn.
Minneapolis, Minn.
May, Albert E.............B. E. M. Ass't Manager North Pole Mine.
Bourne, Baker Co., Oregon.
Tanner, Wallace N........B. E. M.   Smelting Dept. Boston & Mon-
Great Falls, Mont.   tana C. C. & S. M. & S. Co.
Wheeler, Herbert M.......B. E. E.   Fort Wayne Electric Corporation.
Ft. Wayne, Indiana.

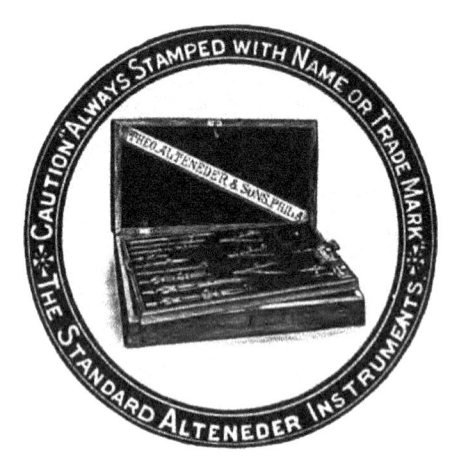

THEO. ALTENEDER & SONS,
945 RIDGE AVE., PHILADELPHIA

Lundell FAN and POWER
Motors

Thousands in Use

Iron and Brass Armored Insulating Conduit

INTERIOR CONDUIT & INSULATION COMPANY

GENERAL OFFICES AND WORKS:
527 West 34th Street
NEW YORK

Meacham & Wright

Manufacturers' Agents for

Utica Hydraulic Cement

and Dealers in

Portland and Louisville Cements, Michigan and New York Stucco.

98 Market Street - - - - - - Chicago

Telephone, Main Express 59.

The Railroad Gazette

PUBLISHED EVERY FRIDAY.

The oldest, best and, in all departments of railroading, the most complete in the world. **$4.20** a year.

Catechism of the Locomotive...... **$3.50**
Car Builders' Dictionary........... **$5.00**

List of Railroad Gazette books sent on application. Unusual inducements for combined subscription and book orders will be furnished on application.

The Railroad Gazette
32 Park Place - - - New York

Pioneer Electrical Journal of America.

Most Popular of Technical Periodicals.

THE ELECTRICAL WORLD is the **Largest**, most **handsomely illustrated** and most **widely circulated** journal of its kind in the world and is read by **Students, Teachers, Electrical Engineers, Professional Men, General Readers**, in short, all who desire to keep informed in this ever advancing branch of **Applied Science**. It is the best **supplement to a course** of study in **Electrical Engineering** that a student can have, for it places him **in actual touch with the profession** he intends to follow.

If you are not already a subscriber direct, through one of our agents, or a local newsdealer, let us send you a **Sample Copy**.

Subscription, postage prepaid, $3.00 per year.

We are the **Largest American Publishers** of and **Dealers in**

ELECTRICAL BOOKS.

There is no work relating to the theoretical or practical application of Electricity that is not either **Published** or **For Sale** by us.

We will be pleased to furnish at any time full information regarding the **latest** and **best works** on any application of electricity in which you may be interested; for which purpose we maintain a **Separate Department,** the manager of which keeps himself at all times **familiar with the contents of every work published,** at home or abroad, on **Electricity** and its **allied branches.**

Any electrical books published, American or foreign, will **be promptly mailed to any address in the world, postage prepaid, upon receipt of price.**

Address and make drafts, P. O. orders, etc., payable to

THE W. J. JOHNSTON COMPANY,
253 Broadway, New York.

Louisville Cement.

The undersigned is agent for the following works:

Hulme Mills,	producing	Star Brand.
Speed Mills,	"	" "
Queen City Mills,	"	" "
Black Diamond Mills, River,	"	Diamond "
Black Diamond Mills, R. R.,	"	" "
Falls City Mills,	"	Anchor "
Silver Creek Mills,	"	Acorn "
Eagle Mills,	"	Eagle "
Fern Leaf Mills,	"	Fern Leaf "
Peerless Mills	"	Crown "
Lion Mills,	"	Lion "
Mason's Choice Mills,	"	Hammer & Trowel "
United States Mills,	"	Flag "

These works are the largest and best equipped in the United States. Orders for shipment to any part of the country will receive prompt attention. Sales in 1892, 2,145,568 barrels.

Western Cement Company,

247 West Main St,, - - Louisville, Ky.

H. C. CUTLER, E. M.,   C. D. WILKINSON, B. E. M.,
Consulting Engineer   Manager
Gold Hill Mine.   Gold Hill Mine.
With American Developing and Mining Co.

CUTLER & WILKINSON,

Mining Engineers and Metallurgists.

Assaying, Chemical Analysis, Ore-testing, Reports on Mining Properties, Installation of Cyanide, Milling and Chlorination Plants.

Special Experience in Cyanide and Chlorination Processes and Development of Mines.

CORRESPONDENCE SOLICITED.

Whitehall and Gaylord, Mont.   528 Lumber Ex., Minneapolis.

Kuhlo & Ellerbe,

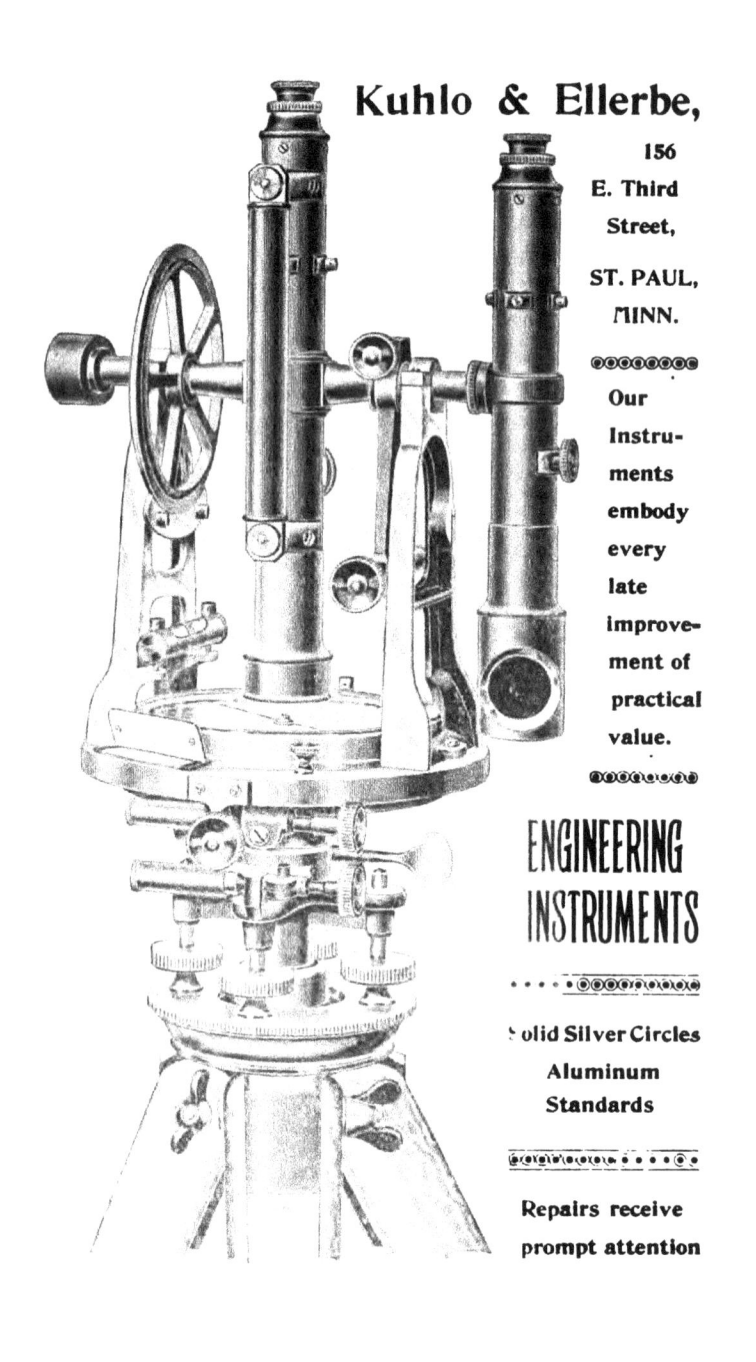

156
E. Third
Street,

ST. PAUL,
MINN.

Our
Instruments
embody
every
late
improvement of
practical
value.

ENGINEERING
INSTRUMENTS

Solid Silver Circles
Aluminum
Standards

Repairs receive
prompt attention

BUFF & BERGER,
IMPROVED
Engineering and Surveying Instruments,

No. 9 Province Court, Boston, Mass.

They aim to secure in their Instruments:

Accuracy of division; Simplicity in manipulation; Lightness combined with strength; Achromatic telescope, with high power; Steadiness of Adjustments under varying temperatures; Stiffness to avoid any tremor, even in a strong wind, and thorough workmanship in every part.

Their instruments are in general use by the U. S. Government Engineers, Geologists, and Surveyors, and the range of instruments as made by them for River, Harbor, City, Bridge, Tunnel, Railroad and Mining Engineering, as well as those made for Triangulation or Topographical Work and Land Surveying, etc., is larger than that of any other firm in the country.

Illustrated Manual and Catalogue Sent on Application.

THE ENGINEERING RECORD
BUILDING RECORD AND THE SANITARY ENGINEER
A JOURNAL for ENGINEER, ARCHITECT, CONTRACTOR, MECHANIC & MUNICIPAL OFFICER

GIVES PROMINENCE TO

MUNICIPAL AND BUILDING ENGINEERING

WHICH INCLUDES

Water-Works (Construction and Operation), Sewerage, Bridges, Metal Construction, Pavements, Subways, Road Making, Docks, River and Harbor Work, Tunneling, Foundations, Building Construction, Industrial Steam and Power Plants, Ventilation, Steam and Hot-Water Heating, Plumbing, Lighting, Elevator and Pneumatic Service.

"The success of this publication has been marked in many ways; not only has it become a source of profit to its projectors but it has been of incalculable value to the general public whose interests it has always served."—*Cincinnati Gazette.*

"It stands as a fine example of clean and able journalism."—*Railroad Gazette.*

Published Saturdays at 277 Pearl St., NEW YORK.

$5.00 PER YEAR   SINGLE COPY, 12 CENTS.

THE ENGINEERING RECORD is the recognized medium for advertisements inviting proposals for all Municipal and U. S. Government Engineering and Public Building Work. Its subscribers include the experienced and reliable Contractors and Manufacturers of Engineering and Building Supplies in all sections of the United States and Canada. The RECORD'S value to secure competition in bids is therefore obvious.

The Deane Steam Pump Co.
HOLYOKE, MASS.

Pumps of **Every Description**.
Single, **Duplex, Triplex,**
Compound, **Triple Expansion,**
Electric.

Deane Triplex Power Pump and Electric Motor.

Fine Mechanical Tools.

STARRETT'S

Milling and Other Cutters.

98=Page Cata= logue Free.

The L. S. STARRETT CO., Box 42 Athol, Mass. U.S.A.

THE ENGINEERS
YEARBOO

UNIVERSITY OF
MINNESOT

MINNEAPOLIS ENGRAVING CO.

We Make a Specialty of Fine Half Tone Engraving.

Designing, Illustrating, Engraving by All Processes, Electrotypes.

16-20 N. Fourth St., MINNEAPOLIS.

FREDK. FRALEY SHARPLESS.   HORACE V. WINCHELL.

SHARPLESS & WINCHELL,
ANALYTICAL CHEMISTS AND ASSAYERS.

Consulting Geologists, Mining Experts and Metallurgists. Complete Chemical and Assaying Outfit with modern equipment; Water, Soil and Food Analyses; Ore Testing and Assaying; Mines Examined and Mining Property Developed. Advice given on all Chemical Questions. Thorough acquaintance with Lake Superior Mineral Lands.

ALL WORK GUARANTEED - - CORRESPONDENCE SOLICITED.

Laboratory and Office, 811 Wright Block, - - -Minneapolis, Minn.

The Denver Fire Clay Co.
Dealers in Assayers' and Chemists' Supplies

Manufacturers of

Crucibles, Scorifiers, and Muffles,

And all other kinds of

FIRE CLAY MATERIAL

Specialties—Strictly c. p. Acids, Iron Sulphide, Test Lead, Bone Ash, Cyanide Potash, Argol, Borax, Litharge, Soda, Mining Fluxes, etc.

1742 to 1746 Champa Street, Denver, Colorado.

J. M. ALLEN, President.   W. B. FRANKLIN, Vice-President.
J. B. PIERCE, Secretary.   F. B. ALLEN, 2d Vice-President.
———————ORGANIZED 1866———————

THOROUGH INSPECTIONS
And Insurance Against Loss or Damage to Property and
Loss of Life and Injury to Persons Caused by
STEAM BOILER EXPLOSIONS

When You Are in Need of

Sewer and Culvert Pipe
Remember that the Best Quality of Ware
is made in large quantities by

EVENS &
HOWARD
St. Louis, Mo.

We Make PROMPT SHIPMENTS and LOW PRICES and our
Goods are Popular Because They Give Satisfaction.

Lightning Source UK Ltd.
Milton Keynes UK
UKHW022320051218
333536UK00016B/1739/P